Petroleum Geology of Africa: New Themes and Developing Technologies

Special Publication reviewing procedures

The Society makes every effort to ensure that the scientific and production quality of its books matches that of its journals. Since 1997, all book proposals have been refereed by specialist reviewers as well as by the Society's Books Editorial Committee. If the referees identify weaknesses in the proposal, these must be addressed before the proposal is accepted.

Once the book is accepted, the Society has a team of Book Editors (listed above) who ensure that the volume editors follow strict guidelines on refereeing and quality control. We insist that individual papers can only be accepted after satisfactory review by two independent referees. The questions on the review forms are similar to those for *Journal of the Geological Society*. The referees' forms and comments must be available to the Society's Book Editors on request.

Although many of the books result from meetings, the editors are expected to commission papers that were not presented at the meeting to ensure that the book provides a balanced coverage of the subject. Being accepted for presentation at the meeting does not guarantee inclusion in the book.

Geological Society Special Publications are included in the ISI Index of Scientific Book Contents, but they do not have an impact factor, the latter being applicable only to journals.

More information about submitting a proposal and producing a Special Publication can be found on the Society's web site: www.geolsoc.org.uk.

It is recommended that reference to all or part of this book should be made in one of the following ways:

ARTHUR, T.J., MACGREGOR, D.S. & CAMERON, N. (eds) 2003. *Petroleum Geology of Africa: New Themes and Developing Technologies.*. Geological Society, London, Special Publications, **207**.

TARI, G., MOLNAR, J. & ASHTON, P. 2003. Examples of salt tectonics from West Africa: a comparative approach. *In*: ARTHUR, T.J., MACGREGOR, D.S. & CAMERON, N. (eds) 2003. *Petroleum Geology of Africa: New Themes and Developing Technologies*. Geological Society, London, Special Publications, **207**, 85–104.

GEOLOGICAL SOCIETY SPECIAL PUBLICATION NO. 207

Petroleum Geology of Africa: New Themes and Developing Technologies

EDITED BY

T. ARTHUR

Burnham, Slough, Berkshire, UK

D. S. MACGREGOR

Sasol Petroleum International, UK

and

N. R. CAMERON

Global Exploration Services, UK

2003
Published by
The Geological Society
London

THE GEOLOGICAL SOCIETY

The Geological Society of London (GSL) was founded in 1807. It is the oldest national geological society in the world and the largest in Europe. It was incorporated under Royal Charter in 1825 and is Registered Charity 210161.

The Society is the UK national learned and professional society for geology with a worldwide Fellowship (FGS) of 9000. The Society has the power to confer Chartered status on suitably qualified Fellows, and about 2000 of the Fellowship carry the title (CGeol). Chartered Geologists may also obtain the equivalent European title, European Geologist (EurGeol). One fifth of the Society's fellowship resides outside the UK. To find out more about the Society, log on to www.geolsoc.org.uk.

The Geological Society Publishing House (Bath, UK) produces the Society's international journals and books, and acts as European distributor for selected publications of the American Association of Petroleum Geologists (AAPG), the American Geological Institute (AGI), the Indonesian Petroleum Association (IPA), the Geological Society of America (GSA), the Society for Sedimentary Geology (SEPM) and the Geologists' Association (GA). Joint marketing agreements ensure that GSL Fellows may purchase these societies' publications at a discount. The Society's online bookshop (accessible from www.geolsoc.org.uk) offers secure book purchasing with your credit or debit card.

To find out about joining the Society and benefiting from substantial discounts on publications of GSL and other societies worldwide, consult www.geolsoc.org.uk, or contact the Fellowship Department at: The Geological Society, Burlington House, Piccadilly, London W1J 0BG: Tel. +44 (0)20 7434 9944; Fax +44 (0)20 7439 8975; E-mail: enquiries@geolsoc.org.uk.

For information about the Society's meetings, consult *Events* on www.geolsoc.org.uk. To find out more about the Society's Corporate Affiliates Scheme, write to enquiries@geolsoc.org.uk.

Published by The Geological Society from:
The Geological Society Publishing House
Unit 7, Brassmill Enterprise Centre
Brassmill Lane
Bath BA1 3JN, UK

(*Orders*: Tel. +44 (0)1225 445046
          Fax +44 (0)1225 442836)
Online bookshop: http://bookshop.geolsoc.org.uk

Project management by Swales and Willis, Exeter, UK
Typeset by Lucid Digital, Honiton, UK
Printed by The Alden Press, Oxford, UK.

Distributors

USA
  AAPG Bookstore
  PO Box 979
  Tulsa
  OK 74101-0979
  USA
Orders: Tel. + 1 918 584-2555
          Fax +1 918 560-2652
          E-mail bookstore@aapg.org

India
  Affiliated East-West Press PVT Ltd
  G-1/16 Ansari Road, Daryaganj,
  New Delhi 110 002
  India
Orders: Tel. +91 11 327-9113
          Fax +91 11 326-0538
          E-mail affiliat@nda.vsnl.net.in

Japan
  Kanda Book Trading Co.
  Cityhouse Tama 204
  Tsurumaki 1-3-10
  Tama-shi
  Tokyo 206-0034
  Japan
Orders: Tel. +81 (0)423 57-7650
          Fax +81 (0)423 57-7651
          E-mail geokanda@ma.kcom.ne.jp

British Library Cataloguing in Publication Data
A catalogue record for this book is available from the British Library.

ISBN 1-86239-128-9

Contents

Acknowledgements

This book is the outcome of the diligence and enthusiasm of the authors, and the generous sponsorship provided by industry. Shell International Exploration and Production made a substantial donation which enabled colour illustrations to be included for the academic papers. WesternGeco Ltd Multiclient funded the production of the CD. The generosity of both these companies is greatly appreciated.

Companies that funded colour figures for papers with which they were associated are Anadarko Algeria Company, PGS Geophysical, Vanco Energy Company and Veritas DGC. Their support, both in financial terms and in permitting their staff to publish, is gratefully acknowledged.

Our gratitude must also be extended to the staff of the Geological Society in London and to the Geological Society Publishing House in Bath. The referees, listed below, gave freely of their time and energy to review the papers and provide help and guidance to the authors.

Jerry Anthony	David Ellis
Ray Bate	Ian Jarvis
David Boote	Gary Karner
John Brewster	Steve Lawrence
Nick Cameron	Duncan Macgregor
Andy Carr	Justin Morrison
Danny Clarke-Lowes	Mike Norton
Chris Cornford	Ed Purdy
Neil Craigie	Bill St John
Mark Davies	Gabor Tari
Ian Davison	Hannah Walford
John Dewey	Chris Walker
Peter Dolan	John Zumberge

Map compilations and synthesis of Africa's petroleum basins and systems

E. G. PURDY[1] & D. S. MACGREGOR[2]

[1]Foxbourne, Hamm Court, Weybridge, Surrey, KT13 8YA, UK (e-mail: egpurdy@mac.com)
[2]Sasol Petroleum International, 93 Wigmore Street, London, W1U 1HJ, UK (e-mail: duncan.macgregor@sasol.com)

Abstract: The purpose of this short contribution is to provide an overview of our current state of knowledge of Africa's petroleum systems as an introduction to the detailed volume accounts that follow. Toward that end we introduce a set of maps on a supplementary CD compiled by Purdy as part of a confidential report provided to industry subscribers some ten years ago. The maps include subsurface as well as surface structural information but, because of their vintage, have not taken account of more recent information. Nonetheless, the regional geological framework apparent on the maps has not changed and, in that sense, the maps are as relevant today as when they were first compiled. Moreover the maps serve as a useful building block on which more recently acquired exploration data can be readily added by others using facilitating computer programs.

Map compilations

Two map compilations are provided on the supplementary CD (Purdy 2003). The first is entitled *Exploration Fabric of Africa* and includes not only conventional outcrop geology but also pertinent subsurface structural trends, including the depths and bottom-hole formations of shallow onshore boreholes that have exploration significance (e.g. in Botswana, Sudan and Nigeria). Bathymetry is also shown along with the location, bottom-hole formations and thickness of section penetrated by the Joint Oceanographic Institutions for Deep Earth Sampling (JOIDES) and Deep Sea Drilling Project (DSDP) holes prior to 1989.

The second compilation map is entitled *Sedimentary Basins of Africa* and illustrates the areal extent and distribution of basins within the general evolutionary framework of Africa described by Burke *et al.* (2003) The basins are classified by name and age of principal hydrocarbon-bearing reservoir play, illustrated against an outcrop background of Precambrian basement, infra-Cambrian sediment, volcanics (mainly Mesozoic and Tertiary) and important subsurface trends. The delineation of infra-Cambrian sediments is viewed as particularly important because there are several basins in Africa that are underlain by significant thicknesses of unmetamorphosed, unfossiliferous sediments lying above Precambrian crystalline basement and below the first age-diagnostic fossils.

The map's basin classification follows the terminology of Kingston *et al.* (1983) with two major modifications. The first is the designation of delta sag (DS) for large basins that are dominated by delta fill. The second is the addition of a W subscript for those basins in which the structural style has been markedly affected by subsequent wrench-fault movements. Some of these may turn out to be expressions of a single divergent wrench couple rather than separate phases of interior fracturing and wrenching. Also indicated is the generalized stratigraphic age of each stage of basin evolution. For example the Niger Grabens designation of IF/IS (K_L-T_P/T_N) translates as an Early Cretaceous to Paleogene Interior Fracture Basin succeeded by an Interior Sag Basin during the Neogene. This classification is important in providing an indication of the type of structures generally present and, in some instances, provides some indication of the size of structures that may be present by comparison to structural sizes known to be generally associated with specific structural styles. It should be noted that the seaward extent of the mapped basins was arbitrarily limited to the 1000 m water-depth contour. In many West African marginal sag basins, however, recent exploration in deeper waters has demonstrated the occurrence of Tertiary reservoir intervals that are demonstrably more important than the older Cre-

From: ARTHUR, T. J., MACGREGOR, D. S. & CAMERON, N. R. (eds) *Petroleum Geology of Africa: New Themes and Developing Technologies.* Geological Society, London, Special Publications, **207**, 1–8. 0305-8719/03/$15
© The Geological Society of London 2003.

taceous reservoir plays identified on the map. These major required revisions are detailed in the following discussion.

Main African reservoir plays

There follows a summary of each of the main reservoir plays identified on the map. These are summarized with respect to age and basin type and are based on innumerable references and personal communications with industry workers, with only the key sources listed as references. For each of the identified reservoir plays, a brief assessment is made both of the proven petroleum provinces discussed by Hemsted (2003) and of frontier basins that may be analogous to these. A view is also taken of the critical exploration factors and technologies applicable to each category. Finally, observations made on key reservoir and petroleum systems are simplified into a number of themes that characterize current hydrocarbon exploration on the continent. For further details of the cited reserves, please refer to the tables and references in the following paper by Burke *et al.* (2003)

Infra-Cambrian reservoirs within interior sag and fracture basins

Widespread areas of interior central and southern Africa are characterized by the occurrence of thick sequences of unfossiliferous sediments overlying Precambrian basement and conformably underlying the earliest fossiliferous beds, which are typically of Silurian age. The largest of these include the Zaire Interior Basin, the Barotse Basin of Angola, the Etosha Basin of Angola and Namibia, the Kalahari Basin of Botswana and the Volta Basin of Ghana.

These beds are typically referred to as 'infra-Cambrian' and are assigned a Vendian–Ordovician age. Such strata are productive on the Arabian plate of southern Oman, where the petroleum system benefits from a thick Mesozoic overburden, but there have been no infra-Cambrian discoveries in Africa. This might reflect a lack of drilling as there are no more than a handful of true exploration wells in the basins concerned and, within this sequence, a number of good source rocks have been mentioned in the literature. However, these are frequently assessed as being overmature. Moreover, burial history relative to phases of trap formation, uplift and potential flushing is difficult to assess. These problems are manifested in the best-documented basin of Zaire (Daly *et al.* 1992). Here, Lawrence & Makazu (1988) have struggled to find source levels in outcrop which are not overmature and additionally could have undergone a Mesozoic reburial phase that post-dated major

uplift and erosion of the infra-Cambrian interval. With these difficulties in mind, the prospectivity of the infra-Cambrian section is considered to be less than encouraging.

Palaeozoic within North African interior sag basins

The North African Palaeozoic play is presented in this volume by Coward & Ries (2003) and in an earlier publication by Boote *et al.* (1998). Reservoirs ranging in age from Cambrian to Early Carboniferous in the Ghadames (Grand Erg), Illizi, Hamra, Murzuk and Ahnet basins contain around 18.5 billion barrels oil equivalent (BBOE), constituting 8% of Africa's petroleum resource. Over half of these reserves are oil in the Cambrian Hassi Messaoud Field. Hydrocarbons were generated from Silurian and Devonian source rocks predominantly during Mesozoic reburial, pre-Hercynian generation having been largely ineffective, although the Ahnet Basin gas reserves could, as Coward & Ries (2003) point out, be exceptions to this general rule. The impact of the secondary Devonian source rocks has been significant for the most recent discoveries and these are described in this volume by Lüning *et al.* (2003) Technologies that are contributing and will continue to contribute to the discovery of more petroleum in these Palaeozoic systems include improved seismic imaging in areas of unfavourable surface conditions, such as shifting sand dunes (Drummond *et al.* 2003)), and shallow carbonate occurrences. Improved petrophysical and engineering practices in some of the tighter reservoirs, particularly the Ordovician fluvioglacial sands, will undoubtedly contribute to increased recovery of *in situ* hydrocarbons.

The Silurian source rock is widespread across most of the frontier basins in this class (Macgregor 1996; Boote *et al.* 1998), parts of the Kufra Basin being an exception. Consequently, the critical factors for frontier areas in this basin category are burial history and preservation. The absence of a significant Mesozoic burial phase in most of the frontier basins seems to be a key difference between these and the productive basins.

Karroo megasequence within interior fracture and foreland basins

The 'Karroo' is a broad lithostratigraphic term applied to sediments extending from the Late Carboniferous 'Dwkya' glacial section up to the Triassic–Liassic 'Stormberg' volcanics. These infill interior fracture and foreland basins developed during or following Permo-Carboniferous tectonism. Despite fairly significant exploration in sediment of this age in Madagascar and South Africa, significant hydrocarbon occurrences are limited to the near-outcropping Bemolanga tar sands of Madagascar, with a reported resource of 22 billion barrels in-place. This oil occurrence is tied to thick Permian shallow marine or lacustrine shales that have not as yet been recorded outside of Madagascar (Raveloson *et al.* 1991) and which seem to be limited to narrow rifts developed at this time. The thick Permian coals that occur elsewhere seem to be predominantly gas-prone (though occasional torbanites are reported) and high maturities are reported from many basins (e.g. Kreiser *et al.* 1998).

The critical factors for Karroo plays thus seem to be presence of an oil-prone source rock, maturation timing relative to structuring, and the potential destructive effects of the numerous phases of uplift that have occurred. The main unexplored areas are the narrow rifts of Zambia, Zimbabwe and Mozambique. A typical exploration case study in one of these is provided by Banks *et al.* (1995).

Triassic and Early Jurassic in interior sag and fracture basins

Triassic sands in Algeria, Tunisia and Libya are one of Africa's most significant plays, with an estimated 31 BBOE of reserves, over half of which is gas in the giant Hassi'R Mel Field. These constitute about 14% of Africa's petroleum reserves (Boote *et al.* 1998). The majority of reserves lie in fluvial sands below Triassic salt and are Palaeozoic-sourced. There are also major reserves at this level in the Sirt Basin, sourced stratigraphically downward from Cretaceous source rocks (Burwood *et al.* 2003). Imaging of undrilled traps below shifting sand dunes and salt is a critical-technology technique that has led to considerable recent success (Drummond *et al.* 2003).

Triassic reservoirs were also developed in widespread Tethyan rifts and plays are developing in areas such as Morocco, where a recent discovery is speculated to be sourced from the Silurian. A separate play is developed in Triassic–Early Jurassic fluvial sands in syn-rift sequences in Somalia and Ethiopia (Tietz 1991), which, at least in the case of Somalia, is an extension of a play developed at this level in rifts in Yemen. Deep burial and complex burial/uplift histories characterize much of this play, and the best possibilities may lie in areas such as the Darror Rift of eastern Somalia, distant from the areas of uplift around the Afar Plume and Gulf of Aden.

Jurassic–Early Cretaceous in interior fracture/marginal sag basins

Reservoir plays of this age are developed along oceanic margins that rifted during the Jurassic, namely the northwestern African margin, the eastern Mediterranean margin and the East African margin. With the exception of a number of Egyptian rifts, such as the Shushan Basin (Keeley *et al.* 1998), some small petroleum systems in Morocco and the small Songo-Songo discovery in Tanzania, these reservoirs have been surprisingly unproductive in exploration to date and have certainly underperformed relative to other basins of this age elsewhere in the world. While it is clearly difficult to generalize over such wide areas, it can be said that the plays most typically comprise either Jurassic carbonates (of variable reservoir quality) or Early Cretaceous clastics (e.g. Slind *et al.* 1988; Jabour *et al.* 2000). Unlike the younger plays described below in younger marginal sag basins, there seems to be no clear regional source level. Lean Late Jurassic source rocks and Early Cretaceous reservoirs are a combination that appears to work in some of the Egyptian basins and in Songo-Songo, but extensive geochemical analyses along many of the margins indicate only sporadic source quality. Many of the East African basins concerned have also undergone upper stage uplift in the Tertiary, which may have caused remigration and flushing. Possibilities may exist along these margins for Triassic–Liassic lacustrine source rocks within the rift sequences (e.g. Jabour *et al.* 2000), as has been locally observed in the Newark rifts of the US Atlantic margin. In many basins, however, these are very deep and may lie, or have previously lain, in the gas window.

Early–Late Cretaceous in interior fracture/marginal sag basins

The continental margins developed during the Early Cretaceous are considerably more productive and also can be shown to contain a series of promising new plays. The margins concerned extend from Senegal to South Africa, with discoveries to date concentrated in the Aptian Salt Basin between Nigeria and Angola (Coward *et al.* 1999). Most traps are anticlinal, related to salt tectonics (Tari *et al.* 2003). A series of petroleum systems in these areas is now well documented (Tiesserenc *et al.* 1989; Katz 2000; Schoellkopf & Patterson 2000), with a total resource base in Cretaceous reservoirs of around 13 BBOE, or 6% of Africa's petroleum. Several petroleum systems can be simplistically subdivided into pre-salt sources charging pre-salt and post-salt reservoirs, and post-salt source rocks charging post-salt reservoirs. Syn-rift Early Cretaceous source rocks charge pre-Aptian salt reservoirs as well as post-rift carbonates and clastic reservoirs in salt-induced structures, while two regional post-salt source levels in the Albo-Cenomanian and Turonian charge primarily clastic turbidite reservoirs. Late Cretaceous deep-water strata comprise an emerging play in these margins, although there is still much information needed about drainage patterns, that were clearly very different from those of today (Taylor 2000). Further exploration successes continue in moderately deep waters in Equatorial Guinea in Late Cretaceous reservoirs with an increasing emphasis on stratigraphic traps and with drilling now heavily driven by direct hydrocarbon indications (DHI) technology (Dailly & Goh 2000; www.tritonenergy. com). This success is stimulating the search for similar plays in areas such as the wrench margin from Cote d'Ivoire to Benin (Macgregor et al. 2003) and as far afield as Senegal and the Gambia.

South of the Walvis Ridge, discoveries are limited to gas in Barremian (Kudu Field) and Albo-Cenomanian (Ibhubezi Field) reservoirs (Jungslager 1999), with the contributing source rock believed to be an anoxic event in the Aptian (van der Spuy 2003). Salt is absent south of the Walvis Ridge, contributing to a deficiency in structural traps and a stratigraphic component seems to be a key element in the discoveries made to date. 3-D seismic imaging of the reservoirs and traps is therefore particularly key in this region.

Early–Late Cretaceous in interior wrench basins

The Early Cretaceous source-rock play extends into the intracratonic rifts of Central Africa (Genik, 1993). The wrench-controlled basins of Chad are a significant developing petroleum province in which only the Doba Basin has been significantly explored to date. Development of an estimated reserve of 1 BBO in Late Cretaceous fluvial reservoirs in this basin is imminent and the existence of a pipeline will undoubtedly stimulate more serious exploration in adjoining basins.

Late Cretaceous and Palaeogene in interior fracture basins

Basins in this set include NW–SE-trending rifts, often initiated in the Late Jurassic–Early Cretaceous, or possibly earlier, but with play systems developed primarily in the Late Cretaceous and Palaeogene (Guiraud & Maurin 1992). In the Sirt Basin (Burwood et al. 2003), which is the predominant hydrocarbon province within the class, the

productive stratigraphic interval extends downward stratigraphically to include the important Albian sandstone play (Gras & Thusu 1998). Other basins in this class include the Gabes Basin of Tunisia and the interior rift basins of Sudan and Niger. With the exception of Sudan, where an Early Cretaceous lacustrine section is the source rock (Schull 1988), the main source level within these basins is the Late Cretaceous marine section, which feeds reservoirs ranging in age from Cambro-Ordovician and Triassic to Neogene (Burwood et al. 2003). Excluding the Palaeozoic reservoirs discussed earlier, this class of petroleum provinces contains around 40 BBOE of reserves, or about 17% of Africa's petroleum.

The majority of these basins have been explored, but many probably still have significant amounts of undiscovered reserves. Large areas of rift basin in Sudan remain untouched by the drill, and several authors have pointed out the low exploration maturity of the Sirt Basin, relative to producing rift-basin analogues, and the limited subtle trap exploration (e.g. Gras & Thusu 1998). The remaining frontier basins in this group include a number of smaller rifts in Sudan and Niger and the Anza Rift of Kenya (Bosworth & Morley 1994), where a number of wells have been drilled without success on valid structures. A source quality or timing problem may well exist in this basin and reservoir risks may also be high.

Tertiary interior fracture basins, northeast and East Africa

This group of basins extends from the Suez Graben in the north through the Red Sea and includes the East African Rift Valleys to the south. All are characterized by thick Neogene fill and active extensional tectonics. To date the only productive basin is the Gulf of Suez, notable for its very high productivity of 7.4 BBOE in an area of only 19 000 km^2 (Alsharan 2002). This productivity relates to its high-quality Miocene and pre-rift reservoirs and the presence of an underlying Late Cretaceous source rock. Future potential of this densely drilled basin would seem to be dependent on improving seismic quality below the obscuring Miocene salt to delineate heretofore unrecognized structural traps.

The Red Sea has long been proposed as an analogue to the Gulf of Suez, yet exploration results have been disappointing (Bunter & Abdel Magib 1989). To some extent, this may be due to poor seismic definition associated with the overlying salt, but it seems clear that the Late Cretaceous source-rock ingredient of the Gulf of Suez petroleum system is missing. The modest discoveries made to date have been of gas, probably reflecting the high heat flow that characterizes the Red Sea and which increases toward the south.

In the East African rift system widespread seeps identify the presence of a petroleum source that may reflect Miocene lacustrine deposition. Good reservoir potential is more speculative, but the main concern is whether or not the size of the extensional structures is commensurate with potential commercial viability, especially in Lakes Malawi and Tanganyika, where water depths exceed 500 m and 1000 m, respectively. In Lake Albert, water depths of less than 100 m offer a more reasonable exploration target, other factors being equal. As most structures are likely to be faulted to surface, petroleum retention is a key issue in such plays.

Tertiary delta sag basins

Nearly half the African petroleum discovered to date has been found within three Tertiary delta sag basins, namely those of the Niger, Nile and Congo. An estimated total of around 96 BBOE has been established to date, and is continuing to increase as new deep-water discoveries are made in turbidite reservoirs. It is the deeper-water parts of these basins that have been the location of most of the new discoveries in recent years (Hemsted 2003).

It is in plays of this class that the appended *Sedimentary Basins of Africa* map must be considered somewhat dated. At the time of map compilation, petroleum plays were only considered as far as the 1000 m water-depth contour, which is now known to be shoreward of several major Tertiary deepwater depocentres. The Tertiary delta play area identified for the Congo Fan, therefore, should now be expanded considerably seaward into deep-water regions. More speculatively, a number of other, lesser order Tertiary depocentres, such as those off Cameroon, Senegal, Mauritania, Tanzania and Kenya, should now be reconsidered with respect to their deep-water potential.

The three major petroleum provinces of this age have some similarities in their petroleum attributes and some key differences. The Niger Delta, with around 80 BBOE of reserves to date, is by far the largest province. The basin contains both deltaic- and turbidite-producing reservoirs ranging in age from Eocene to Pliocene that are charged from Eocene–Oligocene shales. These sources contain a high proportion of terrestrial kerogen that seems to generate gas simultaneously with oil (Haack et al. 2000). Onshore trapping is related to growth faults whereas deep-water trapping is associated with mud diapirism and gravity sliding. Stratigraphic traps formed by pinchouts on diapir flanks have

become important in recent years (e.g. Bonga Field). The key exploration factor, particularly in deep water, is the distinction between oil fill and gas fill, which does not seem, as yet, to be well constrained by either geochemical or geophysical technologies.

The deep-water Congo Fan petroleum system comprises Oligocene–Miocene turbidite reservoirs (Dickson et al. 2003) within salt-induced trapping geometries charged by a primary source in the Late Cretaceous and a secondary source in the Palaeogene (Coward et al. 1999). In contrast, the Nile Delta houses a Miocene–Pliocene gas system, with a mixture of thermogenic gas that is thought to originate from sources ranging in age from Jurassic to Miocene (Doulson & Bucher 2002), and biogenic gas generated within the shallow pro-delta section.

Exploration in these systems continues to move progressively into deeper water where a higher component of stratigraphic trapping must usually be invoked. In the case of the Nile and Niger systems, the deep-water regions have been speculated to be more oil-prone due to lesser burial of the source rock and a lesser proportion of terrestrially derived kerogen within the source bed. The presence of oil seeps in Oligocene quartz sands on the island of São Tomé, hundreds of kilometres off the Niger Delta, provides evidence that these young petroleum systems may extend far beyond both the continental slope and the regions of currently active exploration. DHIs of various types on 3-D seismic have proved crucial in achieving high success rates in all three deltas and, if anything, will prove even more critical as exploration seeks to test more subtle types of traps.

Exploration to date in Tertiary depocentres has concentrated on searching for stepouts within previously established petroleum systems within the largest Tertiary deltas. The only such delta that has not proven productive to date is the Zambezi, probably because of the apparent absence of a regionally developed source rock. Attention is now being given to smaller Tertiary depocentres, including deltaic systems that may have been significant in the past (Taylor 2000). Indeed, the discovery of a new petroleum system in deep-water Mauritania (Brown 2002), albeit one with probable close analogues in terms of reservoirs and source-rock ages to those described above, will continue to stimulate exploration of smaller offshore Tertiary depocentres. Targets include offshore Kenya, Tanzania (Cope, in press), the Mozambique Rovuma Basin and Morocco (Jarvis et al. 1999). The presence of salt or mud diapirism (Tari et al. 2003) provides a structural basis for high-grading such basins.

Summary: future themes and technologies in African exploration

Africa exploration success has continued at high rates in recent years (Hemsted 2003), but the vast majority of new discoveries have been within, or are, deep-water extensions of petroleum systems identified many years ago. Consequently, there have been few new petroleum systems discovered, deep-water Mauritania perhaps providing the main such exception. For example, the discoveries in the Rio Muni Basin of Equatorial Guinea can be argued to be extensions to petroleum systems discovered in the 1960s in offshore Gabon (Katz *et al.* 2000), while the deep-water discoveries in Angola and Nigeria are extensions of the petroleum systems on the adjoining shelves. This conclusion is also supported by the fact that the appended map of reservoir plays requires little modification, despite having been completed many years ago. With this in mind, it seems likely that future discoveries will continue to be concentrated in regions where working plays have already been established or in adjoining regions showing a high degree of analogue to these. More specifically, in North Africa, exploration success will continue to reflect the occurrence of a Mesozoic depositional load that matured Palaeozoic source rocks. Elsewhere in Africa the presence of a Tertiary and, to a lesser extent, a Late Cretaceous depositional load will continue to be a key control on future reserves discoveries, controlling the maturation of Cretaceous and Early Tertiary source rocks. The areal extents of Tertiary and Late Cretaceous turbidite fairways are likely to be expanded further along the West African margin while analogous fairways provide a promising new frontier in the deep-water extensions of the northwestern and East African margins, where Mesozoic plays have underperformed to date on the adjoining shelves.

Additionally, ever-improving technology will aid the identification of traps and reduce risk to the point of exploration acceptability. This is particularly true where seismic trap definition has been hampered by surface and outcrop conditions, or by overlying salt or carbonates. Indeed, increased seismic resolution beneath salt offers the greatest technological potential for significantly increasing the amount of discovered reserves in the Gulf of Suez, the Aptian Salt Basin and Algeria, perhaps even more so than conventional new discoveries. Increased seismic resolution and related direct hydrocarbon recognition technologies also offer hope for continued success in further seaward step-outs from existing deep-water discoveries within regions where stratigraphic traps may dominate. Finally, increased knowledge of basin-timing events relating thermal history, petroleum generation and trap formation to periods of uplift and erosion that may have adversely affected trap integrity will enhance wildcat success ratios and certainly should focus exploration efforts on the frontier basins most likely to provide commercial success.

References

ALSHARAN, A. S. 2002. Petroleum geology and factors characterising the richness and potential plays for hydrocarbon in the Gulf of Suez rift basin, Egypt. *American Association of Petroleum Geologists Annual Convention, Houston, 2002, Abstracts*, A5.

BANKS, N. L., BARDWELL, K. A. & MUSIWA, S. 1995. Karoo rift basins of the Luangwa Valley, Zambia. In: LAMBIASE, J (ed.) *Hydrocarbon Habitat in Rift Basins.* Geological Society, London, Special Publications, **80**, 282–295.

BOOTE, D. R. D., CLARK-LOWES, D. & TRAUT, M. W. 1998. Paleozoic petroleum systems of North Africa. In: MACGREGOR, D. S., MOODY, R. T. J. & CLARK-LOWES, D. D. (eds) *Petroleum Geology of North Africa.* Geological Society, London, Special Publications, **132**, 7–68.

BOSWORTH, W. & MORLEY, R. K. 1994, Structural and stratigraphic evolution of the Anza Rift, Kenya. *Tectonophysics*, **236**, 93–115.

BROWN, L. 2002. Deep water Mauritania, a new petroleum province. *American Association of Petroleum Geologists Annual Convention, Houston, 2002, Abstracts*, A7.

BUNTER, M. & ABDEL MAGIB, A. E. M. 1989. The Sudanese Red Sea, new developments in stratigraphy and petroleum geological evolution. *Journal of Petroleum Geology*, **12**, 145–166.

BURKE, K., MACGREGOR, D. S. & CAMERON, N. R. 2003. Africa's petroleum systems: four tectonic aces in the past 600 million years In: ARTHUR, T. J., MACGREGOR, D. S. & CAMERON, N. R. (eds) *Petroleum Geology of Africa: New Themes and Developing Technologies.* Geological Society of London, Special Papers, **207**, 21–60.

BURWOOD, R., REDFERN, J. & COPE, M. J. 2003. Geochemical evaluation of East Sirte Basin (Libya) petroleum systems and oil provenance. In: ARTHUR, T. J., MACGREGOR, D. S. & CAMERON, N. R. (eds) *Petroleum Geology of Africa: New Themes and Developing Technologies.* Geological Society of London, Special Papers, **207**, 203–240.

CAWLEY, S. J., WILSON, N. P., PRIMMER, T., OXTOBY, N. H. & KHATIR, B. 1995. Palaeozoic gas charging in the Ahnet-Timimoun Basin, Algeria. *American Association of Petroleum Geologists Bulletin*, **79**, 1202.

COPE, M. (In press) The petroleum potential of the Mafia deep marine offshore basin, southern Tanzania. *Oil and Gas Journal*. [Preprint]

COWARD, M. P. & RIES, A. C. 2003. Tectonic development of North African basins. In: ARTHUR, T. J., MACGREGOR, D. S. & CAMERON, N. R. (eds) *Petroleum Geology of Africa: New Themes and Developing Technologies.* Geological Society of London, Special Papers, **207**, 61–83.

COWARD, M. P., PURDY, E. G, RIES, A. C & SMITH, D. G. 1999. The distribution of petroleum reserves in basins of the South Atlantic margins. *In*: CAMERON, N. R., BATE, R. H. & CLURE, V. S. (eds) *The Oil and Gas Habitats of the South Atlantic*. Geological Society, London, Special Publications, **153**, 101–132.

DAILLY, P. & GOH, K. 2000. Rio Muni Basin of Equatorial Guinea – a new hydrocarbon province. *Bulletin of the Houston Geological Society,* **42** (10), 15–17.

DALY, M. C., LAWRENCE, S. R., DIEMU-TSHIBAND, K. & MATOUANA, B. 1992. Tectonic evolution of the Cuvette Centrale, Zaire. *Journal of the Geological Society, London,* **149**, 539–546.

DICKSON, W. G., DANFORTH, A. & ODEGARD, M. 2003. Gravity signatures of sediment systems: predicting reservoir distribution in Angolan and Brazilian systems. *In*: ARTHUR, T. J., MACGREGOR, D. S. & CAMERON, N. R. (eds) *Petroleum Geology of Africa: New Themes and Developing Technologies*. Geological Society of London, Special Publications, **207**, 241–256.

DOLSON, J. S. & BOUCHER, P. J., 2002, The petroleum potential of the emerging Mediterranean offshore gas plays, Egypt. *American Association of Petroleum Geologists Annual Convention, Houston, 2002, Abstracts*, A43.

DRUMMOND, J., KASMI, R., SAKANI, A., BUDD, A. J. L. & RYAN, J. W. 2003. Optimizing 3-D seismic technologies to accelerate field development in the Berkine Basin, Algeria. *In*: ARTHUR, T. J., MACGREGOR, D. S. & CAMERON, N. R. (eds) *Petroleum Geology of Africa: New Themes and Developing Technologies*. Geological Society of London, Special Publications, **207**, 257–273.

DUMESTRE, M. A. & CARVALHO, F. F. 1985. The petroleum geology of the Republic of Guinea Bissau. *Oil and Gas Journal,* **83** (36), 180–191.

GENIK, G. 1993. Petroleum geology of Cretaceous–Tertiary rift basins in Niger and Chad. *American Association of Petroleum Geologists Bulletin,* **77**, 1405–1434.

GRAS, R. & THUSU, B. 1998. Trap architecture of the Lower Cretaceous Sarir sandstone. *In*: MACGREGOR, D. S., MOODY, R. T. J. & CLARK-LOWES, D. D. (eds) *Petroleum Geology of North Africa*. Geological Society, London, Special Publications, **132**, 317–334.

GUIRAUD, R. & MAURIN, J. 1992. Lower Cretaceous rifts of western and central Africa, a review. *In*: ZIEGLER, P. A. *Geodynamics of Rifting*, vol. 2. *Tectonophysics*, **213**, 153–168.

HAACK, R. C., SUNDARARAMAN, P., DIEDJOMAHON, J. O, XIAO, N. J., GANT, N. J., MAY, E. D. & KELSCH, K. 2000. Niger Delta petroleum systems. *In*: MELLO, M. R. & KATZ, B. J. (eds) *Petroleum Systems of South Atlantic Margins*. American Association of Petroleum Geologists Memoirs, **73**, 213–232.

HEMSTED, T. 2003. Second and third millennium reserves in African basins. *In*: ARTHUR, T. J., MACGREGOR, D. S. & CAMERON, N. R. (eds) *Petroleum Geology of Africa: New Themes and Developing Technologies*. Geological Society of London, Special Publications, **207**, 9–20.

JABOUR, H., MORABET AL M. & BOUCHTA, R. 2000. Hydrocarbon systems of Morocco. *In*: CRASQUIN-SOLEAU, S. & BARRIER, E. (eds) *Peri-Tethys Memoir 5: New Data on Peri-Tethyan Sedimentary Basins*. Memoirs Museum National d'Histoire Naturelle, Paris, Serie C Sciences de la Terre, **182**, 143–158.

JARVIS, I., FISH, P. & GARWOOD, T. 1999. Morocco's Tarfaya deepwater prospects encouraging. *Oil and Gas Journal,* **97** (33), 90–94.

JUNGSLAGER, E. H. A. 1999. Petroleum habitats of the Atlantic margin of South Africa. *In*: CAMERON, N. R., BATE, R. H. & CLURE, V. S. (eds) *The Oil and Gas Habitats of the South Atlantic*. Geological Society, London, Special Publications, **153**, 153–168.

KATZ, B. J., DAWSON-WILLIAMS, C., LIRO, L. M., ROBISON, V. D. & STONEBRAKER, J. D. 2001. Petroleum systems of the Ogouee Delta, offshore Gabon. *In*: MELLO, M. R. & KATZ, B. J. (eds) *Petroleum Systems of South Atlantic Margins*. American Association of Petroleum Geologists Memoirs, **73**, 247–256.

KEELEY, M. & MASSOUD, M. S. 1988. Tectonic controls on the petroleum geology of N.E. Africa. *In*: MACGREGOR, D. S., MOODY, R. T. J. & CLARK-LOWES, D. D. (eds) *Petroleum Geology of North Africa*. Geological Society, London, Special Publications, **132**, 265–281.

KINGSTON, D. R., DISHROOM, C. P & WILLIAMS, P. A. 1983. Global basin classification scheme. *American Association of Petroleum Geologists Bulletin,* **67**, 2175–2193.

KREUSER, T., SCHRAMEDEI, R. & RULLKOETER, J. 1988. Gas prone source rocks from cratogene Karoo basins in Tanzania. *Journal of Petroleum Geology,* **11** (2), 169–183.

LAWRENCE, S. R. & MAKAZU, M. M. 1988. Zaire's central basin. Prospectivity outlook. *Oil and Gas Journal,* **86** (38), 105–108.

LÜNING, L., ADAMSON, K. & CRAIG, J. 2003. Frasnian organic-rich shales in North Africa: regional distribution and depositional model. *In*: ARTHUR, T. J., MACGREGOR, D. S. & CAMERON, N. R. (eds) *Petroleum Geology of Africa: New Themes and Developing Technologies*. Geological Society of London, Special Publications, **207**, 165–184.

MACGREGOR, D. S. 1996. Hydrocarbon systems of North Africa. *Marine & Petroleum Geology,* **13**, 329–340.

MACGREGOR, D. S. 2001. Exploration opportunities in Sao Tome-Principe. *Proceedings of the Gulf of Guinea Conference, Paris, 2001.* [CD]

MACGREGOR, D. S., ROBINSON, J. & SPEAR, G. 2003. Play fairways of the Gulf of Guinea transform margin. *In*: ARTHUR, T. J., MACGREGOR, D. S. & CAMERON, N. R. (eds) *Petroleum Geology of Africa: New Themes and Developing Technologies*. Geological Society of London, Special Publications, **207**, 131–150.

PURDY, E. G. 2003. Map compilations for Africa's basins. *In*: ARTHUR, T. J., MACGREGOR, D. S. & CAMERON, N. (eds) *Petroleum Geology of Africa*. Geological Society, London, Special Publications, **207**, supplementary CD.

RAVELOSON, E, ANDRIAMANANTERA, J., ROGER, J. & RAMANAMPISOA, L. 1991. The south Morondava Karoo and its petroleum potential. *In*: PLUMMER, P. S. (ed) *Proceedings of the First Indian Ocean Petroleum Seminar, Seychelles, 10–15 December 1990*, 307–324.

SCHOELLKOPF, N. B. & PATTERSON, B. A. 2000. Petroleum systems of Cabinda, Angola. *In*: MELLO, M. R. & KATZ, B. J. (eds) *Petroleum Systems of South*

Atlantic Margins. American Association of Petroleum Geologists Memoirs, **73**, 361–376.

SCHULL, T. J. 1988. Rift basins of interior Sudan. *American Association of Petroleum Geologists Bulletin,* **72**, 1128–1142.

SLIND, O. L, DU TOIT, S. R. & KIDSTON, A. G. 1988. The hydrocarbon potential of the East African continental margin. *Geotriad '98, Calgary, Abstracts*, 436–437.

TAYLOR, M. 2000. Controls of sedimentation along the African margin of the Atlantic; climate and drainage geometry. *Petroleum Systems and Evolving Technologies in African Exploration & Production, Abstracts of Meeting of 16–18 May 2000*, Petroleum Exploration Society of Great Britain/Geological Society, London, **48**.

TIESSERENC, P. & VILLEMIN, J. 1989. Sedimentary basin of Gabon–geology of oil systems. *In*: EDWARDS, J. D. & SANTOGROSSI, P. A. *Divergent/Passive Margin Basins*. American Association of Petroleum Geologists Memoirs, **48**, 117–199.

TIETZ, H. H. 1991. The Ogaden Basin, Ethiopia, an underexplored sedimentary basin, *Journal of Petroleum Geology*, **14** (1), x–xii.

VAN DER SPUY, D. 2003. Aptian source rocks in some South African Cretaceous basins. *In*: ARTHUR, T. J., MACGREGOR, D. S. & CAMERON, N. R. (eds) *Petroleum Geology of Africa: New Themes and Developing Technologies*. Geological Society of London, Special Publications, **207**, 185–202.

Second and third millennium reserves development in African basins

T. HEMSTED

Independent Consultant, La Billioude, 384 Rue St. Denis, 01170 Ain, France (e-mail: tim.hemsted@wanadoo.fr)

Abstract: This paper discusses the development of hydrocarbon reserves discovered onshore and offshore of the continent of Africa up to the start of the year 2000 and the future potential for new discoveries. Over 2400 hydrocarbon discoveries have been made in Africa, of which 700 were producing in significant quantities at the start of 2000. Initial recoverable reserves on these accumulations stood at 160 billion barrels of oil, 500 trillion cubic feet (TCF) gas and 12 billion barrels of condensate, with estimated cumulative production being 59 billion barrels of oil and condensate and up to 50 TCF gas. A similar proportion of the total discovered oil reserves as gas reserves have been found in cratonic basins while passive/rift basins contain more of the oil reserves than the gas reserves. Despite this mix, considerably more gas has been produced from cratonic basins than from the passive/rift basins. In the last decade, deeper-water exploration using 3-D seismic data has resulted in the discovery of significant reserves in deeper waters off West Africa and the Nile Delta. Future reserves growth is expected to be concentrated in the known petroliferous basins.

There are 179 distinct recognized African geological provinces, of which 102 lie onshore, 22 offshore and the remaining 55 straddle the two (Fig. 1). As of 1 January 2000, over 2400 hydrocarbon discoveries had been made, of which over 700 were producing in significant quantities at that time. These producing fields were located in just 36 geological provinces (Fig. 2), of which only 28 had significant production from more than one field. Eleven of these basins are onshore and the rest are basins straddling onshore and offshore (Table 1).

There are two main geographical concentrations of hydrocarbon fields: the first comprises the North African basins in a swathe from Algeria, across Tunisia, Libya and Egypt, to the Red Sea and Nile Delta; the second lies along the West African coastal seaboard between the Niger Delta and the Congo and Cuanza basins. Other isolated producing areas include the Rharb Prerif Basin in Morocco, the Senegal Basin, the Muglad Basin in Sudan and the Outeniqua Basin offshore South Africa.

Within these producing areas, as a percentage of the total number of producing fields, 26% are in onshore cratonic basins and 13% in pure rift basins. Some 60% of fields are in passive-margin related basins, of which most are in passive/rift basins (53%), the remainder being in pure passive-margin basins and a small proportion in passive/transform

basins (Fig. 3). The reserves contained within cratonic basins are limited to the North African basins of Algeria, Tunisia and Libya, the most prolific being the Sirte (50 billion barrels oil equivalent [BOE]) and the Ghadames (10.6 billion BOE) basins, the Hassi Messaoud High (18 billion BOE) and the Tilrhemt Uplift (22.5 billion BOE). The pure rift basins with significant reserves are the Gulf of Suez (11 billion BOE), Red Sea and Muglad basins. The passive/rift-related basins with significant reserves are found on the western coastal margins and the Mediterranean coastal basins of North Africa (Fig. 4 & Table 1).

The estimated proven plus probable initial recoverable reserves for all African hydrocarbon accumulations at the beginning of 2000 was 160 billion barrels of oil, over 500 trillion cubic feet (TCF) gas and 12 billion barrels of condensate, estimated cumulative production being some 59 billion barrels of oil and condensate and 31 TCF gas. Remaining reserves at this time was therefore estimated to be 102 billion barrels of oil, 11 billion barrels of condensate and 469 TCF of gas (Table 2).

At the start of 2000, there were 25 basins with over 100 million barrels of remaining recoverable liquid reserves (Fig. 5). The majority were located in the North African onshore basins and the West African coastal basins. In North Africa there were 16 basins with more than 100 million barrels of

From: ARTHUR, T. J., MACGREGOR, D. S. & CAMERON, N. R. (eds) *Petroleum Geology of Africa: New Themes and Developing Technologies.* Geological Society, London, Special Publications, **207**, 9–20. 0305-8719/03/$15

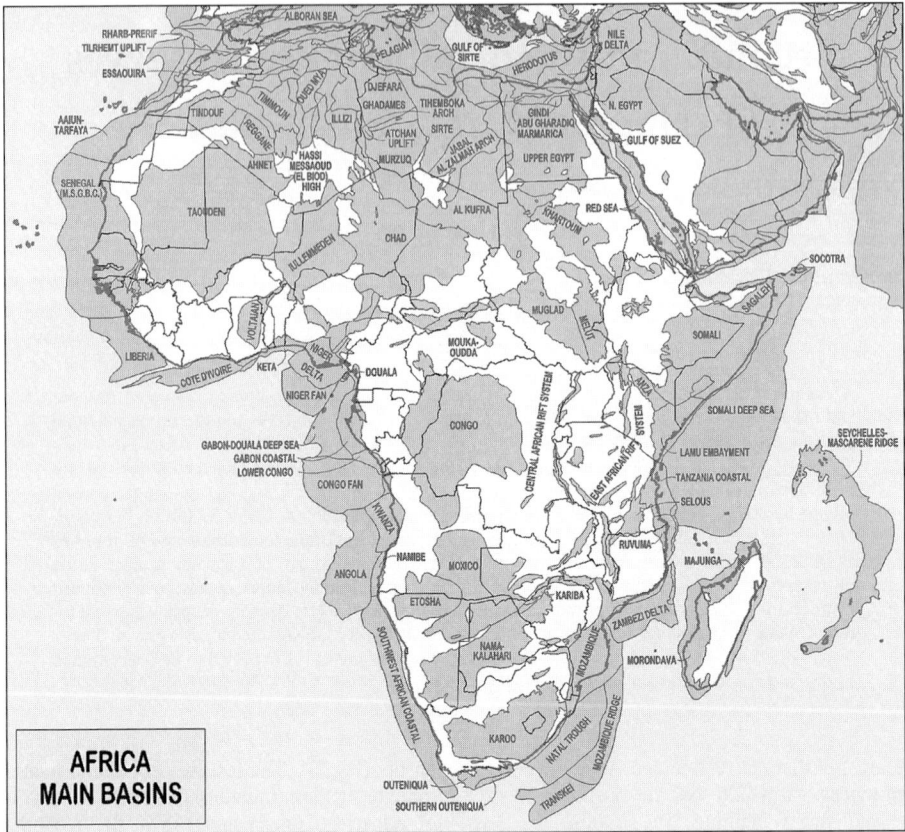

Fig. 1. African main basins map.

remaining reserves, the most notable being the Sirte and Ghadames basins, with some 42 and 8 billion BOE of remaining reserves respectively, and the Hassi Messaoud High with some 14 billion barrels. In West Africa, the Niger Delta (50 billion barrels) and the Lower Congo basins (17 billion barrels) contained the most significant remaining liquids reserves.

There were 25 basins with remaining gas reserves in excess of one TCF (Fig. 6). The basins with the highest remaining proven plus probable gas reserves were the Niger Delta (160 TCF) and elsewhere in North Africa: the Algerian Tilrhemt Uplift (74 TCF), the Sirte Basin (32 TCF), the Nile Delta (39 TCF), and the Hassi Messaoud High (24 TCF).

If reserves and production from the significant producing basins are analysed by basin type, it is observed that the proportion of initial discovered oil reserves located in cratonic basins (45% of the

total) is similar to that of gas (46%). Otherwise, most important are the passive/rift basins, which contain 47% of the oil and 42% of the gas. Despite this mix, considerably more gas has been produced from cratonic basins than from passive/rift basins (55% v. 26% of the total) (Table 3). It is suggested that this is primarily due to the location of the cratonic basins onshore of North Africa relatively close to the main consumer areas and export routes to Europe. This has led to the development of good gas transport infrastructure especially in Algeria. Likewise, both oil and gas production from the rift-basin reserves has been proportionally higher, primarily due to the maturity of rift-basin exploration in Egypt.

Despite a string of well-publicized oil discoveries, primarily in the West African offshore margins in the late 1990s, it is noted that the most successful period for the discovery of both oil and gas reserves was in the early phase of exploration on

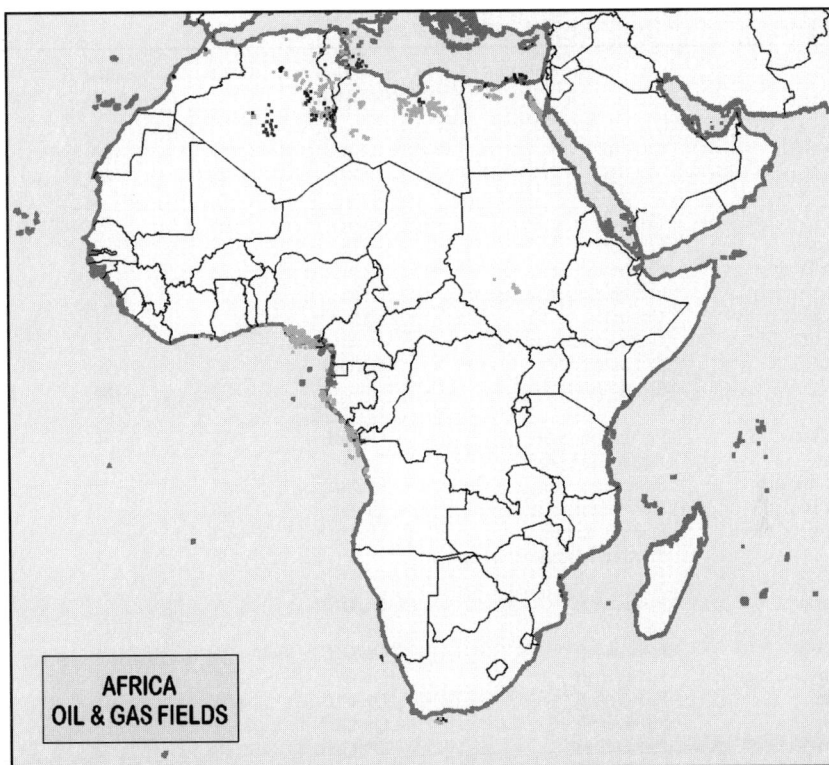

Fig. 2. The distribution of African oil and gas fields.

the continent between 1956 and 1968, when the big onshore fields in Libya and Algeria were discovered (Figs 7 & 8). In Algeria, the Hassi Messaoud Field was discovered in 1956, proving up reserves of 10 billion barrels of oil and 80 TCF of gas. In the following year, the 105 TCF gas giant Hassi R'Mel was discovered on the Tilrhemt Uplift. In the late 1950s and early 1960s, there was a string of oil discoveries in the Libyan Sirte Basin, the most significant fields being the Gialo (5 billion barrels), Sarir (4.5 billion barrels), Amal (4.25 billion barrels), Bu Attifel (2.8 billion barrels) and the Nasser (2.5 billion barrels).

Offshore exploration in Africa commenced in the late 1950s in shallower waters and moved below the 200 m depth contour into the deeper offshore in the late 1970s (Fig. 9). Successes in areas of shallower water were common in the early 1960s in the Gulf of Suez, the Niger Delta and the Gabon offshore basins. In 1966, the first discovery was made in the Lower Congo Basin offshore of Angola and, in 1967, the first offshore discovery in the Sirte Basin, Libya. The first discovery in the

South African Outeniqua Basin was in 1969. The Namibian Kudu Field was discovered in 1974 in 190 m of water and was followed by exploration in deeper waters with the first successes in water depths of more than 200 m occurring in 1977 in offshore Libya and then Tunisia in the Pelagian Basin. In 1980 there were some deep-water gas condensate discoveries offshore of the Cote d'Ivoire in waters of up to 700 m deep. The first deep-water success in the Lower Congo Basin was in 1984 but, in the Nile Delta, it was not until 1995. The first discoveries in Africa in water depths greater than 1000 m were in 1996, both on the Congo Fan and Niger Delta. These were followed by a remarkable run of successes, especially on the Congo Fan and the Nile Delta in the late 1990s.

While the discovery rates for reserve additions in the onshore areas and the shallow offshore show a definite plateau effect in the 1970s and 1980s respectively, it is observed that this plateau had not yet been reached for the deeper offshore discovery rate by the end of the 1990s (Fig. 10). Indeed, more recent results, combined with extremely promising

Table 1. *Summary of main hydrocarbon-producing basins*

Basin name	Classification	Location on/offshore	Fields	Producing fields	Reserves MMBOE
Ghadames Basin	Cratonic	On	151	34	10,637
Hassi Messaoud (El Biod) High	Cratonic	On	26	15	17,769
Ilizi Basin	Cratonic	On	68	23	5,149
Murzuq Basin	Cratonic	On	18	3	2,197
Oued Mya Basin	Cratonic	On	28	12	1,541
Tihemboka Arch	Cratonic/interior sag	On	21	3	1,368
Tilrhemt Uplift	Cratonic	On	3	1	22,503
Atchan Uplift	Cratonic/interior sag	On	16	5	4,006
Sirte Basin	Cratonic/rift	On/Off	267	90	49,937
TOTAL	**Cratonic Basins**		**598**	**186**	**115,107**
Essaouira Basin	Passive margin	On/Off	7	4	24
Gindi Basin	Passive margin	On	7	4	95
Marmarica Basin	Passive margin	On/Off	7	2	124
Nile Delta Basin	Passive margin	On/Off	60	11	7,175
Northern Egypt Basin	Passive margin	On/Off	62	30	1,897
TOTAL	**Passive Margin Basins**		**143**	**51**	**9,315**
Djefara Basin	Passive/rift	On/Off	9	3	80
Gabon Coastal Basin	Passive/rift	On/Off	129	47	4,719
Kwanza Basin	Passive/rift	On/Off	25	4	106
Lower Congo–Congo Fan	Passive/rift	On/Off	266	88	18,993
Niger Delta	Passive/rift	On/Off	684	225	80,331
Pelagian Basin	Passive/rift	On/Off	67	13	5,684
Senegal (M.S.G.B.C.) Basin	Passive/rift	On/Off	10	2	26
TOTAL	**Passive/Rift Basins**		**1190**	**382**	**109,938**
Cote d'Ivoire Basin	Passive/transform	On/Off	29	3	637
Outeniqua Basin	Passive/transform	On/Off	33	3	686
TOTAL	**Passive/Transform Basins**		**62**	**6**	**1,323**
Red Sea Basin	Rift	On/Off	4		265
Abu Gharadiq Basin	Rifts	On	41	22	1,256
Gulf of Suez Basin	Rifts	On/Off	134	72	11,044
Muglad Basin	Rifts	On	28	2	624
TOTAL	**Rifts Basins**		**207**	**96**	**13,189**
Rharb-Prerif Basin	Foredeep faulted	On/Off	29	5	23

Table 2. *African hydrocarbon reserves and production totals*

	Oil	Condensate (billion barrels)	Liquids	Gas (TCF)
Initial recoverable reserves	160	12	172	500
Cumulative production	58	1	59	30–50
Remaining recoverable reserves	102	11	113	450–470

indications from recently acquired 3-D seismic surveys, point to a continued success rate in deeper waters in the Congo Fan, Niger Delta and Nile Delta areas in the first decade of the new millennium.

Drilling and resultant successes in increasingly deeper-water areas has mirrored advances in the quality and use of 3-D seismic data as an exploration tool. This has resulted in the discovery of significant reserves in the West African passive-

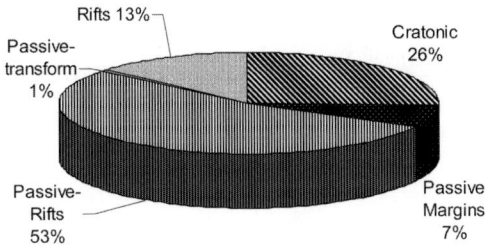

Fig. 3. Breakdown of producing fields by basin type.

margin basins. In parts of these basins, near-total 3-D seismic data coverage over prospective areas has been reached, as in the Congo and Niger Delta basins (Figs 11 & 12).

The first 3-D survey in Africa was acquired in the 1970s in the Cote d'Ivoire offshore basin but it was only in the early 1980s that 3-D surveys become a regular feature on the African Exploration & Production scene, with surveys being shot offshore of West Africa in the Niger Delta, the Gabon Coastal Basin and the Lower Congo Basin.

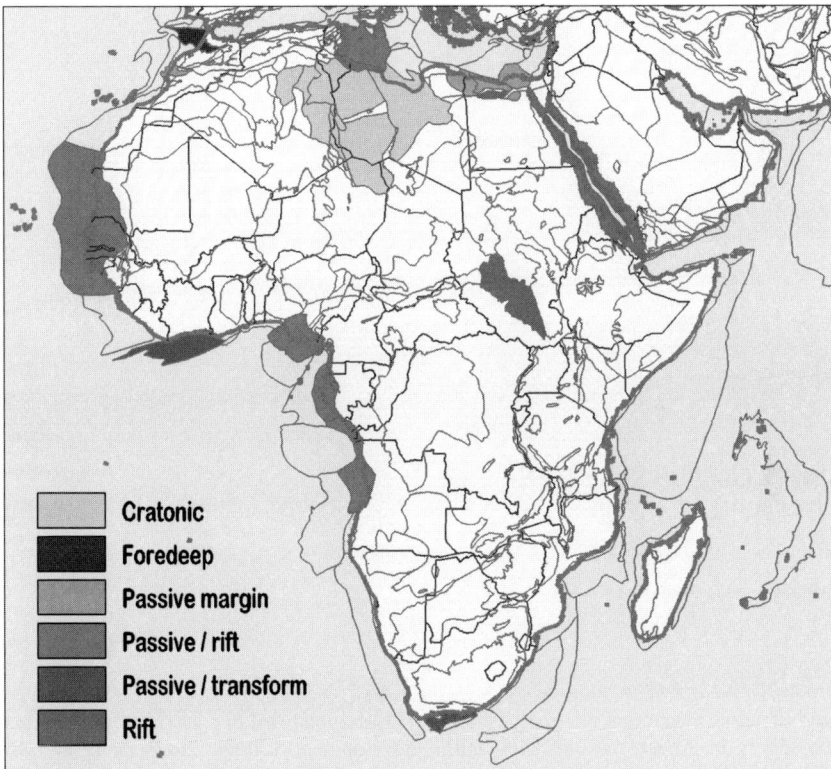

Fig. 4. Significant producing basins by basin type.

Table 3. *Hydrocarbon reserves and cumulative production distribution by basin type*

Basin classification	Oil reserves	Oil production	Gas reserves	Gas production
Cratonic	45%	43%	46%	55%
Passive margin	1%	1%	9%	10%
Passive/rift	47%	43%	42%	26%
Passive/transform	0%	0%	1%	0%
Rift	7%	13%	2%	9%

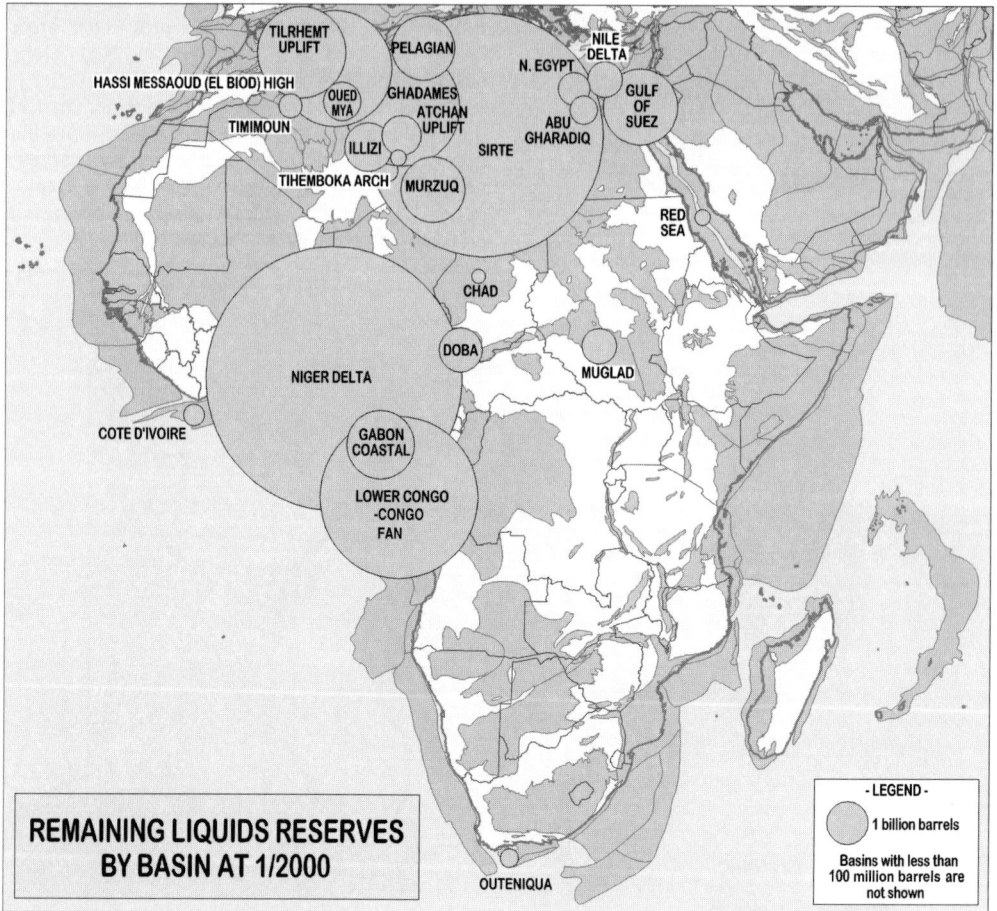

Fig. 5. Remaining liquids reserves by basin.

Operational difficulties meant that the first onshore 3-D seismic survey in Africa was not shot until 1987, on the Niger Delta, where the maturity of the area is marked by a peak in 3-D acquisition in the early 1990s followed by a marked decline in the late 1990s (Fig 13).

The first 3-D survey in North Africa was shot in 1983 in the Gulf of Suez. In contrast, in the North African area, only the Nile Delta shows a similar trend of widespread 3-D coverage (Fig. 14) while, in the petroliferous onshore cratonic basins, 3-D coverage is still relatively sparse and has often only been acquired as a supplement to development operations. This is certainly a function of the cost and difficulties of operating in desert and especially dunefield environments, but is also due to the lack of investment in certain areas of Algeria and Libya (Fig. 15).

Horizontal drilling has been used primarily for development drilling, commencing in 1987 in the offshore Gabon Basin with the Tchibouela Marine 116 well. The use of the technique increased in terms of metres drilled (total depth of well) throughout the 1990s, peaking in 1997, when 300 000 m of horizontal wells were drilled. Drilling declined to some 180 000 m in 1999. Results from the year 2000 indicate that horizontal drilling was close to its peak of 1997 and that, with higher oil prices, further horizontal drilling increases may be expected in the coming years. To date, it has been used primarily in Nigeria, Gabon, Libya and Egypt, but only in Nigeria in significant amounts offshore (Figs 16 & 17).

The future growth of reserves in the coming millennium will no doubt be concentrated initially on drilling in the known petroliferous basins. Seismic

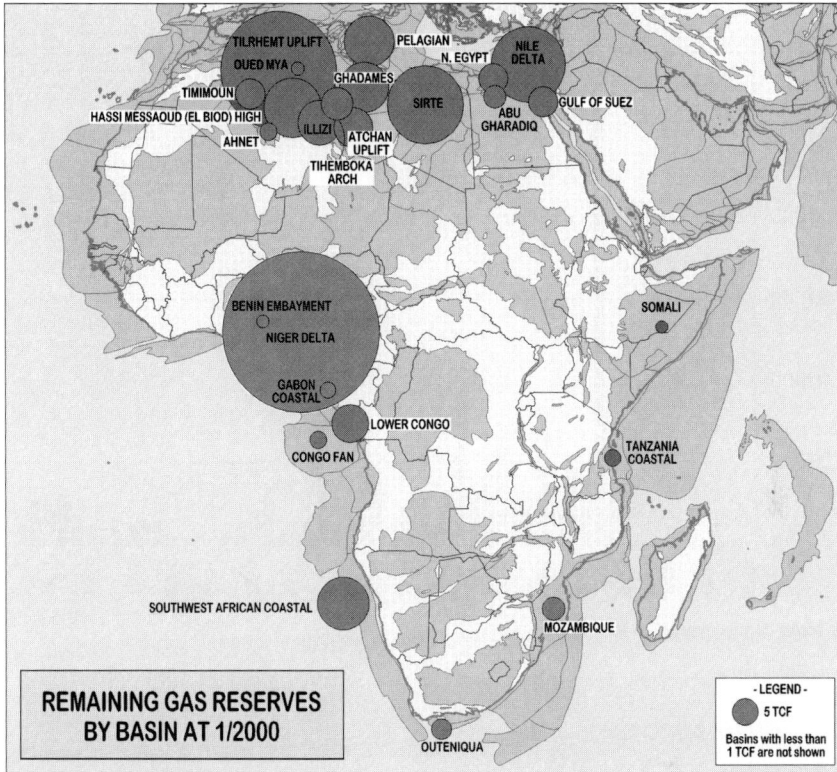

Fig. 6. Remaining gas reserves by basin.

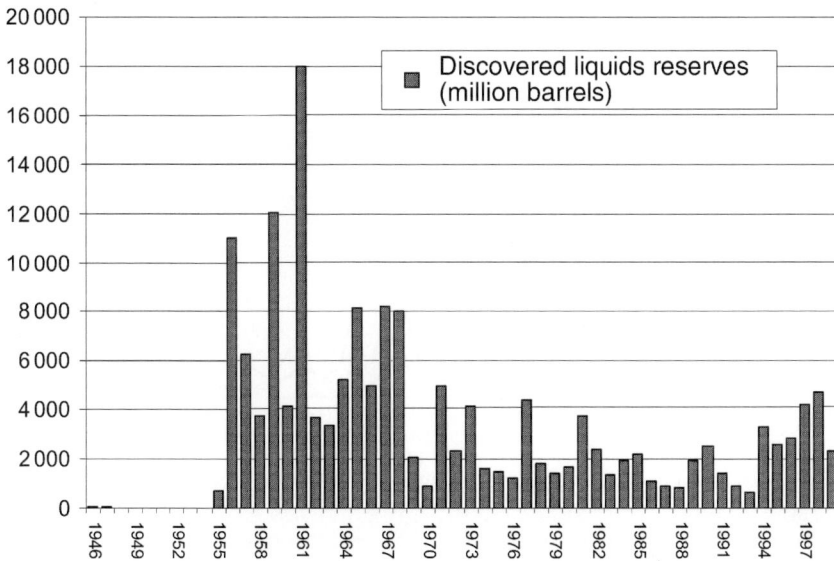

Fig. 7. Liquids reserves discovered 1946–1999.

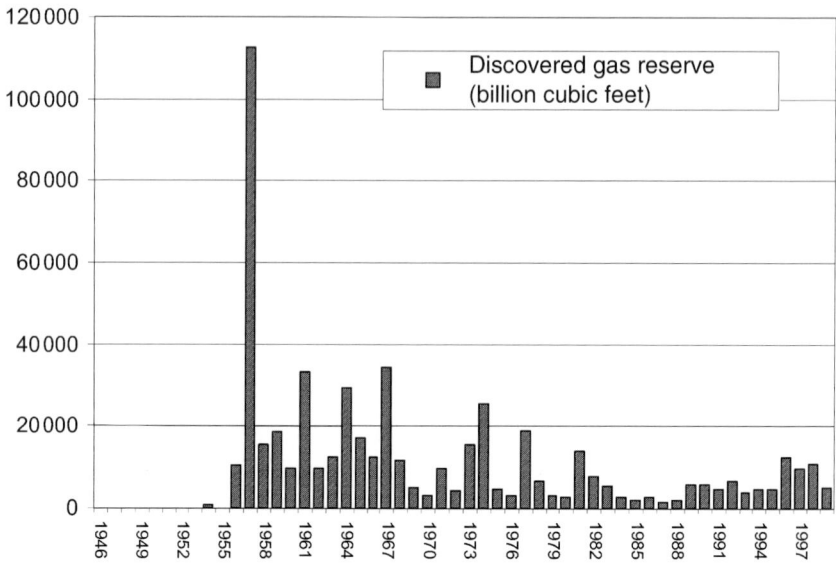

Fig. 8. Gas reserves discovered 1946–1999.

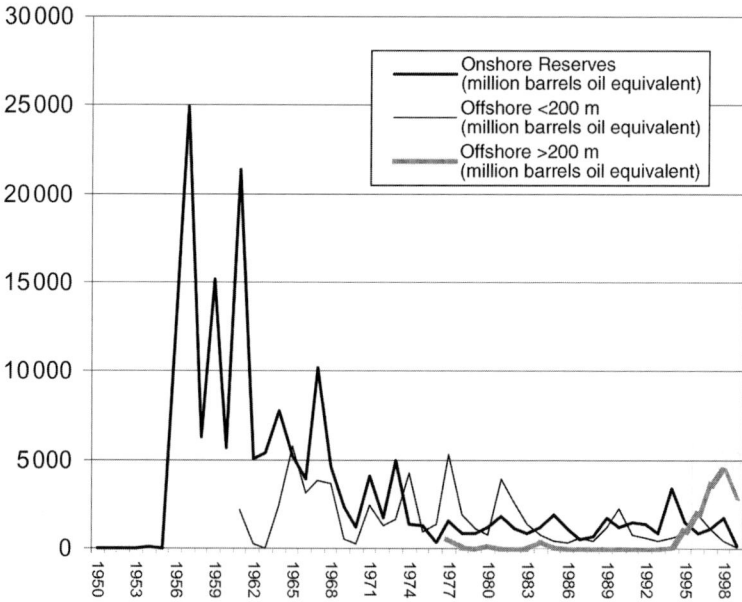

Fig. 9. Reserves development onshore and offshore in shallow and deep water 1950–1999.

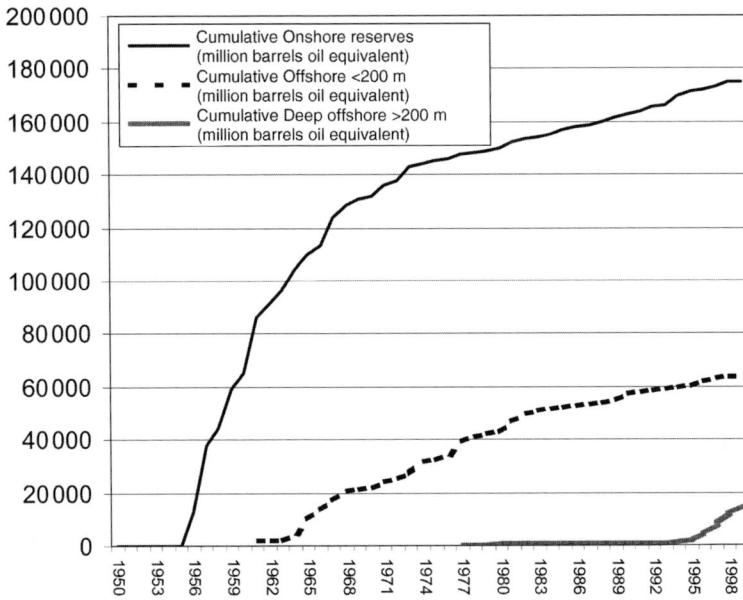

Fig. 10. Cumulative reserves development onshore and offshore in shallow and deep water 1950–1999.

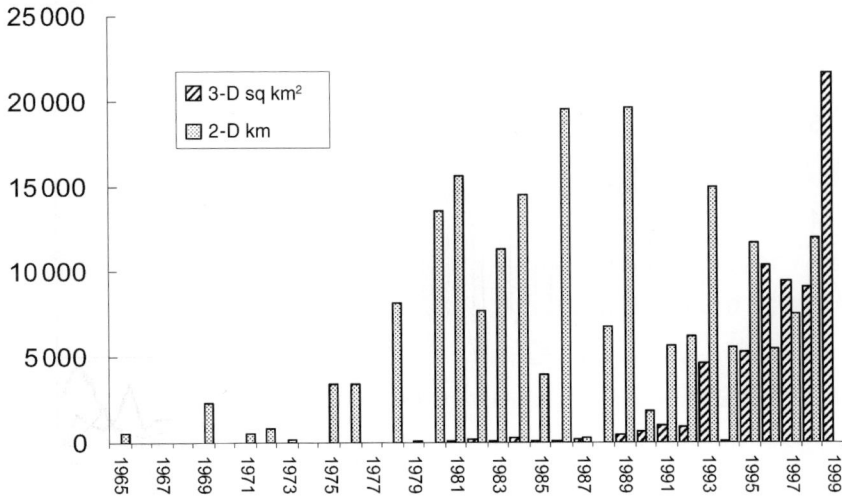

Fig. 11. Congo Basin offshore seismic data acquisition 1965–1999.

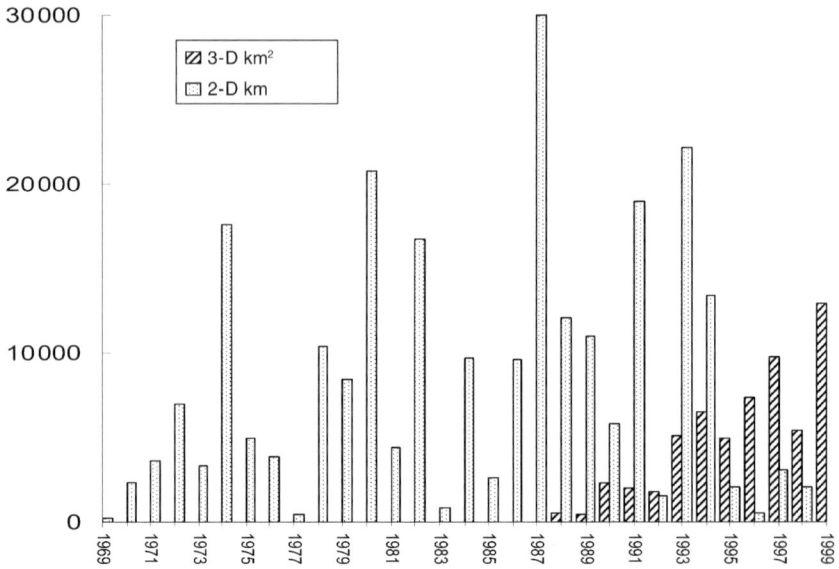

Fig. 12. Niger Delta offshore seismic data acquisition 1969–1999.

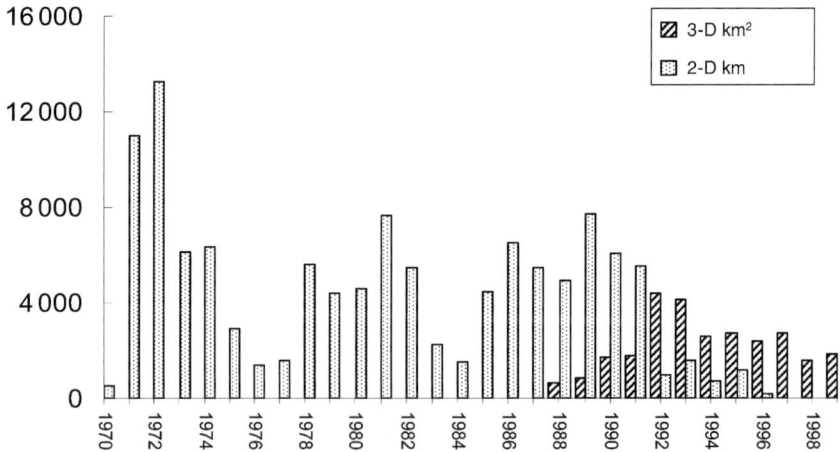

Fig. 13. Niger Delta onshore seismic data acquisition 1970–1999.

data has already given indications of where potential reserves lie, especially in the offshore Nile, Niger and Congo basins. In the more mature onshore North African basins, more reserves will be found by acquiring more 3-D seismic data and identifying the deeper and subtle plays. Elsewhere, there still remain vast areas of speculative potential in other relatively unexplored basins. A recent

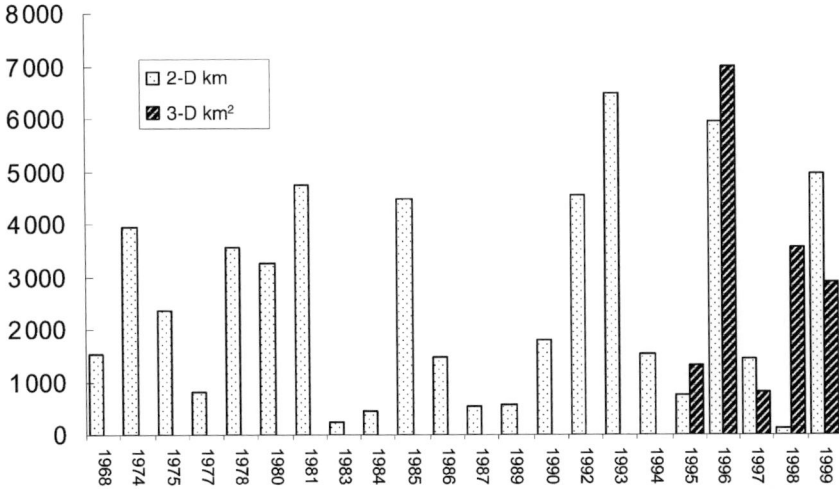

Fig. 14. Nile Delta offshore seismic data acquisition 1968–1999.

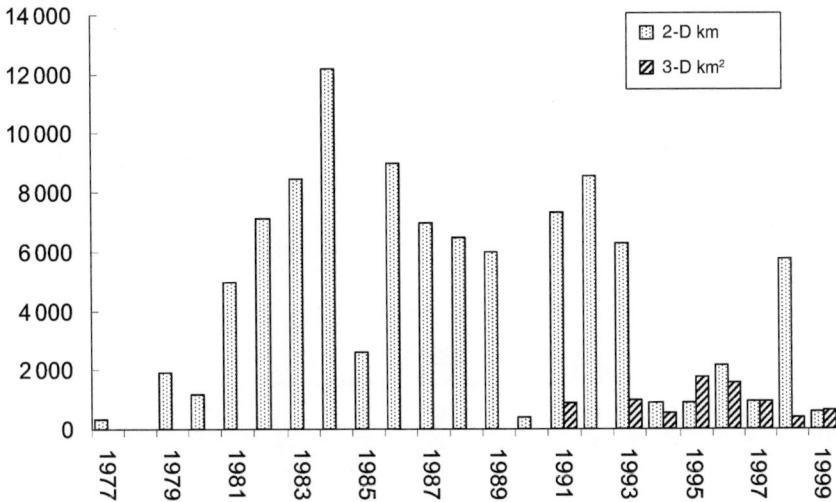

Fig. 15. Sirte Basin onshore and offshore seismic data acquisition 1977–1999. Note that some estimates have been made for 1987–1989 due to incomplete dataset.

example of this nature is the Talsinnt area, where a significant hydrocarbon discovery (Sidi Belka-cem 1) was made in 2000.

Appendix

All data, including the definition of basin outlines and types, referenced in this paper come from IHS Energy's IRIS21 database. Figures given reflect the state of the database on 1 January 2000.

The definition of basin types is derived and modified from systems proposed by Bally & Snelson (1980), Kingston et al. (1983) and St John et al. (1984).

Where not otherwise stated, reserve figures are 'initial recoverable proven plus probable' reserves.

This paper was made possible by the hard work of many who have collected and entered data into the Petroconsultants/IHS Energy database over the past 40

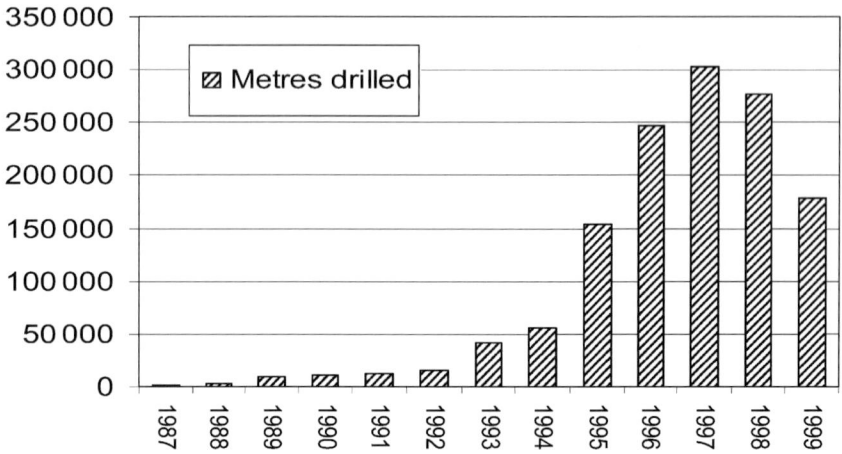

Fig. 16. African horizontal drilling by metres drilled (measured depth of hole) 1987–1999.

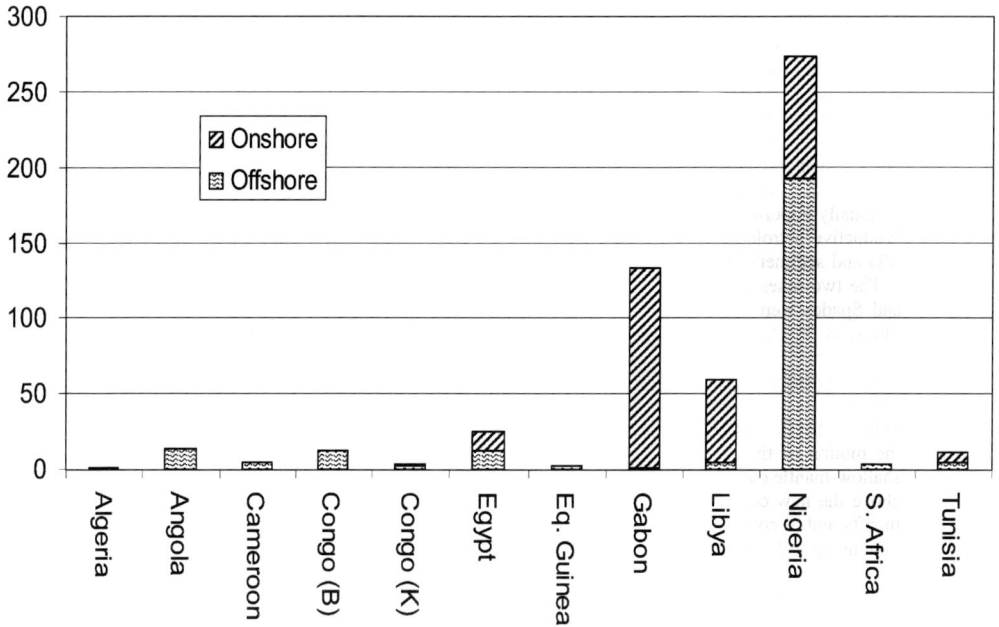

Fig. 17. Numbers of horizontal wells by country, onshore and offshore to end 1999.

years. Particular thanks for assistance and comments go to T. Piperi, J. Pellaux, D. Arbouille and J. Roelofsen.

References

BALLY, A. W. & SNELSON, S. 1980. Realms of subsidence. *In*: MIALL, A. D. (ed.) *Facts and Principles of World Petroleum Occurrence*. Canadian Society of Petroleum Geologists Memoirs, **6**, 9–94.

KINGSTON, D. R., DISHROON, C. P. & WILLIAMS, P. A. 1983. Global basin classification system. *American Association of Petroleum Geologists Bulletin*, **67**, 2175–2193.

Africa's petroleum systems: four tectonic 'Aces' in the past 600 million years

K. BURKE[1], D. S. MACGREGOR[2] & N. R. CAMERON[3]

[1]*Department of Geosciences, University of Houston, Houston, Texas, 77204-5507, USA*
[2]*PGS Reservoir Consultants (UK) Ltd., PGS Thames House, 17–19 Marlow Road, Maidenhead, Berkshire, SL6 7AA, UK*
[3]*Global Exploration Services Ltd., Little Lower Ease, Cuckfield Road, Ansty, West Sussex, RH17 5AL, UK*

Abstract: We relate the depositional and structural histories of the sedimentary rocks containing Africa's primary petroleum systems to four tectonic intervals, which in the light of their widespread and beneficial consequences we designate as 'Aces'. The Ace of Clubs was the assembly of Gondwana by continental collision and the collapse and erosion of the mountains constructed during that assembly, which generated accommodation space through thermal subsidence over a vast area. Africa's oldest great reservoir rocks accumulated in that space during Cambro-Ordovician times (520–440 Ma). After a short-lived glacial interval, Silurian and Devonian source rocks formed parts of a thick section that was deposited as long-term subsidence continued. The Ace of Diamonds consists of the collision of Baltica with Laurentia at *c.*380 Ma and the collision between Gondwana and Laurussia at *c.*310 Ma. It also includes the intracontinental deformation and orogenic collapse associated with the latter event, during the course of which regionally important structures and rifts now containing hydrocarbon-bearing fill were generated. Productive petroleum systems involving older Palaeozoic source rocks are concentrated in the rifts and sedimentary rocks of this phase.

The two other aces relate to the plume-dominated break-up of Pangaea. The Aces of Hearts and Spades were the eruption of the Karroo Plume at 183 Ma and the eruption of the Afar Plume at 31 Ma. These plumes, because they both generated huge volumes of basalt during brief intervals, are considered to have come from the deep mantle where, for more than 200 million years there has been a discrete large volume of hot rock over which Africa has been slowly rotating. Perhaps as many as six other deep-seated plumes have risen from that deep hot volume. The importance of the Karroo and Afar Plumes comes from the fact that they arrested the motion of the African Plate and, on each occasion, fostered the establishment of a new shallow-mantle convective circulation pattern. Intracontinental rifts, basins and swells developed above the new convection pattern after both arrests. Organic-rich sedimentary rocks deposited in rifts and at continental margins that formed in response to the Karroo-Plume-induced plate-pinning episode (K-pippe, 183–133 Ma) are being buried today under piles of sedimentary rock eroded from swells that have been rising since the later Afar-Plume-induced plate-pinning episode (A-pippe) began at 31 Ma. The Afar Plume eruption is designated 'Ace of Spades' because oil and gas generated following source-rock burial by sediments eroded from Africa's active swells during the past 31 Ma together make up three-quarters of Africa's hydrocarbon resource. In addition, half of that petroleum lies in reservoirs deposited during this phase.

Two continental collisions and two episodes of interaction with deep-seated plumes

The complexity of petroleum systems is familiar. Sequential developments of appropriate conditions for the deposition of source sediments, reservoirs and seals, for the maturation of organic material and for the migration of fluids into and through traps, have to be exactly right. As is represented by the concentration of over 90% of African reserves in just seven countries, the probability of just the right combination of processes occurring in just the right sequence is low. Understanding what has happened in the formation of oil and gas resources is much easier after they have begun to be developed than predicting circumstances that are likely to lead to the discovery of new reserves. It is against this familiar background that we have chosen to outline the way in which four tectonic intervals, that we here call 'Tectonic Aces', have

From: ARTHUR, T. J., MACGREGOR, D. S. & CAMERON, N. R. (eds) *Petroleum Geology of Africa: New Themes and Developing Technologies.* Geological Society, London, Special Publications, **207**, 21–60. 0305-8719/03/$15
© The Geological Society of London 2003.

played a critical part in the establishment of Africa's petroleum systems. Our aim is to promote better understanding of these four remarkable intervals in the hope that improved comprehension may help in the search for new resources. We also hope to contribute to better appreciation of the mode of formation of known oil and gas occurrences, wherever they may be located.

Tectonics is defined as 'the large scale evolution of planetary lithospheres' (Basaltic Volcanism Study Project, 1981, p. 804). On the Earth, plate tectonics has been dominant, at least since *c.*4 Ga, because the formation and the aging of ocean floor, the oldest preserved fragments of which are about 4 Ga in age, are the processes by which most heat is removed from the Earth's interior. Africa over the past 200 million years has provided at least a partial exception to the dominance of plate tectonics. Africa's first two Phanerozoic tectonic aces were indeed linked to the inexorable, sequential, plate-tectonic processes of continental collision, related mountain-belt construction and ensuing mountain-belt collapse. That sequence of events characterized first the assembly of the 80 million km^2 Gondwana continent at *c.*600 Ma and later the assembly of the 150 million km^2 continent of Pangaea at *c.*310 Ma. However, the two tectonic aces that were dealt more recently to the hand of Africa were different. They were the result of interaction between plumes of deep-mantle origin and the lithosphere. The first interaction began when the Karroo Plume erupted at 183 Ma, and the second when the Afar Plume erupted at 31 Ma.

Interaction between deep-seated plumes and the African lithosphere appears to have resulted from the persistent presence of a large volume of low-velocity, most probably hot material underlying the continent at depths close to the core/mantle boundary (Grand *et al.* 1997). That material, heated from within and by conduction from the core, has remained hot for at least the past 200 million years, apparently because it has not been penetrated during that time by cold slabs of lithosphere subducted from the Earth's surface. Descending cold slabs provide a principal coolant for the Earth's mantle.

One element contributing to Africa's distinctive behaviour is, therefore, that the hot deep underlying mantle has escaped refrigeration for more than 200 million years. Another important control on Africa's behaviour over the same interval has been that the African lithosphere has experienced very little slab pull. Slab pull, which is generated by the excess weight of cold, relatively dense lithospheric slabs in the mantle, is the dominant plate-moving force (Lithgow-Bertelloni & Richards 1995). Because of a general absence of slab pull, Africa has moved only slowly with respect to the underlying mantle (Fig. 1). That slow motion

appears to have fostered efficient coupling between plumes of deep-seated origin and the lithosphere. Although no heat may have been removed from the deep mantle under Africa by slab refrigeration during the past 200 Ma, the deep mantle has not simply become gradually hotter and hotter. Eight deep-seated plumes: the CAMP Plume at 201 Ma, the Karroo Plume at 183 Ma, the Tristan Plume at 133 Ma, the Rajmahal Plume at 118 Ma, the Kerguelen Plume at 110 Ma(?), the 'Marion' Plume at 90 Ma, the Deccan Plume at 65 Ma and the Afar Plume at 31 Ma, have risen from the region of the deep-seated hot volume. At the time of their eruption, all eight plumes penetrated the African Plate lithosphere or the lithosphere of an antecedent plate (Fig. 2). The Karroo and Afar Plumes have been most important for Africa's petroleum systems because, at eruption, they locally elevated the base of the overlying African lithosphere and arrested the already slow rotation of the continent (Fig. 3). As a result, a new shallow-mantle convective circulation (Fig. 4) was immediately set up under the stationary plate, generating basins, swells, intracontinental rift systems and intraplate igneous rocks (cf. Holmes 1944; Burke & Wilson 1972; McKenzie & Weiss 1975; England & Houseman 1984; Burke 1996). The swells started to be eroded as soon as they had begun to form. It has been the deposition of sedimentary rocks eroded from the swells into rift systems and at Africa's continental margins that has provided the main link between the plumes derived from the deep mantle and petroleum systems which have developed near the surface of Africa.

Considering the effects of the four tectonic episodes together provides something of a cautionary tale about the complexity of Earth systems on the geological time scale. It reveals contrasts as well as similarities in lithospheric response to each of the two continental collisions and to each of the two deep-seated plume eruptions. We will not dwell on that subject further than summarizing the resulting geology in Table 1, but will plunge straight into consideration of Africa's rich hand of petroleum-system cards with the first Ace, which, because it was dealt first, we call 'the Ace of Clubs'.

The Ace of Clubs: collapse of mountains built during Gondwana's assembly as a result of tectonic escape and rifting with ensuing thermal subsidence to accommodate the deposition of reservoir and source rocks in a giant 'steer's-horns' basin overlying those rifts.

Gondwana was assembled from:

- Cratons more than 1 Ga in age.

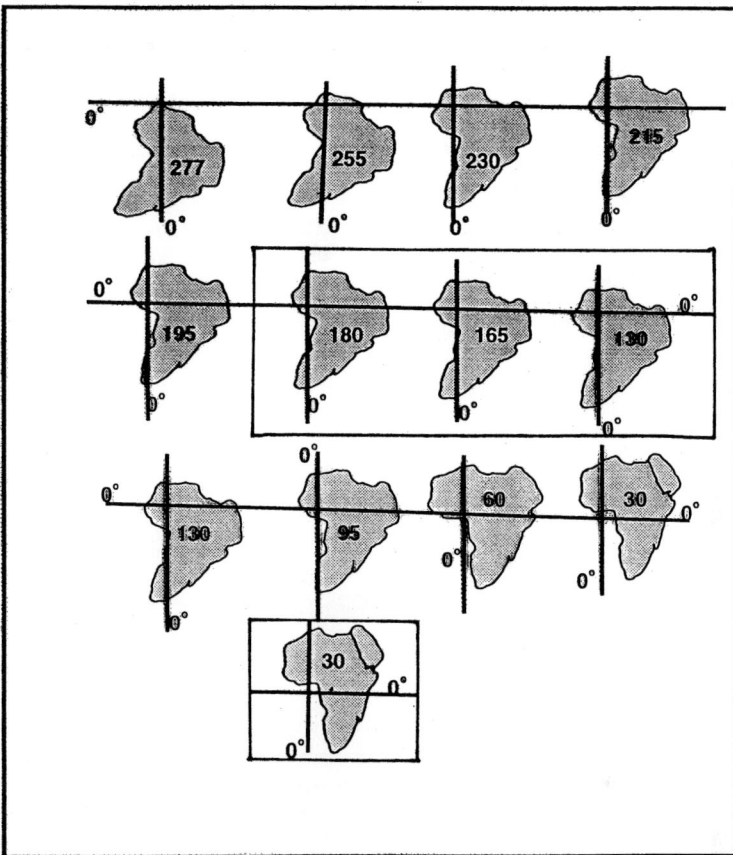

Fig. 1. Sketches indicating that the African continent has rotated only slowly with respect to the zero lines of latitude and longitude over the past 280 Ma. Based on palaeomagnetic and palaeoenvironmental indicators of latitude, the position through time of the '711' plume (Burke 2001) and ocean-floor magnetic anomalies. Boxes have been drawn around the positions of Africa at 30 Ma and its positions between 180 Ma and 130 Ma, since these are times when the African Plate was at rest with respect to the underlying mantle circulation (Burke 1996).

- Island-arc systems that had been active between c.1 Ga and c.0.6 Ga.
- Continental crust more than 1 Ga in age that was reactivated in Tibetan-style collisions during a c.100 million years interval bracketing 0.65 Ga (Fig. 5).

Eastern Gondwana, consisting mainly of Australia, Antarctica and India, was dominantly made up of old cratons that had been assembled at c.1 Ga, but western Gondwana contained an extensive area dominated by collided island-arc systems in the East African Orogen (Stern 1994) and large areas of c. 0.65 Ga reactivated older continental material in much of western Africa and South America (Fig. 5). Low-lying cratonic regions, e.g. the West African craton, which are individually up to 3 million km² in area, were distributed among the newly formed high mountains of western Gond-

wana. There is a resemblance between western Gondwana's mountains constructed at c.0.65 Ga and the mountainous areas of central Asia that have formed during the past 50 million years following India's collision with Asia (Dewey & Burke 1973). A major difference is that the mountains built at the time of Gondwana's assembly were much more extensive, covering an area of more than 20 million km², which is equal to that of North America today. No mountainous area as extensive has been constructed on the Earth since the assembly of Gondwana and there is little evidence that earlier mountain belts, such as those formed at c.1 Ga, were anything like as huge.

An area of c. 5 million km² of the newly formed continent's high mountains, in a region which is now occupied by North Africa and Arabia, collapsed during the final phase of the assembly of

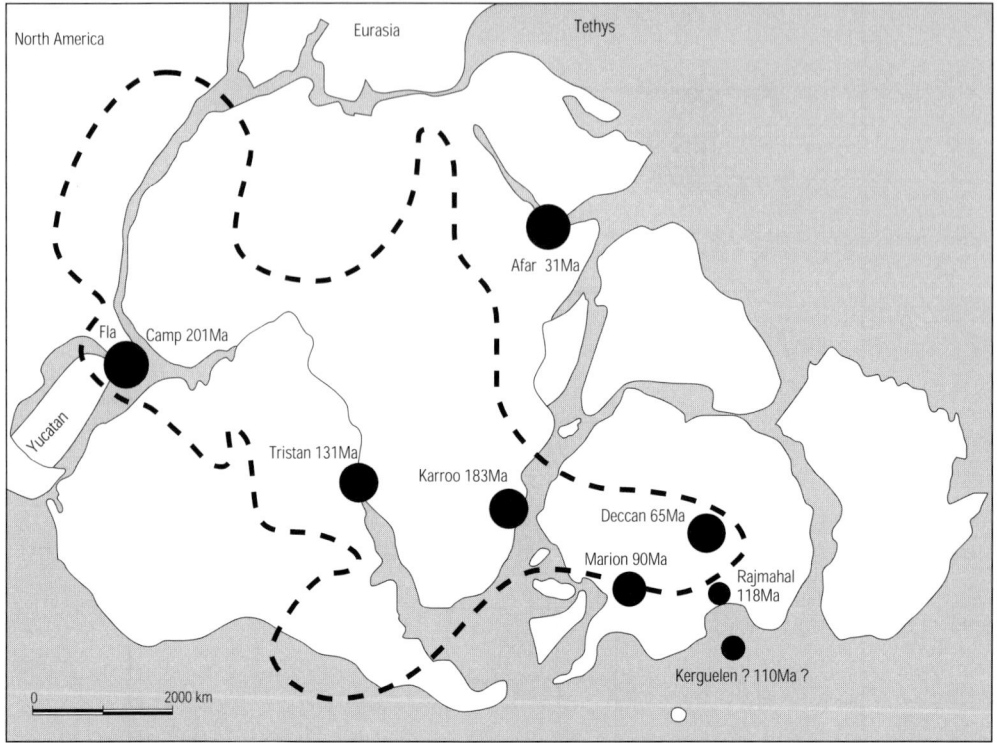

Fig. 2. Part of the Earth's surface showing the continents reassembled into Pangaea superimposed on a map of a deep-seated low-velocity region (dashed line) near the base of the mantle mapped tomographically by Grand *et al.* (1997) and projected radially to the Earth's surface from its position under Africa. The locations of eight plumes erupted over the past 200 Ma that are all regarded as of deep-seated origin (because they gave rise to Large Igneous Provinces (LIPs) in a few million years) are shown in their positions with respect to Africa at their time of eruption. Using the LIP criterion, these plumes represent about half of all the deep-seated plumes to have reached the Earth's surface during the past 200 Ma. This intense concentration of plumes within an area of <20 % of the Earth's surface is regarded as evidence that: (1) the extensive low-velocity volume under Africa today was the source of all those plumes, and (2) the low-velocity volume has been where it is now with respect to Africa throughout the past 200 Ma. The Afar and Karroo Plumes have proved to have been most important for Africa's tectonic evolution because they temporarily arrested the motion of the plate under which they erupted.

Gondwana, beginning at *c.*600 Ma (Dewey 1988). That collapse was accomplished by tectonic escape (Burke & Sengör 1986) much in the same way as South-East Asia is escaping today from the Tibetan collision (Molnar & Tapponnier 1975). The escape was accommodated on strike-slip faults that are well exposed in Arabia, in the eastern deserts of Egypt and Sudan, and in the Ahaggar (see Figs 5 & 6). The faults are best known in the Najd system of Arabia (for a review, see Sengör & Natalin 1996). Evidence of extension is seen in rifts in Oman, Saudi Arabia, Sinai and the eastern desert of Egypt, as well as in core complexes that have been described by Sturchio *et al.* (1983) and Blasband *et al.* (2000). Isotopic ages of igneous rocks in the rifts cluster between *c.*610 Ma and 520 Ma (Stern 1994), giving an indication of the duration of the time over which tectonic escape was in progress.

By the end of the episode of escape, 20% of the area of Gondwana's mountainous continental crust had been thinned from an average thickness of *c.*60 km to a thickness of *c.*30 km and the area of Gondwana had been increased by *c.*5 million km^2. Assuming that the continental area of the world was about the same as today (*c.*150 million km^2), that 3% increase in area would have raised sea level worldwide by *c.*40 m (Flemming & Roberts 1973). That rise would account for all of the well-documented worldwide marine transgressions of the Early and Middle Cambrian.

Thermal subsidence of the rifted lithosphere following the escape did not produce discrete steer's-horns basins over individual rifts of the kind first recognized by McKenzie (1978), but rather a huge composite basin over a complex system of rifts (Fig. 7). That basin extended from Oman, where

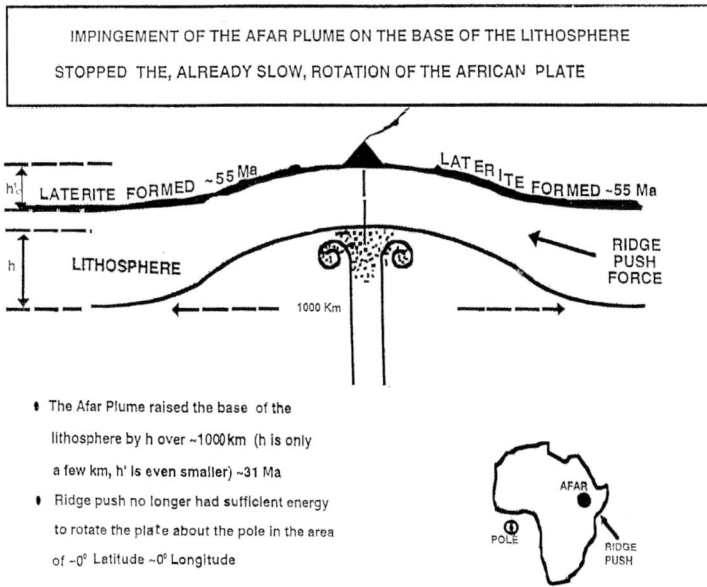

Fig. 3. The Afar Plume is considered to have elevated the lithosphere over an area $c.1000$ km in diameter when it erupted because the widely developed Late Palaeocene ($c.55$ Ma) laterite of Africa is missing from a 1000 km diameter area around the plume eruption site (Burke 1996). Elevation of the lithosphere is suggested to have prevented the weak ridge-push force from continuing to rotate the Afro-Arabian Plate so that the plate came to rest and the characteristic pattern of shallow-mantle convection associated with a stationary plate (McKenzie & Weiss 1975) became established. Because the pattern of rift initiation and intraplate igneous activity that followed the eruption of the Karroo Plume is similar to that which occurred after the eruption of the Afar Plume, that plume too is considered to have arrested plate movement.

the basin overlaid evaporite-bearing rifts (Droste 1997) to the Bové Basin south of Dakar and further, reaching into areas that are now in South America, Pakistan, India and Iran (Fig. 8). Marine and non-marine sandstones were the most abundant rocks deposited in the huge basin. Following Burke (1999), we call those sandstones the 'North African and Arabian Cambro-Ordovician Quartz Rich Sandstones' with the acronym NAACOQRS (which we suggest may be pronounced 'Naycockers'). The sedimentary environments of the NAACOQRS have been wonderfully described by Droste (1997; Fig. 9).

The sandstones are of exceptional extent and great thickness. We consider the sandstones to be generally first-cycle products of the erosion of latest Precambrian mountains. Their overall maturity (10–25% non-quartz minerals) is attributable to repeated reworking on the vast low-lying surfaces on which they were deposited. Our estimated original volume for the NAACOQRS sandstones of Africa and Arabia is 15 ($\pm$5) million km^3. If a comparable volume of sand were to be deposited over Europe today, it would average more than 2 km in thickness everywhere. Most of the NAACOQRS were deposited on top of the rifts

that had been generated during tectonic escape in accommodation space attributable to thermal subsidence between the Late Cambrian ($c.520$ Ma) and the end-Ordovician or Hirnantian glacial episode (at 440 Ma, Burke 1999). Isotopically determined ages cited in this paper are mostly from Harland *et al.* (1989), but later-published estimates have been preferred for Cambro-Ordovician and for some Mesozoic and Cenozoic intervals.

Elevation of the great topographic swells of North Africa and Arabia in the Ahaggar, Tibesti, Djebel Uweinat, Djebel Marra and in the hills on either shore of the Red Sea during the past 30 million years (Burke 1996) has exposed much of the NAACOQRS at outcrop. The sandstones are well known, in Saudi Arabia and Jordan, for example, around the ancient Nabatean city of Petra and in the Algerian Sahara (Beuf *et al.* 1971). Elsewhere producing oil- and gasfields such as Hassi Messaoud have yielded understanding complementary to that obtained at outcrop (e.g. Droste, 1997; Fekerine & Abdallah 1998).

Few other sedimentary basins closely resemble that in which the NAACOQRS were deposited because no comparable episode of orogenic collapse has happened since that of the mountains

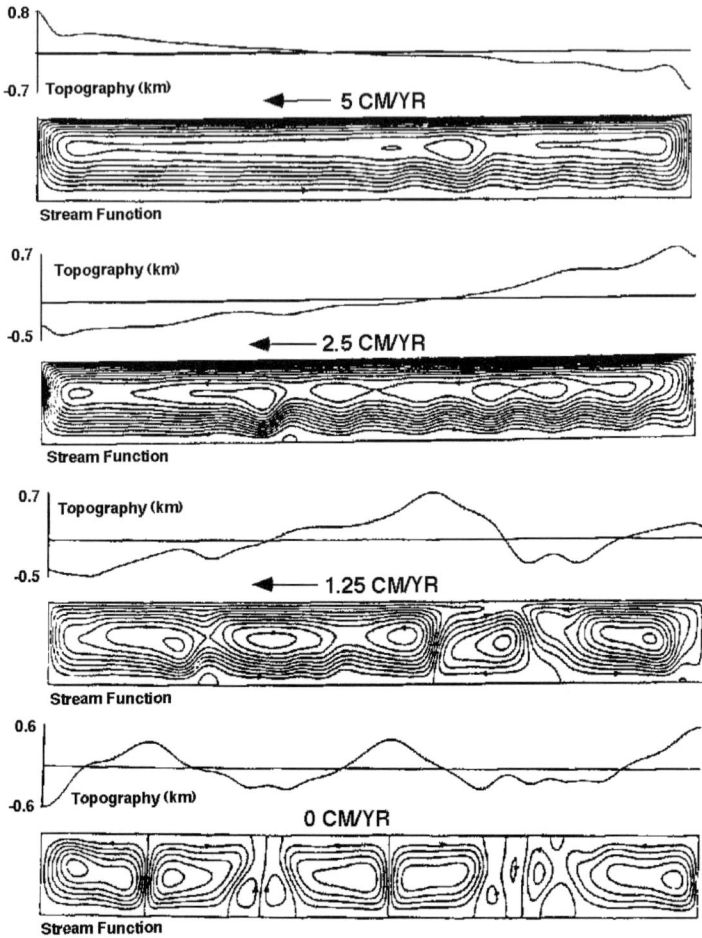

Fig. 4. Sketch (simplified from England & Houseman 1984) indicating the results of a numerical experiment that showed how shallow-mantle flow under a moving plate is streamline flow. Basin and swell structures, reminiscent of those on the surface of the African Plate, only form when the motion of the plate over the shallow mantle stops and a new shallow-mantle convection system forms. The Afar and Karroo Plumes are suggested to have arrested plate motion at 31 Ma and 183 Ma, after which times basin and swell structures were established.

constructed in the final assembly of Gondwana. One sedimentary basin of the later Phanerozoic that does provide something of an analogue to that in which the NAACOQRS were deposited is the much more richly petroliferous western Siberian Basin. The West Siberian Basin differs from that in which the NAACOQRS were deposited in: (1) containing a much larger proportion of shale, and (2) extending over an area of only 4.5 million km², which is less than half the extent of the NAAC-OQRS basin. In Western Siberia, Jurassic and Cretaceous sedimentary rocks overlie, in a composite steer's-horns basin, several individual rifts of a Permo-Triassic rift system that had formed during the final collisional assembly of Pangaea (Sengör 1995; Sengör & Natalin 1996).

Thermal subsidence continued after the end of the Ordovician in North Africa and Arabia, persisting until late in the Palaeozoic. Glaciation at 440 Ma interfered briefly with the record in the subsiding accommodation space. The resumption of subsidence was marked by the deposition of Silurian shale source rocks that are locally developed in low-lying areas on the post-glacial flooding surface. The average total thickness of Palaeozoic carbonate-poor sedimentary rocks deposited in North Africa is almost 5 km (Boote *et al.* 1998, figs 3–8). Nowhere else in the world is comparable sedimentary cover so thick on the continental scale. We regard this exceptional thickness as a measure of how great the extension of the continental crust had been during the tectonic escape and rifting epi-

Table 1. *African geology summarized in terms of the 'Ace' stratigraphy.*

ACE OF CLUBS 600–390 Ma	ACE OF DIAMONDS 390–183 Ma	ACE OF HEARTS 183–133 Ma	THE JOKER in play 133–31 Ma	ACE OF SPADES 31–0 Ma
Collision, collapse, rifting, subsidence	*Two collision episodes with intraplate deformation and an orogenic collapse*	*First interval of plate arrest by a plume: K-pippe*	*Plate growth and relative tectonic quiescence as the African Plate rotated 45° about an internal pole*	*African Plate pinned by the Afar Plume: A-pippe*
600–520 Ma. Orogenic collapse following Gondwana assembly. Rifting and strike-slip faulting extended North Africa and Arabia. Collision continued further south.	390 Ma. In a first collisional episode (between Europe and NOAM). Foreland basins developed south of the Atlas while thermal subsidence continued in the rest of North Africa and Arabia.	183 Ma. Karroo Plume erupted. It attempted to pin Pangaea. 180–165 Ma. The attempt failed. NOAM and Greater Antarctica left Gondwana, initiating formation of the Central Atlantic Ocean and the Arabian Sea.	133 Ma. Tristan Plume erupted unpinning Afro-Arabia and facilitating formation of the South Atlantic Ocean. 126 Ma. Ocean began to open in the South Atlantic, on the Agulhas coast and in the Gulf of Guinea.	31–0 Ma. Shallow-mantle circulation was induced and basin and swell structure immediately set up. Volcanism on swell crests, except on cratons. Red Sea, Gulf of Aden and Ethiopian Rifts form on Afar Swell at 31 Ma. East African rifts propagated rapidly from Afar.
530–440 Ma. Thermal subsidence. Deposition of the NAACOQRS in the accommodation space generated in a giant composite steer's-horn basin over the rifts in North Africa and Arabia.	300–230 Ma. In a second collisional episode (between Gondwana and Laurasia) Pangaea was assembled. Intracontinental deformation became widespread within Africa as rifts and strike-slip fault systems formed, especially in East Africa, Madagascar and Congo.	180–130 Ma. Successful pinning of Residual Gondwana (Arabia, Africa, SOAM) in the K-pippe. Intracontinental rift systems initiated during this episode included: Yemen @ 150, Abu Gharadig @150 Ma, Sirt @ 150, South Atlantic @ 140, Gulf of Guinea @ 140, West African @ 140, Central Africa @ 140 and Anza @140 Ma.	90 Ma. 'Marion' Plume erupted: India left Madagascar. Indian Plate changed. 65 Ma. Deccan Plume erupted: India left Seychelles. Indian Plate again modified 90–31 Ma. Africa low-lying. Great rivers mostly flowing in K-pippe rifts. Episodic flooding over rifts. Organic-rich deposits at highstands.	Erosion of swells generates new (e.g. Nile, Ahwaz) and greatly expanded (e.g. Niger, Zambezi) deltas. Nile, Casamance–Senegal, Volta, Niger, Ogooué and Zambezi deep-water fans developed.
440 Ma. A brief interval of glaciation with a low sea stand.	270 Ma. Cimmeria left Gondwana, initiating formation of the Neo-Tethyan Ocean at a rifted Gondwanan margin which is notably well developed in Arabia where carbonate-dominated accumulation has persisted ever since.		**SEVEN WILD CARDS** (unusual phenomena) were dealt while the Joker was in play:	Four more **WILD CARDS**, all related to Alpine convergence, were dealt after arrest by the Afar Plume:
440–390 Ma. Thermal subsidence continued. Silurian and Devonian source rocks formed as part of an exceptionally thick Palaeozoic siliciclastic succession.	230 Ma. Orogenic collapse of some of the mountains formed in the assembly of Pangaea. New rift systems developed including the Newark, Moroccan and TAGI Rifts.	140–30 Ma. Thick sediments, including organic-rich material, accumulated episodically in these rifts.	(1) 126–112 Ma. South Atlantic lake over ocean floor and evaporite episode. (2) 126–95 Ma. Equatorial Atlantic transforms active until SOAM cleared Africa.	(8) 15 Ma. Zagros suture formed. Ocean floor in the Gulf of Aden (15–5 Ma) and the Red Sea (5–0 Ma) began to form in consequence. Newer Harrats began to erupt (15 Ma). Western Rift initiated (15 Ma). Tectonism in Suez (15 Ma).
	200 Ma. CAMP Plume erupted. Onset of plume-dominated tectonics for Africa. CAMP itself had little tectonic effect, but profound thermal effects on organic-rich sedimentary rocks. Since 200 Ma Africa has moved only slowly over the mantle.	150 Ma. ? Apulia left Residual Gondwana, initiating formation of the East Mediterranean ocean.	(3-7) Convergence and arc-collisional Alpine events at the margin of Afro-Arabia: 100 Ma, 84 Ma (Santonian), 65 Ma, 53 Ma and 40 Ma.	(9) Bitlis suture formed at 9 Ma. Deformation in NE Africa and the Levant. (10) 25 Ma. Maghreb convergence. Atlas Mountains and related foreland structures formed. (11) <5 Ma. Uplift of Jebel el Akhdar on the Libyan coast south of the Mediterranean ridge.

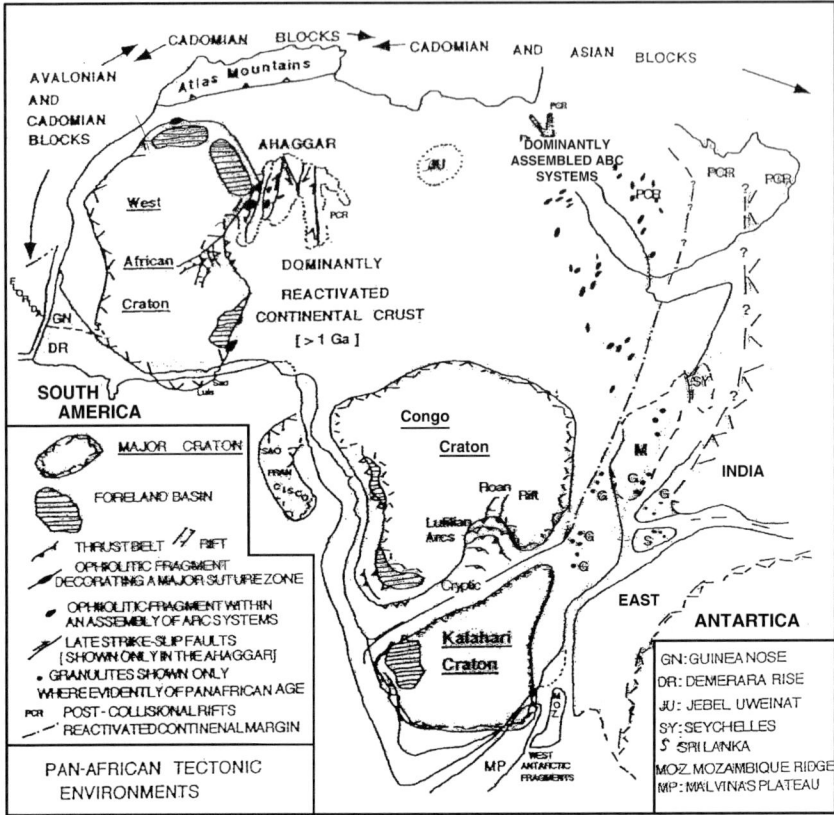

Fig. 5. Africa as it was when first assembled to become a part of Gondwana *c.*600 Ma. The newly assembled continent consisted of: (1) ancient cratons; (2) lately assembled arc systems, recognizable by their numerous contained ophiolitic slivers; and (3) reactivated continental crust. Tectonic escape, toward the top of the map, between *c.*610 Ma and 530 Ma took place on strike-slip faults whose positions are indicated on this map only in the Ahaggar. Similar faults are known in Arabia, Egypt and Sudan and are inferred to exist in the subsurface of Libya.

sode that took place between 610 Ma and 520 Ma. Gradual, and declining, thermal subsidence persisted until Mid-Devonian times (*c.*380 Ma). After that time, continuing subsidence was modified over North Africa as a result of the continental collisions that assembled the great continent of Pangaea.

The North African composite steer's-horns basin that formed by thermal subsidence over the Pan-African rift complex has controlled the formation of some of the world's most significant Palaeozoic-reservoired and sourced petroleum systems (Boote *et al.* 1998; Tables 2 & 3). Silurian source rocks, which contribute 20% of Africa's petroleum, were deposited over wide areas of the basin and have charged extensive reservoirs of the Cambrian, Ordovician, Devonian and Triassic, with the majority of the reserves sealed below Liassic salt. NAACOQRS contain important reserves in Hassi Messaoud, which is Africa's largest oilfield, and in

more recent discoveries in the Murzuk Basin of Libya. The Silurian organic-rich graptolitic shales were deposited over a wide area, extending from Guinea to Egypt and onward into Arabia. A second source-rock depositional event occurred in the Frasnian (*c.*380 Ma), with source rocks of that age charging reservoirs of the Late Devonian and Triassic, most notably in the Berkine Basin of Algeria (Boote *et al.* 1998; Echikh 1998). Both the reservoirs and source rocks of these petroleum systems are unusually extensive. The areal distribution of the systems was controlled by complex burial, generation and destructional histories, with success being largely concentrated in areas where the Palaeozoic source rocks went through the oil and gas windows in Mesozoic times (MacGregor 1998). For that reason, the main hydrocarbon-bearing areas lie in eastern Algeria and western Libya.

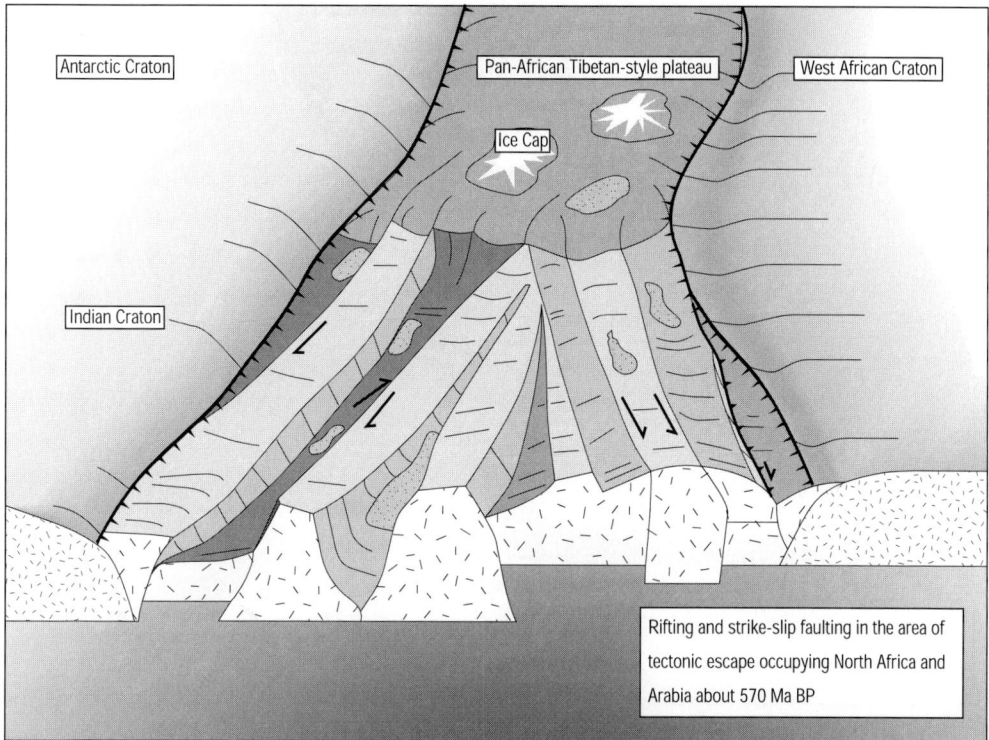

Fig. 6. Sketch indicating how tectonic escape on strike-slip faults and associated rifting thinned the thickened crust of the North African and Arabian part of newly assembled Gondwana. Continuing convergence maintained high areas that were soon to be eroded to provide the sand now embodied into the North African and Arabian Cambro-Ordovician Quartz Rich Sandstones (the NAACOQRS).

The Ace of Diamonds: the collision of Gondwana with North America, the related collisions that together assembled the great short-lived continent of Pangaea and the development of rifts formed by the collapse of the mountains formed during the Pangaean collisions

From an African perspective, the assembly of Pangaea can be seen as falling into two episodes. In the earlier, Mid-Devonian to Namurian (390–320 Ma) episode, much of Europe, then forming the continent of Baltica, collided with North America and Greenland, which at that time together made up the continent of Laurentia. That collision, which built a continent called Laurussia, has left a substantial record in a then adjacent area of northwestern Africa. Foreland basins, e.g. the Tindouf Basin, resulting from the collision formed in the area south of the present site of the Atlas and their depocentres, migrated eastward between *c.*380 Ma and *c.*320 Ma. Over the same interval, deformation

reactivated Pan-African structures far into the continent, e.g. in the Ahaggar (Coward *et al.* 2000) and in the Cuvette Central of the Congo Basin (Daly *et al.* 1992). This deformation has seriously compromised the possibility of preservation of oil that was emplaced in reservoirs during Palaeozoic times.

The second and final episode in the assembly of Pangaea involved the collision of Gondwana with newly assembled Laurussia. That collision contrasted with those that had led to the assembly of Gondwana about 300 million years earlier. Long, relatively narrow mountain belts in the Appalachians, the Hercynian mountains of Europe and the Ural Mountains between Europe and Asia were constructed at the time of the collision. No huge, areally extensive mountainous regions, related to either arc-assembly or to Tibetan-style regional continental thickening comparable to those that had formed during the final assembly of Gondwana, formed during the assembly of Pangaea.

The most distinctive characteristic of the Pangaean-assembling collisions, in relation to the formation of petroleum systems, was the extraordi-

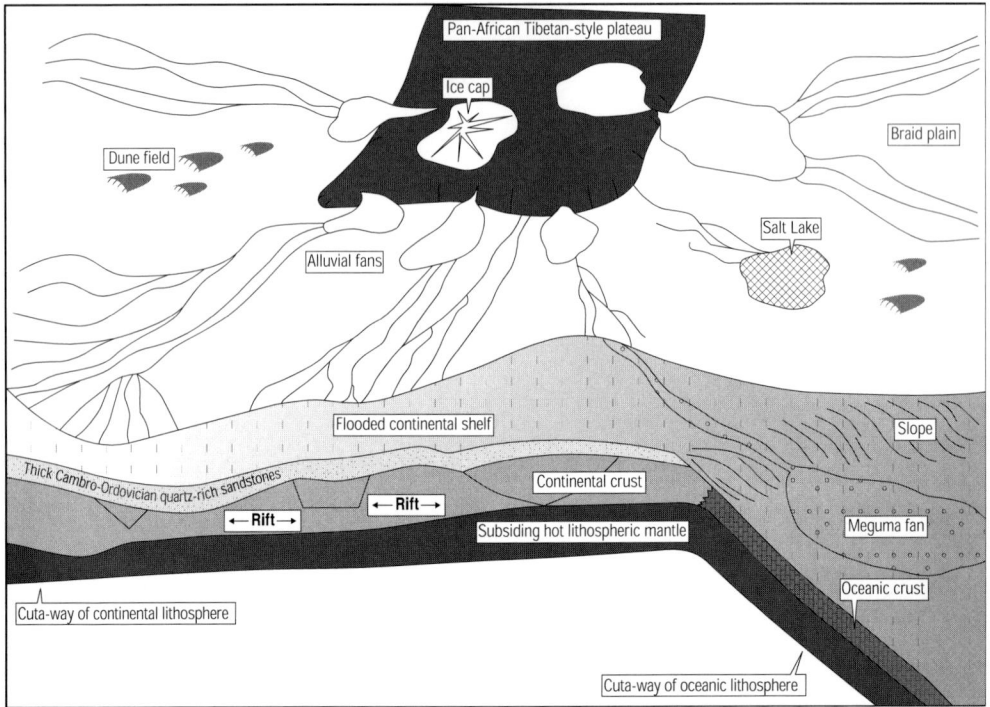

Fig. 7. Sketch indicating how NAACOQRS sandstones of North Africa and Arabia were derived by erosion from a high mountainous region and accumulated in a steer's-horns basin above rifts. The steer's horns are expressed by the distribution of the Cambro-Ordovician sandstones. Marine and non-marine shelf sandstones are represented. A deep-water fan equivalent, the Meguma, that accumulated off the coast of Morocco now forms much of the province of Nova Scotia in Canada.

nary extent of intracontinental deformation and related fluid migration remote from the mountain belts. Some indication of the extent of that deformation is shown in Figure 10, which depicts only the Gondwana continents. In North America the development of the Ancestral Rockies and the inversion of the Southern Oklahoma Rift were contemporary with the Appalachian collision. In Europe, the formation of the Hercynian foreland basin, as well as the Oslo Graben and other rifts of the North Sea rift system, took place at the same time. In Asia, the rifts and strike-slip fault systems underlying the West Siberian Basin began to form in the Permian (Sengör & Natalin 1996). Continental deformation remote from the collisional mountains, but contemporary with the collisions that assembled Pangaea, was distributed over *c.*100 million km^2 or about 60% of the area of the newly assembled great continent.

Permian coals in foreland basins that formed at the time of the final Pangaean collision in Argentina, South Africa and Australia, and in contemporary rifts in Africa, Madagascar, India and Australia (Fig. 10), first began to fuel colonial and commercial economic development about 150 years ago. Permian rifts underlying the Exmouth Plateau off northwestern Australia have played a part in the tectonic evolution of that great gas province. In Africa, the lacustrine-sourced tar sands of Madagascar (Raveloson *et al.* 1991), representing organic-rich materials deposited within part of the extensive rift system shown in Figure 10, are the best known petroleum-related materials. Strike-slip and thrust faults mapped in the interior of Africa, e.g. in the Congo (Daly *et al.* 1992), have been linked to contemporary folding of the Cape Fold Belt (De Wit & Ransome 1992). Cordilleran-type fold belts, of which the Cape folds are a well-exposed example, were active along the coast of Gondwana from Queensland in Australia to Colombia in northwestern South America during the time that the Pangaean collision was in progress (Fig. 10). A similar cordilleran system occupied the west coast of the North American margin of Pangaea, extending as far as the north coast of Greenland. The only substantial part of the margin of Pangaea that was not experiencing shortening was that extending from the Mediterranean to Indone-

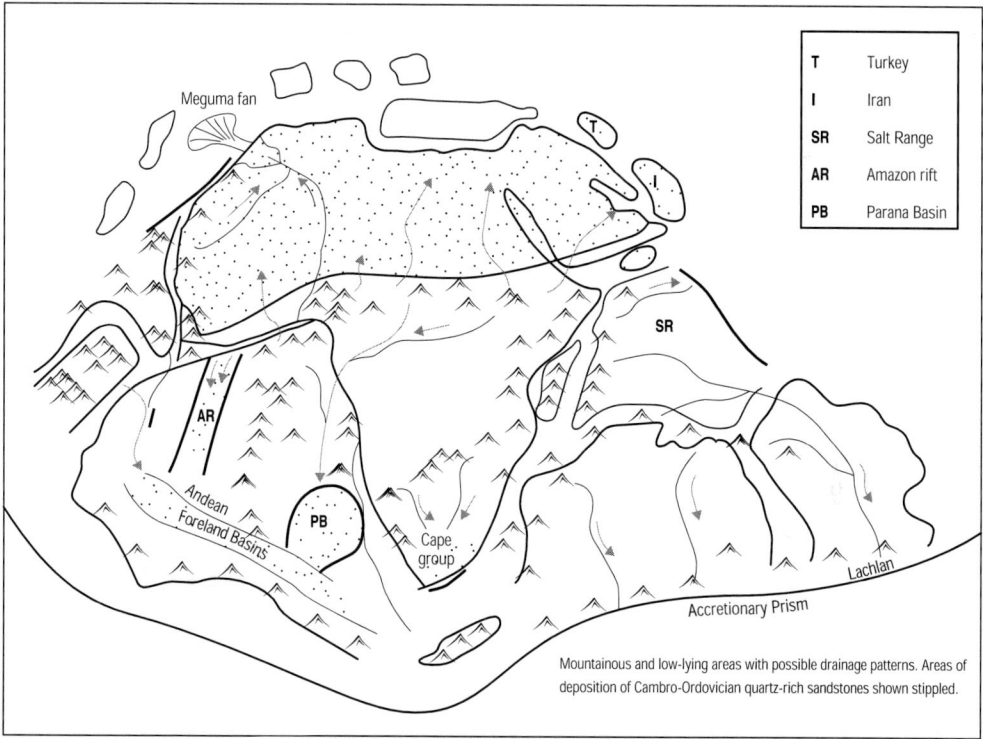

Mountainous and low-lying areas with possible drainage patterns. Areas of deposition of Cambro-Ordovician quartz-rich sandstones shown stippled.

Fig. 8. Gondwana showing the distribution of NAACOQRS sandstones at the time they were deposited over much of North Africa and Arabia. Correlatable sandstones are shown in South America, South Africa, Pakistan, Iran and Turkey. The position of the Meguma deep-sea fan in a marginal basin off the coast of Morocco is shown. Mountainous areas and possible river systems are indicated. Abundant zircons with Pan-African (750–600 Ma) ages in a deep-sea fan of the Lachlan fold belt may have been carried along rivers longer than any flowing in the world today.

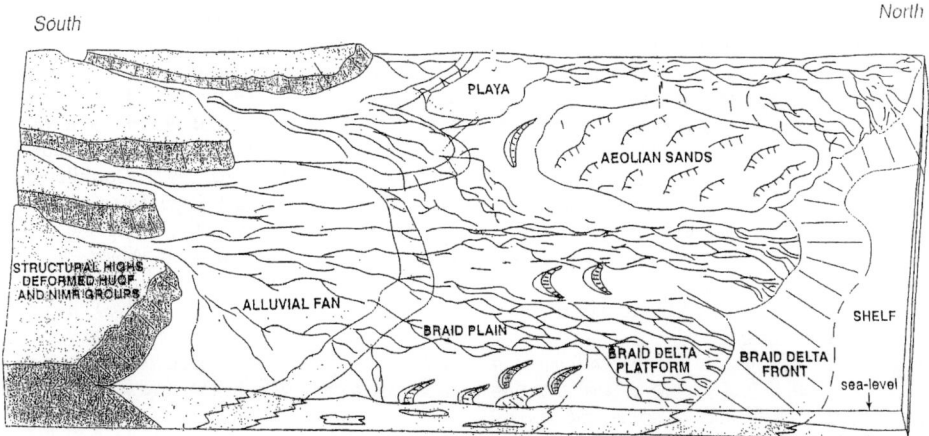

Fig. 9. Sedimentary environments recognized in the NAACOQRS of Oman (simplified from Droste 1997).

Table 2. *African petroleum systems with greater than 2 billion barrels oil equivalent (BBOE) of reserves (in-place volumes are used for tar sands)*

Rank	Predom. source name	Predom. reservoir name	Basin or country	Type field	Reserves – oil (10⁹ BBOE)	Reserves – oil & gas (10⁹ BBOE)	Source age	Reservoir age	Generation age	References
1	Akata	Agbaba	Niger Delta	Bonga	50	80	Pg–Ng	Oligo.–Plio.	Neogene	Haack *et al.* 1998
2	Tanezzuft	Triassic A-G	Algeria	Hassi 'R Mel	0.15	18	E. Silurian	Triassic	L. Cret.–Pg	Boote *et al.* 1998
3	Tanezzuft	Ra (Cambrian)	Algeria	Hassi Messaoud	10	10.5	E. Silurian	Cambrian	E. Cret.	Boote *et al.* 1998
4	Iabe	Malembo	d/water Angola	Girassol	8	9	L. Cret.	Pg–Ng	Neogene	Reynaud *et al.* 1998
5	Sirt	Sarir	Sirt, Libya	Messla	8	8.7	L. Cret.	E. Cret.	Pg–Ng	Burwood *et al.* 2003
6	*Bucomazi*	Pinda-Lucula	Angola, Congo	Malongo	4.5	8.2	Barremian	Albian	L. Cret.–Pg	Burwood 1999
7	Sirt	*Zelten-Sabil*	Sirt, Libya	Zelten	7	7.6	L. Cret.	Palaeogene	Pg–Ng	Burwood *et al.* 2003
8	Brown Lst.-Sudr	Rudeis	G of Suez	Morgan	7	7.4	L. Cret.	Neogene	Neogene	Schutz 1998
9	*Mio-Pliocene*	Mio-Pliocene	Nile Delta	Rosetta	0	7.1	Neogene	Neogene	Neogene	Swallow, pers. comm.
10	Sirt	Amal	Sirt, Libya	Amal	6.5	6.8	L. Cret.	Triassic	Pg–Ng	Burwood *et al.* 2003
11	Sirt	*Waha-Defa etc.*	Sirt, Libya	Waha	5	5.7	L. Cret.	L. Cret.–Dan.	Pg–Ng	Burwood *et al.* 2003
12	Sirt	Gialo	Sirt, Libya	Gialo	3.5	3.8	L. Cret.	Palaeogene	Pg–Ng	Burwood *et al.* 2003
13	Malembo	Malembo	Deep-water Angola	Bengo	3	3.5	Palaeogene	Oligo–Miocene	Neogene	Reynaud *et al.* 1998
14	Tanezzuft	Tadrart	Illizi, Algeria	Stah	1.5	3.5	E. Silurian	E. Dev.	E. Cret.–Pg	Boote *et al.* 1998
15	Tanezzuft	Triassic A-G	Triassic, Algeria	R. El Baguel	1	3.5	E. Silurian	Triassic	E. Cret.–Pg	MacGregor 1998
16	Sirt	Augila	Sirt, Libya	Augila	2.8	3.2	L. Cret.	L. Cret.	Pg–Ng	Burwood *et al.* 2003
17	Sirt	*Intisar*	Sirt, Libya	Intisar	2.8	3.2	L. Cret.	Palaeogene	Pg–Ng	Burwood *et al.* 2003
18	Bahloul	*El Garia*	Tunisia– Libya	Bouri	1.5	3	L. Cret.	Eocene	Neogene	MacGregor & Moody 1998
19	Pic Radioactiv	Triassic A-G	Berkine B, Alg.	Ourhoud	2.2	2.5	L. Dev.	Triassic	E. Cret.–Pg	Boote *et al.* 1998
20	*Sarir*	Sarir	Sirt, Libya	Bu-Attifel	2	2.3	E. Cret.	E. Cret.	L. Cret.–Ng	Burwood *et al.* 2003
21	Azile	Anguille	Gabon	Anguille	2	2.3	L. Cret.	L. Cret.	Neogene	Tiesserenc *et al.* 1989
22	Tanezzuft	Memouinat	Murzuk, Libya	Elephant	2	2.2	E. Silurian	Ordovician	L. Cret.	Boote *et al.* 1998
23	Pic Radioactiv	Aouinet	Illizi, Algeria	Zarzaitine	1	2	L. Dev.	L. Dev.	L. Cret.	Boote *et al.* 1998
TAR 1	*Sakamena*	Isola	Madagascar	Bemolanga	22 in place		L. Permian	E. Triassic	?E. Cret.	Raveloson *et al.* 1991
TAR 2	*Ise*	Abeokuta	W. Nigeria	–	?10 in place		Neocomian	L. Cret.	Neogene	Haack *et al.* 1998

normal font, marine source, clastic reservoir
italics, lacustrine source , carbonate reservoir
underlined italics, biogenic source
E., Early; L., Late; Cret., Cretaceous; Dan., Danian; Dev., Devonian; Ng, Neogene; Pg, Palaeogene

Table 3. *Primary African reserves tabulated by the age of the producing horizon, reservoir facies and source-rock type*

Main oil- and gas-producing horizons				Main contributing source-rock units			
Age	Facies	Region	% of African reserves (BBOE)	Age	Facies	Region	% of African reserves (BBOE)
Oligocene–Recent	Clastics–deltaic and turbidites	West Africa	43	Tertiary	Pro-delta	Niger Delta	38
Triassic	Clastics–fluvial	Algeria, Libya	14	Late Cretaceous	Marine	North Africa	23
Late Cretaceous	Carbonates	North Africa	7	Silurian	Marine	North Africa	20
Cambro-Ordovician	Clastics	Algeria, Libya	6	Early Cretaceous	Lacustrine	West, Central Africa and S. Sirt Basin	6
Palaeogene	Carbonates	North Africa	6	Late Cretaceous	Marine	West Africa	6
Neogene	Clastics	Egypt	6	Late Devonian	Marine	North Africa	2
Early Cretaceous	Clastics–Fluvial	Libya	5	Palaeogene	Marine	North Africa, Angola	2
			(Total = 87%)				(Total = 97%)

BBOE, billion barrels oil equivalent.

sia. Perhaps it was the distribution of compressional forces generated within Pangaea by the cordilleran margins, as de Wit and Ransome (1992) suggested, that led to the departure of the elongated Cimmerian block at *c.*270 Ma (Fig. 10) to form a microcontinent within the Tethys (Sengör & Natalin 1996, fig. 21.41*a*). It was on the newly rifted margin of Arabia from which Cimmeria had departed that the carbonate-rock-dominated successions so important to Middle-Eastern petroleum systems began to accumulate.

Deposition of generally thinner, non-marine sedimentary rock sequences continued into the Triassic (*c.*250–*c.*210 Ma) in the foreland basins and in the rift systems of the interior of Gondwana. During that time a separate more localized and distinct population of rifts developed as a consequence of collapse of the Appalachian Mountains (Dewey 1988) beginning in the Carnian (*c.*230 Ma). Rifting related to the collapse of the Appalachian Mountains is the third element of Pangaean assembly that has influenced Africa's petroleum systems. Three distinct elements – two collisions, the second with great effects in the far field, and one collapse – together make up the Ace of Diamonds. The Appalachian-collapse rift system, which is called the Newark-Rift System in North America, extended beyond the Appalachian Mountains into South America, reaching at least as far as the Takatu Rift of Guyana and Brazil. In the opposite direction it reached into the area presently occupied by the Atlas Mountains in Africa, into Europe in Iberia and onward around the British Isles, terminating in East Greenland. In the North Sea, the Triassic rifts linked with and reactivated the Permian rift system of northern Europe, which had formed part of the intracontinental deformation system associated with the collision between Gondwana and Laurussia (Fig. 10). The rift system initiated in the Carnian extended beyond the area now occupied by the Atlas Mountains into northeastern Algeria and Tunisia and possibly into parts of the Sirt Basin (Gras & Thusu 1998). From there it extended into the Apulia–Menderes–Taurus Block, which was soon to leave North Africa.

An indication of the widespread distribution of rifts active at this time in western Europe, North Africa and the northern Appalachians, and of the amount of extension in those rifts, is provided by distribution of evaporites of the Late Triassic and Early Jurassic, which has been sketched in Figure 11. The evaporites are shown to have been derived from Tethyan sea waters. Our inference is that rifting had extended the continental lithosphere sufficiently for the land surface in all the places in which the evaporites were deposited to have come to lie below sea level. As Arthur Hugh Clough wrote while sitting on an Apulian hill 150 years ago: 'Far off through creeks and inlets passing comes slowly flooding in the main'. Rifts with evaporites derived from Tethyan waters in this way are developed in Africa in three areas: (1) offshore

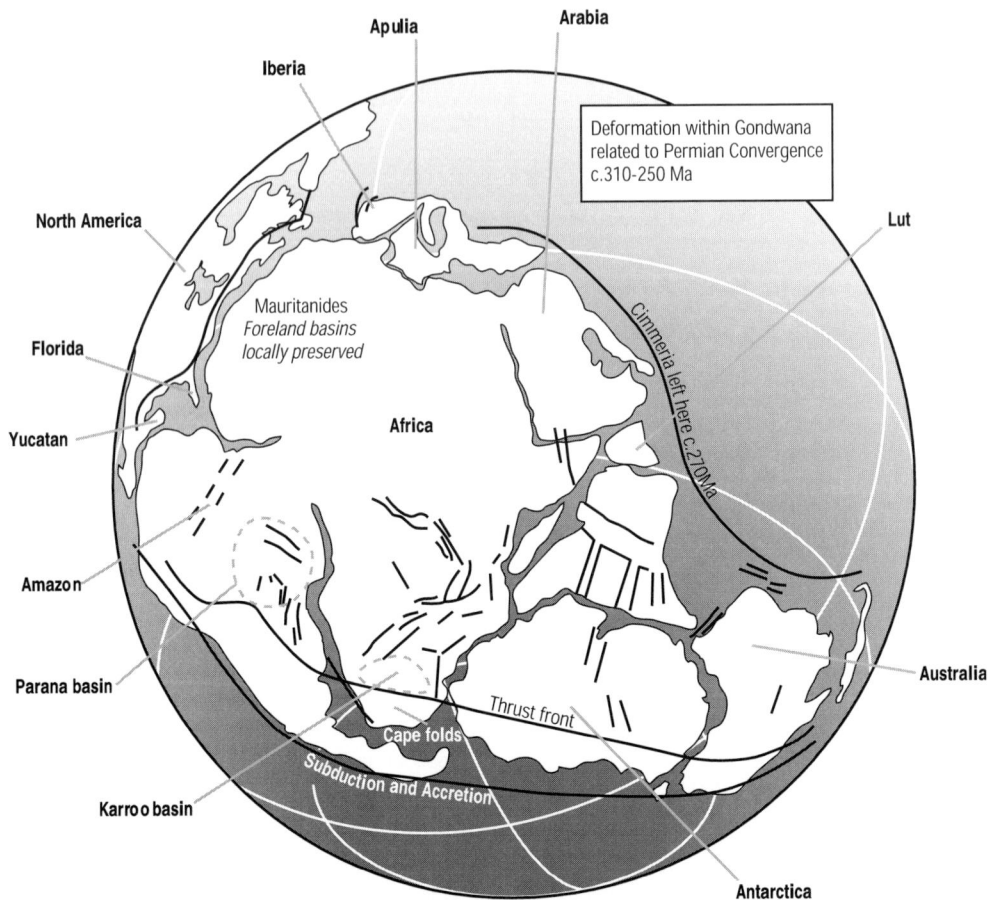

Fig. 10. Part of Pangaea close to the time at which it was assembled, indicating widespread deformation within Gondwana, mainly in the form of rifting and strike-slip movement, that accompanied the final collision of the continents along the Pangaean suture. Comparable intraplate deformation took place at the same time in North America, Europe and Asia.

Morocco; (2) within and south of the Atlas; and (3) in Tunisia, Algeria and Libya. In offshore Morocco, where the evaporites are conjugate with those off Nova Scotia, most appear to overlie rocks deposited in rifts (Jabour *et al.* 2000). They thus appear to be older than Aalenian (180 Ma), at which time the ocean floor in the region first formed (Withjack *et al.* 1998). There is, however, a possibility that some of the Moroccan and Nova Scotian evaporites are younger than 180 Ma and were deposited on ocean floor that had been formed under air.

The rift system of Algeria, Tunisia and Libya extends over about 0.5 million km² and is occupied by non-marine sedimentary rocks with local shallow-marine evaporites which together make up the TAGI or 'Triassique Argilo-Gréseux de l'Intérieur' (Fig. 11; Eschard *et al.* 2000). The extent of

the evaporites, called the 'formations salifères', gives an idea of how the waters of the Tethys, having crossed Apulia, flooded into subaerial, subsealevel depressions on four separate occasions during Late Triassic and earliest Liassic times.

The influence of the Ace of Diamonds on Africa's petroleum systems was two-fold: (1) intracontinental deformation led to destruction of existing petroleum systems during the collisions, and (2) formation of rifts and overlying steer's-horns basins associated with orogenic collapse in which several major petroleum systems would later develop.

The general lack of exploration success in areas where Silurian source rocks are modelled to have gone through the oil window prior to Permo-Carboniferous tectonism (MacGregor 1998) illustrates the major destructive effect of the Permian

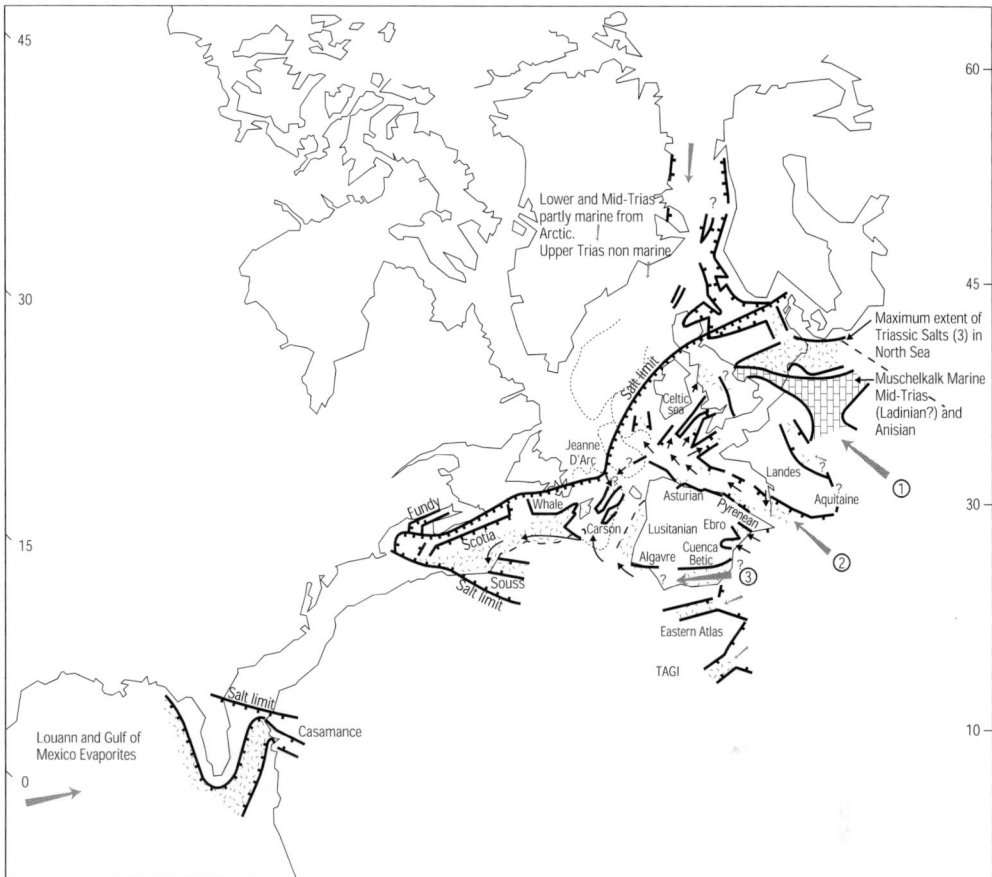

Fig. 11. Sketch-map of Europe, Africa, North America and Greenland on a closed Northern and Central Atlantic showing rifts (defined by stippling) that contain Late Triassic and Early Jurassic evaporites. These rifts, including rifts occupying the northern end of the Central Atlantic Ocean, extend from Agadir and its conjugate shore, the Gulf of Maine, to the Grand Banks, northern England and the North Sea. In Africa, rifts of this population with evaporites occur in and south of the Atlas and in the area in which the TAGI was deposited. The black arrows indicate how the waters of the Tethys spilled across sills (labelled 1, 2 and 3) into subsea-level rifts to form the brines responsible for generation of the evaporites. The evaporites at the mouth of the Casamance Rift are shown in this figure as related to the, much younger, Lou Ann evaporites of the Gulf of Mexico (Burke 1976a). The closed Atlantic has been fitted tightly by making the assumption that Ellesmere Island was about twice the size that it is today. Modified from DOBEX (1976, fig. 6; John Sclater was mainly responsible for the Atlantic reconstruction).

collision on then existing petroleum systems. One possible exception is the Ahnet Basin Petroleum System of central Algeria, where various authors have argued over the timing of the critical moment as being either Late Carboniferous or Triassic–Liassic (Logan 1998; Coward & Ries 2003.). Permo-Carboniferous folding did have a beneficial effect in creating structures that were later filled by oil and gas, including those of the giant Hassi Messaoud oilfield, as well as many of the structures of the fields of the Illizi and Ahnet basins.

The most significant effect of the orogenic collapse was the formation of rifts in which Triassic non-marine sands were deposited that include Africa's second-most significant reservoirs, with 14% of total African reserves. Liassic sealing evaporites were later deposited in the same rifts. These rifts, by deepening the Palaeozoic succession, also delimit the kitchens for the Silurian and Devonian source rocks of North Africa, although generation from the productive areas did not generally occur until the Jurassic–Cretaceous, when there was sufficient post-rift cover to exceed pre-Permian burial (Macgregor 1998). The Triassic is a significant reservoir in areas of salt-seal development in eastern Algeria and also in the Amal Field of Libya. The

Ace of Diamonds interval was not significant for oil-prone, source-rock development, either in Africa or globally. An exception is the Late Permian–Early Triassic lacustrine source rock that feeds the Madagascar tar sands (Raveloson *et al.* 1991).

A change in the style of African tectonics at 201 Ma when the plume-dominated break-up of Pangaea took over

The earliest manifestation of the regime in which Pangaea was to be broken up and in which deeply derived plumes came to dominate Africa's tectonic evolution came at 201 Ma with the eruption of the Central Atlantic Magmatic Province (CAMP) Plume (Fig. 12; Olsen 1999). The CAMP Plume was a huge plume in terms of igneous rock production (Marzoli *et al.* 1999). It generated *c.*3.0 million km^3 of basaltic rocks that are known mainly as sills and dykes in an area extending from the Maranhão Basin of Brazil to the eastern coast

of Greenland. Because the tectonic consequences of the plume's eruption were relatively small, we do not consider it to have constituted an Ace, although its eruption presaged the advent of plume eruption as the dominant control in African tectonics. The CAMP Plume may have had adverse consequences from the petroleum systems viewpoint because it may have raised the temperature of potential source rocks in extensive areas to above those in the oil window. The CAMP Plume generated no new rifts and, in contrast to other plumes that arose from deep beneath the African Plate, it contributed to the establishment of no new plate boundaries. A striking feature of the CAMP Plume's eruption, which it shares with most of the other seven deep-seated plumes that erupted into the African and predecessor plates over the next 170 Ma, is that it erupted into an already existing rift. In the case of the CAMP Plume, that rift was a member of the then active Newark system located in southern Florida (Fig. 12). An explanation of the eruption of plumes into pre-existing rifts has been

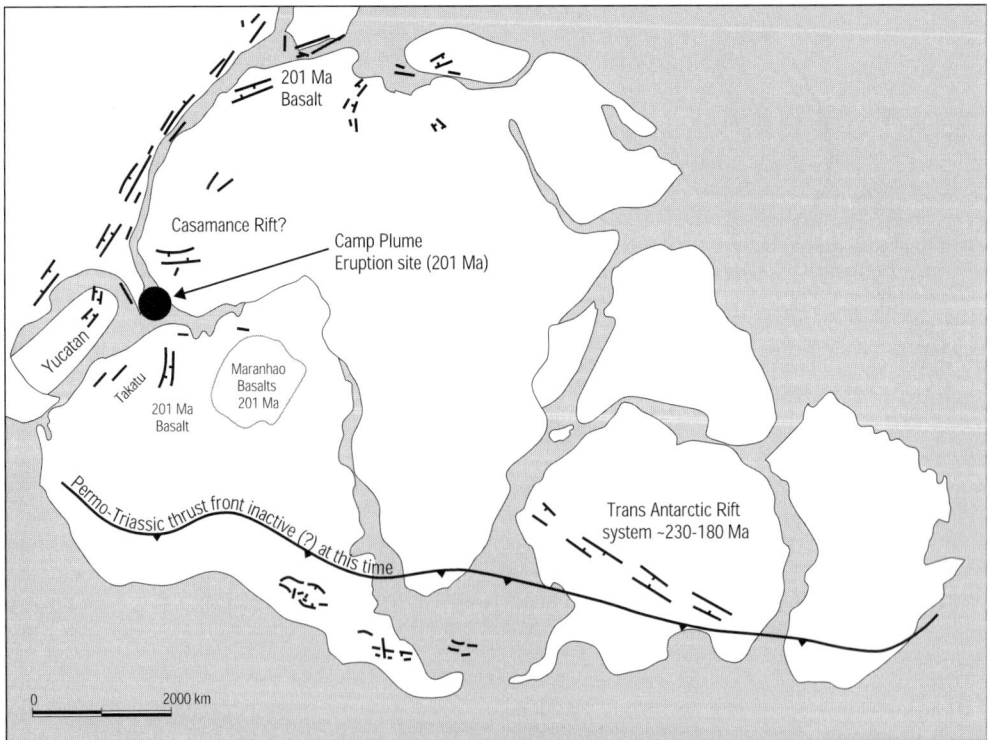

Fig. 12. The CAMP Plume erupted at 201 Ma into the active Newark rift system of North America, generating basalt from Brazil to Greenland (Olsen 2000; Marzoli *et al.* 2000). The plume may have impinged on the base of the plate at a nearby location and found the rift, within which it finally erupted at the surface by 'upside-down drainage' (Sleep 1997). The trans-Antarctic rift system active at this time was soon to be similarly invaded by the Karroo Plume (Elliot *et al.* 1999).

provided by Sleep (1997). He suggested that a plume reaching the base of the lithosphere from depth may there generate buoyant magma which travels by 'upside-down drainage' along the base of the lithosphere until it reaches a place, such as an existing rift, where the lithosphere is thin. Eruption of volcanic rocks is then localized above the place where the lithosphere is thin and not vertically above the place at which the plume impinges on the base of the lithosphere. This explanation works well for the CAMP Plume as well as for the Karroo, Tristan, Kerguelen and Deccan plumes, all of which erupted into pre-existing rifts, but it does not account for the location of the eruption site of the Afar Plume, which is remote from older rifts. The distinctive character of the Afar Plume is that it generated entirely new rifts by horizontal propagation from its eruption site (Burke 1996). For a very different interpretation see Ebinger & Sleep 1998). If the eruption of the CAMP Plume caused relatively little tectonic change, the eruption of the next plume to rise from the hot-deep mantle, the Karroo Plume, provides an adequate contrast. Its eruption was followed by the most significant of

all of the plate reorganizations that were involved in the break-up of Pangaea.

The Ace of Hearts: eruption of the Karroo Plume, its failed attempt to pin Pangaea, and the beginning, in earnest, of the break-up of that supercontinent

Eruption of the Karroo Plume began an episode of crucial importance for the development of Africa's petroleum systems because many rifts, both in the interior of Africa and at the continental margins, that now produce oil and gas were initiated in response to the Karroo Plume's eruption. The Karroo Plume erupted on what is now the southeastern coast of Africa at one end of an active rift system stretching across Antarctica (Fig. 13). Elliot et al. (1999) suggested that the rift crossing Antarctica, which had been active since c.230 Ma, was formed as a result of back-arc extension in the Gondwana continent parallel to the Cordilleran or Andean volcanic arc occupying the continental margin. The widespread initiation of intracontinental rift sys-

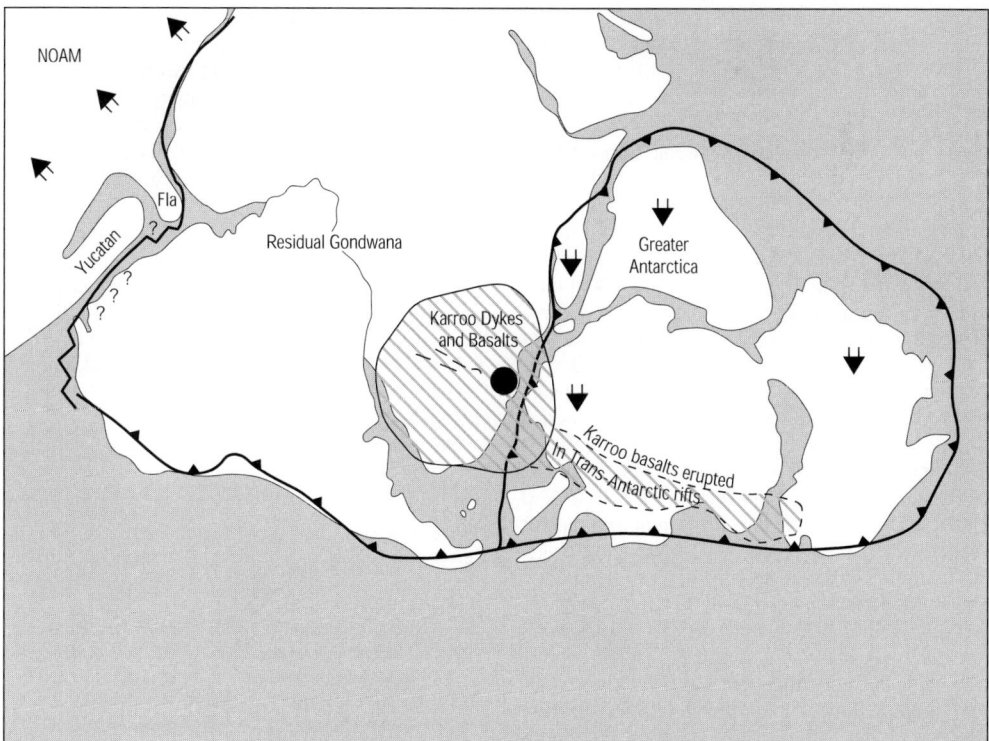

Fig. 13. The Karroo Plume erupted on the southeast coast of Africa at 183 Ma at one end of the then active trans-Antarctic rift, pinning Residual Gondwana. The rapid departure of North America (NOAM) at 180 Ma (Withjack et al. 1998) and Greater Antarctica at 180–160 Ma in the directions indicated by the arrows is interpreted as showing that the Karroo Plume may have attempted, and failed, to pin the motion of a larger area: perhaps all of Pangaea.

tems and the widespread eruption of intraplate igneous rocks between 183 Ma and 133 Ma in Residual Gondwana (Africa, South America and Arabia) closely resembles the initiation of rifts and intraplate igneous events that have characterized Africa since 31 Ma. Because that was the time when the Afar Plume erupted and pinned its overlying plate (Fig. 3), the Karroo Plume is inferred to have also pinned its overlying plate. Our inference is that the Karroo Plume set up an earlier version of the shallow-mantle convection-induced basin and swell regime that exists today in Africa (Burke & Wilson 1972; McKenzie & Weiss 1975; England & Houseman 1984).

The Karroo Plume may have failed in an attempt to make an even greater arrest: that of the whole of Pangaea. Evidence for this attempt is that North America and Greater Antarctica (a continent consisting of India, Antarctica, Madagascar, the Seychelles and Australia, including Greater New Zealand) both left Africa abruptly between 180 Ma and 160 Ma, i.e. they left immediately after the Karroo Plume had erupted at 183 Ma. Pangaea at the time was subject to little slab pull except that on the southern shore of Eurasia, where a reversal of subduction polarity had followed the beginning of the involvement of Cimmeria in a microcontinent/continent collision (Sengör & Natalin 1996, fig. 21.41c). Slab pull is likely to have been rapidly moving the floor of the world ocean (Panthalassa) occupying the site of today's Pacific Ocean, but only the relatively weak slab-rollback force and, on most of the margins of Tethys, the ridge-push force, were being applied to Pangaea from plate boundaries. For both the North American and the Greater Antarctic continents which promptly left Africa (Fig. 13), slab rollback at their cordilleran margins appears to have been a force sufficient to overcome the Karroo Plume's attempt at the pinning of Pangaea. Coupling of slab rollback with the South American cordillera appears to have been less efficient because South America remained with Africa until the plate-moving forces were again perturbed at 133 Ma (Fig. 14). The activity during much of the Jurassic (201–140 Ma) of the La Quinta arc-crestal rift system and the rift system on which the Rocas Verdes marginal basin was to form (Burke 1988, fig. 4) is indicative of extensional forces along the 6000 km Andean arc crest (cf. Molnar & Dalmayrac 1981). The Jurassic arc-crestal rift of the Andes appears to have interfered with the coupling of a Pacific Ocean floor plate to South America and thereby diminished the influence of the slab-rollback force on the South American part of Residual Gondwana.

The part of Pangaea successfully pinned by the Karroo Plume, the continent of Residual Gondwana (Fig. 14), consisted of Africa, South America and Arabia. Apulia appears to have left Residual Gondwana between 180 Ma and 130 Ma, but exactly when is unclear. An extensive intracontinental rift-system network was initiated within Africa, South America and Arabia between 180 Ma and 133 Ma, which was the time when the Tristan Plume erupted, Afro-Arabia was unpinned and South America departed (Fig. 15). Establishing the time of initiation of an intracontinental rift is difficult because the basal sediments in the rift are typically unfossiliferous and the igneous rocks that might indicate age are sporadically distributed. The latter are also hard to distinguish from rocks erupted later into the rift and are often altered in ways that make age determination difficult. In spite of this difficulty, there are concentrations of rift-initiation ages between Toarcian–Aalenian (*c.*180 Ma) and Barremian (*c.*133 Ma) in Central Africa and Sirt (Gras & Thusu 1998). Identical ages are present in the Benue Valley (Burke 1976*a*; Maluski *et al.* 1995) and in rifts disposed along what were later to become the Gulf of Guinea (De Matos 1999) and the South Atlantic (Bate 1999) (Fig. 14). It is on this evidence of timing that all these rifts are here treated together as elements of a population initiated during the Karroo-Plume-induced plate-spinning episode (K-pippe).*

It is intriguing to speculate as to whether there was an episode of basin and swell development over shallow upper-mantle circulation, comparable to that of today, during the Karroo-Plume-induced interval of plate-arrest (K-pippe). Because it was so long ago and because so much has happened at the surface of Africa since, the existence of former basins and swells can only be inferred indirectly. Erodable high ground was clearly present because thick sediments were deposited during the Late Jurassic and the Early Cretaceous (*c.*150–133 Ma) in the Central African, West African and Sudan rifts and at the mouths of the Anza, Casamance (Dombrowski *et al.* 2000) and the Sirt Rifts, which all reached the continental margin.

The rifts shown in Figure 14 have been ident-

---

* We are sensitive to the fact that the name 'Karroo' is given to a Supergroup of rocks, with a type locality in the Great Karroo of South Africa, that was deposited between the end of the Carboniferous (*c.*295 Ma) and the Toarcian–Aalenian (183 Ma) (De Wit & Ransome 1992). The eruption of the Karroo Plume generated only the topmost unit of the Karroo Supergroup: the Stormberg lavas. On the other hand, the Karroo-*plume-induced plate-pinning episode* (K-pippe), which has proved so important for Africa's petroleum systems, began with the last event recorded in Karroo Supergroup rocks and lasted over the next 50 million years. The acronym 'pippe' is used from here onwards for 'plume-induced plate-pinning episode'.

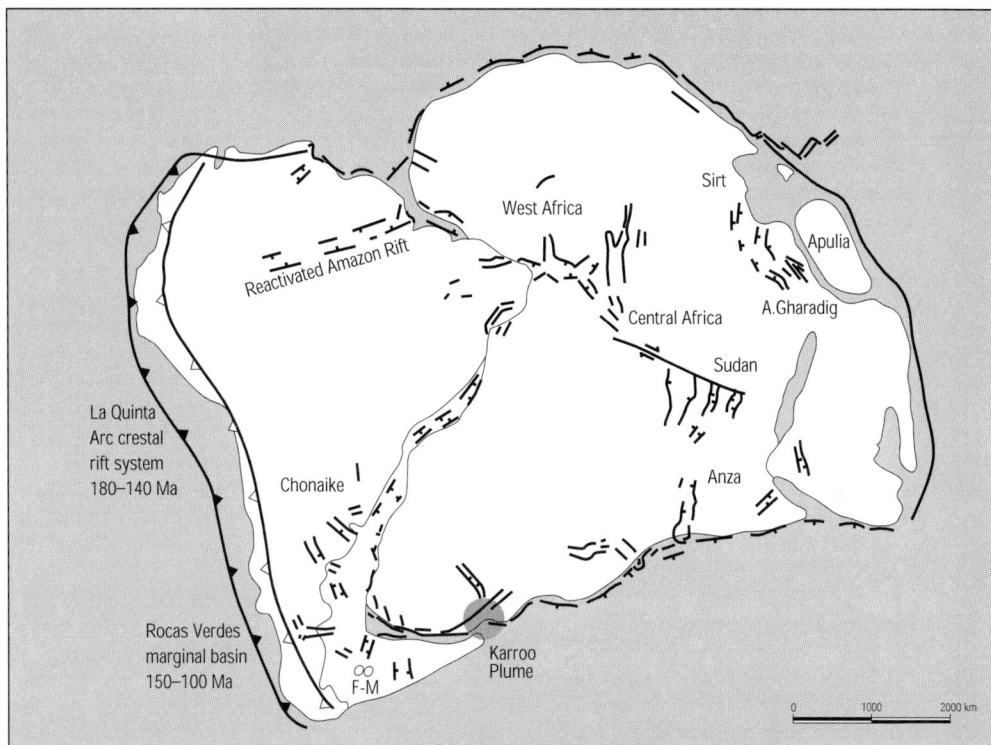

Fig. 14. Residual Gondwana as it was between 180 Ma and 130 Ma. Abundant intracontinental rifting is, by analogy to Africa's more recent response to the eruption of the Afar Plume, interpreted to have been a result of plate-pinning. Many of the rifts formed during this 50 Ma interval have developed significant petroleum systems. Some of those rifts still lie within the continents, but some are now at continental margins. The presence of a well-developed arc-crestal rift system in the Andes at this time may have been the reason that South America did not leave Gondwana when the Karroo Plume erupted.

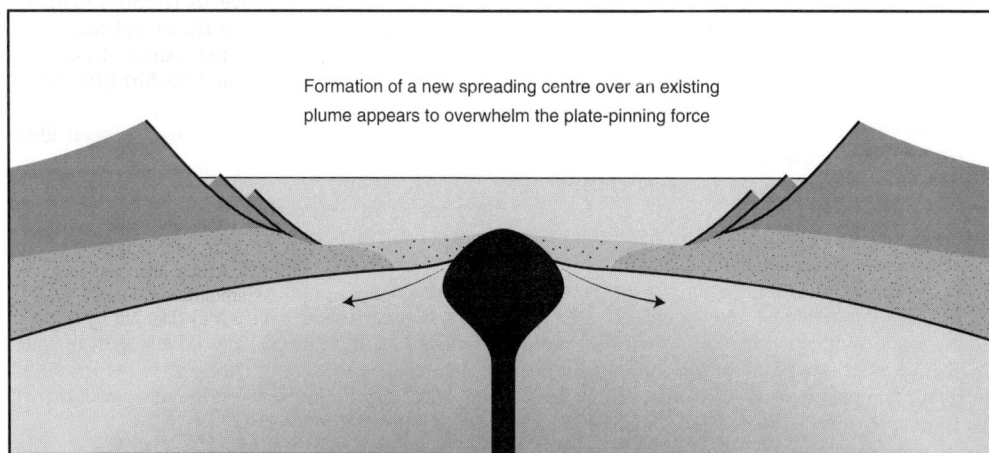

Fig. 15. Almost as soon as the Tristan Plume erupted at 133 Ma, the South Atlantic Ocean began to form (at 126 Ma). Once a spreading centre had formed over the plume it could not pin either of the flanking plates and served to augment the ridge-push force.

ified as a coherent population because they were all initiated during the time of the K-pippe, but many of them continued to receive sediments, in most cases episodically, long after pinning had ceased at 133 Ma. This is not surprising. Once a rift has been formed in a continent, the mass distribution beneath it represents such a perturbation of the surrounding continental structure that it is likely to be repeatedly reactivated either as a topographic high or as a depression when the stress field within the continent changes. Guiraud and Bosworth (1997, 1999) have analysed individual rifts of the African population showing what kinds of activity were in progress in which rifts at what times. This provides important information about rift evolution and the tectonics of the continent, but that information can only be fully appreciated by defining the timing and characteristics of the rift-initiation episode and distinguishing what happened during that critical interval from later reactivation events. A distinction can be made between intracontinental rifts that formed during the K-pippe (*c.*180–*c.*133 Ma), which are still located within continents, and rifts which were intracontinental when they formed during the K-pippe but are now in places where Atlantic-type margins have later developed (Fig. 14).

Considering first the Atlantic-type margins, Madagascar and Antarctica left the east coast of Africa between 180 Ma and 160 Ma on a plate boundary that established an Atlantic-type margin along a chain of existing intracontinental rifts. These were mainly Permian in age having been initiated during the final Pangaean collision (*c.*310–250 Ma) (Fig. 10). Slab rollback acting on Greater Antarctica appears to have localized extension within Pangaea along that rift chain. Slab rollback pulling on North America appears similarly to have localized extension at *c.*180 Ma to form the Central Atlantic oceanic margin all round the bulge of Africa (Withjack *et al.* 1998). That ocean floor began to form as soon as the Karroo Plume erupted at 183 Ma. Ocean-floor formation was concentrated on rifts of the active Newark Rift population, which had been initiated by orogenic collapse at *c.*231 Ma.

An existing Atlantic-type margin had been rejuvenated on part of the northern coast of Gondwana by the departure of Cimmeria during the Permian (*c.*270 Ma). Further west the Apulia–Menderes–Taurus microcontinent left North Africa at some time between 180 Ma and 130 Ma, perhaps during the time of initiation of the Sirt rifts (*c.*150 Ma). The Moroccan north coast of Africa became active as a mainly transform boundary with ocean-floor development only in pull-apart basins along a chain of Newark rifts as the Central Atlantic Ocean began to open at 180 Ma (Sengör & Natalin 1996,

fig. 21.42). South Atlantic ocean floor nucleated on rifts formed during the K-pippe. The oldest ocean floor of the South Atlantic formed virtually simultaneously from southernmost Mozambique to the Guinea Nose at *c.*126 Ma (Austin & Uchupi 1982). Ocean floor was nucleated along a chain of rifts that had formed within Residual Gondwana which lie between what is now Africa and South America (Fig. 14). The Guinea coast and the southern coast of South Africa began to develop at this time mainly as transform margins.

Rifts initiated between 180 Ma and 130 Ma, which now lie at Atlantic-type margins of Africa and which have already proved significant in petroleum production, include the Gabon, Congo and Kwanza basins. Basins in the Equatorial Atlantic and on the Agulhas shelf may also be included. The source rocks and reservoirs to these petroleum systems were deposited during the infill of these rifts. The source rocks are mainly post-133 Ma in age, but Jurassic sources are present in eastern and southern Africa.

Rifts that formed during the 180–133 Ma episode which today remain within Afro-Arabia include: rifts of the West African system, rifts of the Central African system, the Yemen Rift, the Anza Rift, the Sirt and Abu Gharadiq rifts and the Shire or Urema Rift (Fig. 14). This brief list indicates, and Figure 14 illustrates, the great importance of the K-pippe in the subsequent history of Residual Gondwana. The complexity of Africa's petroleum systems is nicely illustrated by the observation that some of these rifts are major producers, some are prospective, some appear unlikely to become productive and some are as yet very poorly explored. Overall it is worth emphasizing that basins formed as consequences of this rifting episode contain two-thirds of Africa's petroleum.

The Joker: an interval of tectonic quiescence from 133 Ma to 31 Ma, during which time growth of the Afro-Arabian plate dominated over plume-induced behaviour

The reason for referring to this interval as a Joker is that, although it began with the eruption of a plume, the Tristan Plume, it was also a time when the influence of plumes was slight in comparison with the previous 50 million years and the subsequent 31 million years. The joke is that, in spite of the relative absence of plume activity, this interval was very important for the development of Africa's petroleum systems.

Eruption of the Tristan Plume at 133 Ma marked the beginning of the Joker's time. In contrast to the Karroo and Afar Plumes, its effect was not to pin the plate, but the reverse (Fig. 15). Soon after it erupted, by 126 Ma, the South Atlantic spreading centre formed over the Tristan Plume. From its position on the spreading centre, the plume augmented the ridge-push force. The South American Plate, newly formed at 126 Ma, rotated quite rapidly away from Afro-Arabia in a fixed plume reference frame while, in the same reference frame, the simultaneously formed Afro-Arabian Plate began to rotate slowly about an internal pole close to 7°N 11°E. That location is the site of the shallow-sourced '711' Plume above which igneous activity has been continuous since c.250 Ma (Burke 2001). The Afro-Arabian Plate rotated about 45° anticlockwise about a pole close to '711', which is now at Mount Oku in Cameroon between 133 Ma and 31 Ma (Burke 1996).

This slow rotation appears to have been sufficient to suppress the shallow mantle convection pattern that had been active during the K-pippe because no new rifts were initiated in Afro-Arabia between 133 Ma and 31 Ma. Deposition continued episodically in existing rifts and at the continental margins, but the Afro-Arabian continent was rapidly eroded and soon became low-lying with little igneous activity. The rift shoulders of the South Atlantic supplied abundant siliciclastic sediment to the continental margin between the time when the ocean floor first began to form, at 126 Ma, and in the Albian at 100 Ma. Subsequently, as accommodation space increased, carbonate environments characterized the South Atlantic margins of Africa until the rift shoulders were first bypassed by siliciclastic sediments c.95 Ma and then completely eroded by 75 Ma. From that time on siliciclastic sedimentary rocks have dominated the margins. The mature topographic state of the Afro-Arabian continent at this time, i.e. during the later Cretaceous and earlier Cenozoic (80 Ma in the Campanian to 31 Ma or the Early Oligocene), is illustrated in Figure 16. This sketch-map indicates the carriage of siliciclastic sediment to the continental margin as dominated by a small number of relatively large river systems, most of which flowed along rifts that had been formed between 180 Ma and 130 Ma. This observation serves to further emphasize the importance of the K-pippe during which those rifts were initiated.

Wild cards that were dealt while the Joker was in play

The interval during which the Joker was in play extended over 100 million years and affected an area of around 30 million km^2. There is a good record of how petroleum systems developed over this large area and long time interval because these changes have taken place during the relatively recent past. We have here chosen to emphasize

events that we consider to have been particularly important for Africa's petroleum systems during the 133–31 Ma interval.

The first wild card The first of these events, which we call 'wild cards' to emphasize their unusual character, was the development of the northern South Atlantic oceanic basin between the Walvis Ridge and Cameroon after 126 Ma and before 112 Ma during the Barremian and Aptian. This northern South Atlantic basin (Fig. 17) has long been recognized to be one of those places, like the Red Sea, the Mediterranean, the North Caspian depression, the Central Atlantic and possibly the Gulf of Mexico, where thick evaporites have been laid down directly on oceanic crust (Burke 1975, 1977; Burke & Sengör 1988). The evaporites in the first two areas were deposited during ocean-closing and the others during ocean-opening episodes (Wilson 1968, table 1).

The Tristan Plume erupted at 133 Ma into the South Atlantic part of the K-pippe rift system. Rifts along what is now the continental margin had been active there for about 10 million years. Ocean floor began to form soon after the Tristan Plume erupted at 126 Ma (Austin & Uchupi 1982). The last few M-series magnetic anomalies record short time intervals and, using maps of the distribution of those anomalies south of the Walvis Ridge–Rio Grande Ridge, the short-term progression in the opening of the South Atlantic between Cape Town and the Walvis Ridge has been discerned. North of the Walvis Ridge–Rio Grande Ridge off both the Brazilian and African coasts, no persistent M-series magnetic anomalies have been mapped. This is readily understandable by comparison with conditions in the Afar. The Afar, also formed under air, has no persistent magnetic anomalies; they appear to develop only where circulating ocean waters and sheeted dykes are present (Hall 1970).

A high volcanic rampart generated by the Tristan Plume and stretching along the oldest parts of the Walvis Ridge and the Rio Grande Rise appears to have sufficed to keep oceanic waters out of the northern South Atlantic basin for c.14 million years between 126 Ma and 112 Ma. During that time, ocean floor was constructed under air in a youthful ocean basin with a spreading centre, which, by analogy with those active today, was about 2.8 km below sea level. Thermal subsidence over 14 million years would have produced a subaerial basin by 112 Ma that lay at an average depth of 3 km below sea level. The location of the northern South Atlantic during the Barremian and Aptian (Burke 1975; Burke & Sengör 1988) helped to prevent the establishment of a permanently filled deep lake in the basin. Intermittently balance-filled lakes of varying salinity would have occupied the deep

Europe passing by

~ 140 →

Arc collides 84

Atlas

Syrian Arcs

Euphrates rifts

?

180

Tan Tan

Sirt R. Estuarine

Kuwait ~ 75 SSTS.

Thin Sporadic Continental Sandstones

Casamance & Senegal R.

Benue R.

Sudan Rifts Rivers

Abidjan

Anza

R.

125

Gabon Rivers

Rift R.

50AM Passing

R. Rufiji Delta

~ 160

125

Kwanza Rivers

Shire R.

Tswana ?

Z

Walvis

Orange River

L

125

African Low-Lying

Flooded at highstands

~ 125

Main Rivers:

Sirt, Casamance, Senegal

Benue, Orange, Shire-Zambesi, Anza

Z	Zambezi
L	Limpopo
125	Age of Oldest Ocean Floor

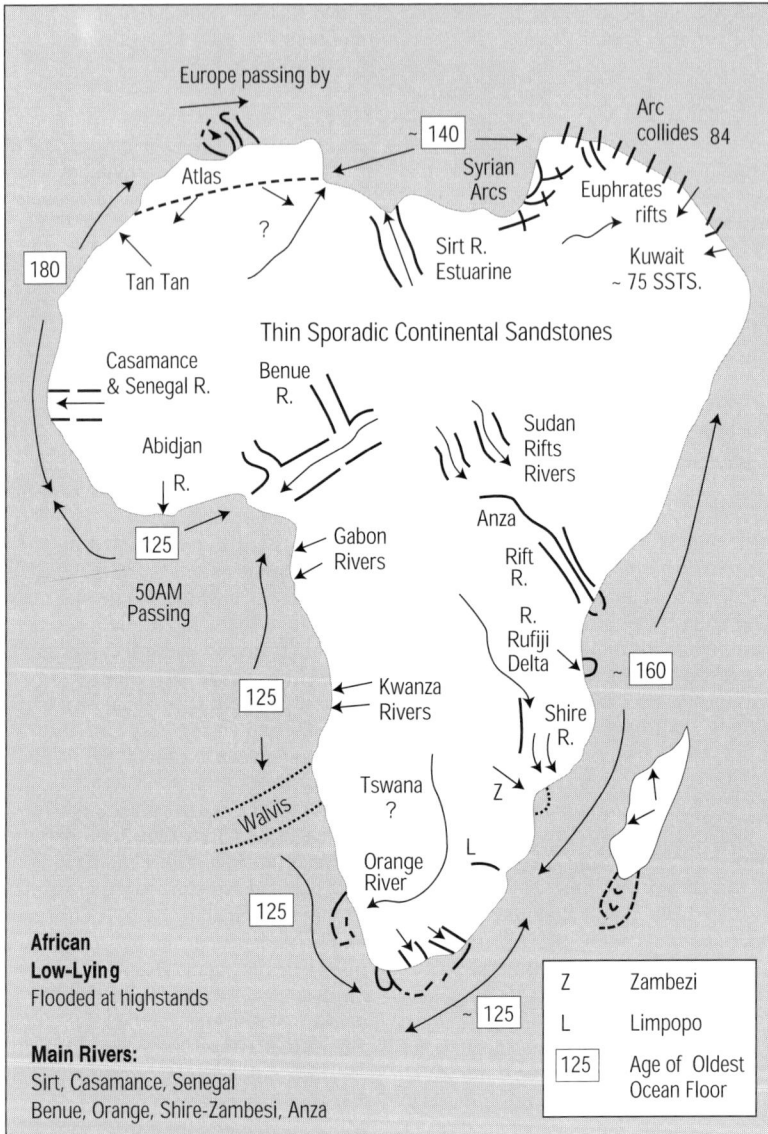

Late Cretaceous ~ 80–65 Ma

Fig. 16. Dominant river systems of Afro-Arabia as they were from *c*.80 Ma to 40 Ma after rift shoulders at the continental margins had subsided and been eroded and before the Afar Plume erupted at 31 Ma. Several of the largest rivers flowed along rifts initiated during the Karroo-Plume-pinning episode (183–133 Ma). Only the Orange River and the Tan Tan Delta River among major rivers appear not to have occupied rifts at this time.

subaerial basin and reached a maximum extent of about 1 million km². By comparing the geochemistry of evaporites in the Sergipe Basin and in Gabon, Wardlaw & Nicholls (1972) made a strong case that, when ocean waters did spill across the Walvis Ridge into the northern South Atlantic, both the Brazilian and the African shores lay within a single giant saline lake of *c*.1 million km² area.

This is about three times the size of the present Caspian Sea.

Cameron *et al.* (1999, fig. 5) showed that there was a change from active faulting in the then *c*.15 Ma-old rifts of the margin of the South Atlantic at about 126 Ma with the establishment of an overlying steer's-horns or sag environment. That change coincided with the time when active fault-

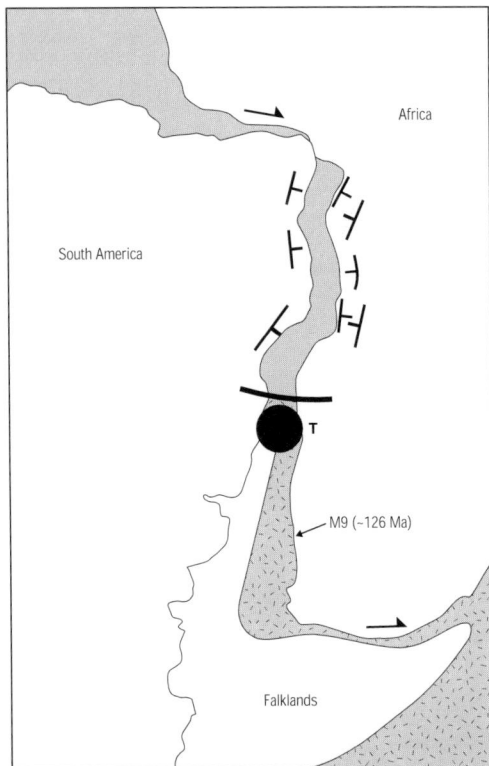

Fig. 17. South Atlantic at about 115 Ma. Normal ocean floor with magnetic anomalies (M9–M0) had begun to form south of the Walvis Ridge from *c.*126 Ma (Austin & Uchupi 1982). Tristan-Plume-derived volcanic rocks formed a barrier to oceanic waters and the northern part of the South Atlantic was occupied by a subaerial basin overlying ocean floor at a depth of *c.*3 km below sea level. That basin, which lay in desert latitudes, was occupied by balance-filled lakes, many of which were the site of organic-rich accumulations. In the Late Aptian (*c.*112 Ma) oceanic waters spilled into the deep basin and evaporites were deposited on the ocean floor, on transitional crust and in intracontinental rifts. Similar lakes on thin slivers of ocean floor probably occupied the Gulf of Guinea, but no evaporites are known from that area.

ing moved to the newly formed oceanic spreading centre to the west. We have extended figure 5 of Cameron *et al.* (1999) to the west in Figure 18 to give an idea of how lake sediments and evaporites were deposited on ocean floor, the transition zone and the rifts of the K-pippe population. The geometry of the drift section is also illustrated.

Because of the great thickness of sedimentary rock that has accumulated across the continent/ocean boundary, a challenging question in the interpretation of the geology of the South Atlantic margin of Africa is: 'What is the structure of the continent–ocean transition and how wide is

the zone of transitional crust?' In our sketch (Fig. 18), we have followed Rosendahl and Groschel-Becker (1999, fig. 1) who found, using deep-penetration seismic techniques, that the transition zone, away from fracture zones, varied from 150 km to 200 km in width. Fault structure above the Moho is considered to be dominated by detached rotational blocks with geometries analogous to those reported over a comparable width transition zone in the northern Bay of Biscay (De Charpal *et al.* 1978). The observations of De Charpal *et al.* (1978), although dated, remain exceptionally valuable because the Biscay coast has been more starved of sediment since its rifted margin first formed than any Atlantic-type margin of comparable age anywhere in the world. We have depicted the extension and thinning of the continental lithosphere as having ended by the Late Barremian (126 Ma) when ocean floor of normal thickness first began to form. Resolving the time at which this critical change happened north of the Walvis Ridge is difficult for three reasons:

- The sedimentary rocks of both the rift facies and the overlying steer's-horns or sag facies below the evaporite are non-marine.
- The sections are thick and deeply buried.
- Well/corehole penetrations have been few.

Workers in Gabon and further south have contributed to the understanding of the structure of the continental margin by distinguishing inboard faulting at an Eastern Hinge from more oceanward faulting at an Atlantic Hinge (Karner & Driscoll 1999, fig.1). Analysis of the structure across the two hinges making use of potential field data (gravity, magnetic and heat flow) has proved understandably popular because of the great thickness of sedimentary rock. Karner and Driscoll (1999) did not estimate a width for the transition zone, but mapped a line as the ocean/continent boundary. Comparison between Rosendahl and Groschel-Becker (1999) and Karner and Driscoll (1999) for the 450 km-long zone in which their studies overlap shows that the linear boundary of the latter lies somewhere close to the outer edge of the broad transition zone mapped by the former. By analogy with De Charpal *et al.* (1978), the Atlantic Hinge of Karner and Driscoll (1999, fig. 1) and some other, less well-defined structural highs shown on their map inboard of their ocean/continental boundary are here suggested to mark the crests of rotated and tilted blocks in the transition zone. Karner and Driscoll (1999, fig. 8) showed faulting at their Eastern Hinge as most active at the beginning of the Cretaceous (*c.*140 Ma). They also indicate the Atlantic Hinge, 100 km to the west and within the transition zone, to have been most active in a separate episode *c.*10

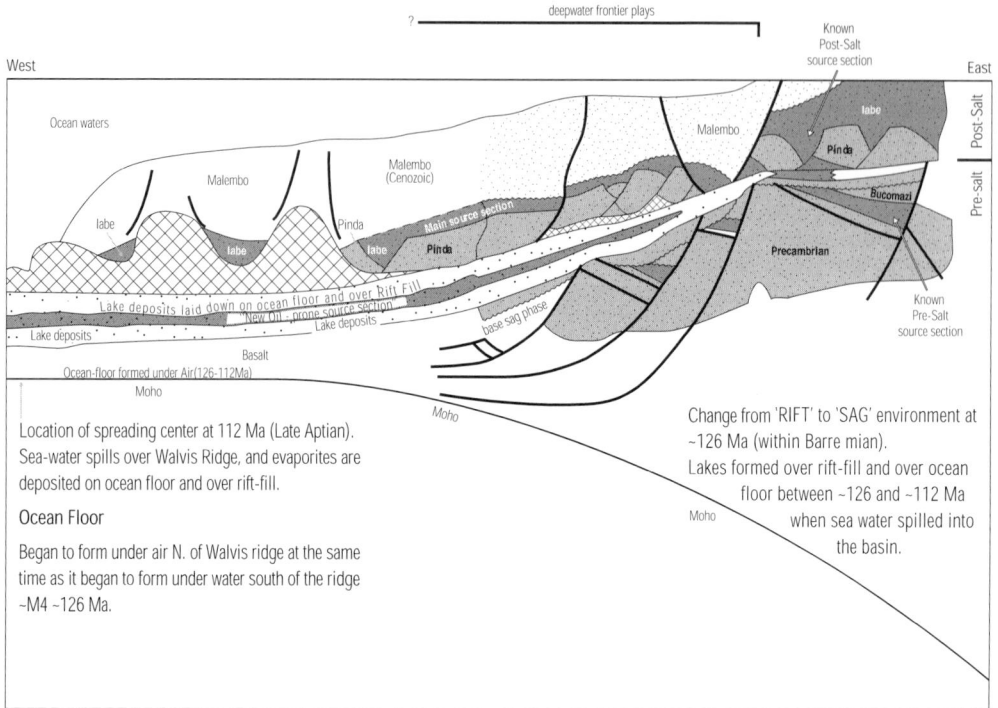

Fig. 18. Sketch cross-section (extended west from fig. 5 of Cameron *et al.* 1999) from the Angola coast reaching westward as far as ocean floor formed at 112 Ma. The transition from continental to oceanic crust is emphasized and the lake deposits of the beginning of the sag phase are shown as extending across the entire section.

million years later. We interpret both these faulting episodes as recording extension within a rift on the site of the future South Atlantic Ocean of the population that formed during the K-pippe. Final, less intense, faulting is indicated on their figure 8 at both hinges at the end of the Barremian (*c.*126 Ma). We interpret this as coinciding with the beginning of the formation of ocean floor of normal thickness further to the west, as we depict in Figure 18 for Angola. In their figure 9, Karner and Driscoll (1999) show no faulting at either hinge in the Mid-Aptian (*c.*118 Ma) when, we have suggested, faulting had moved westward to the active spreading centre (Fig. 18). In summary, although the great thickness of sedimentary rock and the complex later history combine to make working out the structural evolution of the Atlantic margins north of the Walvis Ridge early in Cretaceous times a difficult and complicated task, it is clear that progress is beginning to be made.

The lithologies deposited between Nigeria and the Walvis Ridge at this time have played a key role in the development of petroleum systems in this highly oil-productive area of Africa. One of the key source environments of Africa and Brazil is that of the organic-rich shales deposited within

the Great Lake around the Barremian (*c.*124 Ma). Rift-fill lacustrine source rocks were also deposited in the southernmost rifts of the Sirt Basin at that time (Gras & Thusu 1998) and in the Central African rifts (Genik *et al.* 1991), establishing a rift-associated source-rock fairway that extends the length of Africa.

The lacustrine rift-fill shales of the South Atlantic have charged interbedded non-marine sandstones, as well as limestones and sandstones of the Albian and Cenomanian post-salt successions of Angola, Gabon and Congo. Of even greater importance was the deposition of salt over the Cameroon to Angola region during the Aptian. Later salt deformation, stimulated largely by Tertiary overburden, has set up the trapping styles of some of Africa's most prolific reservoirs in the post-Salt section, and also led to fracturing of the section that allowed vertical oil migration into Tertiary reservoirs through salt-evacuation windows.

The second wild card The evolution of the Gulf of Guinea, between 140 Ma and 95 Ma, which was the time when the northern coast of South America cleared the westernmost part of the Guinea coast in transform motion, constitutes our second wild card. Rifts of the Karroo-related population had formed along the Guinea coast (De Matos 1999) and are best known in the Potiguar Basin of Brazil, where the oldest rift-fill sediments are Berriasian (*c.*140 Ma) (Fig. 19a). As Figure 19b indicates, deformation of the Potiguar rift fill began in the Barremian (*c.*126 Ma) with the formation of fault blocks that we interpret to have developed in

strike-slip motion. That movement was part of a broad zone of transform motion which was active as South America began to slide to the west past the Guinea coast. In Potiguar, faulting was largely, but not completely, over by the Mid-Aptian (Fig. 19c), but faulting must have continued in the offshore area where ocean-floored pull-apart basins were beginning to develop during the Barremian

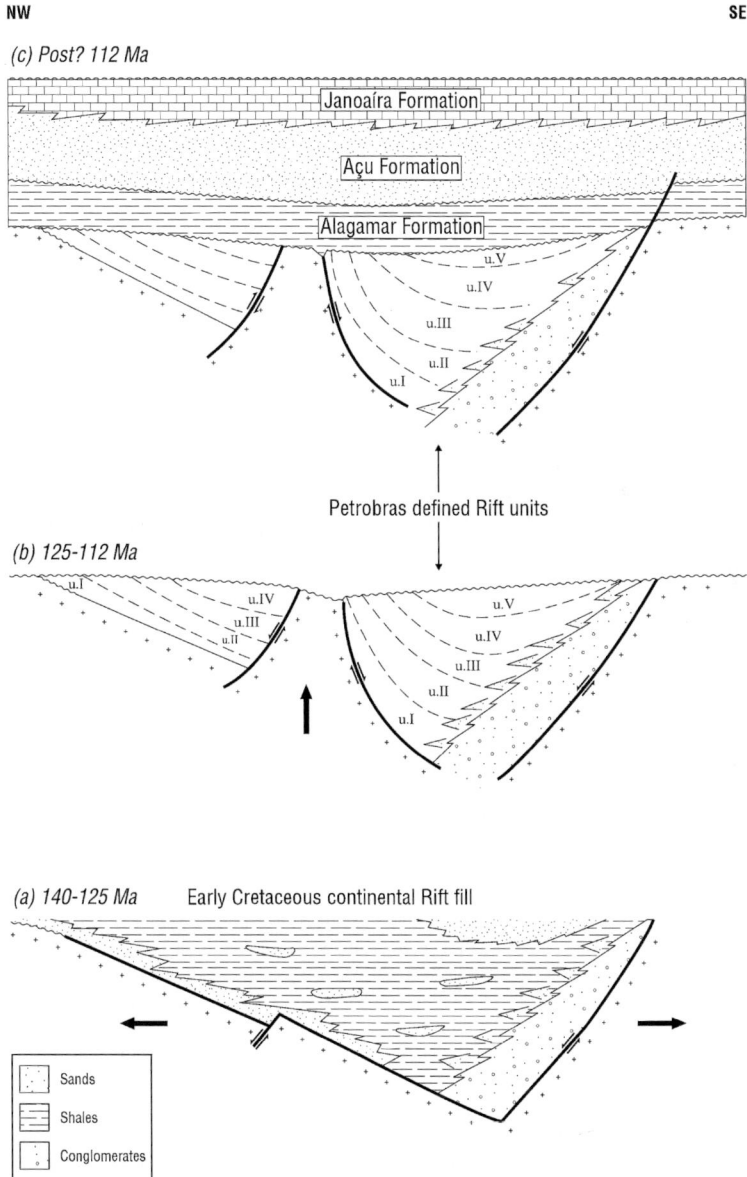

Fig. 19. Cross-sections of the Potiguar Basin in Brazil modified from Neves (1989, fig. 6) indicating how: (**a**) the basin was occupied by a half-graben from 140 Ma to 125 Ma when faulting (**b**), which we suggest was dominated by strike-slip movement began. Most faulting had ended by Late Aptian time (**c**).

and Aptian (127–112 Ma). Lacustrine sedimentary rocks, including organic-rich rocks related to those of the Potiguar Basin, may be widely distributed at depth along the Guinea coast, but they do not appear to have yet been penetrated offshore by the drill bit. A lacustrine source is, however, reported from the Nigerian onshore well Ise-2 (Haack *et al.* 2000). Because there is no preservation of Aptian evaporites in the Gulf of Guinea or under the Niger Delta, it can be inferred that there was no more than limited continuity with the Great Barremian–Aptian lake of the South Atlantic during Late Aptian times (112 Ma).

The development of the so far largely unproven petroleum systems of the Guinea coast may be largely governed by variations in the thickness and character of the later Cretaceous and Tertiary cover and the role of that cover in the burial of potential source rocks of the Early Cretaceous. The main indications of a lacustrine-fed petroleum system come from western Nigeria and Benin, where the large reported volumes in a tar belt to the north are typed to a Neocomian lacustrine source (Haack *et al.* 2000). Mid- to Late Cretaceous clastics are known in the subsurface in shelf, slope and deep-sea fan facies (MacGregor *et al.* 2003). Variation along the Guinea coast in the thickness of these rocks, and more particularly in their Tertiary successors, may be one of the more important controls on the maturation of source rocks and the evolution of oil and gas resources of the Guinea coast.

The third wild card During the past 100 million years the internal stress field of the African Plate has been repeatedly subjected to modification as a result of convergent and collisional events on the northern margin of Afro-Arabia which record the progressive closing of the Neo-Tethyan ocean. Folding, faulting, uplift and erosion on the continent and on the ocean floor have been the characteristic responses to these Alpine events. New sediments have been deposited and new fluid-flow patterns have been established in association with the events. The Afro-Arabian Plate, rotating slowly about an internal pole, and subject, along most of its boundary, only to the weak ridge-push force appears to have proved very susceptible to these plate-margin induced stress modifications. To change our analogy, Afro-Arabia can be thought of as facing a Eurasian bruiser who has unleashed onto the gently rotating continent a series of straight rights (in the Atlas) and left hooks (in Oman, Iran, Iraq, Syria and Turkey). Altogether, we distinguish nine episodes in Afro-Arabia related to Alpine convergence and collision. Individually these constitute the third to the eleventh wild cards dealt to Africa's hand.

The events were not of uniform intensity, e.g.

that at 84 Ma (Santonian) appears to have been particularly widespread and intense (Guiraud & Bosworth 1997). We recognize that arguments can be made for distinguishing more events and alternatively for abandoning the idea of discrete wild-card events and treating the Alpine convergent and collisional phenomena as continuous. Fully defining individual events must depend on mapping sets of phenomena that took place over extensive areas during short intervals. Because we have not attempted that kind of analysis, the nine episodes that we distinguish are perhaps best thought of as only a preliminary explanation of a complex problem.

The first of the nine events took place at *c.*100 Ma, at the end of the Albian. It is quaintly and regrettably sometimes referred to as generating the Austrian unconformity, although it has nothing whatever to do with anything that happened in Austria. Rather it seems to have propagated into Africa from the Atlas and to have been most intense in its consequences in Morocco, Algeria, Tunisia and Libya. Folding, faulting, erosion, the establishment of a major unconformity and new fluid-flow patterns have been well documented as having developed at this time in North Africa (Boote *et al.* 1998).

A possible cause for the 100 Ma event is the beginning of extension in the Bay of Biscay. A comparison of Sengör and Natalin's (1996) figures 21.46 and 21.47 for 110 Ma and 83 Ma shows that the beginning of extension in the Bay of Biscay during that interval rotated Iberia onto the transform fault that was carrying Iberia and France eastward past Morocco. We suggest that the convergence induced in the Atlas as a result of that rotation was the trigger for the widespread 100 Ma events.

This tectonism has significant implications for petroleum systems in Algeria, where it was a major trap-forming event. Inversion occurred along normal faults formed during the Ace of Diamonds interval, establishing anticlinal structures, e.g. on the El Biod Ridge. Structures were probably also created or enhanced at this time in the Illizi Basin and possibly at Hassi 'R Mel. Many traps of this age are gas-bearing, as it was around this time that Silurian source rocks entered the gas window (MacGregor, 1998). Earlier filled structures, such as Hassi Messaoud, retained their oil charge.

The fourth wild card This card was an episode of deformation that is recorded in structures extending over a 20 million km^2 area of northern Afro-Arabia from Oman to the Atlas and from Kenya to Equatorial Guinea. This surprisingly widespread episode made its presence felt as an unconformity over an even larger area (Guiraud & Bosworth 1997, fig.

3). Where it can be accurately dated, e.g. in the southern Benue Valley of Nigeria, this event has been shown to have taken place over a brief interval, perhaps in less than 2 Ma, during the Santonian (84 Ma).

The brevity of the episode was a key clue for Guiraud and Bosworth (1997), who, in a very thorough analysis, confirmed an earlier speculation of Dewey and Burk [sic] (1974) that this widespread episode of deformation formed part of the Syrian arc events of c.84 Ma (Krenkel 1925) which are now recognized (see Sengör & Natalin 1996, fig. 21.47) to have coincided with the obduction of the Oman and Kermanshah-Neyriz ophiolites onto the northeastern margin of Afro-Arabia.

Few episodes of intracontinental deformation are recognized as being as extensive or as intense as the 84 Ma Syrian arc event. Most of those that are, such as the current deformation of Eurasia or the deformation of much of Pangaea during the Permian, can be linked to major continental collisions. A possible explanation of the intensity of this, the initial Syrian arc event, is that, when it happened, the Afro-Arabian Plate was subject to few forces other than the weak ridge-push force. Spreading centres were symmetrically distributed around the plate, which was rotating slowly about an internal pole close to 7°N 11°E. The plate, perhaps because forces on it were small and balanced, appears to have responded dramatically to the obduction of ocean floor onto its northeastern corner. At the time of the obduction, slab pull along c.2000 km of convergent plate boundary was eliminated from the margin of the Afro-Arabian Plate, but, as Sengör and Natalin show (1996, fig. 21.47), a comparable length of convergent boundary further to the northeast immediately replaced the eliminated boundary. Over the interval straddling 84 Ma the Afro-Arabian Plate was rotating about the internal '711' pole in a direction that was tangential to that of the limited amount of slab pull on Afro-Arabia at both the earlier and the later sets of convergent plate boundaries. As Guiraud and Bosworth (1997, fig. 2) illustrate, the azimuth of relative convergent motion between the African and Eurasian plates changed abruptly at 84 Ma. What had been motion of Africa toward the ENE rotated through c. 60° in an anticlockwise direction to an azimuth only slightly east of north.

The influence of the 84 Ma deformational episode on Africa's petroleum systems has been varied. Large anticlines were developed within several areas of earlier extension at this time, notably in the Doba Basin of Chad (Genik 1991), Abu Gharadig Basin of Egypt (Awad 1984), the Benue Trough of Nigeria, offshore Rio Muni (Equatorial Guinea) and offshore Togo (MacGregor et al. 2003). The largest fields in the Central African rift system, in the Doba Basin, are in inversion anticlines formed at this time. A prominent unconformity formed along the entire northwestern African margin, from Guinea-Bissau to Morocco, as Jurassic and Early Cretaceous continental-margin carbonate-bank deposits were buried by siliciclastic sediments. The La Ceiba discovery in Rio Muni is associated with sands delivered to the deep water following this event (Dailly & Goh 2000), as are most of the offshore Ogooué Fan fields (Teisserenc & Villemin 1990).

The fifth, sixth and seventh wild cards Although the major Syrian Arc deformation event was at 84 Ma, several later, less intense, episodes appear to have been later manifestations of similar deformation originating from the northeastern margin of Afro-Arabia. Guiraud and Bosworth (1999) distinguished an end-Cretaceous (65 Ma) event as well as an end-Palaeocene (53 Ma) event and a Bartonian–Priabonian (40 Ma) event. The exact cause of these deformations is unclear. Sengör and Natalin (1996, fig. 21.48) indicate that only for the last episode is there evidence of the impingement of a convergent plate boundary onto the northeasternmost corner of Afro-Arabia. However, other authors have suggested earlier onset of the convergence at the continental margin that culminated in the final progressive collision from east to west between Arabia and Asia along first the Zagros and then the Bitlis sutures at 15–10 Ma (Burke 1996).

A return to quiescence In spite of these minor tectonic disturbances, the latter part of the interval during which the Joker was in play, from 84 Ma to 31 Ma, was a quiet time for the Afro-Arabian Plate. The continent was low-lying with widespread laterite development (Burke 1996, p. 348 & fig. 3). Some of the rivers shown in Figure 17 continued to flow and to carry siliciclastic material to the coast. Others, e.g. the Shire River at its Zambezi mouth and what had been the Sirt estuarine system, were replaced in the Early Cenozoic (after 65 Ma) by mainly carbonate shelves. Igneous activity within the African Plate was persistent only:

- where the Tristan Plume was erupting;
- on the track of the Deccan Plume between 65 Ma and 31 Ma;
- in Cameroon where the '711' Plume has been responsible for continuous igneous activity throughout the Cretaceous and Cenozoic (Burke 1996, figs 3 & 10).

A 10 million years interval of igneous activity between 45 Ma and 35 Ma in southern Ethiopia was clearly related to a shallow-sourced, rather than to a deeply sourced plume (George et al.

1998) because it produced only 0.2 million km³ of basalt as it migrated southwestward *c.*300 km while the African Plate rotated about the '711' pole. This shallow-plume-generated volcanism reached an area west of Lake Turkana by 35 Ma, where it was soon overwhelmed by pressure-relief volcanism related to the initiation of the eastern rift at 31 Ma (Morley *et al.* 1992; Burke 1996). Sporadic and episodic igneous activity of small volume occurred in several areas of Africa, such as the eastern desert of Egypt and coastal southwestern Africa. In spite of these minor events, the scene in Afro-Arabia was quiet and set for its most recent great tectonic perturbation as the Afar Plume rose from the deep mantle to reach the base of the lithosphere.

Late Cretaceous to Eocene source rocks The 100–50 Ma interval was a key period for the development of source rocks on both the northern and West African rifted margins. At this time global sea levels were high and, because Africa was low-lying, large areas were frequently and extensively inundated. Those source rocks were to be later brought to maturity by burial under Tertiary overburden. Important source-rock depositional events occurred in West Africa, centring on the Cenomanian–Turonian (e.g. from Angola to Mauritania), and in North Africa, centring on the Santonian (e.g. in the Sirt Basin and the Gulf of Suez). Source rocks of these ages are significant on a global scale, reflecting, beside high sea levels, warm conditions, restricted ocean circulation and, possibly on western coasts, a high element of upwelling, as evidenced by the nature of the microfauna associated with many of these occurrences (Arthur & Sageman 1994).

Similar source rocks occur in Eocene strata, particularly in Tunisia, in western offshore Libya and in deep-water Angola. Maturity is a critical concern for these younger source rocks which, particularly in Libya and Angola, commonly feed shelf sedimentary rocks of their own or younger age in both carbonate and siliciclastic reservoirs.

The Ace of Spades: elevation and erosion of swells and related phenomena, including massive deposition at the continental margins following the eruption of the Afar Plume at 31 Ma

From the petroleum systems viewpoint, the most important consequence of the A-pippe (Fig. 3) was the elevation of the continent of Africa. That elevation has shown itself primarily in the development of 75 swells that are actively rising over upwelling parts of a new shallow-mantle convection system

(Burke 1996). These swells are distributed over both the oceanic and the continental parts of the plate (Fig. 20). Even the low points of the newly formed great basins of the continent lying among the swells, such as the Kalahari, Congo, Chad, Niger, Taoudeni and Sudd basins, are at a few hundred metres above sea level.

By one of those coincidences that have characterized the evolution of the Earth, the formation of swells in Africa (Holmes 1944; Burke & Wilson 1972; McKenzie & Weiss 1975; England & Houseman 1984; Burke 1996) was almost simultaneous with a global lowering of sea level and consequent further relative elevation of Africa. That global sea-level lowering, which may have been by as much as 50 m, followed the establishment of the Eastern or Greater Antarctic ice sheet at the time of the Eocene–Oligocene boundary (*c.*34 Ma). Ocean drilling in Prydz Bay on the rim of the Antarctic ice sheet has shown that a grounded ice sheet developed at the continental margin at that time. Sequence stratigraphers have long recognized an episode of sea-level lowering at *c.*34 Ma (Haq *et al.* 1987) which can be identified on the margins of all the continents. Because it is very hard to define the magnitude of sea-level changes from sequence stratigraphic data, the distinct and separate tectonic elevation of Africa has not always been recognized. It is easy to suggest that the Early Oligocene (31–34 Ma) unconformity around Africa, as seen in seismic reflection records, is more prominent than that around other continents, but quantifying that statement is hard. An easier statement to defend is that submarine canyon development beginning at 31–34 Ma around Africa has been more spectacular than contemporary canyon activity around other continents. The best evidence of the distinction of the two nearly simultaneous events would perhaps come from places in which two unconformities could be recognized: one, of glacial origin, at 34 Ma and a second, of tectonic origin, at 31 Ma. We suspect that there are deep-water well logs in which two unconformities can be distinguished. In most cases the two unconformities are likely to be telescoped into one.

These unconformities may mark, in different areas, either destructive or constructive events in petroleum system development. An Oligocene erosional event was responsible, for instance, for the near-unroofing of the Cap Juby oilfield in Morocco. The same event removed large volumes of Cretaceous source rocks from wide areas of the northwestern African shelf from Morocco and Senegal. More positively, a major pulse of reservoir sand input followed the establishment of the Oligocene unconformity in many deep-water areas, e.g. in Angola, Mauritania (Frost 2000) and Benin/Togo (MacGregor *et al.* 2003). In a third style of devel-

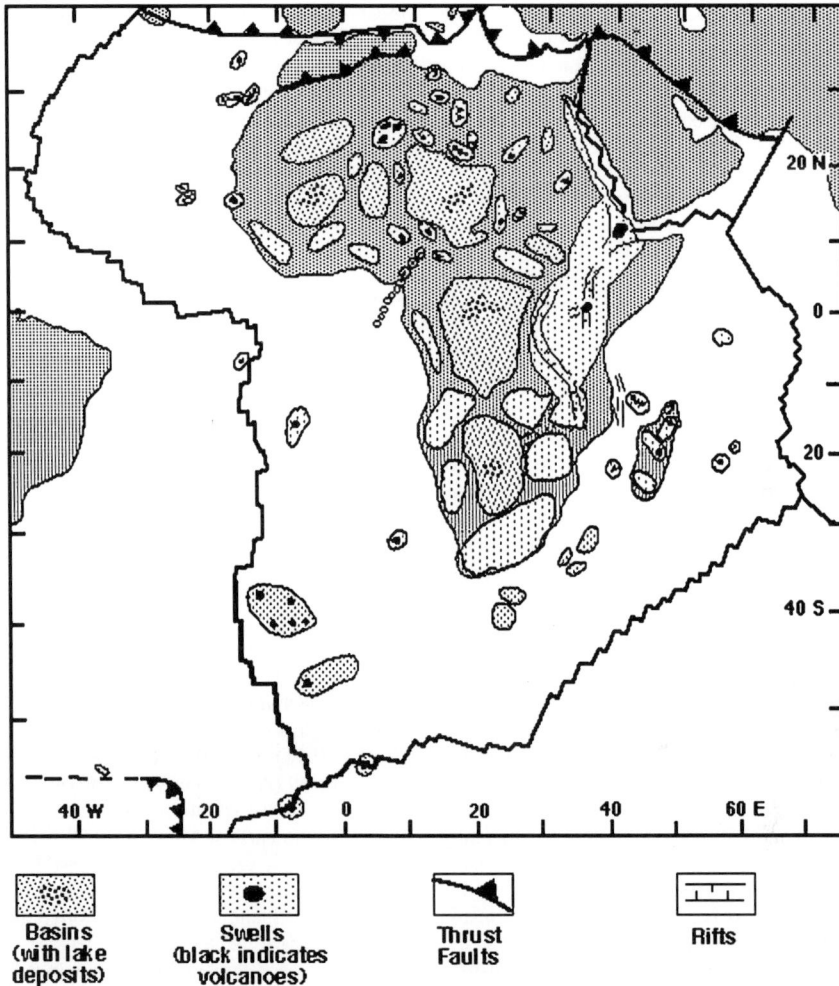

Fig. 20. Basins and swells of the African Plate that are interpreted as representing the plan-form of a pattern of shallow-mantle convection (McKenzie and Weiss 1975) set up in response to pinning of the plate at the eruption of the Afar Plume (Burke 1996). Swells on the continent are elliptical and range from 100 km to 2000 km in greatest length. Erosion from these swells has yielded vast amounts of sedimentary material that has been deposited during the past 31 Ma near the continental margin, especially in the Nile and Niger deltas and in the Congo deep-sea fan. Apart from the Cameroon line (Burke 2001) no simple pattern characterizes the distribution of the swells. Basins among the swells are irregular, simply filling intervening spaces. Offshore swells are hard to identify and have only been indicated where they are capped by young volcanoes or where there is strong evidence (as in the Seychelles) of recent elevation. Swells referred to in the text include the South African Swell at 31°S 25°E, the Ahaggar Swell at 23°N 10°E, the Jos Swell at 10°N 9°E, the Guinea coast swell at 6°N 3°E., the Congo-Gabon coastal swell at 5°S 12°E, the East African swell at 3°S 35°E and the Afar Swell at 10°N 40°E. The South African and East African Swells are particularly extensive and high, dominating what has traditionally been called 'High Africa'. This dominance is suggested to be more because they have been eroded less than swells like the Ahaggar Swell and not because they have been elevated more.

opment, erosion patterns on the Oligocene unconformity have set up the geometry for the traps in the Seme and Aje fields of Benin and Nigeria (MacGregor *et al.* 2003).

Africa's swells Because the general geological development of Africa in response to the arrest of the plate at *c.*31 Ma has lately been reviewed at some length (Burke 1996), discussion is here concentrated on topics closely linked to petroleum systems. The first of these is the elevation of the swells themselves. Clearly, elevation has been a mixed blessing for Africa's petroleum systems. The most obvious consequence of the elevation has been that Africa is now the continent with the largest proportion of basement outcrop.

Erosion removed huge volumes of rock from the high ground of Africa. For example, some 100 km to the north and the east of the 0.5 million-km^2 Ahaggar crystalline outcrop in the central Sahara, the Cambrian /Precambrian unconformity lies 3 km below sea level (Boote *et al.* 1998, figs 4 & 5). Precambrian rocks are exposed everywhere within the Ahaggar, even at the highest point, which is at 2918 m above sea level. The amount of rock eroded from the Ahaggar over the past 31 million years must have been prodigious. A conservative estimate shows that a minimum of 1 million km^3 of rock is likely to have been eroded from the Ahaggar during the interval since the elevation of Africa's swells began. The Ahaggar (Fig. 20) is not one of Africa's biggest swells, but it has clearly been greatly eroded.

Undoubtedly many oilfields were destroyed at this time by uplift. An example of such an extinct system may be the seep-rich but unproductive Karroo Foreland Basin of South Africa, which has been eroded to the level of its reservoirs and source rocks. The uplift and degradation of tar belts such as those in the Kwanza Basin may also have occurred at this time. Boote *et al.* (1998, especially figs. 13–18) also show how elevation of Palaeozoic rocks on the flanks of swells in the Sahara over the past 31 million years has caused gas-cap expansion and flushing. Furthermore, flushing by meteoric waters from the newly rising swells has redistributed hydrocarbons over huge areas.

Without doubt the most positive role of the rising 31 Ma and younger swells on the African continent from the petroleum systems viewpoint is as providers of siliciclastic sediment to the continental margins. A good question is: 'Where has all that rock gone?' It has not accumulated to any great extent in any of the interior basins of Africa. The Chad Basin, for example, contains only 600 m of 31 Ma or younger sediments (Burke 1976*b*). The reason that relatively little 31 million years or younger sediment has accumulated in the interior basins of Africa is that rivers have everywhere, at least intermittently, traversed the basins and transported sediment to the continental margin (Burke 1976*b*). Sediments eroded from the Ahaggar are probably mainly in the Niger Delta, having been carried either directly to the Niger along the Dallol Bosso and its tributaries or indirectly to the Chad Basin and on through the Bongor spillway to the Benue (Burke 1976*b*). Today, the Niger Delta

receives its sediments mainly from the Benue, but that has only been so since the Sahara first became a desert at 2.8 Ma (DeMenocal 1995). Introduction of consideration of rivers draining Africa's swells leads to the more general question: 'How have the river systems of Africa changed in response to the development of the new swells starting at 31 Ma?'

River systems and their deltas An indication of the importance of this question is that it is beginning to be widely addressed (e.g. Purdy *et al.* 2000; Taylor 2000). A simple way to tackle the issue is to compare a map of Africa's modern rivers with Figure 16 to see the resemblances and differences between the rivers of today and those active during the later Cretaceous and the earlier Cenozoic. The map of the active swells of Africa (Fig. 20) complements the two maps of rivers because it shows where disturbances to the older river systems began to develop at 31 Ma.

Drainage in northeastern Africa was dominated during the Late Cretaceous by rivers feeding the Sirt estuaries. During the earlier Cenozoic, a mixed carbonate-siliciclastic bank occupied that region, being particularly well known at Fayum in Egypt. As soon as the Ethiopian Swell began to rise at 31 Ma, two completely new and very large rivers began to flow from the Afar Swell crest: the Nile (Burke 1996) and the river feeding the Ahwaz Delta in Iran (Al Sharhan & Nairn 1997, fig. 9.53). The great deltas of both rivers were well established by the earliest Miocene (*c.*23 Ma).

Another new great river of the past 31 Ma is the Congo, which rises on the western flank of the East African Swell and crosses the Congo Basin, the most extensive of the interior basins that have developed since the swells began to rise (Fig. 20). The Congo has cut through a rising swell close to and parallel to the western coast of Africa. An 18 Ma volcano located offshore of Libreville, Gabon, caps this swell. The Congo, although it has not always used the same passage through the swell, has been able to feed its deep-sea fan throughout the past 31 Ma by keeping pace with uplift. Before there was a Congo River, the offshore basins of Gabon and Angola received relatively small amounts of siliciclastic sediment from relatively short rivers, perhaps comparable to the Ogooué today, which are labelled 'Gabon rivers' and 'Kwanza rivers' in Figure 16.

The Benue is the most prominent of the rivers flowing along rifts that had formed during the K-pippe (Fig. 16) to persist as a major river during the past 31 million years. It was joined by the lower reaches of the Niger draining southward from the headwater tributaries of the Dallol Bosso in the Ahaggar and flowing seaward between the active swells of the Guinea coast and the Jos Plateau (Fig. 20; Burke 1996, figs 29 & 31). The Senegal–Casamance river system has continued to flow along its rift during the past 31 million years (Fig. 16), but the Anza Rift river feeding the Lamu Delta (Figs 16 & 20) died at 31 Ma when the new swell system began to develop. The Zambezi Delta area was occupied by a carbonate bank during the Early Cenozoic (Burke 1996), but a new delta has developed as neighbouring rising swells have been eroded during the past 31 million years (Fig. 20). The Orange River is one of the few major rivers that, in its lower reaches, continues to flow with relatively little change, although its drainage area has been drastically reduced (Burke 1996). Offshore deposition near the mouth of the Orange River is much less than it was in Late Cretaceous times, when growth faults were active in its delta (Jungslager 1999, fig. 14).

Discussion of the tectonic influence of the newly forming swells of Africa on drainage, and thus indirectly on petroleum systems, is incomplete without also considering the influence of climatic change on that drainage. The formation of the ice sheet of East Antarctica that lowered global sea level at c.34 Ma had another immediate effect on Africa by changing atmospheric circulation. Since 34 Ma, something approaching desert conditions has dominated in southern Africa because the Hadley Cell circulation has been driven northward and the flow of the Benguela Current has been initiated as a result of the presence of the ice sheet. As a consequence, the great rising topographic swell of South Africa (Fig. 20) has been little eroded compared, for example, with the Ahaggar. It is unusually elevated not because, as has been suggested, it overlies the hottest and most voluminous part of the anomalous deep mantle (Lithgow-Bertelloni & Silver 1998), but because it has received less rainfall. The hot volume in the deep mantle has been where it is now for 200 million years. The elevation of the South African Swell cannot be directly related to the deep hot volume because that elevation only began at c.31 Ma. Limited erosion in the region also explains the preservation of the distinctive Great Escarpment around Southern Africa (Burke 1996).

The limited erosion of the great swells of South and East Africa is important from the petroleum systems viewpoint. It accounts for the observation that the post-31 Ma depositional piles at the edge of the continent off southern and eastern Africa are generally much less thick than those developed along the west coast from the Walvis Ridge and the western Niger Delta. Exceptions are the Zambezi Delta, the Nile Delta and the Suez Rift.

Although the most important climatic change of the Cenozoic for African petroleum systems came at 34 Ma, when the East Antarctic ice sheet for-

med, the consequences of a more recent change are much more obvious. That change was the onset of Northern Hemisphere glaciation and, in Africa, the establishment of the Sahara as a desert for the first time at 2.8 Ma (DeMenocal 1995). That change produced spectacular results. During times of Northern Hemisphere glaciation, as now, no permanently flowing river links the Ahaggar with the Niger Delta; the Casamance–Senegal river system is, as now, relatively insignificant and, at glacial maxima, the Nile fails to flow north of the Sudd. By contrast, during full interglacial summers, when the Inter Tropical Convergence Zone (ITCZ) reaches well into the Sahara, the Ahaggar is directly drained to the Niger as well as through the Chad Basin to the Benue. Furthermore, the Senegal–Casamance river system has proved powerful enough to have kept the surface of the rising Dakar Swell close to sea level (Burke 1996, fig. 13) and to carry abundant siliciclastic sediment to deep water (Jacobi & Hayes 1982). Finally, as now, the Nile reaches to the Mediterranean. Conditions during Northern Hemisphere interglacials replicate for Africa the conditions that existed between 34 Ma and 2.8 Ma, when the Antarctic ice sheet was a dominant climatic influence on the continent and there was no Northern Hemisphere glaciation. The persistence of conditions of that kind through most of the time since the swells of Africa began to rise from 31 Ma has to be borne in mind when interpreting the role of the present drainage system of Africa in relation to deltaic and offshore sedimentary deposits. As an example, the Congo Basin has lain for the past 31 million years in latitudes at which its relation to the ITCZ has ensured that it has been likely to receive rainfall in both January and July, whether or not Northern Hemisphere glaciation is in progress (Burke 1996, fig. 49).

Rifting related to the A-pippe Rifting in Africa over the past 31 million years has been concentrated around the Afar, where a plume from depth erupted and arrested the plate. That rifting is therefore regarded as a part of the Ace of Spades tectonic episode. The many plumes of the shallow-mantle convection pattern set up as a result of plate arrest have, with the exception of the Samburu Plume (Burke 1996), played no such obvious role in rift development. Three radially disposed rifts formed on the crest of the Afar, or Ethiopian, Swell within temporal resolution, as soon as the swell began to rise at 31 Ma. Two of those rifts, the Red Sea and Gulf of Aden rifts, developed from cracks that rapidly propagated toward sharp bends in the continental margin.

We interpret this to have been because the elevation of the Afar Swell had perturbed the stress distribution within the lithosphere on the swell crest. That perturbation was relieved by the rapid propagation of cracks toward nearby points of existing anomalous stress. Sharp bends in the continental margin, particularly the right-angled bend at the Levant corner of Egypt and Israel, were apparently the closest places where stress distribution was unusual (Burke 1996). Extension to develop intracontinental rifting followed crack propagation and was probably initially accompanied by dyke emplacement.

The Suez Rift, a part of the Red Sea Rift close to the Levant corner, is, from the petroleum systems standpoint, the most important part of the active East African Rift System. This is defined here, following Burke (1996), as comprising the Red Sea, the Gulf of Aden, the Suez Basin, the Eastern and Western Rifts, the Dead Sea transform and the rift system offshore of northern Mozambique (Fig. 20). Suez formed, with the rest of the Red Sea Rift, close to 31 Ma and has evolved since as an intracontinental rift flooded by the sea.

The eighth and ninth wild cards The most significant tectonic change in the Suez Rift took place when the Zagros collision modified stress distribution within Afro-Arabia and led to the initiation of the first phase of Dead Sea transform motion (Bosworth & McClay 2001). Other effects of the Zagros collision on Afro-Arabia include the beginning of Newer Harrat igneous activity and initiation of the rifting in the western rift (Burke 1996). Although the collision in the Zagros and that further to the east along strike in the Bitlis of Turkey are commonly thought of as continuous and progressive from 15 Ma to 10 Ma, Guiraud and Bosworth (1999) found evidence in the Levant and in northeastern Africa that enabled them to distinguish an episode of intraplate deformation concentrated around 15 Ma (Burdigalian) from an event at 9 Ma (Tortonian). For this reason, we have distinguished the Zagros collision at 15 Ma as our eighth Alpine-induced collisional wild card from the Bitlis collision at 9 Ma, which we distinguish as our ninth. Under the influence of these two events, the past 15 Ma have seen radical change in Afro-Arabia. These included the beginning of ocean-floor formation, first in the Gulf of Aden from 10–15 Ma and then propagation along the Gulf of Aden to reach the Afar by 5 Ma. From 5 Ma, northward propagation began along the Red Sea, which is about the time that separate African, Arabian and Nubian plates began to be discernible (Burke 1996).

The tenth wild card While the intense tectonic activity of northeastern Africa and Arabia represented by the A-pippe and the dealing of eight

and ninth wild cards were in progress, renewed convergence was beginning far away in Morocco. Extension in the Valencia Trough and extension and transform motion in the Alboran Sea beginning at 25 Ma carried the two Kabylies southward onto the Atlas (Sengör & Natalin 1996, fig. 21.49). The effects of this collision, the second straight right, are clearest in the Atlas and immediately to the south in the Moroccan Sahara. Whether and how any more remote effects can be separated from those generated more locally within North Africa by the active elevation of Africa's swells is very doubtful. These relatively recent Atlas-linked events seem to have had many negative effects for the petroleum systems of that area, unroofing traps and stimulating leakage that is now represented by widespread seeps.

The eleventh wild card Alpine influence is, for what appears to be the first time, making itself felt directly in Libya where convergence at the Mediterranean ridge is causing uplift on the north coast at Jebel al Akhdar.

Impact of the Afar Plume on petroleum systems The prime importance of the Afar Plume comes from the huge amount of reserves that have developed as indirect consequences of its eruption. Nearly all those reserves are in reservoir sands eroded from high ground and supplied to the continental margins over the past 31 million years. Most of those reserves were generated over the same interval by the thick overburden imposed by deltas and deep-water fans that were active at this time. It is thus no accident that the deltas and deep-water fans of the Niger, Congo and Nile rivers are Africa's main petroleum provinces.

The development of the Nile Delta and the huge expansion of the Niger Delta during the past 31 million years are linked to the erosion of swells formed in consequence of the Afar Plume's eruption (Burke 1996). All the components – reservoir, source and seal – of Africa's largest petroleum system in the Niger Delta are linked to the huge influx of Tertiary sediment to the delta. The source organic matter is predominately terrestrially derived. The development of the Congo Fan with its turbidite reservoirs that are now proving so prolific in offshore Angola and Congo is another consequence of erosion of newly elevated swells. The Nile Delta is again a self-sourcing petroleum system, in this case primarily of biogenic gas, with a key control being the very rapid burial of the organic-rich deltaic sediments.

The formation of the Suez Rift, originally as a part of the Red Sea Rift, and the entire history of its petroleum systems relate directly to the elevation of the Afar Swell. In summary, 75% of

Africa' s petroleum has been generated in the past 30 million years and 50% of that petroleum lies in reservoirs of the Oligocene to Holocene. These reserves lie almost entirely in the four delta and deep-sea fan systems discussed above.

Summary of Africa's primary petroleum systems

This discussion attempts briefly to demonstrate the control that the tectonic aces and wild cards described in this paper have exerted on Africa's petroleum systems. A compilation of Africa's 25 largest petroleum systems, which contain nearly 90% of Africa's total reserves, is presented in Table 2. Each of the systems contains at least 2 billion barrels of recoverable oil equivalent (or 10 billion barrels of in-place tar sand) tied to a literature-referenced petroleum system. The predominant reservoir and source rocks are indicated. We do not yet have sufficient overall information to present the results in terms of the formal nomenclature for petroleum systems developed by Magoon and Dow (1994). Several additional petroleum systems, for instance the syn-rift petroleum system of Gabon, the Iabe–Pinda system of Angola and Congo, the Sudanese and Chadian lacustrine-sourced systems and the Kudu gas system of Namibia, just fail to meet our volume qualifications and are not included. The onshore asphalts east of Luanda are excluded due to the uncertainty of their size (we suspect they are more than big enough). For each of the systems in Table 2, ages have been assigned for the source rock, the predominant reservoir and, more tenuously, the timing of generation (Fig. 21). Plotting the data against time, and in this case including the many smaller petroleum systems not listed in Table 2, reveals that the vast majority of African petroleum is associated with a limited number of discrete sources and reservoirs. These are classified by geographical region, age and facies, and are presented in Table 3 and Figures 22 and 23.

As is evident in Table 2, the most prolific petroleum systems of Africa are limited to two relatively small regions of the continent. These are:

- North Africa, as represented by the countries of Algeria, Tunisia, Libya and Egypt ('North Africa' on Figs 22 & 23);
- the coastal/offshore zone from Nigeria south to Luanda in Angola ('West Africa' on Figs 22 & 23).

These two regions contain 95% of African reserves. No petroleum systems outside these regions qualify for Table 2.

Three distinct settings characterize the successful provinces of North Africa (MacGregor 1996a).

- The so-called 'Palaeozoic' provinces of eastern Algeria and western Libya (Boote et al. 1998). These originated within a single steer's-horns basin formed as a consequence of the Ace of Clubs orogenic collapse. Within that basin, the extensive NAACOQRS reservoirs and Silurian and Devonian source rocks were deposited. Ace of Diamonds orogenesis had destructive effects on any early-formed petroleum systems, but deformation at that time set up structures, which were charged and modified in later events, such as those of the third wild card. Equally critical to the petroleum systems in the area were the sedimentary rocks deposited in and over the rifts that developed during the Ace of Diamonds collapse episode. The rifts controlled the deposition of the Triassic reservoirs and the Liassic salt with its critical preservative effects. Overburden deposited in thermal subsidence basins in late post-rift times served to stimulate generation from the Palaeozoic source rocks. There is a very close correlation between reserve distribution and areas of expected Cretaceous generation (Macgregor 1998). It is largely where Clubs and Diamonds basins coincide that petroleum systems involving Palaeozoic source rocks are effective at the present day.
- The Sirt Basin of Libya. This is a composite basin within and above rifts that were initiated in the Late Jurassic and are thus related to the K-pippe phase (rifting following pinning by the Karroo Plume). An earlier rift of the Triassic, associated with the Ace of Diamonds collapse, is locally represented. Both source rocks and the multiple reservoirs were deposited in the rift fill and cover of the Sirt Basin.
- The Gulf of Suez and Nile Delta provinces, both of which are young provinces originating from Afar Plume-related events. Rift-fill load and high heat flow have matured Late Cretaceous source rocks, which, in turn, have charged rift-fill and pre-rift reservoirs that are sealed by late rift-fill salt. The Nile Delta biogenic gas system lies in even younger reservoirs and is associated, as are most other biogenic systems, to rapid deposition.

The West African region of success from Nigeria to Angola is in an area with a common tectonic history. This ocean margin records rift formation at c.140 Ma during the K-pippe and ocean-margin evolution on the rift system subsequent to the eruption of the Tristan Plume at 133 Ma. The four critical elements responsible for the exceptional richness of this region are:

- the deposition of lacustrine source rocks and salt during the first wild-card event;

Pet. System rank	Age (Ma)	625	600	570	550	525	510	480	450	430	390	370	350	325	300	275	245	225	208	183	160	133	105	80	55	31	0
			Vendian		Cambrian			Ordovician		Sil.		Devonian	Carboniferous			Permian		Triassic		Jurassic			Cretaceous			Tertiary	
			ACE OF CLUBS									ACE OF DIAMONDS									A/HEARTS		JOKERS				SPADES
	Source Fm.	Reservoir Fm.	Collision	Rifting			thermal subsidence			glaciation		First Collision				Main collision		Rifting			Karroo Plume	Rifting	Salt Basin	Alpine Impact Events			Afar Plume
1	Akata	Agbada																					B			S	S,R,G
2	Tanezzuft	Triassic A-G								S								B	R					G→			
3	Tanezzuft	Ra (Cambrian)		B			R			S												G→					
4	Iabe	Malongo																				B	S				R,G
5	Sirt	Sarir																				B	R	S		G→	
6	*Bucomazi*	*Pinda*																				B,S	R			G→	
7	Sirt	*Zelten-Sabil*																					B	S		R,G→	
8	Brown Lst.-Su	Rudeis																						S			R,G
9	*Mio-Pliocene*	Mio-Pliocene																B									S,R,G
10	Sirt	Amal																B					B	S		G→	
11	Sirt	*Waha-Defa etc.*																					B	S,R		G→	
12	Sirt	Gialo																					B	S		R,G→	
13	Malongo	Malembo																				B				S	R,G
14	Tanezzuft	Tadrart		B						S		R											G→				
15	Tanezzuft	Triassic A-G		B						S								B2	R				G→				
16	Sirt	Augila																					B	S,R		G→	
17	Sirt	*Intisar*																					B	S		R,G→	
18	Bahloul	*El Garia*																B								S,R	G
19	Pic Radioactiv	Triassic A-G		B									S					B2	R				G→				
20	*Sarir*	Sarir																				B,S,R			G→		
21	Pic Radioactiv	Aouinet		B									S,R											G→			
22	Tanezzuft	Memouinat		B					R	S														G→			
23	Azile	Anguille																				B	S	S,R			G
TAR 1	*Sakamena*	Isola															B,S	R					G				
TAR 2	*Ise*	Abeokuta																					B,S		R		G

Figure 21. The relationship between the 'ace' tectono-stratigraphy elements outlined in Table 1 and the evolution of the primary Petroleum Systems of Africa.

Key

B	Basin Forms by rifting (thermal subsidence thereafter)	R	Reservoir
B2	Second phase of rifting	G	Generation (spread over arrowed period)
S	Source rock		

Fig. 21. The relationship between the 'Ace' tectonostratigraphy elements outlined in Table 1 and the evolution of the primary petroleum systems of Africa. Note the concentration, particularly of basin-forming events in petroliferous basins (B, Basin Formation), related to the 'Ace' tectonostratigraphy discussed in this paper and the concentration of Generation Episodes (G) in recent times, following the Afar Plume. S, source rock deposition; R, reservoir deposition.

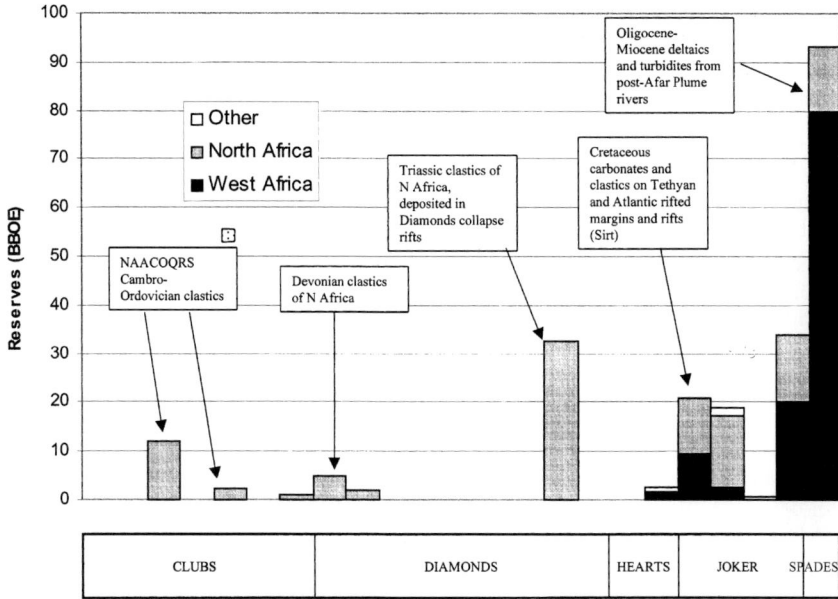

Fig. 22. The main oil- and gas-producing horizons of Africa. Note the very high proportion of reserves associated with Tertiary reservoirs, particularly in West Africa.

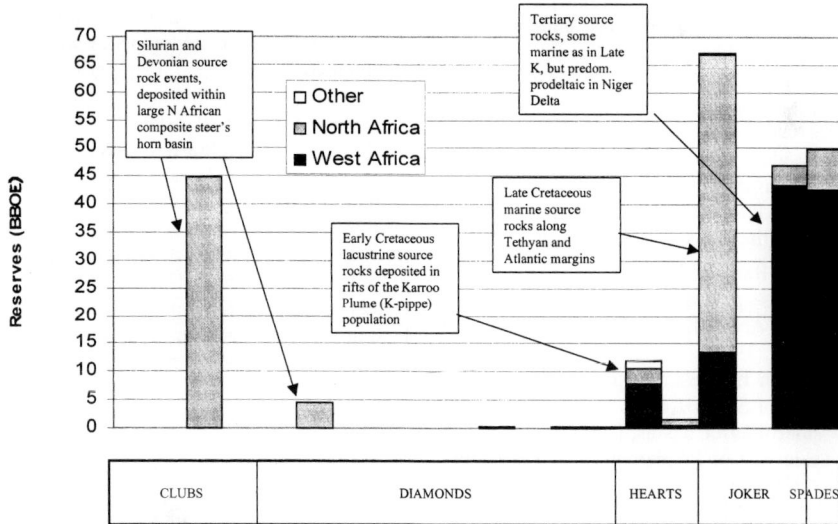

Fig. 23. The main source-ock units of Africa. The most important intervals are the Silurian, the Late Cretaceous and the Tertiary. Most of the source rocks lie within 'post-rift' sequences. As the main source-rock episodes are fixed in time, the relative timing of basin formation is critical as basins must be formed a sufficient time prior to source-rock deposition. Sufficient time must also elapse following source-rock deposition to allow maturation.

- the influence of rift systems on focusing sediment out of the Niger Delta. The flow of the Benue River has been localized for the past 140 Ma within the Benue Rift, one of the rifts formed in response to the K-pippe;

- deposition of marine source rocks from the Mid-Cretaceous (Albian) onward and most critically of all;
- the increase in sediment supply to West African deltas and deep-sea fans from erosion

of the uplifts that are rising in response to the A-pippe. Those sediments: (1) formed the main reservoirs, and (2) controlled, by burial, the areas in which source rock maturity has been attained, both within the syn-rift and drift succession. The increased rate of sedimentation also created the traps within the drift section. These are salt-related in the south and mud-diapir or toe-thrust hosted in the Niger Delta.

In all these successful areas, there is a discernable favourable order of events, most of which are tectonic led. Some of the trends are apparent in Figure 21. For instance, there is a set of petroleum systems identifiable in North Africa that formed in basins initiated in the Late Cambrian 'Ace of Clubs' thermal-subsidence phase and is characterized by source rocks deposited in the Silurian–Devonian and the generation during Cretaceous times of hydrocarbons. Another set lies in the West African basins formed in the K-pippe phase (Early Cretaceous) with reservoirs predominantly of the Tertiary (post-Afar Plume eruption at 31 Ma) fed by a variety of Mid-Cretaceous to Tertiary source rocks that were matured by Afar Plume-associated overburden. It is particularly interesting here to consider the importance of Oligocene–Holocene reservoirs (Fig. 22, Table 2), containing nearly half of Africa's petroleum in a land mass generally considered to be the 'oldest continent'.

It is equally important to the explorer to consider areas without exploration success to date and to suggest possible reasons for that lack of success. Many can be linked to the events discussed in this paper. Some petroleum systems were probably formed, but were later destroyed, by the tectonics associated with the Aces. Examples are:

- the Tindouf Basin of Algeria, whose Petroleum System was probably destroyed by uplift related to Ace of Diamonds convergence (Boote *et al.* 1998);
- the seep-rich but reserves-poor Karroo Foreland Basin of South Africa, which, along with many other interior basins, has been subject to large-scale Ace of Spades uplift and erosion;
- 31 Ma and younger uplift, which has postdated modelled periods of oil generation, has been a downgrading factor over areas of East Africa;
- other areas have not benefited from deposition of the thick Ace of Spades interval overburden that has controlled the maturity of the main regional source rocks, e.g. the immature Cretaceous source rocks over much of the Moroccan shelf.

We are unable to provide cast-iron criteria for predicting new petroleum provinces. However, several commonalities have been illustrated between the largest working system that can be used to help predict areas of likely future success. Positive factors for the prospectivity of any region include: recent overburden and generation; presence of Tertiary delta/turbidite reservoirs, rifts and basins formed at favourable times relative to the main source-rock episodes; and an absence of a major uplift or tectonic episode following generation. On this basis, examples of high-graded plays include Early Cretaceous lacustrine-sourced plays in the Gulf of Guinea area, and those associated with large Tertiary deltas or palaeodeltas, including the Tan-Tan (Morocco), Casamance (Senegal), Orange (South Africa), Zambezi (Mozambique) and Lamu (Kenya; prior to capture by Nile following Afar Plume uplift).

Conclusions

Our analysis presented in terms of 16 episodes, characterized as four Aces, one Joker and eleven wild cards, emphasizes the similarities between the dealing of a hand of cards and the evolution of petroleum systems. Both involve complex, close to random sequences of events that, to produce successful outcomes, have to play out in a particular order with correct timing. There are rarely alternative pathways. A clear example of such a sequence of favoured 'cards' over a long period of geological history is that of the history of the North African Palaeozoic source-rock related systems.

The most petroleum-rich areas of the African continent have proved to be those which record a favourable interplay between two or more of the events that we have discussed in this paper. Thus three-quarters of Africa's petroleum lies at margins or in rifts which formed as a result of the Ace of Hearts event (the K-pippe) and were later subject to significant overburden deposition resulting from Ace of Spades (A-pippe) uplift and erosion. The widespread Palaeozoic source rocks deposited in the extensive Ace of Clubs steer's-horns basin are generally effective only where an additional overburden has been imposed by deposits in later-formed rift basins of the Ace of Diamonds interval. It is equally important to consider the destructive effects of many of the main tectonic events that have left much of interior Africa of questionable prospectivity. Future analyses of the tectonic evolution of Africa, more elaborate than that which we have attempted, are unlikely to be able to find a use for our simple card-playing analogy. We think of it as providing no more than a set of scaffolding rods, ready to be discarded as the full structure of the edifice begins to emerge. We prefer that analogy to thinking of our construction as a house of cards.

References

ALSHARHAN, A. S. & NAIRN, A. E. M. 1997. *Sedimentary Basins and Petroleum Geology of the Middle East.* Elsevier, Amsterdam.

AMARAL, J., BITEAU, J. J., ZAROLINSKA, P. & DECOSTA, L. 1998. The Lower Congo Basin Tertiary petroleum system; Hydrocarbon distribution in relation with the structural and stratigraphic evolution. *American Association of Petroleum Geologists International Conference/Exhibition, November 8–11, 1998, Rio de Janeiro, Brazil, Abstracts*, 924.

ARTHUR, M. A. & SAGEMAN, B. B. 1994. Marine black shales. *Annual Reviews of Earth and Planetary Sciences*, **22**, 499–521.

AUSTIN, J. A. & UCHUPI, E. 1982. Continental–oceanic transition off southwest Africa. *American Association of Petroleum Geologists Bulletin*, **66**, 1328–1347.

AWAD, G. M. 1984. Habitat of oil in Abu Gharadiq and Faiyum basins, Western Desert, Egypt. *American Association of Petroleum Geologists Bulletin*, **68**, 564–573.

BASALTIC VOLCANISM STUDY PROJECT. 1981. *Basaltic Volcanism on the Terrestrial Planets.* Pergamon Press, Inc., New York.

BATE, R. H. 1999. Non-marine ostracod assemblages of the Pre-Salt basins of West Africa and their role in sequence stratigraphy. *In*: CAMERON, N. R., BATE, R. H. & CLURE, V. S. (eds). *The Oil and Gas Habitats of the South Atlantic.* Geological Society, London, Special Publications, **153**, 283–292.

BEUF, S., BIJU-DUVAL, B., DE CHARPAL, O., ROGNON, P., GARIEL, O. & BENNACEF, R. 1971. *Les Grès du Paléozöique Inférieur au Sahara.* Publications de l'Institut Français du Pétrole, Science et Technique du Pétrole, Editions Technip, Paris, **18**.

BLASBAND, B., WHITE, S., BROOIJMANS P., DE BOORDER H. & VISER W. 2000. Late Proterozoic extensional collapse in the Arabian-Nubian Shield. *Journal of the Geological Society, London*, **157**, 625–629.

BOOTE, D. R. D., CLARK-LOWES, D. & TRAUT, M. W. 1998. Paleozoic petroleum systems of North Africa. *In*: MACGREGOR, D. S., MOODY, R. T. J. & CLARK-LOWES, D. D. (eds) *Petroleum Geology of North Africa.* Geological Society, London, Special Publications, **132**, 7–68.

BOSWORTH, W. & MCCLAY, K. 2001. Structural and stratigraphic evolution of the Gulf of Suez rift Egypt. *In*: ZIEGLER, P. A., CAVAZZA, W. & ROBERTSON, A. H. F. (eds) *Peri-Tethys Memoir 6: Peri-Tethyan Rift/Wrench Basins and Passive Margins. Memoires Musée National d'Histoire Naturelle, Paris, Serie C Sciences de la Terre*, **186**, 735–754.

BRETTHAUER, H., CASIMIRO, L., ORSOLINI, P. & DE CAIVALHO, A. 1998. Regional evaluation of the Post-Salt system of Angola. *American Association of Petroleum Geologists International Conference/Exhibition, November 8–11, 1998, Rio de Janeiro, Brazil, Abstracts*, 920.

BURKE, K. 1975. Atlantic evaporites formed by evaporation of water spilled from Pacific Tethyan and Southern oceans. *Geology*, **3**, 613–616.

BURKE, K. 1976a. Development of graben associated with the initial ruptures of the Atlantic Ocean. *Tectonophysics*, **36**, 93–112.

BURKE, K. 1976b. The Chad Basin: an active intra-continental basin. *Tectonophysics*, **36**, 198–206.

BURKE, K. 1977. Aulacogens and continental breakup. *Annual Reviews of Earth and Planetary Sciences*, **5**, 371–396.

BURKE, K. 1988. Tectonic evolution of the Caribbean. *Annual Reviews of Earth and Planetary Sciences*, **16**, 201–231.

BURKE, K. 1996. The African Plate. *South African Journal of Geology*, **99**, 339–410.

BURKE, K. 1999. Tectonic significance of the accumulation of the voluminous early Palaeozoic reservoir-containing quartz-rich sandstones of North Africa and Arabia. *Bulletin of the Houston Geological Society*, **41** (7), 11–13.

BURKE, K. 2001. The origin of the Cameroon Line of volcano-capped swells. *Journal of Geology* **109**, 349–362.

BURKE, K. & SENGÖR, A. M. CA. 1986. Tectonic escape in the evolution of the continental crust. *In*: BARANZANGI, M. & BROWN, L. (eds) *The Continental Crust.* American Geophysical Union, Geodynamics Series, **14**, 41–53.

BURKE, K. & SENGÖR, A. M. CA. 1988. Ten metre global sea-level change associated with South Atlantic Aptian salt deposition. *Marine Geology*, **83**, 319–312.

BURKE, K. & WILSON, J. T. 1972. Is the African plate stationary? *Nature*, **239**, 448–449.

BURWOOD, R, REDFERN & J. COPE, M. J. 2003. Geochemical evaluation of East Sirte Basin (Libya) petroleum systems and oil provenance. *In*: ARTHUR, T. J., MACGREGOR, D. S. & CAMERON, N. R. (eds) *Petroleum Geology of Africa: New Themes and Developing Technologies.* Geological Society of London, Special Publications, **207**, 203–240.

CAMERON, N. R., BATE, R. H., CLURE, V. S. & BENTON, J. 1999. Oil and gas habitats of the South Atlantic: Introduction. *In*: CAMERON, N. R., BATE, R. H. & CLURE, V. S. (eds) *The Oil and Gas Habitats of the South Atlantic.* Geological Society, London Special Publications, **153**, 1–9.

COWARD, M. P. & RIES, A. C. 2003. Tectonic development of North African basins. *In*: ARTHUR, T. J., MACGREGOR, D. S. & CAMERON, N. R. (eds) *Petroleum Geology of Africa: New Themes and Developing Technologies.* Geological Society of London, Special Publications, **207**, 61–83.

COWARD, M. P., PURDY, E. G, RIES, A. C & SMITH, D. G. 1999. The distribution of petroleum reserves in basins of the South Atlantic margins. *In*: CAMERON, N. R., BATE, R. H. & CLURE, V. S. (eds) *The Oil and Gas Habitats of the South Atlantic.* Geological Society, London Special Publications, **153**, 101–132.

COWARD, M. P., BROWN, M. *et al.* 2000. Structural and tectonic evolution of SW Algeria: the results of regional tectonic and detailed 3D/4D structural modelling. *In*: *Petroleum Systems and Evolving Technologies in African Exploration and Production, Abstracts of Meeting of 16–18 May 2000*, Petroleum Exploration Society of Great Britain/Geological Society Meeting, London.

DALY, M. CA., LAWRENCE, S. R., DIEMU-TSHIBAND,

K. & MATOUANA, B. 1992. Tectonic evolution of the Cuvette Centrale, Zaire. *Journal of the Geological Society, London*, **149**, 539–546.

DAILLY, P. & GOH, K. 2000. The Rio Muni Basin of Equatorial Guinea – a new hydrocarbon province. *In: Petroleum Systems and Evolving Technologies in African Exploration and Production, Abstracts of Meeting of 16–18 May 2000*, Petroleum Exploration Society of Great Britain/Geological Society Meeting, London.

DEMENOCAL, P. B. 1995. Plio-Pleistocene African climate. *Science*, **270**, 53–59.

DEWEY, J. F. 1988. Extensional collapse of orogens. *Tectonics*, **7**, 1123–1129.

DEWEY, J. F. & BURKE, K. 1973. Tibetan, Variscan and Precambrian basement reactivation: products of continental collision. *Journal of Geology*, **81**, 406–433.

DEWEY, J. F. & BURK, K. 1974. Two plates in Africa during the Cretaceous. *Nature*, **249**, 313–316.

DE CHARPAL, O., GUENNOC, P., MONTARDET, L. & ROBERTS, D. G. 1978. Rifting, crustal attenuation and subsidence in the Bay of Biscay. *Nature*, **275**, 706–711.

DE MATOS, R. M. D. 1999. History of the northeast Brazilian rift system: kinematic implications for the break-up between Brazil and West Africa. *In:* CAMERON, N. R., BATE, R. H. & CLURE, V. S. (eds) *The Oil and Gas Habitats of the South Atlantic*. Geological Society, London, Special Publications, **153**, 55–73.

DE WIT, M. J., & RANSOME, I. G. D. (eds) 1992. *Inversion Tectonics of the Cape Fold Belt, Karroo and Cretaceous Basins of Southern Africa*. Balkema, Rotterdam.

DOBEX. 1976, *The Petroleum Potential of Atlantic Continental Margins North of 40 degrees N*. DOBEX International Ltd, Nyack, New York. [Unpublished report]

DOMBROWSKI, J., FAYE, M., BATE, R. H., CAMERON, N. R. & CARR, A. D. 2000. Evidence for a newly recognised petroleum system in the deep water portion of the Senegal sedimentary basin. *In: Petroleum Systems and Evolving Technologies in African Exploration and Production, Abstracts of Meeting of 16–18 May 2000*, Petroleum Exploration Society of Great Britain/Geological Society Meeting, London.

DROSTE, H. H. J. 1997. Stratigraphy of the Lower Paleozoic Haima Supergroup of Oman. *GeoArabia*, **2**, 419–472.

EBINGER, CA. J. & SLEEP, N. H. 1998. Cenozoic magmatism throughout East Africa resulting from impact of a single plume. *Nature*, **395**, 788–791.

ECHIKH, K. 1998. Geology and hydrocarbon occurrences in the Ghadames Basin, Algeria, Tunisia, Libya, *In:* MACGREGOR, D. S. MOODY, R. T. J. & CLARK-LOWES, D. D. S. (eds) *Petroleum Geology of North Africa*. Geological Society, London Special Publications, **132**, 109–130.

ELLIOT, D. H., FLEMING, T. H., KYLE, P. R. & FOLAND, K. A. 1999. Long-distance transport of magmas in the Jurassic Ferrar large igneous province, Antarctica. *Earth and Planetary Science Letters*, **167**, 89–104.

ENGLAND, P. & HOUSEMAN, G. 1984. On the geodynamic setting of kimberlite genesis. *Earth and Planetary Science Letters*, **67**, 109–122.

ESCHARD, R., BEKKOUCHE, D., DESAUBLIUAUX, G. & HAMEL, A. 2000 stratigraphic architecture of the Triassic reservoir in the Saharan province. *Petroleum Systems and Evolving Technologies in African Exploration & Production, Abstracts of Meeting of 16–18 May 2000*, 53.

FEKERINE, B. & ABDALLAH, H. 1998. Paleozoic lithofacies correlatives and sequence stratigraphy of the Sahara platform. *In:* MACGREGOR, D. S. MOODY, R. T. J. & CLARK-LOWES, D. D. S. (eds) *Petroleum Geology of North Africa*. Geological Society, London, Special Publications, **132**, 97–108.

FLEMING, N. C. A. & ROBERTS, D. G. 1973. Tectono-eustatic changes in sea level and sea-floor spreading. *Nature*, **243**, 19–22.

FROST, B. 2000. Deep water Mauritania – a frontier petroleum system. *In: Petroleum Systems and Evolving Technologies in African Exploration and Production, Abstracts of Meeting of 16–18 May 2000*, Petroleum Exploration Society of Great Britain/Geological Society Meeting, London.

GENIK, G. 1993. Petroleum geology of Cretaceous–Tertiary rift basins in Niger and Chad. *Bulletin of the American Association of Petroleum Geologists*, **77**, 1405–1434.

GEORGE, R., ROGERS, N. & KELLEY, S. 1998. Earliest magmatism in Ethiopia; evidence for two mantle plumes in one flood-basalt province. *Geology*, **26**, 923–926.

GRAND, S. P., VAN DER HILST, R. D. & WIDYANTORO, S. 1997. Global seismic tomography: a snapshot of convection in the Earth. *GSA Today*, **7**, 1–7.

GRAS, R. & THUSU, B. 1998. Trap architecture of the Early Cretaceous Sarir sandstone. *In:* MACGREGOR, D. S., MOODY, R. T. J. & CLARK-LOWES, D. D. (eds) *Petroleum Geology of North Africa*. Geological Society, London, Special Publications, **132**, 317–334.

GUIRAUD, R. & BOSWORTH, W. 1997. Senonian basin inversion and rejuvenation of rifting in Africa and Arabia: synthesis and application to plate scale tectonics. *Tectonophysics*, **282**, 39–82.

GUIRAUD, R. & BOSWORTH, W. 1999. Phanerozoic geodynamic evolution of NE Africa and the NW Arabian platform. *Tectonophysics*, **315**, 73–108.

HAACK, R. C., SUNDARARAMAN, P., DIEDJOMAHON, J. O, XIAO, N. J., GANT, N. J., MAY, E. D. & KELSCH, K. 2000. Niger Delta petroleum systems. *In:* MELLO, M. R. & KATZ, B. J. (eds) *Petroleum Systems of South Atlantic Margins*. American Association of Petroleum Geologists Memoirs, **73**, 213–232.

HALL, S. A. & GIRDLER, R. W. 1970. An aeromagnetic survey over the junction of the Ethiopian rift with the Gulf of Aden and Red Sea rifts. *Proceedings of the Geological Society of London*, **166**.

HAQ, B., HARDENBOL, J. & VAIL, P. R. 1987. Chronology of fluctuating sea levels since the Triassic. *Science*, **235**, 1156–1167.

HARLAND, W. B., ARMSTRONG, R. L., COX, A. V., CRAIG, L. E., SMITH, A. G. & SMITH, D. G. 1990. *A Geologic Time Scale 1989*. Cambridge University Press, Cambridge.

HOLMES, A. 1944. *Principles of Physical Geology*. Thomas Nelson and Sons, Edinburgh.

JABOUR, H., MORABET, AL M. & BOUCHTA, R. 2000. Hydrocarbon systems of Morocco. *In:* CRASQUIN-SOLEAU, S. & BARRIER, E. (eds) *Peri-Tethys Memoir*

5: *New Data on Peri-Tethyan Sedimentary Basins. Memoires Musée National d'Histoire Naturelle, Paris, Serie C Sciences de la Terre*, **182**, 143–158.

JACOBI, R. D. & HAYES, D. E. 1982. Bathymetry, microphysiography and reflectivity characteristics of the West African margin between Sierra Leone and Mauritania. *In*: VON RAD, U., HINZ, K., SARNTHEIN, M. &. SEIBOLD, E. (eds) *Geology of the Northwest African Continental Margin.* Springer-Verlag, Berlin, Heidelberg and New York, 182–212.

JUNGSLAGER, E. H. A. 1999. Petroleum habitats of the Atlantic margin of South Africa. *In*: CAMERON, N. R., BATE, R. H., & CLURE, V. S. (eds) *The Oil and Gas Habitats of the South Atlantic.* Geological Society, London, Special Publications, **153**, 153–168.

KARNER, G. D. & DRISCOLL, N. 1999. W. Tectonic and stratigraphic development of the West African and eastern Brazilian Margins: insights from quantitative basin modelling. *In*: CAMERON, N. R., BATE, R. H.& CLURE, V. S. (eds) *The Oil and Gas Habitats of the South Atlantic.* Geological Society, London, Special Publications, **153**, 11–40.

KRENKEL, E. 1925. *Geologie der Erde, Geologie Afrikas.* Gebrüder Borntrager Verlagsbuchhandlung, D-70176, Stuttgart.

LITHGOW-BERTELLONI, C. A. & RICHARDS, M. A. 1995. The dynamics of Cenozoic plate driving forces. *Geophysical Research Letters*, **22**, 1317–1320.

LITHGOW-BERTELLONI, CA. & SILVER, P. G. 1998. Dynamic topography, plate driving forces and the African superswell. *Nature*, **395**, 269–272.

LOGAN, P. & DUDDY, I. 1998. An investigation of the thermal history of the Ahnet and Reggane basins, Central Algeria and the consequences for hydrocarbon generation and migration ; *In*: MACGREGOR, D. S., MOODY, R. T. J.& CLARK-LOWES, D. D. (eds) *Petroleum Geology of North Africa.* Geological Society, London, Special Publications, **132**, 131–156.

MALUSKI, H., COULON, CA., POPOFF, M. & BAUDIN, P. 1995. 40 Ar/39Ar chronology, petrology, and geodynamic setting of Mesozoic to Early Cenozoic magmatism from the Benue Trough, Nigeria. *Journal of the Geological Society, London*, **152**, 311–326.

MARZOLI, A., RENNE, P. R., PICCIRILLO, E. M., BELLIENI, G., DE-MIN, A. 1999. Extensive 200-million-year old continental flood basalts of the Central Atlantic Magmatic Province. *Science*, **284**, 616–618.

MACGREGOR, D. S. 1996a. Hydrocarbon systems of North Africa. *Marine & Petroleum Geology*, **13** (3), 329–340.

MACGREGOR, D. S. 1996b. Factors controlling the destruction and preservation of giant light oilfields. *Petroleum Geoscience*, **2** (3), 197–217.

MACGREGOR, D. S. 1998. Giant fields, petroleum systems and exploration maturity of Algeria. *In*: MACGREGOR, D. S., MOODY, R. T. J. & CLARK-LOWES, D. D. (eds) *Petroleum Geology of North Africa.* Geological Society, London, Special Publications, **132**, 79–96.

MACGREGOR, D. S. & MOODY, R. T. J. 1998. Mesozoic and Cenozoic petroleum systems of North Africa., *In*: MACGREGOR, D. S., MOODY, R. T. J. & CLARK-LOWES, D. D. (eds) *Petroleum Geology of North Africa.* Geological Society, London, Special Publications, **132**, 201–216.

MACGREGOR, D. S. & CAMERON, N. R. 2000. An overview of the petroleum systems of Africa. *In*: *Petroleum Systems and Evolving Technologies in African Exploration and Production, Abstracts of Meeting of 16–18 May 2000*, Petroleum Exploration Society of Great Britain/Geological Society Meeting, London.

MACGREGOR, D. S., ROBINSON, J. & SPEAR, G. 2003. Play fairways in the Gulf of Guinea transform margin. *In*: ARTHUR, T. J., MACGREGOR, D. S. & CAMERON, N. R. (eds) *Petroleum Geology of Africa: New Themes and Developing Technologies.* Geological Society of London, Special Publications, **207**, 131–150.

MAGOON, L. B. & DOW, W. G. 1994. The Petroleum System. *In*: MAGOON, L. B. & DOW, W. G. (eds) *The Petroleum System – From Source to Trap.* American Association of Petroleum Geologists Memoirs, **60**, 3–23.

MCKENZIE, D. P. 1978. Some remarks on the development of sedimentary basins. *Earth and Planetary Science Letters*, **40**, 25–32.

MCKENZIE, D. P. & WEISS, N. 1975. Speculations on the thermal and tectonic history of the Earth. *Geophysical Journal of the Royal Astronomical Society*, **42**, 131–174.

MOLNAR, P. & DALMAYRAC, B. 1981. Parallel thrust and normal faulting in Peru and constraints on the state of stress. *Earth and Planetary Science Letters*, **55**, 473–481.

MOLNAR, P. & TAPPONNIER, P. 1975. Cenozoic tectonics of Asia: effects of continental collision. *Science*, **83**, 5361–5375.

MORLEY, CA. K., WESCOTT, W. A., STONE, D. M., HARPER, R. M., WIGGER, S. T. & KARANJA, F. M. 1992. Tectonic evolution of the northern Kenya rift. *Journal of the Geological Society, London*, **149**, 333–348.

NEVES, CA. A. DE O. 1989. Hydrocarbon generation, migration and accumulation in the Eo-Cretaceous continental sequence of the intracontinental Potiguar Rift Basin, northeastern Brazil. *Boletim de Geociências da Petrobras*, **3** (3), 131–145.

OLSEN, P. E. 1999. Giant lava flows, mass extinctions, and mantle plumes. *Science*, **284**, 604–605.

PURDY, E., CAMERON, N. R., MILSOM, J. & OEHLERS, M. 2000. Implications of South Atlantic coastal drainage for deepwater exploration. *In*: *Petroleum Systems and Evolving Technologies in African Exploration and Production, Abstracts of Meeting of 16–18 May 2000*, Petroleum Exploration Society of Great Britain/Geological Society Meeting, London.

RAVELOSON, E., ANDRIAMANANTERA, J., ROGER, J. & RAMANAMPISOA, L. 1991. The South Morondava Karoo and its petroleum potential. *In*: PLUMMER P. S. (ed.) *Proceedings of the First Indian Ocean Petroleum Seminar, Seychelles, 10–15 December 1990*, 307–324.

REYNAUD, F. & DRAPEAU, D. 1998. Post-salt petroleum systems in the Lower Congo Basin (Congo-Angola). *American Association of Petroleum Geologists International Conference/Exhibition, November 8–11, 1998, Rio de Janeiro, Brazil, Abstracts*, 838.

ROSENDAHL, B. R. & GROSCHEL-BECKER, H. 1999. Deep seismic structure of the continental margin in the Gulf of Guinea: a summary report. *In*: CAMERON, N. R., BATE, R. H. & CLURE, V. S. (eds) *The Oil and Gas*

Habitats of the South Atlantic. Geological Society, London, Special Publications, **153**, 75–83.

SCHOELLKOPF, N. B. & PATTERSON, B. A. 2000. Petroleum systems of Cabinda, Angola. *In*: MELLO, M. R. & KATZ, B. J. (eds) *Petroleum Systems of South Atlantic Margins.* American Association of Petroleum Geologists Memoirs, **73**, 361–376.

SCHUTZ, K. I. 1994. Structure and stratigraphy of the Gulf of Suez, Egypt. *In*: LANDON, S. (ed.) *Interior Rift Basins.* American Association of Petroleum Geologists Memoirs, **59**, 57–96.

SENGÖR, A. M. C. A. 1995. Sedimentation and tectonics of fossil rifts. *In*: BUSBY, CA. J. & INGERSOLL, R. V. (eds) *Tectonics of Sedimentary Basins.* Blackwell Science, Oxford, 53–117.

SENGÖR, A. M. C. A. & NATALIN, B. 1996. Paleotectonics of Asia. *In*: YIN, A. & HARRISON, T. M. (eds) *The Tectonic Evolution of Asia.* Cambridge University Press, Cambridge, 486–640.

SLEEP, N. H. 1997. Lateral flow and ponding of starting plume material. *Journal of Geophysical Research*, **102**, 10.001–10.012.

STERN, R. J. 1974. Arc assembly and continental collision in the Neoproterozoic East African orogen. *Annual Reviews of Earth and Planetary Science*, **22**, 319–35.

STURCHIO, N., SULTAN, M. & BATIZA, R. 1983. Geology and origin of the Meatiq Dome, Egypt. *Geology*, **11**, 72–76.

TAYLOR, M. 2000. Controls of sedimentation along the African margin of the Atlantic; climate and drainage geometry. *In*: *Petroleum Systems and Evolving Technologies in African Exploration and Production, Abstracts of Meeting of 16–18 May 2000*, Petroleum Exploration Society of Great Britain/Geological Society Meeting, London.

TEESSERENC, P. & VILLEMIN, J. 1989. Sedimentary basin of Gabon – geology of oil systems. *In*: EDWARDS, J. D. & SANTOGROSSI, P. A. (eds) *Divergent/Passive Margin Basins.* American Association of Petroleum Geologists Memoirs, **48**, 117–199.

WARDLAW, N. CA. & NICHOLS, G. D. 1972. Cretaceous evaporites of Brazil and West Africa and their bearing on the theory of continental separation. *Proceedings of the 24th International Geological Congress, Montreal 1972, Section 6: Stratigraphy and Sedimentology*, **2**, 43–55.

WILSON, J. T. 1968. Static or mobile Earth; the current scientific revolution. *In*: *Gondwanaland Revisited – New Evidence for Continental Drift.* Proceedings of the American Philosophical Society, Philadelphia, **112**, 309–320.

WITHJACK, M. O., SCHLISCHE, R. W. & OLSEN, P. E. 1998. Diachronous rifting, drifting and inversion on the passive margin of central eastern North America: an analog for other passive margins. *American Association of Petroleum Geologists Bulletin*, **82** (5A), 817–835.

Tectonic development of North African basins

M. P. COWARD & A. C. RIES

Ries-Coward Associates Limited, 70 Grosvenor Road, Caversham, Reading, RG4 5ES, UK

Abstract: Mostly the Palaeozoic and Mesozoic basins of North Africa have generally followed, and reworked, earlier basement trends formed by: (1) the NW–SE accretion of continental and oceanic terranes onto a Pan-African nucleus in northeastern Africa, and (2) the collision of this amalgam of accretionary terranes with the West African Craton. During the Upper Precambrian Pan-African Orogeny, the West African Craton formed a rigid block which indented this amalgam of accreted mobile belts to form much of North Africa. Intrusion of this indentor into North Africa caused the expulsion of narrow, triangular-shaped blocks of lithosphere to the north and south in a tectonic style very similar to the Miocene–Pliocene deformation of Tibet. Expulsion reactivated the earlier shear zones to form an anastomosing pattern of steeply dipping shears with left and right lateral sense of displacement. Left lateral shear also affected the northern edge of the West African Craton during this process of indentation.

Subsequent rifting of the Pan-African mountain belt resulted in a series of grabens, which were infilled with Upper Precambrian–Cambrian molasse. These are the precursor basins for the Palaeo-zoic sediments which cover much of North Africa. The effects of rifting continued into the Cambro-Ordovician in the western basins. During the Silurian–Devonian many of the rifts were reworked. A new basin formed in the Atlas and Anti-Atlas, related to the growth of the proto-Tethyan Ocean.

Basin inversion characterizes the Palaeozoic structures of the western Atlas and Anti-Atlas, pro-ducing thickened crust and a large mountain belt during the Carboniferous. Foreland basins formed on either side of this mountain belt and both the mountains and the adjacent basins were compart-mentalized by WNW–ESE-trending transfer zones. Pan-African structures, within the African Plate, were reworked with further indentation of the West African Craton into Pan-African crust. The craton was pushed eastward, generating a left lateral shear couple along its northern margin. NW–SE-trending faults were reworked as dominantly left lateral strike-slip faults and N–S-trending fault blocks were rotated slightly in a clockwise sense. There was probably further lateral expulsion of lithosphere, ahead of the NE–SW-trending front of the indentor, reworking earlier N–S-trending shear zones.

The North African Palaeozoic basins were inverted during the Hercynian–Appalachian Orogeny. In the Ahnet Basin the shortening was approximately NNE–SSW, perpendicular to the trend of the structures. This inversion was particularly marked in the Ougarta–Ahnet Basin where it produced a series of open to closed, north–south to NW–SE-trending folds above reactivated basement faults.

During the Mesozoic, the Hercynian–Appalachian mountain belt underwent extension to produce deep rift basins infilled with continental sediments and some volcanics. The High Atlas formed as an arm to the Atlantic Basin. Transfer zones have a WNW–ESE trend, indicating that this was the main extension direction, similar to that in western and southwestern Europe.

In northeastern Algeria, the orientation of the Mesozoic grabens suggests reworking of the basement fabric formed by Pan-African accretionary tectonics. The structures appear to die out toward the southwest into a broad transfer zone with some NW–SE-trending faults. The northeastern edge of the basin is obscured by later rift basins in the eastern Mediterranean. The Palaeozoic faults of the Amguid Spur, overlying one of the major shear zones of the Hoggar, formed a structural high through-out the Mesozoic with probably several pulses of inversion. An important episode of inversion occurred during the Aptian–Albian with the development of anticlines and associated reverse faults.

Crustal extension associated with block faulting occurred in the Sirte Basin of Libya during the Mid- and Late Cretaceous. The block faults trend NNW–SSE to NW–SE, cross-cutting earlier Palaeozoic fold structures at a high angle but possibly parallel to some of the basement shear zones. The faults form the tips of a rift basin which opened between Sicily and Tunisia in the central Mediterranean. The Cretaceous faults have a component of right lateral displacement as well as normal fault movements.

The Mesozoic basins of the High and Middle Atlas were inverted during the Late Cretaceous–Early Oligocene. The displacement direction, as seen from the transfer systems, was NW–SE, almost perpendicular to the Middle Atlas, but at a lower angle with the High and Sahara Atlas, which must have had components of oblique or right lateral movement. Minor effects of this inversion are reported from the Saharan basins.

From: ARTHUR, T. J., MACGREGOR, D. S. & CAMERON, N. R. (eds) *Petroleum Geology of Africa: New Themes and Developing Technologies.* Geological Society, London, Special Publications, **207**, 61–83. 0305-8719/03/$15

Introduction

This paper discusses the tectonic framework of that part of North Africa which extends from Morocco through Algeria to Libya. The important features which control the tectonic history are: (1) the effect of the Pan-African basement on subsequent rift and inversion structures, (2) the variation in the time of rifting and subsidence across the Palaeozoic basins, (3) the relationship between the development of the Mesozoic basins and the opening of the Atlantic, and (4) the effects of Alpine tectonics on the interior basins.

Precambrian tectonics

West African Craton

The oldest rocks exposed in North Africa form part of the West African Craton in the west (Archean), and the Touareg Shield in the east (Eburnean, 2000–1500 Ma) (Fig. 1) (Black & Fabre 1980). The gneisses and granites of the craton, which has been stable since c.1700 Ma, are overlain by Proterozoic calcareous and clastic sediments (Taoudeni Basin) that grade northward into mafic and volcaniclastic rocks dated at 780 Ma (Leblanc & Lancelot 1980). The Reguibat Shield, which forms the northern part of the West African Craton (Fig. 1), consists of Archean rocks in the western and central parts and Early Proterozoic rocks in the east; radiometric dates (mainly Rb–Sr isochrons) indicate events at 2700 Ma and 2400 Ma.

Caby (1970) and Leblanc (1972) recognized collisional deformation processes along the northern and eastern margins of the West African Craton during the Pan-African and related these to the collision of the passive margin of the West African Craton with the active margins of the Touareg Shield to the north and the Benin–Nigeria Shield to the south during the Pan-African Orogeny. In Morocco, along the northern margin of the West African Craton, a zone of ophiolitic blocks (Bou Azzer), representing a collisional suture, is marked by an elongate zone of positive gravity anomalies (Crenn 1957) extending from Morocco to the western Hoggar. On the eastern margin of the West African Craton, the suture zone which extends for >2000 km, is marked by positive gravity anomalies with amplitudes of 30–80 mgal.

Touareg Shield

The striking feature of the Touareg Shield is the approximate N–S-trending pattern of anastomosing Pan-African shear zones (Figs 1 & 2), which has been compared to a similar pattern of Cenozoic

faults in northeastern Asia (Caby *et al.* 1981), related to the collision of India with Asia (Fig. 3). In both areas many of the shear zones formed late in the collisional process and may be due to lateral continental escape. Furthermore, the Pan-African collisional belts of North Africa formed by the accretion of numerous microplates, as in Central Asia. Each block can be identified by differences in stratigraphy, although the bounding sutures may be gently dipping, folded and consequently difficult to trace. On a broad scale the blocks and block boundaries are defined by different ages of tectonics and sometimes by preserved remnants of oceanic crust or lower continental crust. From west to east these blocks are (Fig. 2):

1. the Pharusian Belt, with a platform sequence of lower Upper Proterozoic (1000–800 Ma) quartzites and marbles (Série a Stromatolites), overlain by upper Upper Proterozoic volcaniclastic rocks, characteristic of modern island arcs and active continental margins;
2. the central Hoggar–Aïr Domain, with pre-Pan-African basement, intruded by Pan-African granitoids;
3. the eastern Hoggar–Ténéré Domain, stabilized at c.730 Ma and bounded on its western margin by an upper Pan-African reactivated shear zone, the Tiririne Belt.

The above domains contain large bodies of rocks with high-grade, often granulite-facies metamorphic assemblages of Mid-Proterozoic Eburnean age (Bouillier *et al.* 1978). In the western Hoggar, the Eburnean granulite blocks are large and have a dominant ENE–WSW fabric. The blocks are surrounded by N–S-trending Pan-African amphibolite-facies gneisses and mylonite zones. There are also bodies of Lower Proterozoic high-grade amphibolite-facies gneisses in central Hoggar, giving ages of 2250 (±100) Ma. The presence of high-grade Mid-Proterozoic Eburnean rocks in the Touareg Shield contrasts to the Archean ages of the West African Craton.

The earliest accretionary terrane developed in the east, associated with magmatism dated at c.750 Ma, and is related to a magmatic arc developed above a SE-directed subduction zone. The boundary is uncertain; its western tectonic boundary is represented by the late tectonic Tiririne or 8° 30′E Lineament, a left lateral strike-slip shear zone which now forms the eastern margin of the Hoggar–Aïr Domain (Fig. 2). This early collisional belt was part of a Pan-African accretionary mass of magmatic arcs and oceanic/continental terranes, accreted onto an Upper Proterozoic nucleus prior to 700 Ma. Upper Proterozoic tectonics across much of the Precambrian terrane of the Eastern Desert of Egypt and Sudan involved accretion of

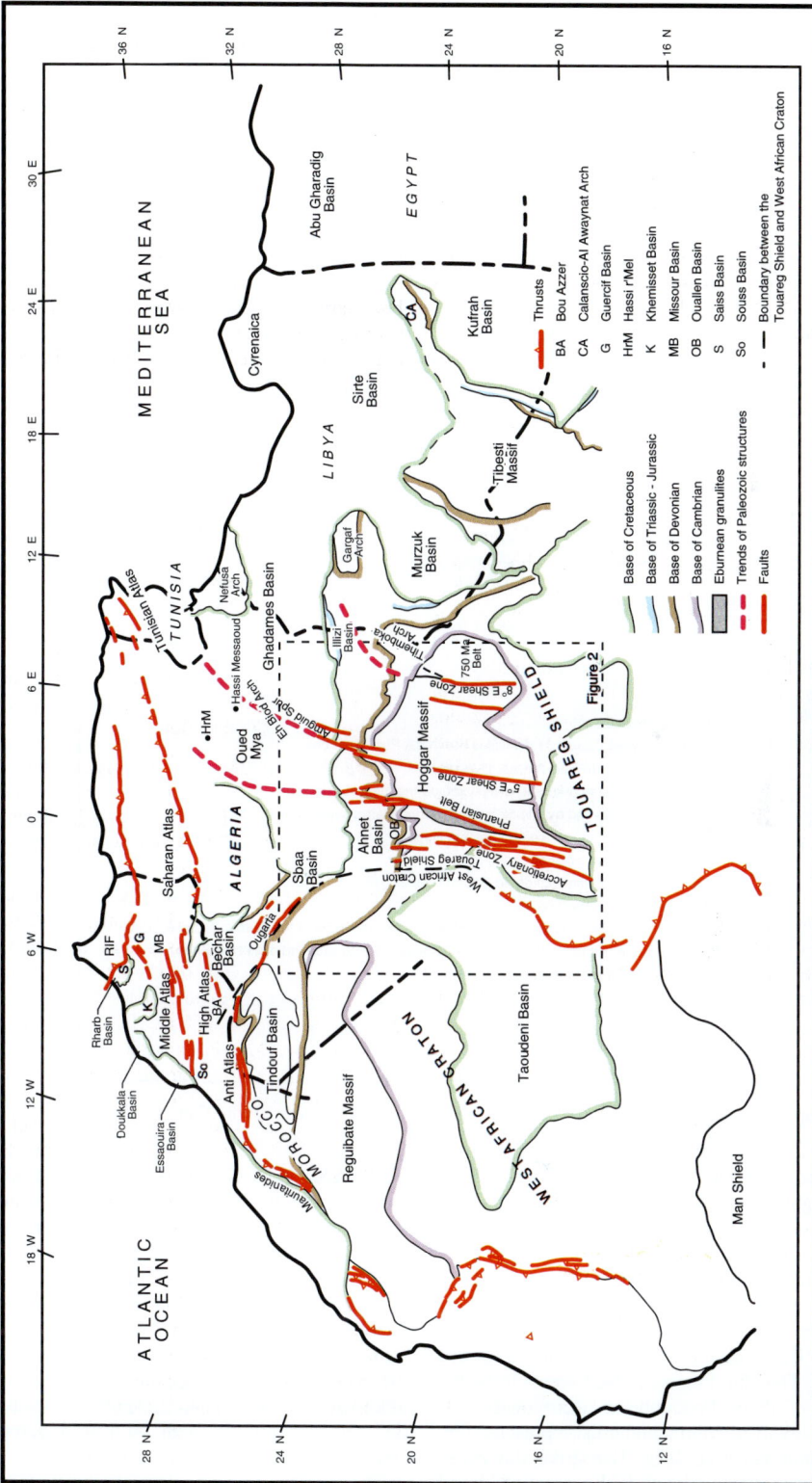

Fig. 1. Map of the northern part of Africa showing the main Precambrian structures, Phanerozoic basins and locations mentioned in the text.

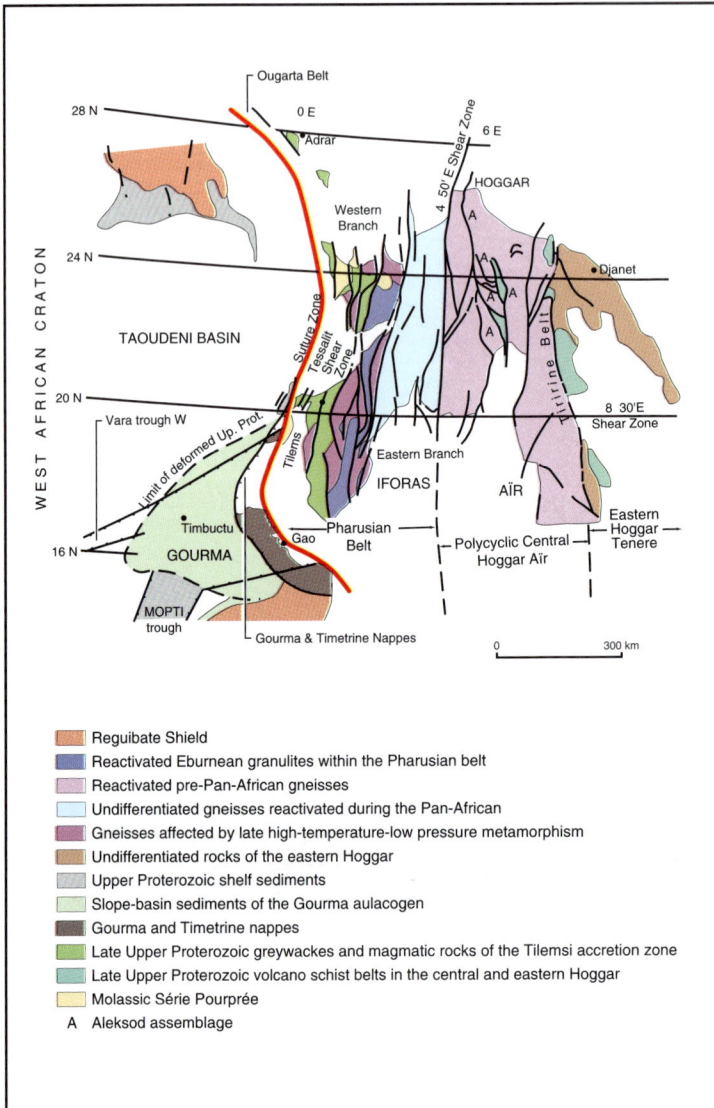

Fig. 2. Simplified geological map of the Touareg Shield and adjacent areas. (From Caby *et al.* 1981)

material from the northwest. Associated thrust belts trend roughly NE–SW, although there are many local variations, while strike-slip faults trend NNW–SSE to NW–SE (Shackleton 1994).

According to Black & Fabre (1980), a similarly directed collisional event (*c.*700 Ma or earlier) affected the Upper Proterozoic rocks of the western Pharusian Belt with the transport of nappes toward the NNW. Thus the Pharusian Belt seems to be the youngest of these Precambrian accretionary terranes and involved continental crust represented by Mid-Precambrian granulites. The strike-slip zones (transfer zones), associated with accretion, trend

NNW–SSE to N–S. The 4° 50′E Lineament is a major N–S- to NNE–SSW-trending shear zone which extends through the Hoggar southward to Benin and forms the boundary between the Pharusian Belt to the west and the Hoggar–Aïr Domain to the east (Fig. 2). The Pan-African in this zone is characterized by intensely deformed N-S-trending schist belts and shear zones with granitic intrusions. The amount of displacement is unknown but only the latest events, associated with the collision process, can be traced across this zone.

The western Hoggar seems to be underlain by

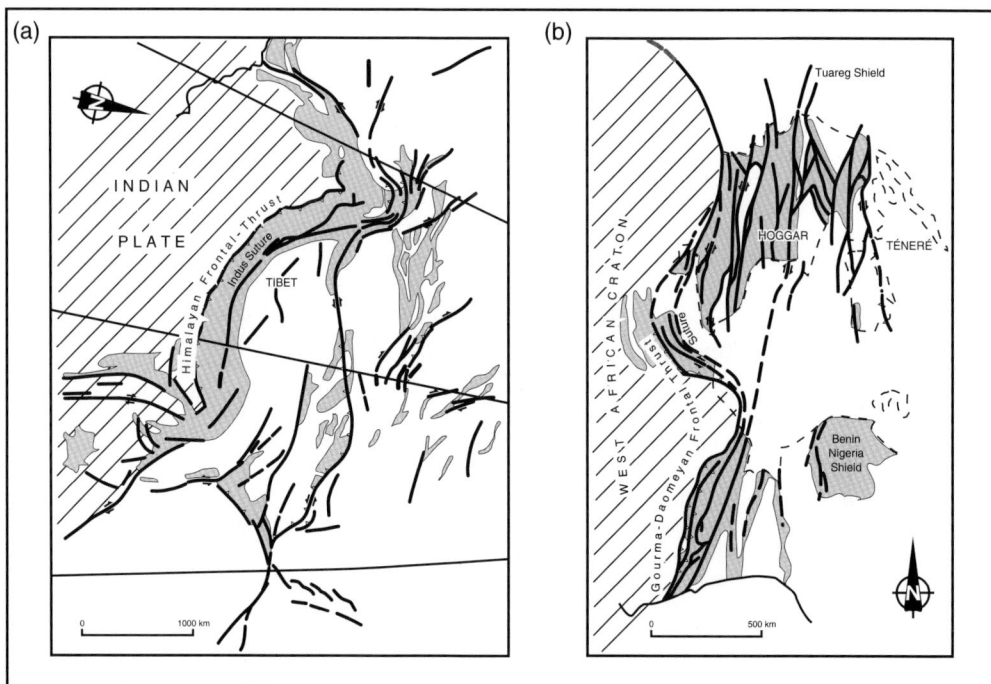

Fig. 3. Comparison between (**a**) the Cenozoic–Recent fault patterns in northeastern Asia and (**b**) the Pan-African structures of northeastern Africa. Shaded areas show the extent of the Tertiary and Upper Pan-African deformation, respectively. (From Black & Fabre 1980)

accretionary sediments and a magmatic arc, dated at 700–650 Ma. This amalgam of accretionary terranes was then accreted onto the West African Shield (Bouillier *et al.* 1978). This late east–west collisional event (*c.*600 Ma) affected the entire Touareg Shield.

The latest Pan-African structures involve shear zones and normal faults. Many of the faults bound Upper Precambrian molassic basins (Série Pourprée). In the Ouallen Basin (northwestern Hoggar–southern Ahnet area), conglomerates and fanglomerates occur close to the master fault and pass westward into shaly sequences. The basal sediments, away from the fault, are aeolian and fluvial arkoses. The bounding fault is currently steeply dipping, suggesting back-rotation, and the half-graben fill has been uplifted and eroded, probably largely during Upper Palaeozoic deformation, but possibly also during late pulses of Pan-African deformation.

Anti-Atlas Belt

The basement exposed in the Anti-Atlas Belt in southern Morocco comprises Eburnean rocks forming the northern part of the West African Craton to the south and rocks affected by Pan-African oro-

genic events, dated at 680–570 Ma (Leblanc & Lancelot 1980), to the north. These two areas of different aged basement are separated by the Bou Azzer ophiolite zone (Leblanc 1972; Leblanc & Lancelot 1980). This is the only outcrop of ophiolitic rocks exposed along the eastern and northern edge of the West African Craton, although the trend of the suture can be seen from the gravity data (Crenn 1957).

Basement fabrics

The Pan-African structures in North Africa (Fig. 1) form the basement fabric of the Upper Palaeozoic and Mesozoic basins. In the west this fabric is governed by the Pan-African suture between the North African accretionary complex and the West African Shield. Pan-African fabrics trend NW–SE and dip moderately to the SE.

In Central and Eastern Algeria, the basement fabrics are related to the growth of the Touareg accretionary complex. In the Hoggar the basement fabric trends N–S and is steeply dipping to the east, with dominantly oblique-slip zones generated as transfer zones during accretion. The intervening thrust structures trend ENE–WSW to NNE–SSW and have a gentle to moderate dip to the ESE. To the north and east, the thrust structures seem to dominate. There the basement fabric probably largely has a NE trend and dips to the SE.

In Libya and Egypt, the basement structures also trend NE–SW to ENE–WSW and have gentle to moderate dips to the southeast. However there are also large shear zones, trending NW–SE, parallel to the accretion direction. These shear zones form prominent features in the basement of East Africa. They may be present under parts of the Murzuk and Sirte basins.

Palaeozoic basin development

The structural framework and general stratigraphy of the Palaeozoic basins of North Africa are shown in Figures 4 and 5. Basement fabric appears to control the trend of the Palaeozoic isopachs and fault trends. This is demonstrated by the changing orientation of the Palaeozoic basins from west to east (Fig. 4). In the western Sahara–northwestern Hoggar and Ougarta, these basins trend NNW–SSE to NW–SE, mimicking the Pan-African suture and adjacent basement beneath. Further east, in the northern and eastern Sahara of Algeria, the basins trend N–S to NNE–SSW and NE–SW and reactivate basement fabrics and shear zones related to Pan-African accretionary tectonics. In Libya, the Palaeozoic structures trend NE–SW to ENE–WSW (Figs 4 & 5).

Pan-African tectonics ended in the Late Precambrian followed by regional subsidence, local intracratonic rifting and a synchronous erosion period. This levelled many of the earlier structures into a northerly dipping pediplane, forming a flat depositional basin with little regional differentiation. The basin was infilled with paralic to shallow marine clastics. There are also occasional marine limestone intercalations, continental clastics and igneous rocks. The Palaeozoic stratigraphy of the several different basins is summarized in Figure 6 while Figure 7 illustrates the main phases of basin development.

There were three main cycles of sedimentation during the Palaeozoic, as described below.

1. Cambro-Ordovician cycle

Isopachs for Cambrian strata mimic the basement structures and presumably define areas of Upper Pan-African rifting. The basement is overlain unconformably by a molassic sequence of pebbly sheet sandstones, quartzites and conglomerates deposited by fast, northerly flowing braided rivers. This was followed by cyclic sequences of thick, regionally extensive, transgressive, fluvial and estuarine sandstones passing up into shallow-water sandstones and shales, deposited during marine incursions from the north. Throughout the Cambrian, sediment influx onto the northern Sahara followed the main graben trends, which had initially developed at the end of the Pan-African Orogeny. The Upper Pan-African and Cambrian faulting favoured the development of half-grabens with the master faults dipping toward the ENE, although there are important WSW-dipping faults. Some of the major Pan-African shear zones, such as the 4° 50′ and 8° 30′ shear zones, may have experienced some strike-slip displacement at this time. The thickness of the Cambrian sediments also increases in the Anti-Atlas and Atlas. Thicknesses of several kilometres are reported from the Atlas, but some of this may be tectonic thickening rather than stratigraphically controlled.

Similar rift structures occurred in other mid-Gondwana basins. In the Paranaiba and Parana basins of Brazil there were extensional events during the Cambrian and Silurian, forming N–S-trending grabens which followed the trends of earlier Brazilian (Pan-African equivalent) structures.

During the Ordovician the direction of the sediment supply continued from south to north, along the axes of the older basins. The Hamra Quartzite (Fig. 6) represents a low-stand sequence of Arenig age, with abundant *skolithus*-type worm burrows and planar cross-bedding. Stacked regressive sequences characterize the overlying Ordovician sequence with marine graptolitic shales, coarsening up into delta-front and delta-top sandstones. During the Ashgill the North African Platform was situated near the South Pole and an icecap developed over much of Africa and South America. The regressions are probably associated with cycles of ice-sheet advance and retreat (Fig. 8a). Falling sea levels encouraged locally deep incisions into the underlying strata which were subsequently infilled with fluvioglacial and glaciomarine shelf deposits, during final retreat of the icecap.

During the Early Ordovician the relatively uniform thickness of the Tremadocian suggests that the region was dominated by post-rift subsidence with only minor rifting along the Ougarta Zone and more active rifting along the fault zones bounding the Precambrian Pharusian Belt (Fig. 2). These rift basins trend N–S to NNE–SSW and appear to be asymmetric. The intervening blocks had little to no sedimentation. Rifting was renewed on all the major Pan-African structures during the Arenig, with elongate depocentres developing in rift basins, south of El Golea and Hassi Messaoud. The NW-trending basins of southern Libya may have been active at this time, opening the western Murzuk Basin. Unconformities, which occur at the base and top of the Arenig–Llandeilo sequence, are largely confined to the footwall uplifts, reflecting pulses of minor fault-controlled extension, particularly along

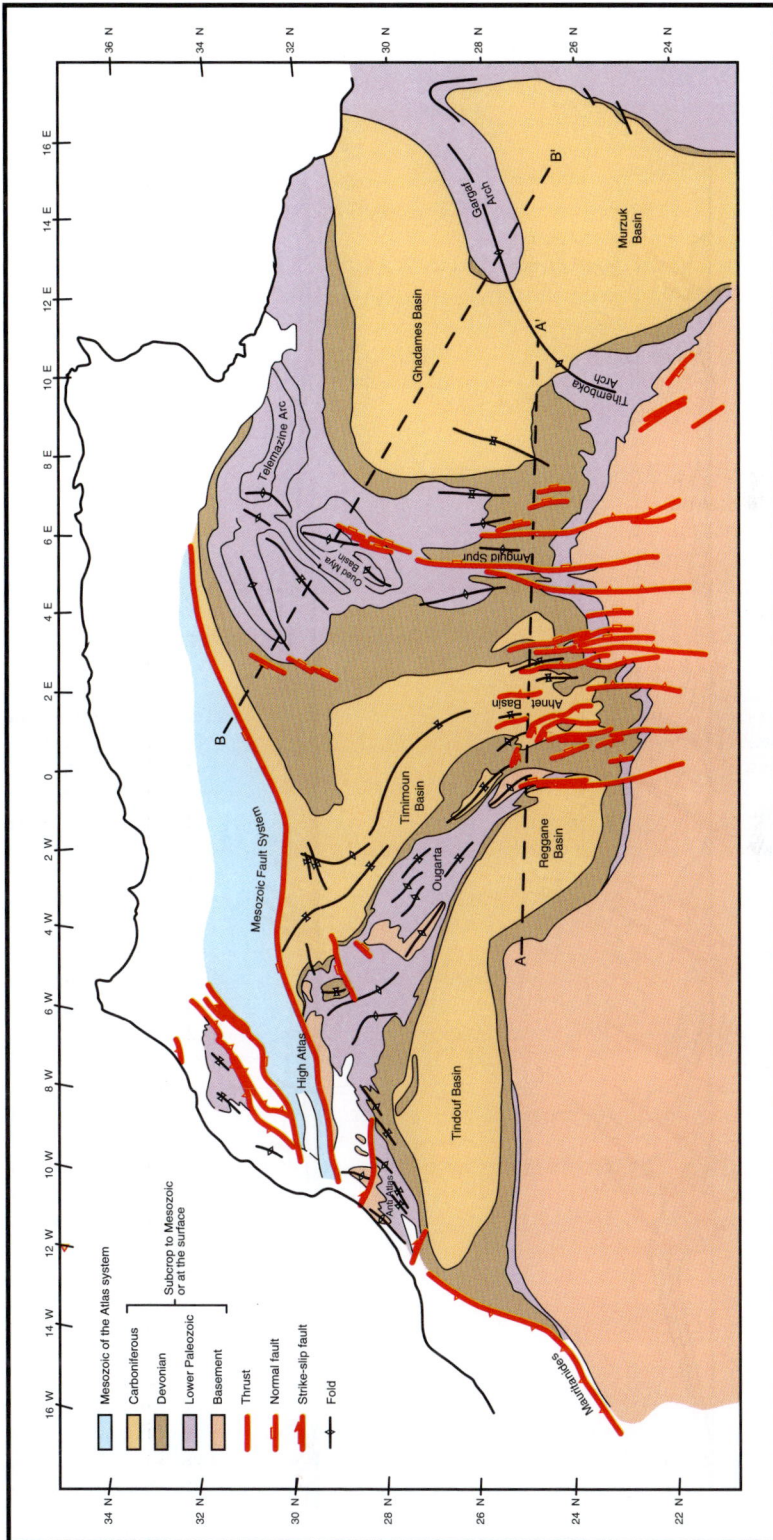

Fig. 4. Map showing the distribution and trend of Palaeozoic structures in North Africa.

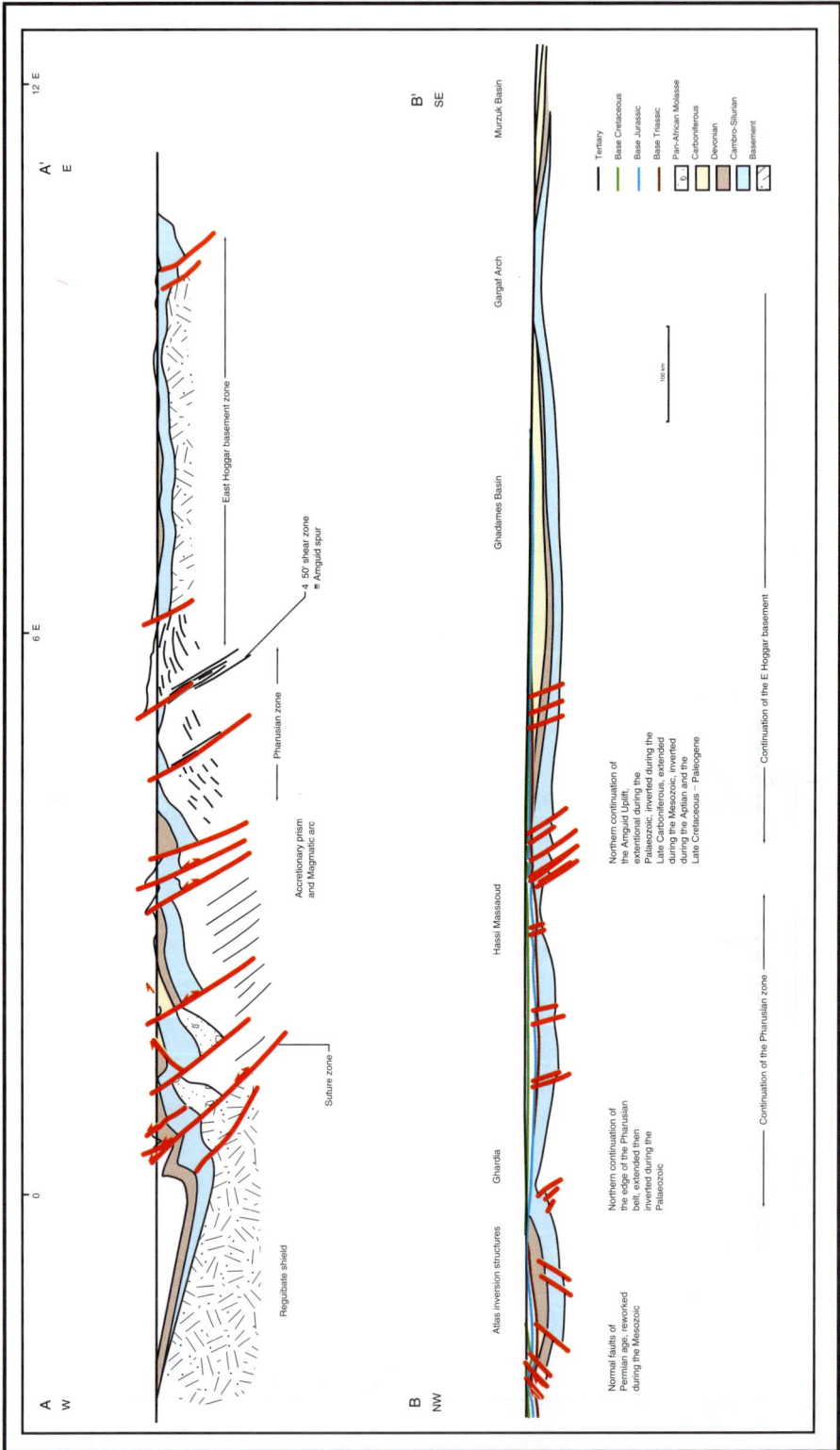

Fig. 5. Cross-sections through the Palaeozoic and Mesozoic structures of Algeria. Location shown on Figure 4.

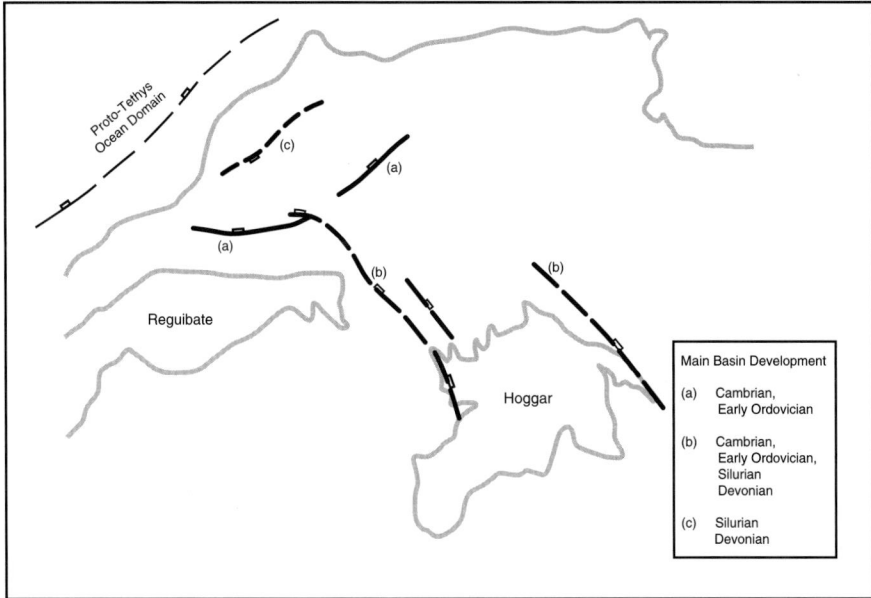

Fig. 7. Map of North Africa showing the main Palaeozoic fault trends.

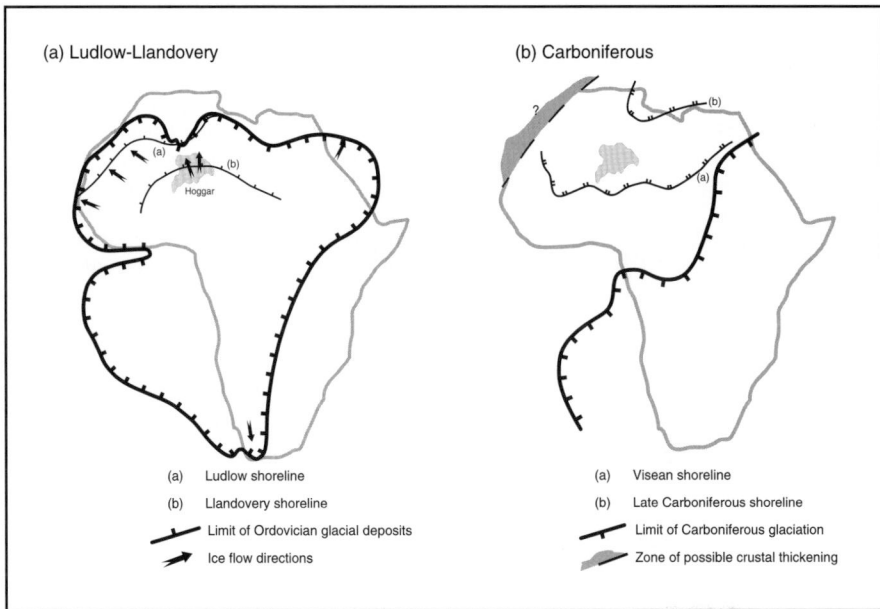

Fig. 8. Maps of Africa showing: (**a**) the limit of Ordovician glacial deposits and the shoreline during the Ludlow and Llandovery, and (**b**) the extent of the Carboniferous glaciation and the limits of the Visean transgression and the Upper Carboniferous regression.

the Ougarta, Amguid–El Biod, Tihemboka, Gargaf, Calanscio–Al Uwaynat and Al Uwaynat–Bahariyah arches. Isopachs for the later Ordovician (Fig. 9a) indicate subsidence over these basins with only minor rifting. However, faulting continued to affect sedimentation and appears to have controlled the development of canyon networks which were feeding fan deposits in the Illizi Basin (Abdallah 1999).

2. Silurian–Devonian cycle

Regional subsidence at the beginning of the Silurian was marked by a marine transgression over the western part of the Sahara, with the deposition of black graptolitic shales and siltstones (Tannezuft Formation) and sandstones. This transgression reached its peak during the Mid–Late Llandovery and may be related to the melting of the Ordovician icecap over central Gondwana. Lower Silurian bituminous shales, deposited during the transgression, are the main Palaeozoic source rock in North Africa (Klitzsch 1968; Boote et al. 1998; Luning et al. 2000). They are overlain by an Upper

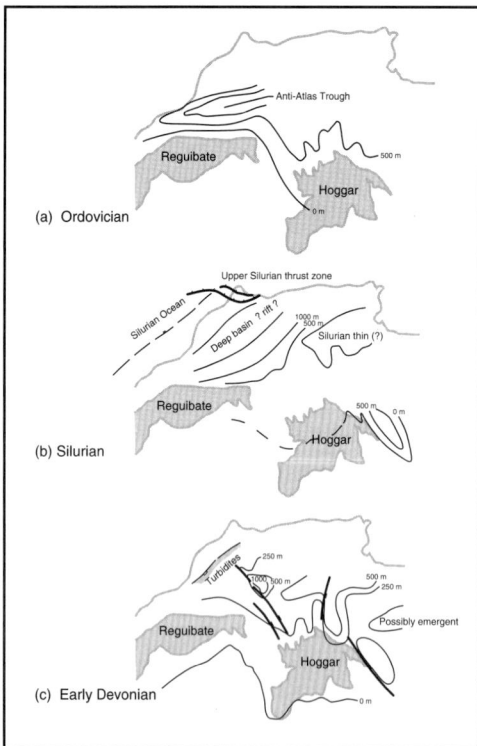

Fig. 9. Maps of North Africa showing the approximate distribution of: (**a**) Ordovician isopachs, (**b**) Silurian isopachs, and (**c**) Lower Devonian isopachs and associated tectonic features.

Silurian regressive marine sandstone unit (Acacus Formation) which is present in most of the Sahara except where removed by later erosion. There is a regionally extensive unconformity at the end of the Silurian, representing an episode of minor inversion.

Renewed rifting in the Silurian is suggested by the increase in the thickness of Silurian sediments on the hangingwall of many of the faults in, for example, the Ahnet and Timimoun basins (Figs 9b & 10). Following structural inversion in the Late Carboniferous, the top Silurian generally shows net contraction while the base Silurian shows net extension. A deep, regionally extensive Silurian basin formed in western Algeria and southern Morocco, probably due to rifting along the northern edge of Gondwana. Farther to the northwest, on the Moroccan Coastal Block, there are only thin platform equivalents. In southern Libya the moderately thick Silurian sequence in the Murzuk Basin may simply reflect regional thermal subsidence following Cambrian rifting. However, more rapid thickness changes in the western part of this basin suggest that there may have been renewed local rifting during the Silurian.

The unconformity at the top of the Silurian Acacus Formation represents a pulse of minor, but regionally extensive, inversion which reworked earlier normal faults. The inversion is most important at the edges of the Illizi and Murzuk basins (Echikh 1998). Basin inversion occurred along the eastern edge of the Hoggar Massif and across the Gargaf Arch. In the Murzuk Basin, the depocentre was shifted eastward, creating a new sub-basin.

The Lower Devonian Tadrart Formation shows seaward progradation and basinward thinning and a transition from fluviocontinental to estuarine to marine neritic conditions toward the northwest. Lower Devonian sediments were deposited in a dominantly transgressive sequence, with cycles of basal fluvial sands passing upward into a distal tidal offshore facies. The sands were derived from the southeast. Thickness variations indicate a strong tectonic control, related to basin inversion and reactivation of underlying structures.

Mid- and Upper Devonian cycles are made up of regressive, paralic delta systems, each with an erosional upper surface, in places incised and capped by extensive transgressive marine shales, limestones and iron oolites. Devonian rocks become more marine to the north and northwest. Givetian reefs developed in southwestern Algeria. There is a regionally developed unconformity of Frasnian age. Devonian thickness variations are difficult to reconstruct because of later erosion in much of north-central and northeastern Algeria, northwestern Libya and the Sirte Basin and are best

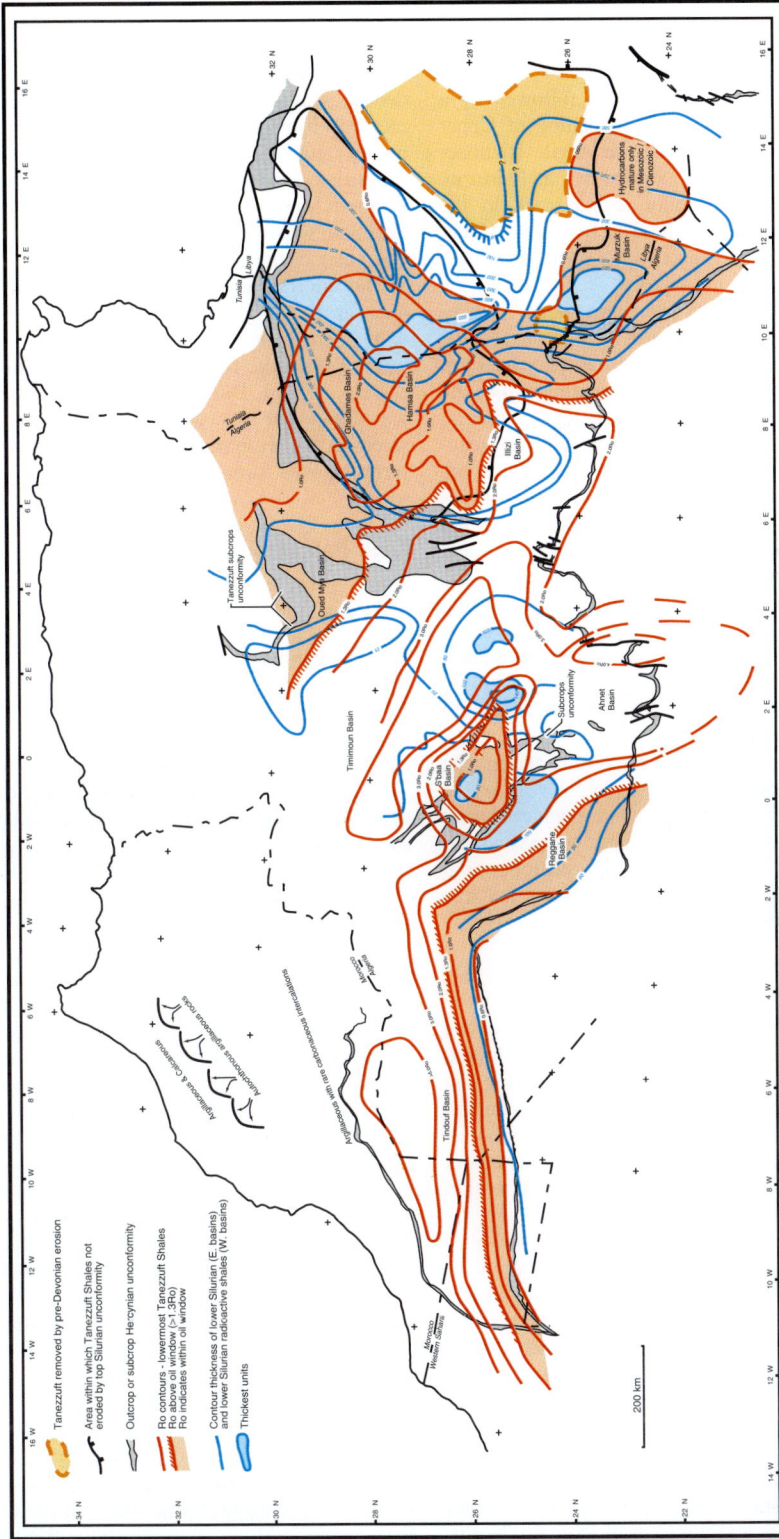

Fig. 10. Map of North Africa showing the outcrop/subcrop and the detailed isopachs for the Early Silurian areas where Lower Silurian shales have been removed by pre-Devonian erosion and maturity values for the Lower Silurian source rocks. Isopachs extrapolated over erosional areas.

preserved in the Illizi, Ahnet, Bechar and Tindouf basins of Algeria and in the Ghadames Basin of Libya (Fig. 9c).

The Frasnian unconformity was followed by a marine transgression and the deposition of Upper Frasnian organic-rich shales, the Argile Radioactive, across much of the North African Platform. These shales form a major source rock in the Illizi and Ghadames basins. The overlying Upper Frasnian–Tournasian succession comprises thin, cyclic, platform deltaic deposits.

During the Devonian, there was active movement on the 4° 50' Lineament in central Hoggar and its continuation into northern and northeastern Algeria. A basin formed on the hangingwall to the east of the shear zone. This basin is asymmetric, deepening to the west, and was formed by the reactivation of several E-dipping normal faults. Some faults were also active in the Ahnet Basin at this time and there was renewed movement on normal faults at the edge of the Murzuk Basin, with local mild inversion. Mild uplift also occurred in the southeastern part of the Saharan Platform during the Late Devonian. There is evidence for erosional truncation over the Ougarta and Gargaf arches. These structures presumably reflect the initial stages of collision tectonics to the west and north of the African Plate.

In the western part of the Atlas, from the Rehamna Massif to the Palaeozoic block in the High Atlas, there was a zone of emergence during the Early and Mid-Devonian. This was associated with the deposition of terrestial sandstones and conglomerates. During the Late Devonian the whole domain emerged.

3. Carboniferous

The Carboniferous marine transgression spread across the Sahara from the west and northwest. Erosion occurred on many of the uplifts during this period and Carboniferous marine facies are unconformable on Devonian rocks. The Early Carboniferous transgression reached its maximum extent during the Visean, followed by a transgression partly caused by Hercynian tectonics and partly by growth of the Carboniferous icecap (Fig. 8b).

The Late Carboniferous is shallow marine to lacustrine, with a predominant deltaic clastic component in the eastern Sahara. In the western Sahara, lagoonal evaporitic, reefal or open marine carbonates pass into thick clastics in the central part of the basins. Basic igneous rocks occur locally in the southern and western Sahara.

During the Late Devonian to Early–Mid-Carboniferous the sediments of the Moroccan Coastal Block were clastic, derived from the western part of the High Atlas and the Coastal Domain. North-

central Morocco became a subsiding flexural basin to the Acadian and Appalachian collision zones and was filled with several thousand metres of detrital sediments. These sediments include Fammenian olistostromes, turbidites and their contemporary equivalents, followed by Visean and Namurian sandstones and shales. Elsewhere in southern Morocco and Algeria, Late Devonian and Carboniferous sediments are represented by shales and pelagic limestones.

The Hercynian–Appalachian Orogeny culminated in the Mid- to Late Westphalian with uplift of the Atlas and Anti-Atlas and the development of the Palaeozoic rift basins in southern and southeastern Algeria. During this period sedimentation was dominated by continental molassic sands and shales in the western Sahara while carbonate platform deposits formed to the east.

Late Palaeozoic tectonics

Morocco

The Hercynian orogenic event, dated at 350–280 Ma, marked the collision of Gondwana and the North American continent (Fig. 11). At least two oceanic domains were closed during this process, ending with the disappearance of the eastern proto-Tethyan Ocean. The suture runs along the eastern part of the Appalachian fold belt in the USA and across Spain to the northeast (Fig. 11). A diagrammatic cross-section through the Hercynian belt (Fig. 12) shows a series of NNW-verging thrusts and sutures which form a thin-skinned thrust zone on the western side of the Appalachians, propagating westward onto the North American Craton. The Atlas lay on the hinterland side of the orogenic belt and ESE-dipping subduction zone. The North African basins are therefore viewed as hinterland basins, with relatively little structural propagation toward the east. This Hercynian orogenic belt is directly analogous to the Himalayan collision zone of Central Asia, and the Carboniferous structures and tectonic fabric of North Africa can be compared to those in Central Asia, north of Tibet. Based on this comparison the Palaeozoic deformation in the Atlas appears closely analogous to the Kun Lun or Tien Shan of western China.

Figures 13a and 13b show the areas affected by Lower Carboniferous and Upper Carboniferous deformation, respectively. There is no evidence either for a Palaeozoic suture or accreted terranes in the Algerian Sahara or most of central-southern Morocco. Instead the sedimentary record suggests that the different Hercynian terranes all formed part of the same epicontinental shelf, which extended from northern Morocco, south to the West African

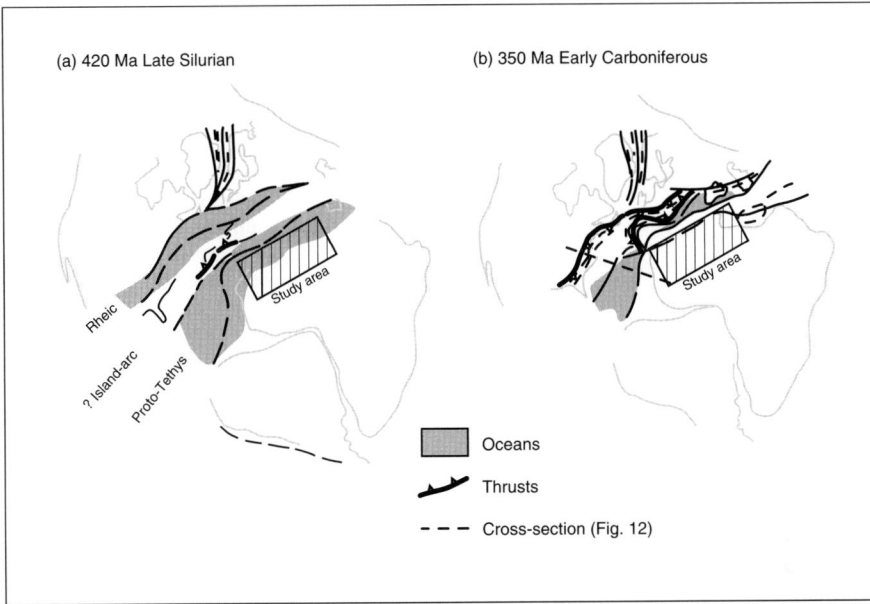

Fig. 11. Maps showing the possible reconstruction of Gondwana and the North American continent for (**a**) Late Silurian and (**b**) Early Carboniferous. (From Windley 1995, modified from Matte 1986)

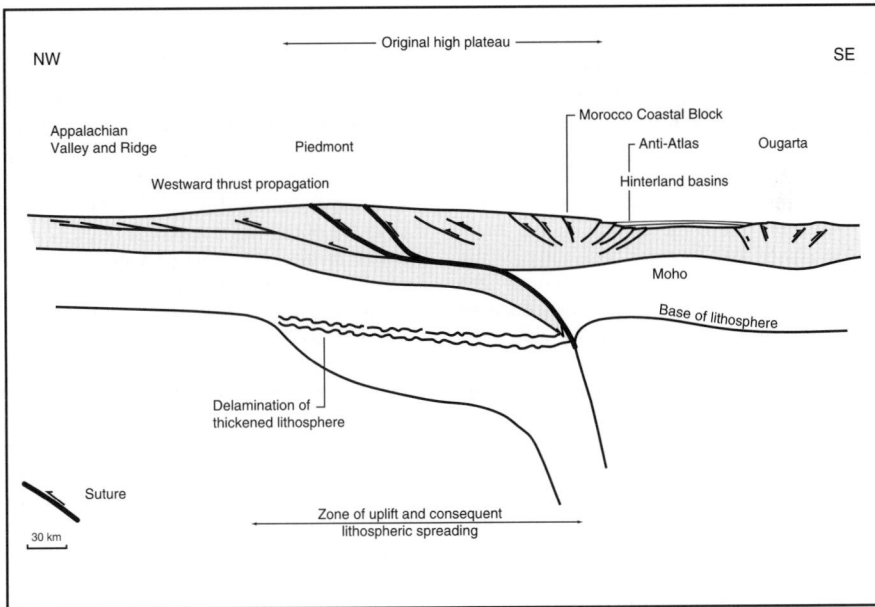

Fig. 12. Schematic section through the Appalachian–African Hercynian mountain belt.

Shield. This shelf was rifted during the Devonian to produce fault-bounded grabens, locally infilled with chaotic material (Pique 1989). Subsequently these basins were inverted and earlier basement fabrics reactivated during the Late Palaeozoic.

The western part of the Sahara comprises three structural domains: (1) the Coastal Block of Morocco and the Atlas, (2) the Anti-Atlas, and (3) the Mauritainide Belt of southern Morocco, separated by continental transform zones.

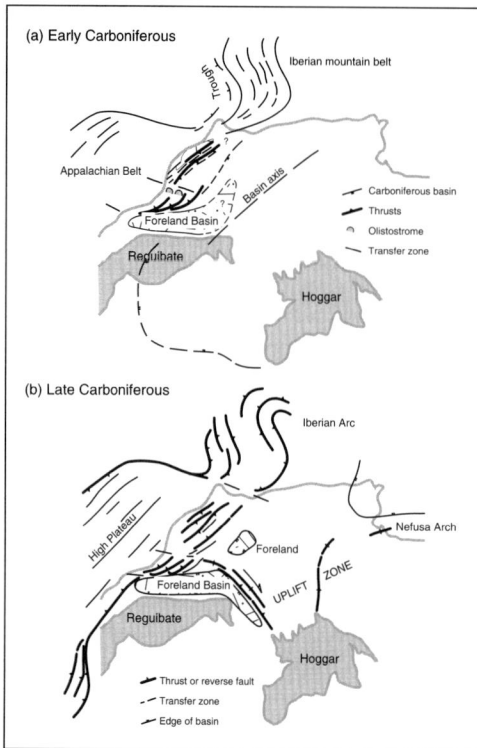

Fig. 13. Maps of North Africa showing the extent of: (**a**) Early Carboniferous and (**b**) Late Carboniferous deformation and basin development.

The Coastal Block of Morocco and the Atlas The Palaeozoic stratigraphy of coastal Morocco and the Atlas is essentially the same as that in the Anti-Atlas to the south, with general continuity of the Hercynian structural fabric. The structures are dominated by NNE-trending, large-scale, open to moderately tight folds, with some thrusts and without associated cleavage or metamorphism. However, the vergence is toward the west and the intensity of deformation increases eastward, with cleavage and metamorphism both more conspicuous toward the eastern limit of the Coastal Block.

The Coastal Block appears to extend beneath the African continental margin toward the west. Mylonitic granodioritic gneisses, which give mica and feldspar ages of *c.*455 Ma, overprinted by a 360–315 Ma event (Pique 1989), have been recorded from the continental margin. These dates differ from those known onshore and suggest that the offshore block has a different tectonometamorphic history from that onshore. A granulite-facies rock, dated at *c.*1000 Ma, was dredged from the western limit of the offshore area.

Hercynian deformation in the High Atlas began during the Visean and ended after the Early West-

phalian. The thrusts on the northwest side of the Atlas are steeply dipping structures, which probably reworked earlier basin-bounding normal faults, together with NW-vergent shortcut thrusts. During the Early Carboniferous, the Coastal Block formed a foreland basin to the inverted Palaeozoic Atlas Basin. However, by the Mid- and Late Carboniferous, the deformation had spread to the Coastal Block, uplifting and folding the foreland basin sediments, and incorporating them within the Hercynian mountain belt. The general trend of the folds is NNE–SSW, with mainly WNW vergence, except in the more metamorphosed zones where the axial surfaces are steep to vertical. The earlier normal faults/basin boundaries were reworked as later semi-ductile shear zones, associated with the inversion of the Palaeozoic Atlas Basin. These inversion-related shear zones and faults separate the High Atlas from the Anti-Atlas and the Coastal Block from the Middle Atlas.

In the western part of the High Atlas, the main cleavage-forming phase of deformation has been dated as post-Westphalian B/pre-Westphalian C (315–310 Ma) with a second phase of pre-Stephanian age (295–290 Ma) (Choubert & Faure-Muret 1971). This deformation generally obscures any older structures. However geochronological work near the junction of the High and Middle Atlas suggests that an important thermal event occurred earlier in the Late Devonian with a subsequent tectonic event in the Early Carboniferous.

Syntectonic granites in the western part of the High Atlas are dated as 316–312 Ma, while post-tectonic granites give ages of 290–280 Ma (Choubert & Faure-Muret 1971). Bossiere & Peucat (1985) obtained an age of 271 (±12) Ma from granites in the Grand Kabylie Massif. Ages from detrital zircons, obtained from the Petite Kabylie in Northern Algeria, suggest a late-stage Hercynian event at 270–280 Ma.

There may have been thrust development on the southeastern side of the High Atlas during the Late Carboniferous, but this zone is now largely obscured by Mesozoic sediments. The eastern part of Morocco and western Algeria would have formed a hinterland basin to the Hercynian orogenic belt, with major thrust structures facing towards the northwest, away from the basin.

2. The Anti-Atlas Two structural trends interfere in the Anti-Atlas and the western part of the High Atlas:

- the north–south to NE–SW-trending structures which control the margin of the Mauritanides;
- the WNW–ESE-trending structures which mark the reworked Pan-African suture zone and associated basement fabrics in the Anti-

Atlas. This zones swings to NW–SE in the Ougarta.

Structurally the Anti-Atlas is a broad anticlinorium with a core of Precambrian and Lower Palaeozoic rocks. Along its southern flank the Visean and Namurian sequences are essentially undeformed and hence the major tectonic activity must have been Early Carboniferous or older, related to inversion of earlier Palaeozoic rift basins, largely predating the tectonic activity in the High Atlas. Lower Carboniferous olistostromic sediments, derived from uplifted areas to the north, occur along the northern flank of the Anti-Atlas. However, south-facing folds and thrusts of Westphalian age occur along the southern edge of the Anti-Atlas, indicating renewed thrust activity at the end of the Carboniferous. The Tindouf Basin (Fig. 1) formed as a foreland basin to this tectonic event. The degree of deformation increases from east to west. Along the western edge of the Tindouf Basin, in the western Anti-Atlas and in Zenmour, the folds are tight and the Palaeozoic rocks are cut by E-verging thrusts, related to the folded foreland of the Mauritanides.

3. Mauritanides

The western part of the Reguibat Shield is overthrust by Mauritanide allochthonous units displaced to the ESE. The overthrust units comprise a lower sheet of quartzites, phyllites and deformed conglomerates of probable Cambro-Ordovician age. These are overlain by the Mauritanide klippe with chloritoid schists, recrystallized brown carbonates and an allochthonous unit of granulite-facies rocks farther to the west.

Southern Algeria

A zone of Carboniferous folding can be traced from the eastern part of the Anti-Atlas, across the Ougarta and into the Sbaa and Ahnet basins. The Ougarta is characterized by high-amplitude broad open folds. Anticlinal structures in the Sbaa Basin generally verge to the SW, with long gentle limbs dipping to the northeast and steeper limbs dipping to the southwest. The steeper limbs form the hangingwalls of reverse faults, some of which originated as Palaeozoic extensional structures. The Ahnet Basin is dominated by thick-skinned structures that rework Pan-African basement fabrics. These inversion anticlines are discrete structures bordered by relatively flat broad synclines. They have gentle back limbs and short steeper forelimbs. Master faults underlie the steeper forelimbs, but smaller reverse faults occur on both limbs. In many areas, the dip of the master fault appears to vary

along strike, reflecting variations in original normal fault geometry. The folds and original normal faults are linked by WNW-trending left lateral strike-slip faults. The earliest growth of the inversion structures appears to be Late Visean but the strongest deformation occurred in the Westphalian.

Field and seismic studies suggest that the sense of shear along the NW–SE-trending faults of the Ougarta and western part of the Ahnet Basin was left lateral, related to further indentation of the West African Craton into the Pan-African crust of North Africa. In the Ahnet Basin and the northern Hoggar, the faults trend NW–SE and N–S. Three-dimensional structural modelling suggests that there was clockwise rotation of the earlier fault blocks associated with the left lateral shear. There was also some right lateral shear along the inverted N–S-trending faults accompanying this rotation.

Eastern Algeria and Libya

Broad open folds characterize the Hercynian deformation of northern Algeria and Libya forming prominent NE–SW-trending arches. The trend of these structures suggests reworking of Pan-African crustal fabrics or heterogeneities. The broad anticlines are cut by smaller but important inversion structures, reworking Palaeozoic normal faults with oblique compression. The Nafusa and Gargaf arches in eastern Algeria and western Libya were uplifted during the Late Palaeozoic, as part of the regional Hercynian basin inversion of North Africa, to separate the eastern Ghadames and Murzuk basins. Smaller-scale basin inversion occurred along steep basement-involved faults along the western side of the Murzuk Basin.

Mesozoic basin development

The Permian and Early Triassic was a period of erosion. The Late Triassic consists of fluvial sandy shales and lacustrine deposits conformably overlain by Jurassic marine and lacustrine deposits and lagoonal evaporites (Popescu 1995). The depositional history of the Triassic was largely controlled by the interaction of global sea-level changes and local fault reactivations. Figure 14 shows the depth to base Mesozoic in eastern Algeria and Libya.

Morocco and Northern Algeria: High Atlas Basin

The Missour Basin and the High and Middle Atlas (Fig. 1) developed as rift basins during the Mesozoic. Extension and rifting began at the end of the Permian and Early Triassic. Sedimentation rates

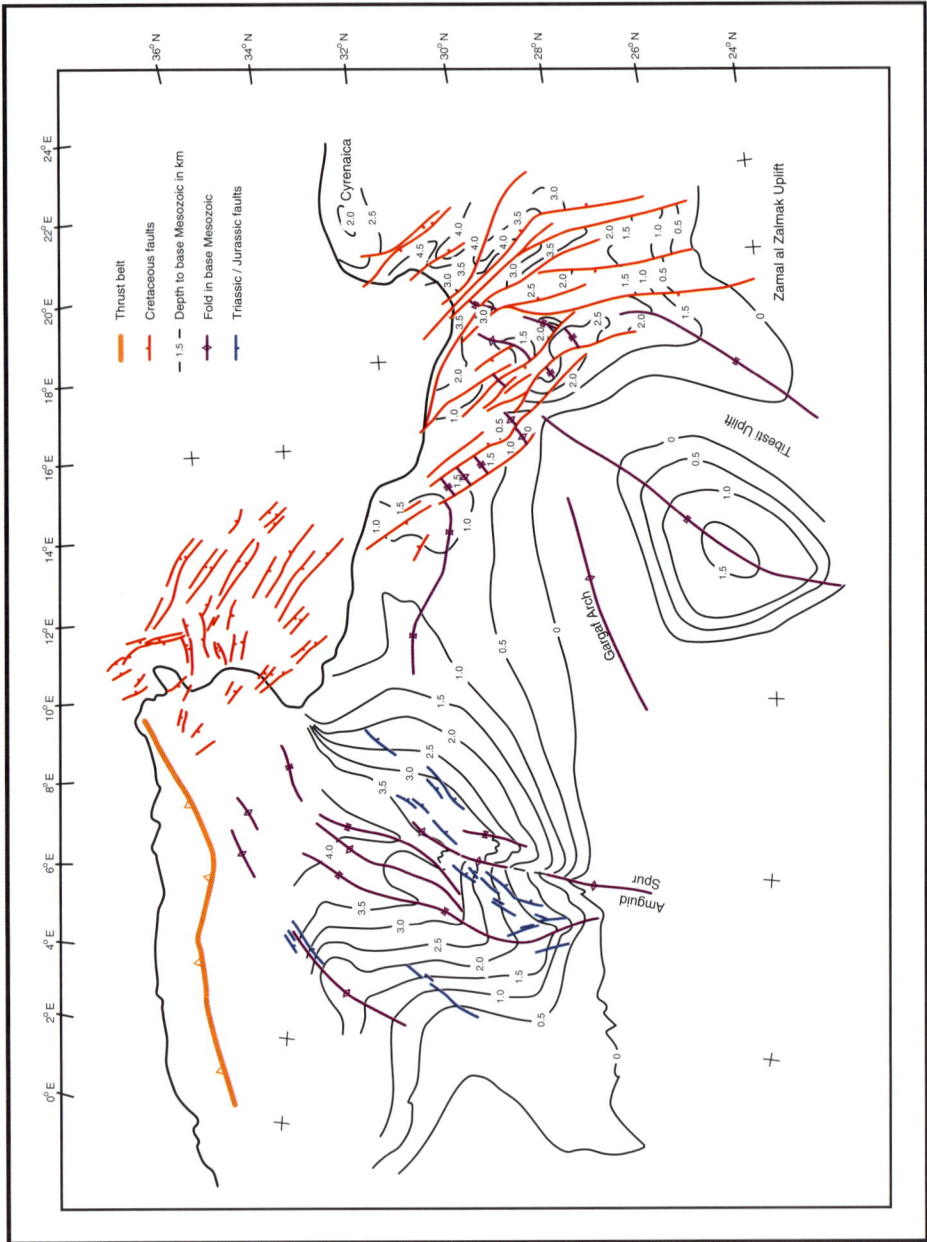

Fig. 14. Map showing the depth to base Mesozoic in kilometres, Triassic faults in Algeria and Cretaceous faults in Libya.

increased during the Jurassic as rifting continued with the opening of the western Tethys Ocean and the North Atlantic. Faults in the High Atlas may have reactivated earlier Hercynian structures. The High Atlas continued eastward into the Saharan Atlas of Algeria and Tunisia while the Middle Atlas trends northeastward beneath the Alpine Rif. The Anti-Atlas remained as a structural high throughout the Mesozoic and provided the source for the sediments in the adjacent troughs.

The basal Triassic sequence comprises non-marine conglomerates, sandstones and sandy mudstones overlain by gypsiferous mudstones with basalt flows which form a thin veneer on the eroded Hercynian basement. Subsequently, thick sequences (1–1.5 km) of reddish brown paralic-facies conglomerates, sandstones and overlying mudstones infilled ENE- and NE- trending troughs and basins. Toward the western and northern continental margin this clastic sequence grades laterally into thick salt deposits (Van Houten 1977) in the Rharb, Souss, Khemisset and Boufekane basins and in the Essaouira Basin and its offshore continuation. The upper part of the red bed/evaporite sequence is overlain by, or intercalated with, basaltic lava flows 30–500 m thick. The lavas have been dated at 180–200 Ma, suggesting a Lower Liassic age (Van Houten 1977).

The distribution of Liassic rifts is shown in Figure 15. The Jurassic sequence thickens dramatically into the Middle and High Atlas basins (Stets & Wurster 1982) and is dominated by mar-ine carbonates, deposited as thick turbidite flows, alternating with basinal marls. Thick olistostromic breccia deposits occur on some fault scarps. Basin-margin carbonate facies, including reef build-ups, platform sequences and intertidal sediments are present along the southeastern margin of the rift basins. Thick Liassic reefal carbonates were deposited on the footwall highs. Evaporitic basins developed on the West African Craton, pinching out farther south. Illite crystallinity in Jurassic sediments indicates that 6–8 km of syn-rift and post-rift sediments were originally present in the deepest part of the rift basin (Brechbuhler et al. 1988). As the Triassic rocks in the High Atlas are 4–4.5 km thick (Beauchamp 1988), the composite thickness of syn-rift sediments may be 10–12 km.

Normal faulting continued through the Mid-Jurassic but to a lesser degree. By the Mid- to Late Jurassic the post-rift thermal subsidence basin extended out onto the flanks of the High Atlas, where further deposition occurred during the Early Cretaceous. Lower Cretaceous sediments rest unconformably on Jurassic strata along the flanks of the High Atlas Graben where erosion, associated with rift-flank uplift and possibly weak basin inversion, has removed a significant thickness of Jurassic rocks. This base Cretaceous unconformity is often easily recognizable on seismic dip lines. Subsidence slowed or ended during the Late Cretaceous, when Cenomanian–Turonian sediments were deposited uniformly across the rift basin and margins.

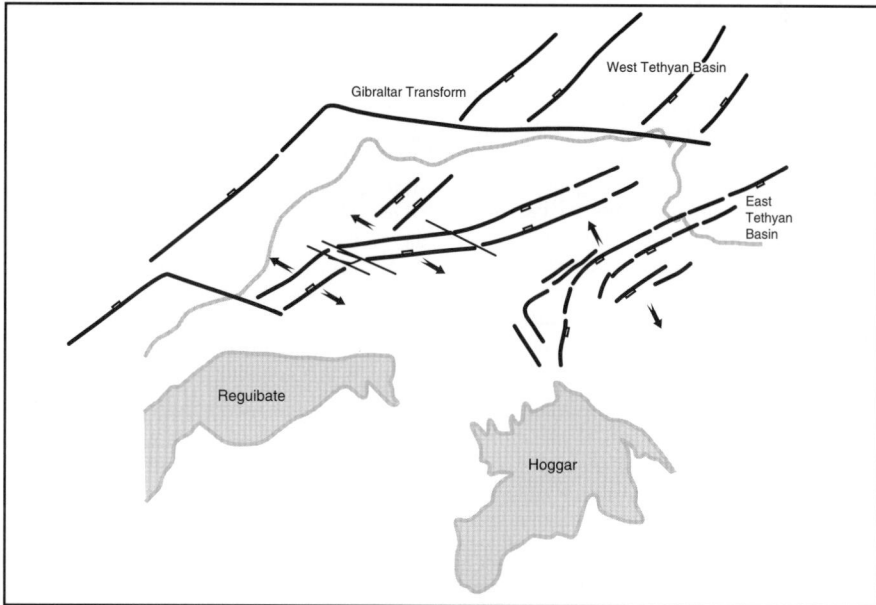

Fig. 15. Map of North Africa showing the distribution of Liassic rifts.

Saharan Platform: Oued Mya & Ghadames basins

A thick succession of Triassic–Lower Cretaceous sediments, deposited in a large sag basin, forms the 'Triassic Basin' of eastern Algeria, southern Tunisia and western Libya. These strata rest unconformably on the eroded Palaeozoic basins and show striking continuity and consistency of thickness over many thousands of square kilometres.

Syn-rift Permian clastics and reefal carbonates were deposited in northernmost Algeria and central Tunisia in the eastern Atlas rift (Rigo 1995), while clastics were being shed southward onto the Sahara Platform in response to rift-shoulder uplift. Permian sediments pass upward into Triassic fluvial sandstones and shales, which transgressed southward across the peneplaned Hercynian unconformity. The Mid- to Upper Triassic continental clastic unit (Trias Argilo-Greseux Inferieur – TAGI) was deposited in northern and central Algeria. These beds, which vary in thickness from 0 m to 1000 m, are represented by continental braided-fluvial clastics in the south, passing northward into lacustrine and estuarine sediments. They form an important reservoir unit in the Ghadames Basin. A similar fluvial sand and shale sequence, roughly equivalent in age, developed to the north and northwest of the Hassi Massaoud Arch and now provides reservoir rocks for the Oued Mya and Hassi R'Mel fields. The lowermost sand member is confined to erosional irregularities on the unconformity surface, formed by minor rifting. The succeeding sand members are more widespread, passing up into alluvial muds and evaporites of the Upper Triassic saline member.

The basal Triassic clastic sequence is overlain by a thick cyclic Liassic succession of cyclic anhydrites, salts and interbedded muds with sandstone and shale equivalents toward the south, which onlaps the Hercynian unconformity to the south and north. This was followed by an Upper Liassic marine transgression with marine clastics and carbonates. A carbonate facies is present in northeastern Libya while Jurassic continental sandstones occur in southern Libya. Regression followed in the Early Cretaceous with the development of an extensive delta system. This is dominated by the Nubian sandstone facies in southern Libya and southern Algeria which grades northwards into a nearshore marine facies, with some marine limestones in northern Algeria, Tunisia and northern Libya.

The established pattern of Triassic extensional faults suggests reworking of basement heterogeneities. The basin appears to die out to the southwest into a broad transfer zone with some NW–SE-trending faults. The northeastern edge of the basin is obscured by later rift basins (Fig. 16).

Basin inversion occurred during the Aptian with the growth of inversion anticlines along the normal faults bounding the eastern and southeastern sides of the Pharusian Belt in northeastern Algeria. The N–S-trending Amguid–Hassi Touareg Axis was inverted at this time, possibly with some strike-slip movement. The Palaeozoic fault blocks of the Amguid Spur, overlying the Palaeozoic 4° 50′ Lineament, formed structural highs throughout the Mesozoic and experienced several pulses of inversion. However, the most important episode occurred during the Aptian–Albian, with the production of inversion-related anticlines and associated reverse faults.

Similar but less dramatic movements occurred along the Tihemboka Arch between the Illizi and Murzuk basins. The Upper Palaeozoic Nafusa and Gargaf arches were further uplifted during the Mid- to Late Cretaceous and the Lower Cretaceous rocks of the Sirte Basin were folded into NE–SW-trending anticlines and synclines, synchronous with the extension that produced the Sirte Basin. The inversion may be related to changes in relative plate motion between Africa and Eurasia at this time. There was also mild inversion in the Atlas basins, Cyrenaica and western Egypt (Boote *et al.* 1998).

The Triassic Basin deepened to the north during the Late Cretaceous, contemporaneously with rifting in the Sirte Basin, and regional extension, along NW–SE-trending faults, continued from eastern Tunisia to central Libya. This basin was associated with the opening of the eastern Mediterranean and the separation of Italy from Tunisia. In northern Algeria, the Late Cretaceous is represented by marine carbonates and clastics with some evaporites, including a thin salt layer.

Elsewhere along the eastern part of the Afro-Arabian plate margin several basins, belonging to an early Eastern Tethys system, were active during the Permo-Triassic, with localized renewed rifting in the Jurassic and/or Early Cretaceous. These basins include the Palmyra Basin of Syria (Lovelock 1984), the Erez Graben along the coast of Israel and the half-grabens of the Western Desert of Egypt, which continue into the southeastern part of the Sirte Basin (Thusu & Mansouri 1995).

Late Cretaceous rifting and basin development: the Sirte Basin

The horst and graben province of the Libyan Sirte Basin developed in response to crustal extension during the Mid- and Late Cretaceous (Figs 14 & 16). The larger structural elements exhibit a pronounced NNW–SSE to NW–SE trend, changing to

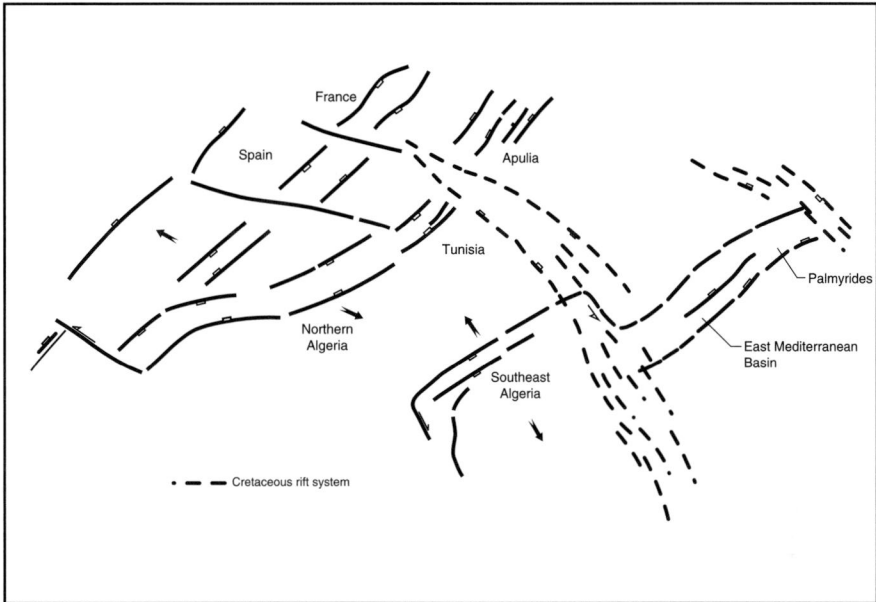

Fig. 16. Sketch-map showing the pattern of Mesozoic rift basins in North Africa, the Mediterranean and the western part of the Middle East.

ENE and WNW toward the southeast. The faults form the tips of a rift basin which opened between Sicily and Tunisia in the central Mediterranean. This rift system can be part of the East Tethyan rift system which propagated into the eastern Mediterranean and the Middle East (Fig. 16).

The faults, which show both right lateral and normal fault displacements, offset the Cretaceous fold axial traces by several kilometres. The Upper Cretaceous rift-fill sequence is dominated by basal clastics passing up into basinal shales and muds with transgressive shallow-water carbonates on footwall highs. By the end of the Cretaceous most of the topography had been buried and the entire rift province was blanketed by a north-facing carbonate platform during a period of rapid Lower Tertiary regional subsidence. Cretaceous subsidence in the Sirte Basin occurred at the same time as mild inversion in Cyrenaica and western Egypt. Some of these inversion structures may have formed by block rotations associated with the change in extension direction from NW–SE to NE–SW, similar to inversion kinematics in the northern North Sea, rather than simply a change in plate motion.

Upper Cretaceous and Tertiary basin inversion

Atlas Mountains

The Mesozoic basins of the High and Middle Atlas were inverted during the Late Cretaceous–

Paleogene and again during the Miocene–Pliocene (Stets & Wurster 1982; Vially et al. 1994). The western and the central High Atlas reach >4 km in elevation (Jebel Toubkal, Jebel Mgoun) and their width ranges from 50 km to 100 km. The displacement direction, obtained from trends of transfer structures, was NW–SE, almost perpendicular to the trend of the Middle Atlas, but at a lower angle in the High and Sahara Atlas, suggesting some oblique-slip or right lateral strike-slip movement. As is usually the case, the geometries of structures generated by inversion are controlled by the fault orientations, fault polarities and fault distributions present within the original extensional basin (Beauchamp et al. 1996). The amount of shortening in the High Atlas and the Middle Atlas is estimated to have been up to 20% and 10%, respectively.

The Atlas structures are far from the continental boundary and the main zone of Alpine collision tectonics. A step in the Moho between the High and Middle Atlas and the presence of gently dipping high-conductivity layers is suggested by the presence of upper mantle and lower crustal shear zones, linking the deformation in the High Atlas to the collisional margin, north of the Rif (e.g. Giese & Jacobshagen 1992; Beauchamp et al. 1996). According to this model, the shear zone would have had the kinematics of a low-angle basement thrust sheet.

The High Atlas is flanked by shallow flexural foreland basins. The Ouarzazate and Tadla basins to the west contain only up to 1 km of Tertiary sediments. However, the adjacent Atlas Mountains are >4 km high and, hence, the shallow depth of the basins suggests either that loading of the lithosphere, due to thickening of the Atlas Mountains, is minimal or that the underlying lithosphere is strong.

The intermontane Missour Basin, between the High and Middle Atlas, has only a thin cover of Tertiary sediments. This basin underwent Mesozoic extension, similar to that in the adjacent Atlas rift basins, but later tectonic inversion was minimal. The Guercif Basin developed along strike from the Middle Atlas and evolved as an extensional Mesozoic basin with relatively mild inversion during the Paleogene. In the Late Miocene it experienced further extension with the development of narrow grabens and deep marine sedimentation. This extension and associated subsidence was probably due to the onset of the Rif compression farther north. The basin itself was affected by the Rif compression during the latest Miocene. Miocene–Pliocene Alpine tectonics are largely confined to the northern edge of the African continent and comprise a thin-skinned thrust zone which can be traced from northern Tunisia into the Rif of northern Morocco. However, several of the shear zones bounding the Atlas rifts were reactivated as strike-slip faults at this time, possibly due to right lateral strike-slip movements.

Farther southwest in central Libya, the Hoggar was uplifted during the Tertiary, presumably by hot-spot development and igneous underplating.

Alpine tectonics across Algeria and Libya

Many of the Mesozoic grabens were inverted between the Late Cretaceous and the Miocene as a result of the Alpine collision in southern Europe (Wildi 1983). Figure 17 lists some of the main unconformities associated with this inversion.

The arcuate Rif mountain belt in Morocco developed during Alpine tectonics. In the Rharb foreland basin, sedimentation began with a thick sequence of Eocene and Oligocene turbidites passing up into Lower–Mid-Miocene molasse. Frequent sea-level fluctuations characterize the Mid-Miocene–Pliocene. A marine incursion occurred during the Mid-Miocene, with shales to shaley sand sedimentation, followed by marls with turbiditic lenses, deposited during a subsequent sea-level fall.

Inversion occurred in eastern Libya and western Egypt during the latest Cretaceous, continuing until the Late Oligocene (Keeley & Massoud 1998). There was renewed inversion during the Oligocene associated with opening of the Red Sea.

Conclusions

During the Pan-African Orogeny, oceanic and small continental terranes were accreted from NW to SE onto a Pan-African nucleus in East Africa. The age of the accretion becomes younger to the northwest. This amalgam of accreted terranes eventually collided with the more rigid Archean–Proterozoic West African Craton. The first event produced large-scale steep shear zones which vary in trend from NW–SE in northeastern Africa to north–south in Algeria. The second event produced a NW–SE-trending fold and thrust belt (Ougarta).

Subsequently the West African Craton formed a rigid indentor protruding into the amalgam of accreted mobile belts making up much of North Africa. This indentor into North Africa caused expulsion of narrow, triangular-shaped blocks of lithosphere to the north and south, in a tectonic style very similar to that of Tibetan tectonics in the Miocene–Pliocene. Expulsion reactivated earlier-formed shear zones to form an anastomozing pattern of steeply dipping shear with left and right lateral sense of displacement. Left lateral shear occurred along the northern edge of the West African Craton at this time. A rifting event occurred during the Late Precambrian and associated grabens were infilled with Upper Precambrian–Cambrian molasse.

During the Cambrian and the Silurian–Devonian many of these Upper Pan-African rift systems were reactivated and developed along an ENE–WSW to NE–SW trend in southeastern Morocco. These Moroccan rifts represent failed systems on the northwestern margin of Gondwana along the flank of the proto-Tethyan ocean. The rift basins of southern Algerian and western Libya represent failed arms to the Moroccan system (Fig. 7), with a triple junction in western Algeria.

There was a pulse of minor inversion during the Ordovician, with further extensional rifting in the Silurian. Presumably the rifting was initially associated with collapse of the Pan-African mountain belt in the Late Precambrian–Cambrian, causing perturbations in the mantle lithosphere. There was a renewed attempt to break up Gondwana during the Silurian along pre-existing structural trends. However, this proved to be of insufficient strength to form a major rift depression or oceanic basin.

Rifting was accompanied, or immediately followed, by subsidence leading to a rapid post-glacial sea-level rise. Silurian graptolitic shales directly overlie Ordovician sandstones and the erosion surface appears generally smooth except for the fluvioglacial incisions infilled by locally coarse clastic rocks. Transgressive sands did not develop in this basin. The lowest deposits are therefore part of a transgressive systems tract, with black shales

		ATLAS	GHADAMES	SIRTE	NW EGYPT	REMARKS
		Inversion		Subsidence	Folding of Syrian Arc	Opening of the Red Sea
	Top Oligocene					
	Top Eocene			Subsidence		
50		Inversion			Folding of Syrian Arc	Early Alpine tectonics
Ma	Top Cretaceous		Subsidence	NW Rifts		
	Santonian	Inversion			Folding of Syrian Arc	First stage of collision between Apulia and Eastern Europe
100	Albian			NW Rifts		
	Aptian					
			Left-Lat. strike-slip on N/S faults		Strong subsidence	
		Subsidence	Subsidence			
	Top Jurassic					
150						

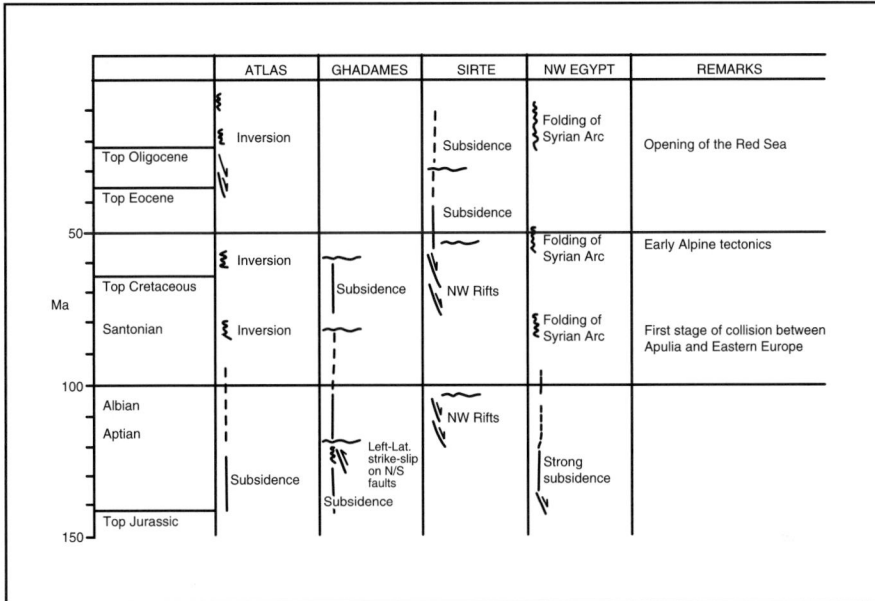

Fig. 17. Chart showing the timing of uplift and subsidence across North Africa.

rich in pyrite. The deposits become progressively sandier upward and the sandstones pinch out basinward (Fekerine & Abdallah 1998). These pass up into a regionally developed highstand sequence, prograding toward the north during the Mid- and Late Silurian.

In northern Morocco the ENE–WSW-trending rift system began to close during the Late Silurian to form a thrust zone. Elsewhere the rift system was fully inverted at the end of the Devonian at the onset of the Hercynian deformation. Hercynian deformation was associated with collision tectonics between North America, Europe and northwestern Africa, and the Hercynian structures along the western edge of Africa formed along the hinterland of this collisional mountain belt. The Hercynian is characterized by: (1) strongly inverted structures with thin-skinned deformation related to soft-linked tectonics or gravity gliding above the deeper inverted basins, as exemplified by the Hercynian deformation in Morocco; (2) moderately inverted Palaeozoic basins, associated with variable degrees of strike-slip movement and block rotation, with the development of hangingwall folds above the original half-graben, as observed in the Ahnet and Sbaa basins; and (3) weakly inverted Palaeozoic basins with very large-wavelength folds cut by short-wavelength hangingwall anticlines above inverted normal faults, as seen in northern and eastern Algeria.

Basin inversion is very characteristic of the Palaeozoic structures of the western Atlas and

Anti-Atlas, forming large mountainous terranes with thickened crust. Foreland basins developed along the flanks of these inverted zones and both the mountains and the basins were compartmentalized by WNW–ESE-trending transfer zones.

Within the interior of the African Plate, the Pan-African fabrics were reworked during the Hercynian with further indentation of the West African Craton into Pan-African crustal collage. The craton was pushed eastward, generating a left lateral shear couple along its northern margin. NW–SE-trending faults were reworked as dominantly left lateral strike-slip faults and north–south-trending fault blocks were rotated slightly with a clockwise sense. There was probably further lateral expulsion of lithosphere from the NE–SW-trending front of the indentor, reworking earlier north–south-trending shear zones.

Away from the indentor boundaries, the deformation intensity was far less. In northern Algeria and Libya the Hercynian structures are characterized by very long-wavelength, low-amplitude folds associated with NE–SW-trending arches. The trend of these structures suggests reworking of Pan-African crustal fabrics or weaker lithosphere. The broader anticlines are cut by smaller but important inversion structures, reworking Palaeozoic normal faults with oblique compression.

Two important basinal systems developed on the North African platform during the Triassic–Jurassic (Fig. 17).

(1) The High Atlas graben system, extending from

western Morocco to Tunisia, which is offset by several WNW-trending transfer fault zones and links with spreading in the Atlantic to the west (Fig. 16). In these western basins there were minor extension events during the Jurassic, with a pulse of inversion, possibly resulting from the onset of Atlantic Ocean spreading and associated rift push. (2) The northeastern Algerian Basin, which formed by renewed rifting of the Pharusian basement and its bordering faults and terminates in a tip zone or possibly a minor transfer zone to the southwest. This basin links with East Mediterranean basins and the Palmyrides (Fig. 16).

In Libya the second of these systems is crossed by the NW–SE-trending Sirte Basin. The latter forms the onshore section of a rift system which can be traced from north of Tunisia and was responsible for the separation of the Apulian Platform from Africa. The two basin systems interfere in the northern and eastern parts of the Sirte Basin, probably generating local inversion of the earlier faults and complex block rotation. The larger-scale uplift of the Cyrenaica Massif may be partly due to rift-flank uplift or to Hellenide tectonics farther north.

Uplifts in the Atlas resulted from: (1) rift-flank uplift during the Triassic–Liassic, (2) local inversion during the Mid-Jurassic, (3) renewed rift-flank uplift during the Cretaceous, (4) major inversion during the Paleogene, and (5) Neogene–Quaternary deformation, associated with strike-slip movements along some of the bounding faults of the Atlas graben.

Collisional events between Africa and Europe are recorded by unconformities and mild tectonic inversion in the various basins of North Africa (Fig. 18).

The authors are grateful to LASMO plc for funding some of this work and for discussions over the regional geology. We thank D. Boote for an excellent job in refereeing the paper and correcting some of the geological inaccuracies. The figures were drafted by ap graphics.

References

ABDULLAH, H. 1999. Submarine fan and turbidite deposits dominated reservoirs in the Upper Illizi Basin. *Association of American Petroleum Geologists, Extended Abstracts, Birmingham*, 1.

BEAUCHAMP, J. 1988. Triassic sedimentation and rifting in the High Atlas (Morocco). *In*: Manspeizer, W. (ed.) *Triassic–Jurassic Rifting*. Elsevier, Amsterdam, 477–497.

BEAUCHAMP, W., BARAZANGI, M., DEMNATI, A. & EL ALJI, M. 1996. Intracontinental rifting and inversion: Missour Basin and Atlas Mountains, Morocco. *American Society of Petroleum Geologists Bulletin*, **80**, 9, 1459–1482.

BLACK, R. & FABRE, J. 1980. A brief outline of the geology of West Africa. *Episodes*, **4**, 17–25.

BOOTE, D. R. D, CLARK-LOWES, D. D. & TRAUT, M. W. 1998. Palaeozoic petroleum systems of North Africa. *In*: MACGREGOR, D. S., MOODY, R. T. J. & CLARK-LOWES, D. D. (eds) *Petroleum Geology of North Africa*. Geological Society, London, Special Publications, **132**, 7–68.

BRECHBUHLER, Y. A., BERNASCONI, R. & SCHAER, J. 1988. Jurassic sediments of the central High Atlas of Morocco: deposition, burial and erosion history. *In*: JACOBSHAGEN, V. (ed.) *The Atlas System of Morocco – Studies on its Geodynamic Evolution*. Lecture Notes in Earth Science, Springer-Verlag, Berlin, **15**, 201–215.

BOSSIERE, G. & PEUCAT, J. -J. 1985. New geochronological information by Rb-Sr and U-Pb investigations from the pre-Alpine basement of Grande Kabylie (Algeria). *Canadian Journal of Earth Sciences*, **22**, 675–685.

BOULLIER, A. M., DAVISON, I., BERTRAND, J. M. L. & COWARD, M. P. 1978. Les granulites du mole des Iforas: une nappe de socle d'age Pan African precoce. *Bulletin de la Societé Geologique de France*, **20(6)**, 877–882.

CABY, R. 1970. La chaine Pharusienne dans le nord-ouest de l'Ahaggar (Sahara Central, Algerie); sa place dans l'orogenese du Precambrien superieur en Afrique). These, University of Montpellier, 335.

CABY, R., BERTRAND, J. M. L. & BLACK, R. 1981. Pan-African ocean closure and continental collision in the Hoggar-Iforas segment, Central Sahara. *In*: KRONER, A. (ed.) *Precambrian Plate Tectonics*. Elsevier, Amsterdam, 407–434.

CHOUBERT, G. & FAURE-MURET, A. 1971. Epoque hercynienne. *In*: *Tectonique de l'Afrique*. UNESCO, *Sciences de la Terre*, **6**, 353–359.

CRENN, Y. 1957. *Mesures gravimetriques et magnetiques en African occidentale*. ORSTOM, Paris.

ECHIKH, K. 1998. Geology and hydrocarbon occurrences in the Ghadames Basin, Algeria, Tunisia, Libya. *In*: MACGREGOR, D. S., MOODY, R. T. J. & CLARK-LOWES, D. D. (eds) *Petroleum Geology of North Africa*. Geological Society, London, Special Publications, **132**, 109–129.

FEKIRINE, B. & ABDALLAH, H. 1998. Palaeozoic lithofacies correlatives and sequence stratigraphy of the Saharan Platform, Algeria. *In*: MACGREGOR, D. S., MOODY, R. T. J. & CLARK-LOWES, D. D. (eds) *Petroleum Geology of North Africa*. Geological Society, London, Special Publications, **132**, 265–281.

GIESE, P. & JACOBSHAGEN, V. 1992. Inversion Tectonics of intracontinental Ranges: High and Middle Atlas, Morocco. *Geologische Rundschau*, **81/1**, 249–259.

KEELEY, M. L. & MASSOUD, M. S. 1998. Tectonic controls on the petroleum geology of NE Africa. *In*: MACGREGOR, D. S., MOODY, R. T. J. & CLARK-LOWES, D. D. (eds) *Petroleum Geology of North Africa*. Geological Society, London, Special Publications, **132**, 265–281.

KLITZSCH, E. 1968. Die Gotlandium-Transgression in der Zentral-Sahara. *Zeitschrift der Deutschen Geologischen Gesellschaft*, **117**, 492–501.

LEBLANC, M. 1972. Un complexe ophiolitique dans le Precambrien II de l'Anti-Atlas central (Maroc); Description, interpretation et position stratigraphique.

Notes et Memoires du Service Geologique du Maroc, **236**, 119–144.

LEBLANC, M. & LANCELOT, J. 1980. Interpretation geodynamique du domaine panafricain (Precambrien terminal) de l'Anti-Atlas (Maroc) a partir de donnees geologiques et geochronologiques. *Canadian Journal of Earth Sciences,* **17**, 142–155.

LOVELOCK, P. 1984. A review of the tectonics of the northern Middle East region. *Geological Magazine*, **125**, 577–587.

LUNING, S., CRAIG, J., LOYDELL, D. K., STORCH, P. & FITCHES, W. 2000. Lower Silurian "hot shales" in North Africa and Arabia: regional distribution and depositional model. *Earth Science Review,* **49**, 121–200.

MATTE, PH. 1986. Tectonics and plate tectonics for the Variscan belt of Europe. *Tectonophysics*, **126**, 329–374.

PIQUE, A. 1989. Hercynian terranes in Morocco. Geological Society of America Special Papers, **230**, 115–129.

POPESCU, B. M. 1995. Algeria/Algerien. *In*: KULKE, H. (ed.) *Regional Petroleum Geology of the World. Part II*, 11–34.

RIGO, F. 1995. Overlooked Tunisia reef play may have giant field potential. *Oil and Gas Journal,* **93**, 56–60.

SHACKLETON, R. M. 1994. Review of Late Proterozoic sutures, ophiolitic melanges and tectonics of eastern Egypt and north-east Sudan. *Geologische Rundschau,* **83**, 537–546.

STETS, J. & WURSTER, P. 1982. Atlas and Atlantic – Structural Relations. *In*: VON RAD, U., HINZ, K., SARNTHEIN, M. & SEIBOLD, E. (eds) *Geology of the Northwest African Continental Margin*. Springer-Verlag, Berlin, 69–85.

THUSU, B. & MANSOURI, A. 1995. Reassignment of the Upper Amal Formation to Triassic and its implications for exploration in southeast Sirte, Libya. *First Symposium on Hydrocarbon Geology of North Africa, London, Abstracts*, 48.

VAN HOUTEN, F. B. 1977. Triassic-Liassic deposits of Morocco and Eastern North America: Comparison. *American Association of Petroleum Geologists Bulletin,* **61**, 79–99.

VIALLY, R., LETOUZY, J., BENARD, F., HADDAD, N., DESFORGES, G., ASKRI, H. & BOUDJEMA, A. 1994. Basin inversion along the North African Margin. The Sahara Atlas (Algeria). *In*: ROURE, F. (ed.) *Peri-Tethyan Platforms*. Editions Technip, Paris, 79–118.

WILDI, W. 1983. La chaîne tello-rifaine (Algérie, Maroc, Tunisie: structure stratigraphique et évolution du Trias au Miocene. *Revue de Géologie Dynamique et de Géographie Physique, Numéro spécial, Chaîne Tello-Rifaine*, **24**, 201–298.

WINDLEY, B. F. 1995. *The Evolving Continents*. John Wiley & Sons, London.

Examples of salt tectonics from West Africa: a comparative approach

G. TARI, J. MOLNAR & P. ASHTON

Vanco Energy Company, One Greenway Plaza, Sixth Floor, Houston, Texas, 77046, USA

Abstract: Exploration experience gained in specific salt basins of West Africa may not be directly applicable to other salt basins along the entire passive margin. To conduct a comparative structural analysis, regional reflection seismic transects were constructed across the salt basins of Morocco, Senegal, Guinea-Bissau, Equatorial Guinea, Gabon and Angola.

Regional-scale similarities of the salt basins include the progressive complication of salt-related structures basinward, the change from an extensional domain on the shelf to a compressional domain on the slope and the presence of a toe-thrust front at the oceanward edge of the basins. Regional-scale differences are partly attributed to the relative stratigraphic position of the salt in relation to the rift history.

In the better-known post-rift salt basins of Equatorial Guinea, Gabon, Congo and Angola updip extension is represented by a broad rafted domain balanced by downdip contraction in the form of salt tongues, canopies and a toe-thrust zone. The efficiency of this gravity sliding/spreading across the whole margin is due to the more or less uniform original distribution of Aptian salt in the post-rift succession forming a continuous detachment level.

In contrast, the typically uneven original distribution of the Late Triassic and Early Jurassic syn-rift salt in Morocco, Mauritania, Senegal, The Gambia and Guinea-Bissau is due to basement highs separating rift half-grabens and creating a different structural pattern. Individual salt structures, such as pillows and diapirs, originated from isolated patches of the autochthonous salt. In the case of syn-rift salt, updip extension may not always be the ultimate driving force for the contractional salt-deformation downdip.

There are several salt basins along the West African passive margin (Fig. 1). The somewhat isolated salt basins of northwestern Africa, in Morocco, Mauritania, Senegal, The Gambia and Guinea-Bissau, are characterized by salt of Late Triassic–Early Jurassic age. The salt was deposited in the rift phase of the Central/North Atlantic Basin. In contrast, the salt basin of the southwestern margin of Africa, including southernmost Cameroon, Equatorial Guinea, Gabon, Congo and Angola, is Aptian in age and was deposited during the post-rift thermal-subsidence phase after the opening of the South Atlantic Basin.

Primarily as a result of ongoing exploration success in the southwestern, post-rift salt basin of Africa, a large number of papers have been published during the last decade discussing certain aspects of salt tectonics (e.g. Duval *et al.* 1992; Lundin 1992; Liro & Coen 1995; Spathopoulos 1996; Dailly 2000; Jackson *et al.* 2000; Marton *et al.* 2000). In contrast, significantly less has been published on the northwestern, syn-rift salt basins of Africa (e.g. Heyman 1988; Hafid 2000; Hafid *et al.* 2000).

The aim of this paper is two-fold: (1) to provide a regional structural synopsis of all the salt basins of West Africa, and (2) to give some insights into the similarities/differences between these basins. Current efforts in the underexplored syn-rift salt basins of Morocco, Mauritania, Senegal, The Gambia and Guinea-Bissau may result in some significant discoveries in the near future. Therefore the application of exploration experiences and approach developed in African post-rift salt-basin settings to the syn-rift salt basins is important.

Regional structure transects across West African salt basins

The best way directly to compare the basin-scale geometry of different salt basins of West Africa is to compile regional seismic transects. As the available data quality is extremely variable, even along the same composite seismic transect line drawing, interpretations were prepared with the same vertical exaggeration applied to maintain a uniform presentation style. Furthermore, to emphasize the role of salt tectonics in a comparative manner, as opposed to the large differences in pre- and post-salt stratigraphy along the West African passive

From: ARTHUR, T. J., MACGREGOR, D. S. & CAMERON, N. R. (eds) *Petroleum Geology of Africa: New Themes and Developing Technologies.* Geological Society, London, Special Publications, **207**, 85–104. 0305-8719/03/$15
© The Geological Society of London 2003.

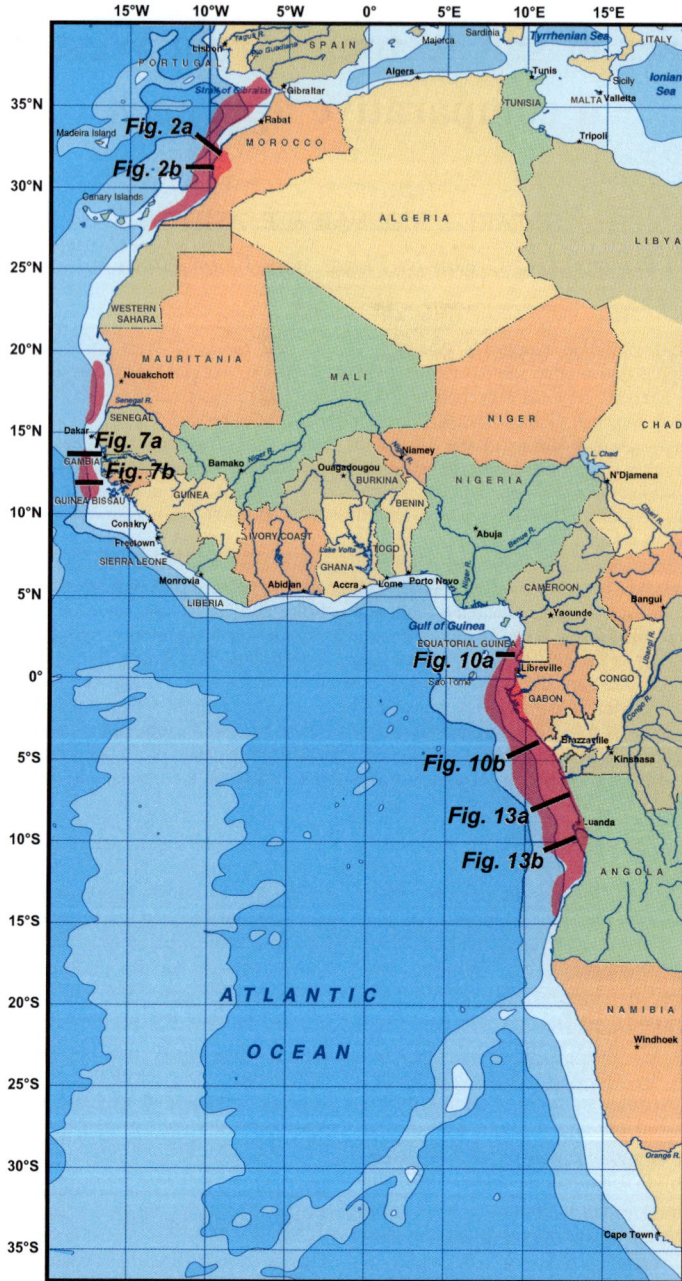

Fig. 1. Index map of West African salt basins. Approximate trace of regional transects shown in Figures 2, 7, 10 and 13 are highlighted in black; salt is shown in magenta.

margin, only the salt is highlighted in these transects. For a regional treatment of the stratigraphy of the West African continental margin the reader is referred to the works of von Rad *et al.* (1982) and Coward *et al.* (1999).

Morocco

Figure 2 shows a line-drawing version of two regional composite seismic profiles across the Safi Haute Mer and Ras Tafelney permits in offshore

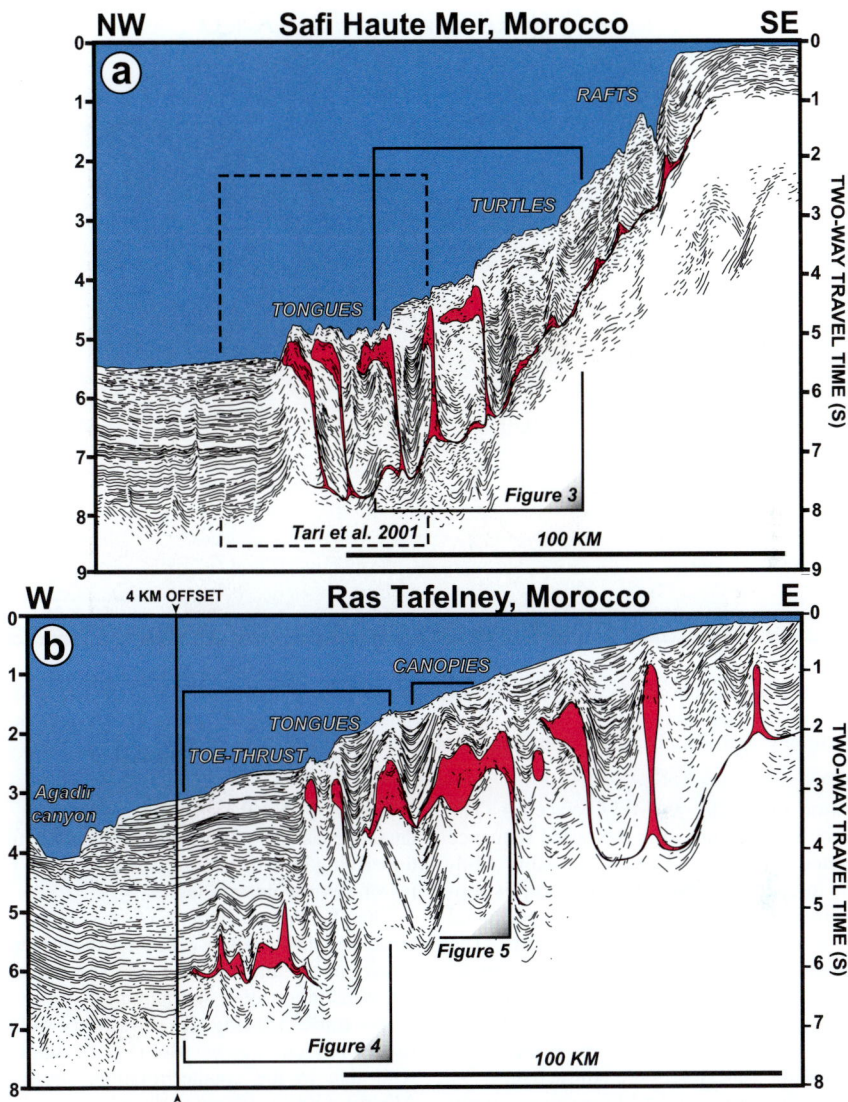

Fig. 2. Line-drawing interpretations of composite, regional reflection seismic transects across the Moroccan passive margin. Salt is highlighted in magenta. (**a**) Regional transect A through the Safi sub-basin. (**b**) Regional transect B through the Tafelney Plateau (modified from Tari *et al.* 2001). For location see Figures 1 and 16.

Morocco. In this northwest African basin the Late Triassic–Early Jurassic salt was deposited during the last stage of rifting prior to the continental break-up (Hafid, 2000).

The northern transect across the Safi sub-basin (Fig. 2a) shows the characteristic halokinetic elements of a passive margin, such as extensional salt structures beneath the shelf and the upper slope, and compressional features downdip on the lower slope. A nearby seismic profile (Fig. 3) illustrates a critical part of the regional transect where

updip extension transitions to downdip compression. The upper slope shows several raft-like features sliding downdip on the salt, accommodating significant extensional strain. Based on the updip correlation of the Mesozoic stratigraphy encountered in the Deep Sea Drilling Project (DSDP) well 416 (Lancelot & Winterer 1980), the inferred stratigraphic sequence within the rafts includes Jurassic shallow- and then deeper-water carbonates and Early Cretaceous shales. The overlying beds display progressive growth which dates

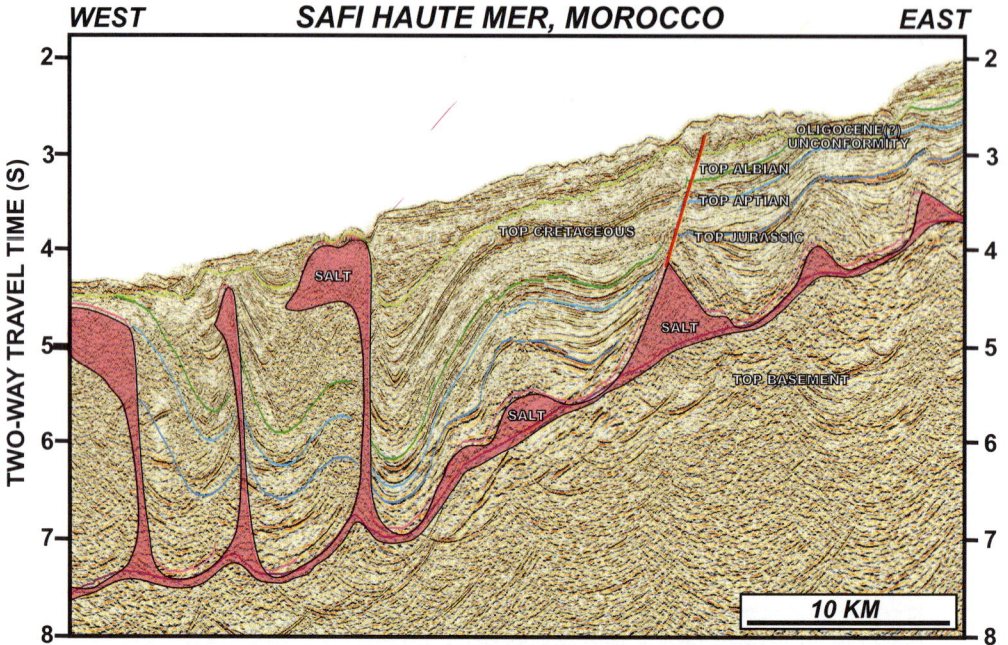

Fig. 3. Reflection seismic illustration of a portion of transect A across the Safi sub-basin of offshore Morocco (for location see Fig. 2*a*). Note that the actual location of this seismic example is about 4 km to the NE from the regional transect. Vertical exaggeration is about five-fold at 4 km/s velocity.

the inception of rafting as Mid-Cretaceous. The major unconformity truncating the rafted sequence is interpreted to be intra-Tertiary, possibly Mid-Oligocene in age. The rafted domain updip is separated from the allochthonous salt tongues and sheets by a narrow zone of turtle structures. The most important observation is that all the salt structural domains appear to be linked by the salt detachment. Based on the expression on the present-day sea floor, the salt tongues at the basinward edge of the salt system are still active.

The southern transect in the Moroccan salt basin over the Tafelney Plateau shows a different picture (Fig. 2b). Whereas the width of the salt basin is about the same, in this area the post-salt basin fill is much thicker. Nonetheless, the structures beneath the shelf can be interpreted as rafts sliding basinward on a salt décollement surface. Basinward, a large salt canopy can be observed ramping up from the level of the autochthonous salt to the Tertiary section. Further out, several allochthonous salt sheets and tongues are developed. There are several large toe-thrust anticlines at the oceanward edge of the salt basin.

The actual seismic reflection data from the western edge of the regional transect is shown in Figure 4. Note that, on this seismic example, the pronounced truncation within the Early Tertiary

succession, over the crest of the anticline, directly dates the last growth period of these structures. The underlying pre-rift basement is dominated by a series of syn-rift normal faults.

The Tafelney segment of the Moroccan salt basin is fairly mature based on the general complexity of the allochthonous salt structures. A seismic illustration of a salt canopy from the central part of transect B (Fig. 2b) shows very well-imaged subsalt reflection geometries (Fig. 5) dipping to the west due to the counter-rotation, *sensu* Hart & Albertin (2001), related to the adjacent salt canopy. Also note the clear expression of the ramp-flat geometry at the base of the allochthonous salt.

Senegal, The Gambia and Guinea-Bissau

The syn-rift salt basins of Mauritania and Senegal/The Gambia/Guinea-Bissau are located in the central Atlantic segment of the West African passive margin (Fig. 1). Generally, these salt basins are much smaller (Fig. 6) and less developed than their Moroccan counterpart. The regional transect from the northernmost part of the salt basin (Fig. 7a) shows just one salt feature. This salt pillow is located just oceanward from a basement hinge zone (Fig. 8). Note that, on the opposite side of the Atlantic Ocean, in the southern Carolina

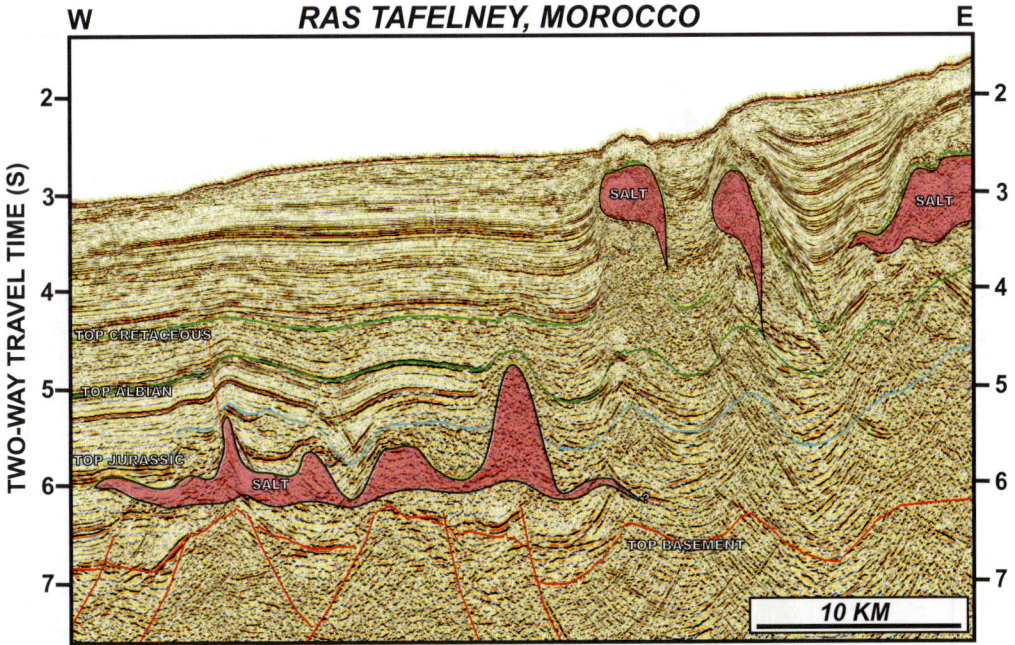

Fig. 4. Reflection seismic illustration of the western end of transect B across the Ras Tafelney Plateau of offshore Morocco (for location see Fig. 2b). Vertical exaggeration is about five-fold at 4 km/s velocity. (Modified from Tari *et al.* 2001).

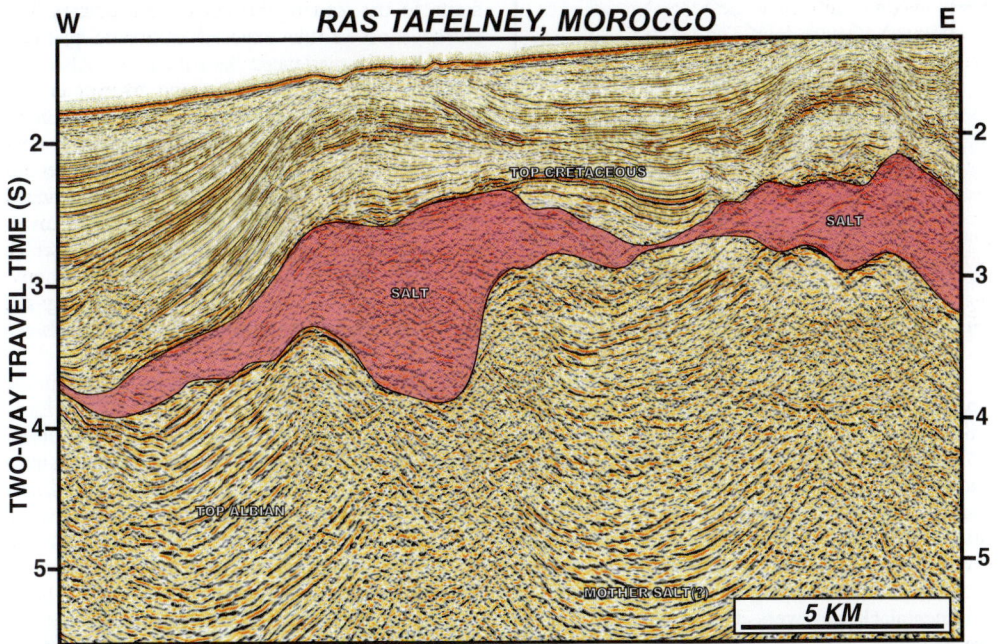

Fig. 5. Reflection seismic illustration of the western end of transect B across the Ras Tafelney Plateau of offshore Morocco (for location see Fig. 2b). Vertical exaggeration is about five-fold at 4 km/s velocity.

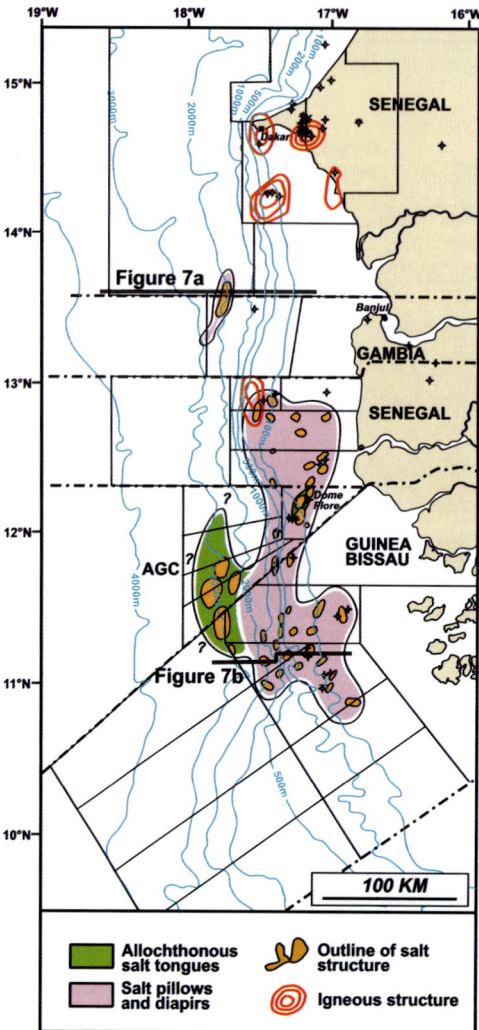

Fig. 6. Simplified salt tectonic map of the southern part of the central Atlantic segment of the West African passive margin between Senegal and Guinea-Bissau. Note the location of transects shown in Figure 7. Also compare Figure 16.

Trough, very similar salt geometries exist (e.g. Dillon *et al.* 1983).

To better illustrate the structural/stratigraphic evolution of this salt structure, the middle part of transect C (Fig. 7a) was sequentially restored (Fig. 9). During the Late Triassic and Early Jurassic a large rift shoulder developed in the footwall of a major west-dipping normal fault. Whereas continental clastics were deposited to the east of the rift shoulder in small extensional grabens, salt deposition occurred in the main rift graben in shallow-water conditions. The lateral extent of the

salt basin was constrained by the fault-related escarpment to the east and stratigraphic pinchout to the west. This rift system heralded the opening of the Central Atlantic.

By the Mid-Jurassic, bathyal conditions prevailed in the west due to thermal subsidence. To the east, where subsidence was not so dramatic, a carbonate platform was established on top of the continental clastics. On top of the basement hinge zone, at the western edge of the platform, a reef system was formed which was able to catch up with the ongoing subsidence. The carbonate platform shed some debris on the steep proximal slope of the deep basin, whereas calciturbidites were deposited in the distal position.

During the Mid-Cretaceous the reef system above the hinge zone was drowned and the edge of the carbonate platform was back-stepping to the east. By this time the circulation pattern in the Central Atlantic caused strong upwelling on this eastern margin of the ocean. Evidenced by the nearby DSDP 367 well (Lancelot & Seibold 1977), the upwelling resulted in the deposition of world-class source rocks in the deep-water region.

The Late Cretaceous Santonian saw the rapid uplift and subaerial exposure of much of the region east of the hinge zone. This regional uplift is well documented around the African continent and is attributed to plate reorganization. The subsequent erosion on the Senegalese passive margin might have removed as much as 1.5 km of the Mesozoic strata in the area of the hinge zone. During this period most of the exposed carbonate platform must have experienced karstification. Also, the regional uplift in the continental interior caused renewed clastic influx which bypassed the platform area through incised valleys. The clastics were deposited in the deep basin onlapping onto the steep submarine escarpment of the hinge zone. This sudden increase in the sedimentary overburden initiated the mobilization of the salt, which started to grow as a salt pillow. The subtle anticlinal topography created by the underlying salt may have caused some ponding of clastics between the structure and the escarpment.

By the Mid-Tertiary, the dramatic bathymetric difference that has been persisting since the Early Jurassic rifting until the Late Cretaceous has disappeared due to the clastic input. On the unstable slope over the palaeoshelf, slumping occurred, creating several debris flow units further basinward. Due to the continuing load of sediments, the salt pillow and the associated anticline kept growing, although at a decreasing rate.

There is no bathymetric expression of the salt-cored anticline on the present-day sea floor and it is considered to be dormant. Deposition on the slope continues to be characterized by slumps and debris

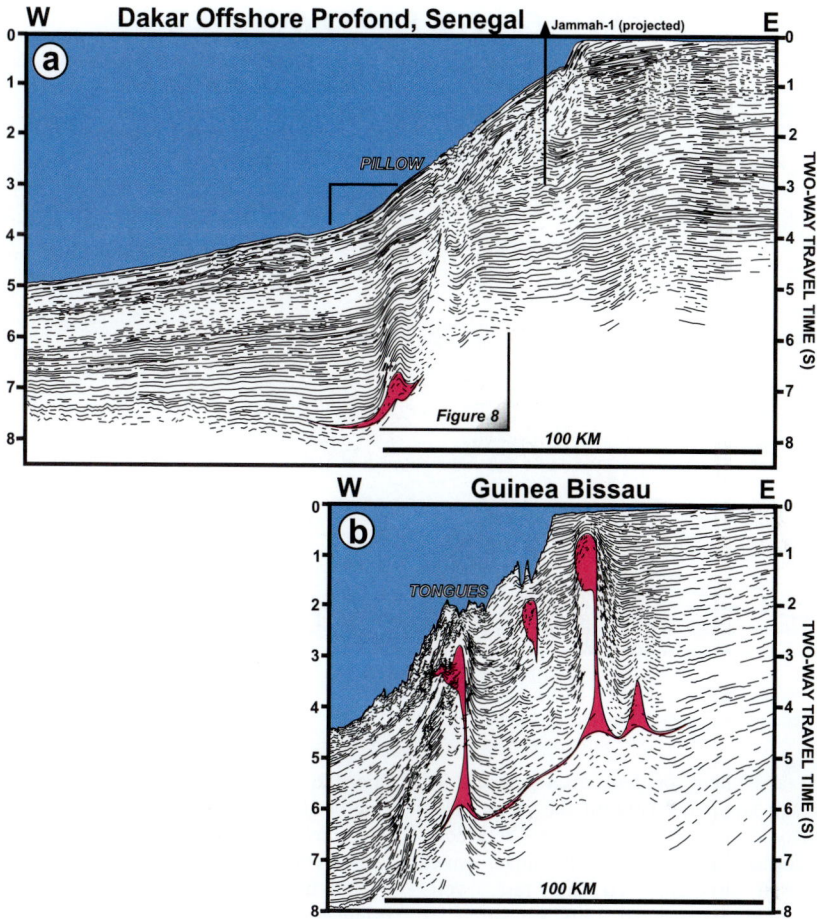

Fig. 7. Line-drawing interpretations of composite, regional reflection seismic transects across the passive margins of Senegal and Guinea-Bissau. Salt is highlighted in magenta. (**a**) Regional transect C through offshore Senegal, just north of The Gambia. (**b**) Regional transect D through offshore Guinea-Bissau (for location see Fig. 6).

flows. Due to coast-parallel currents, sediments also appear to be redepositing on the slope in forms of large sediment waves.

Further to the south, the Senegal/The Gambia/Guinea-Bissau salt basin has more complex salt structures (Fig. 7b), including diapirs and allochthonous salt tongues. The salt structures can be subdivided into a shelf and a deep-water domain, based on their map-view distribution. Note that the oceanward outline of the salt basin is poorly constrained due to the limited seismic reflection data base available for this paper.

Equatorial Guinea and Gabon

The Rio Muni Basin of Equatorial Guinea is located close to the northern termination of the large Aptian salt basin of the South Atlantic

(Fig. 1). The regional transect (Fig. 10a) shows a fairly narrow passive margin, quite similar to the Moroccan examples. The close-up of the toe-thrust anticline, however, shows ongoing deformation creating an escarpment on the present-day sea floor (Fig. 11). Well-developed allochthonous salt sheets and large raft blocks beneath the shelf (Fig. 10a) provide further evidence for ongoing salt deformation using a very efficient salt décollement within the post-rift succession.

The regional transect from the Lower Congo Basin of southern Gabon displays a wider passive margin (Fig. 10b), mostly due to enormous sedimentary influx from the Congo River during the Tertiary. However, the considerable present-day width of the salt basin also reflects the original basin shape. The most striking feature of this part of the Aptian salt basin is the high-frequency

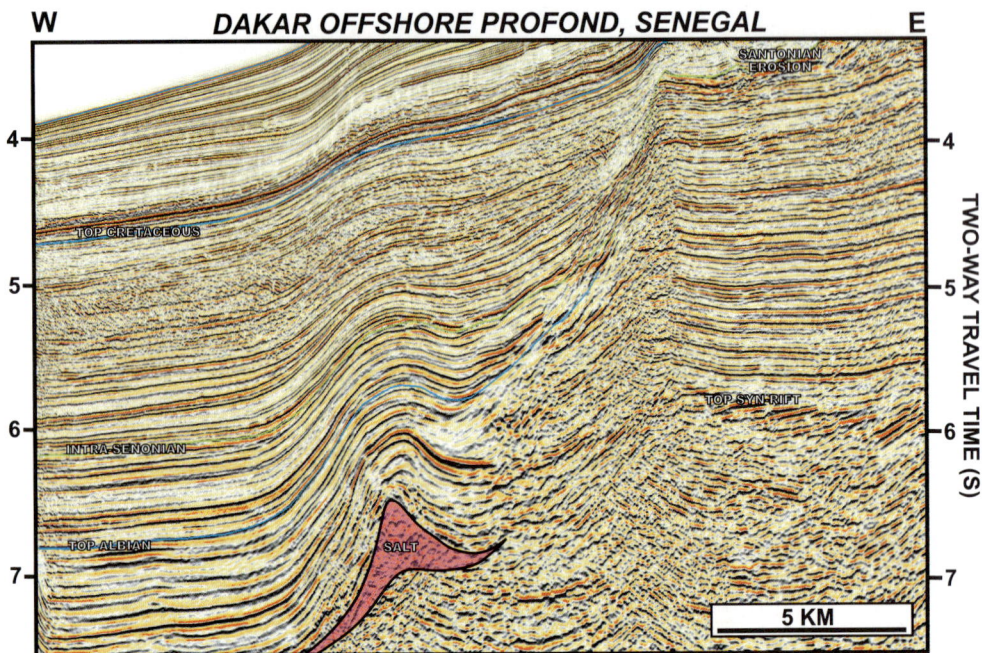

Fig. 8. Reflection seismic illustration of the middle part of transect C across Senegal (Fig. 7a; for location see Fig. 6). Vertical exaggeration is about five-fold at 4 km/s velocity.

occurrence and more or less even spacing of numerous salt structures beneath the slope. This may be partly inherited from an earlier, Cretaceous stage of deformation which set the dominant wavelength of salt structures. At any rate, well-developed updip extensional rafts transition to diapirs, and compressive allochthonous salt tongues and sheets downdip. Turtle structures are notably absent downdip from the raft domain. The edge of the salt basin may have thrust basinward as much as 10–15 km (Fig. 12).

Angola

Further south, in Angola, the original Aptian salt basin was even wider (Fig. 13). In the Angolan part of the Lower Congo Basin (Fig. 1), the salt basin stretches out for more than 250 km from the shoreline (Fig. 13a). The first-order difference between the Gabonese and Angolan sectors of the Lower Congo Basin is the overall amount of salt. The Gabonese transect has about one-third of the salt as the Angolan section. Examination of the topography of the present-day sea floor shows that the Angolan salt features have not yet completely exhausted their Aptian source layer because almost all the salt structures are still moving. In contrast, some of the Gabonese salt features are dormant and now buried deep below the thick Miocene–

Pliocene sediments blanket of the Congo Fan. Another important difference is that the salt structures have a much smaller average wavelength in Gabon compared with those in Angola.

Both transects shown in Figure 13 are quite representative of the Angolan salt basin (Marton *et al.* 2000). Several different names have been suggested to describe the same major salt structural domains (cf. Spathopoulos 1996; Jackson *et al.* 2000; Marton *et al.* 2000). Here we follow the terminology proposed by Marton *et al.* (2000), subdividing the transects into four salt tectonic domains. From east to west these domains are: (1) a raft domain, which developed to the east of the Atlantic hinge; (2) a diapir domain, which developed in the central part of the basin; (3) a canopy domain with allochthonous salt sheets further outboard, which must be regarded as a transition between the diapirs to the east and the inflated salt mass to the west; and (4) a massive outboard salt domain with internal deformation. For actual seismic examples of these domains see Marton *et al.* (2000).

The eastern raft domain is well developed along most of the Angolan passive margin. Generally, rafts appear to be confined to the shelf and upper slope area (e.g. Eichenseer *et al.* 1999). Although raft tectonics (*sensu* Burollet 1975) have been more or less continuous since the deposition of the

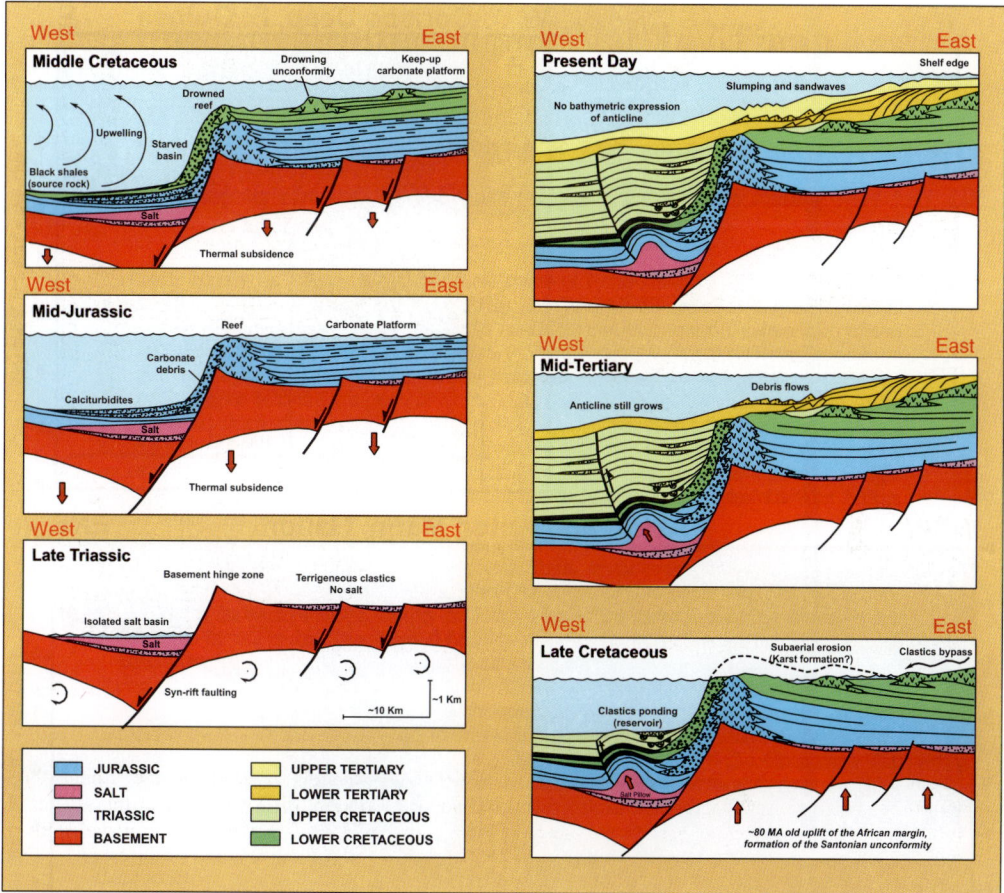

Fig. 9. Sequential restoration sketch of regional reflection seismic transect shown in Figure 7a. For detailed explanation see text.

Albian carbonates, two main phases were recognized: (1) Albian–Cenomanian formation of 'mini-rafts' or pre-rafts, characterized by small block size, close fault spacing of planar fault and high rotation of the intervening blocks; and (2) mainly Late Tertiary 'mega-rafts', characterized by large fault heaves, more listric type faults, larger fault spacing and block size (Duval et al. 1992; Lundin 1992). The formation of these large-scale blocks was triggered by the sudden influx of Tertiary clastics and by the basinward tilt of the entire margin during the Early Miocene.

Basinward, the diapir domain (Baumgartner & van Andel 1971; Leyden et al. 1976) has widely spaced diapiric walls with large intervening turtle structures. Individual diapirs may show signs of an early stage of reactive diapirism (sensu Vendeville & Jackson 1992) but most of them are actively growing and currently compressional in nature. However, changes between extension and

compression in the evolution of individual diapirs may often be observed. Also, the diapirs in the Lower Congo Basin have more expression on the sea floor than their counterparts in the Kwanza Basin (Fig. 13). The canopy domain has a large number of salt tongues and allochthonous sheets coalesced to various degrees. These must be very recent features as they are only covered by very thin probably Pliocene–Pleistocene sediments (Marton et al. 2000). Similarly, the outermost massive salt domain is undergoing recent deformation. This salt domain is being actively inflated by salt from the updip direction (Jackson et al. 2000). A major bathymetric step, the Angola escarpment, marks the basinward extent of the deforming salt where the salt has overthrust 5–15 km west of its original depositional limit (Marton et al. 2000).

The southern Angolan transect (Fig. 13b) shows the same overall salt domains in the Kwanza Basin.

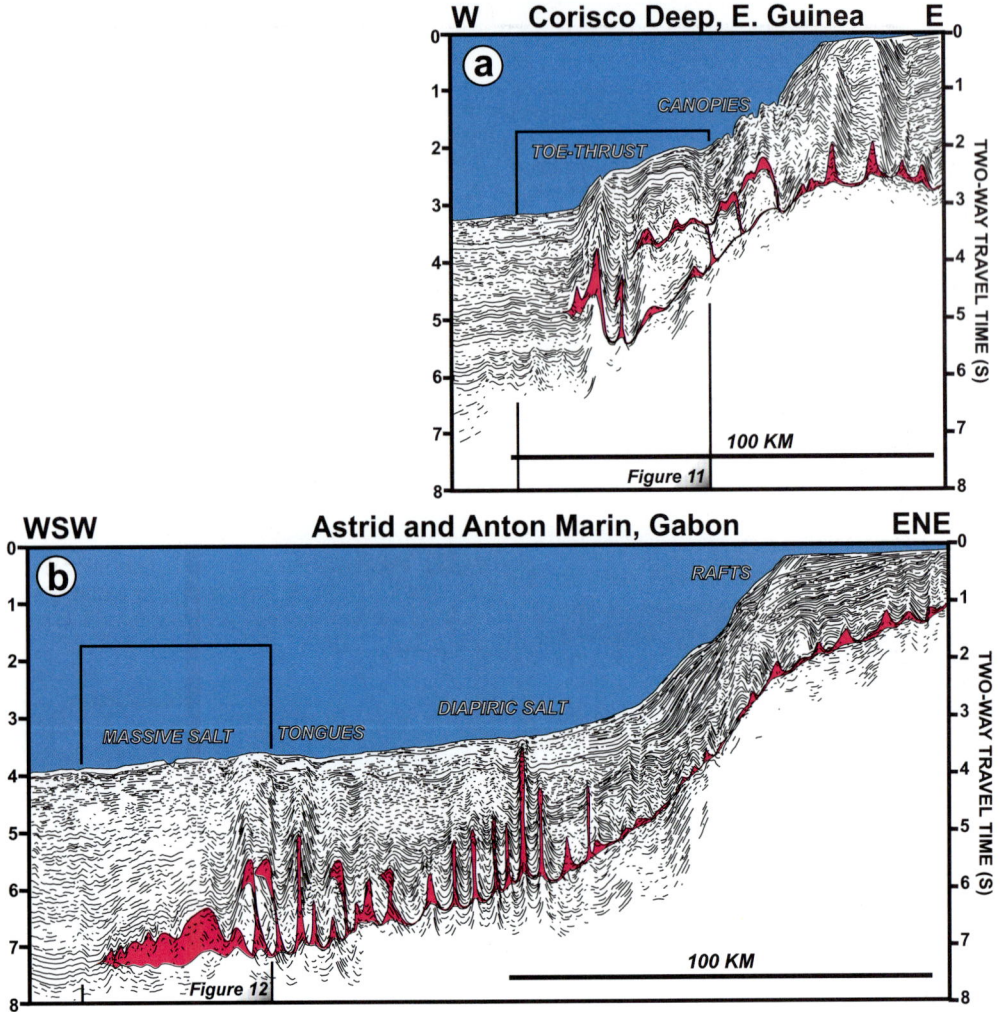

Fig. 10. Line-drawing interpretations of composite, regional reflection seismic transects across the Equatorial Guinean and Gabonese passive margins. Salt is highlighted in magenta. (**a**) Regional transect east through the Rio Muni Basin. (**b**) Regional transect F through the Gabonese part of the Lower Congo Basin.

One of the differences, however, compared to the Lower Congo Basin transect is the presence of a pronounced basement step under the slope: the Atlantic Hinge Zone. Even though the salt is post-rift in age and therefore not expected to be influenced by the underlying syn-rift basement structure, the Atlantic Hinge Zone appears to be a notable exception.

The critical area of the Hinge Zone (see Fig. 13b) was sequentially restored (Spencer *et al.* 1998) in order to demonstrate its role (Fig. 14). As it turns out from the restored geometry of the Tertiary–Mid-Cretaceous sedimentary units, the Hinge Zone has always had an impact on the sedimentation (see also the knick-point on the present-day sea floor). Since the Albian, *c.* 27 km of basinward displacement can be demonstrated using the Hinge Zone as a fixed reference point. Thus this amount gives a minimum estimate of the amount of cumulative extension by rafting, including the onshore Kwanza Basin, updip from the Atlantic Hinge Zone. This result appears to be consistent with the map-view reconstruction of Albian rafts in the Lower Congo Basin by Eichenseer *et al.* (1999) and the sequential restoration of the entire transect shown in Figure 13a by Marton *et al.* (2000).

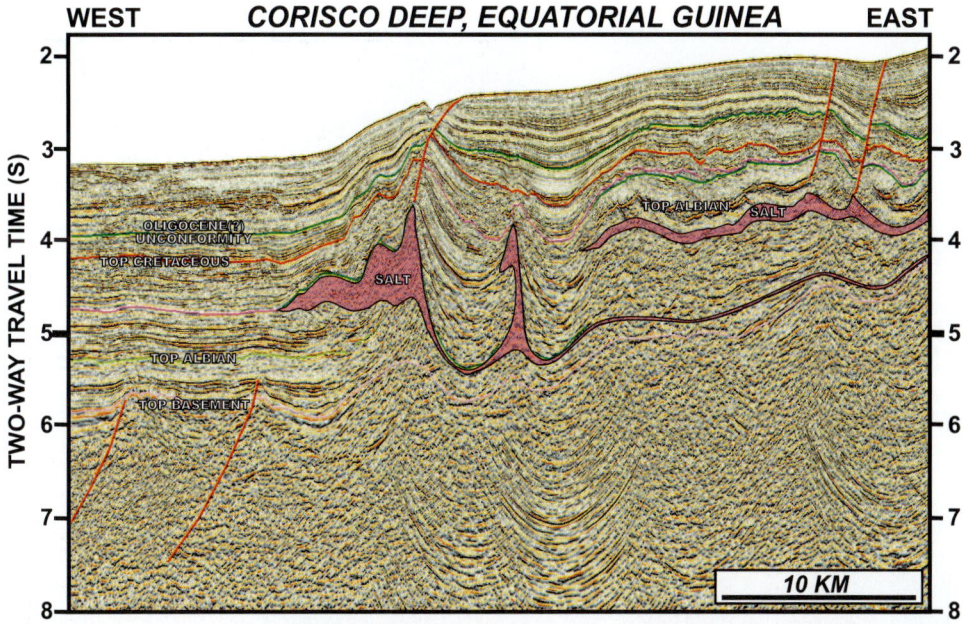

Fig. 11. Reflection seismic illustration of the western end of transect E across the Rio Muni Basin of Equatorial Guinea (for location see Fig. 1). Vertical exaggeration is about five-fold at 4 km/s velocity. (Modified from Tari *et al.* 2001)

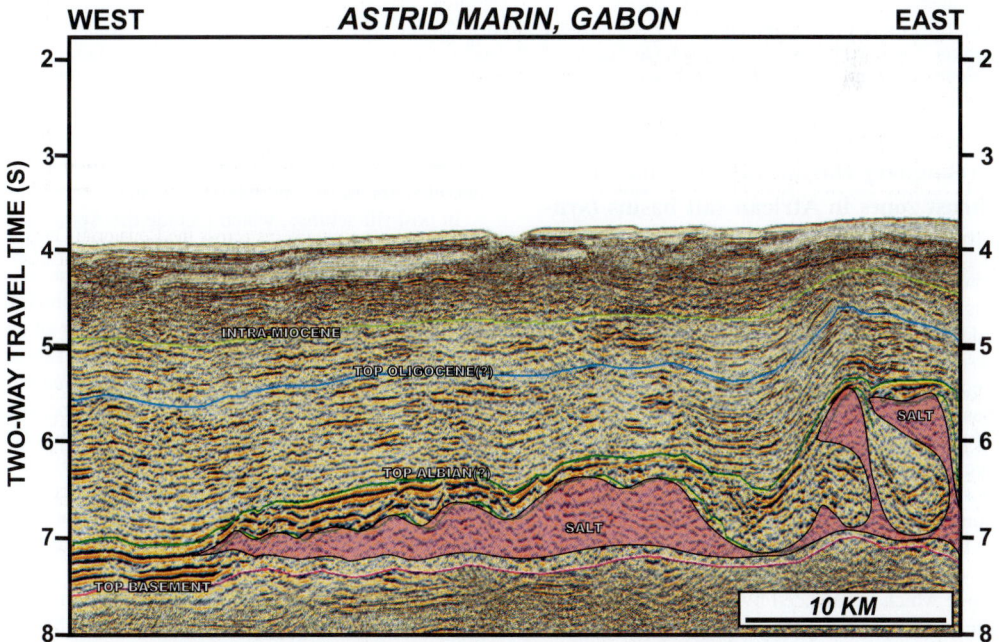

Fig. 12. Reflection seismic illustration of the western end of transect F across the Gabonese sector of the Lower Congo Basin (for location see Fig. 1). Vertical exaggeration is about five-fold at 4 km/s velocity.

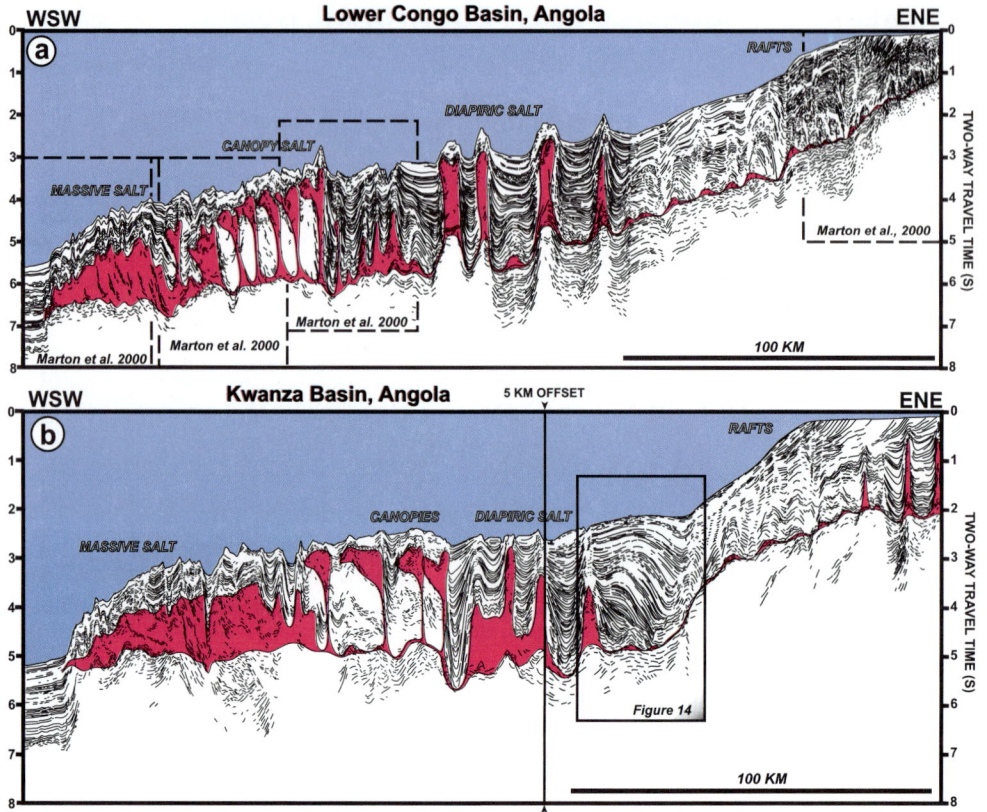

Fig. 13. Line-drawing interpretations of composite, regional reflection seismic transects across the Angolan passive margin. Salt is highlighted in magenta. (**a**) Transect G through the Lower Congo Basin, modified from Marton *et al.* (2000). (**b**) Transect H through the Kwanza Basin (modified from Marton *et al.* 2000). For location see Figures 1 and 16.

Preliminary classification scheme of toe-thrust zones in African salt basins (syn-rift v. post-rift salt)

The speculative classification scheme shown in Figure 15 is a preliminary attempt to qualitatively address some of the major factors which are responsible for the formation of very different salt geometries at the deep-water edge of African salt basins. Most of the deep-water salt-cored fold belts of Africa are plotted in the context of their cross-sectional geometry as it may relate to the tectonic position of the salt and the overall steepness of the continental margin (Tari *et al.* 2001).

In continental margin settings, salt basins always have some sort of extension on the shelf and upper slope which is balanced by downdip contraction in the form of salt tongues, sheets, canopies and, most of the time, a fold belt at the basinward edge of the salt (e.g. Letouzey *et al.* 1995). The importance of the syn-rift v. post-rift setting of the salt is criti-

cal as it directly affects the efficiency of the salt décollement across the whole margin.

In post-rift settings, which include the Aptian salt basin of West Africa, the more or less uniform original distribution of Aptian salt translates to a continuous and therefore very efficient detachment level. It is the primary reason for the presence of an enormous salt structure at the edge of the Lower Congo salt basin in ultradeep-water Angola, regardless of its internal complexity (Fig. 13a, b). The fold belt in Gabon has been inflated very significantly too, although in a different manner (Fig. 10b). The toe-thrust structure in Equatorial Guinea is related to a fault-bounded edge, but it has been vigorously supplied by salt from updip via a continuous and relatively steep detachment surface.

In contrast, one should infer an uneven original distribution for the syn-rift salt in Morocco, Madagascar and Kenya/Somalia (Coffin & Rabinowitz 1988), due to the basement highs separating rift half-grabens. In this setting, individual salt struc-

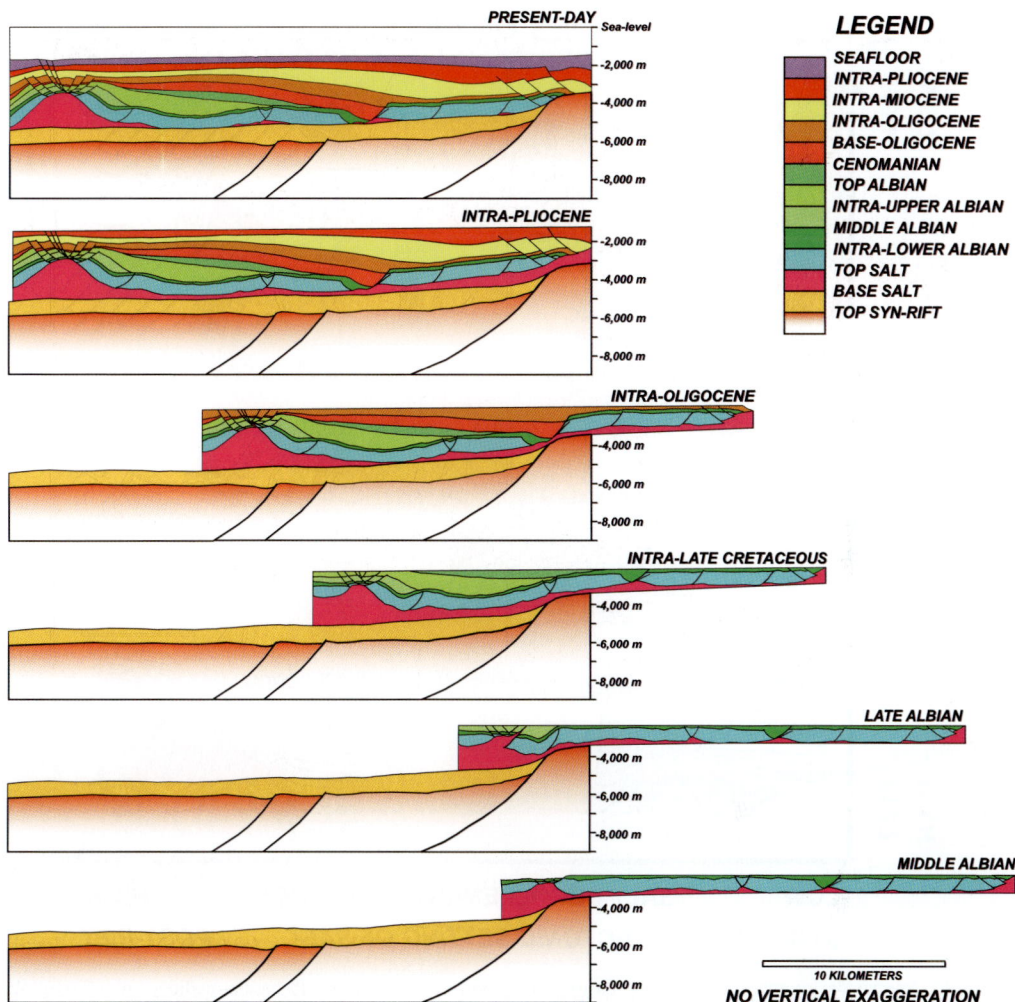

Fig. 14. Computer-aided restoration of a depth-converted part of a regional Angolan seismic transect H (modified from Spencer *et al.* 1998). For location see Figure 13*b*. Note that the restoration has been performed without taking into account the effects of differential compaction, isostasy and regional tilting of the passive margin.

tures, such as tongues, sheets and canopies, might have originated from autochthonous salt in isolated half-grabens. Therefore, updip extension may not be the ultimate driving force for the contractional salt deformation downdip. Nonetheless, a fairly steep margin, like the Safi segment of offshore Morocco, shows structures surprisingly similar to its post-salt counterparts in the Lower Congo Basin, e.g. rafts and turtles on the slope and allochthonous salt tongues at the toe of the slope (Fig. 3). However, the absence of a salt-cored fold belt in the Safi example suggests a sharp, fault-bounded edge to the original syn-rift salt basin (Fig. 2a).

Similarly, in the northern part of Senegal/The Gambia/Guinea-Bissau, the original distribution of

the salt is clearly confined landward by a major hinge zone. The growth of the associated anticline is entirely the function of the sedimentary load and it is not driven by any regional décollement (Fig. 9). Therefore, the salt-cored anticlinal trend in Senegal and The Gambia (Fig. 8) cannot be interpreted as a toe-thrust structure and has been omitted from the compilation shown in Figure 15.

Comparison of salt tectonics in Morocco and Angola

Even though the cross-sectional comparison of West African salt basins already shows some major

TOE-THRUST ZONES OF AFRICAN SALT BASINS

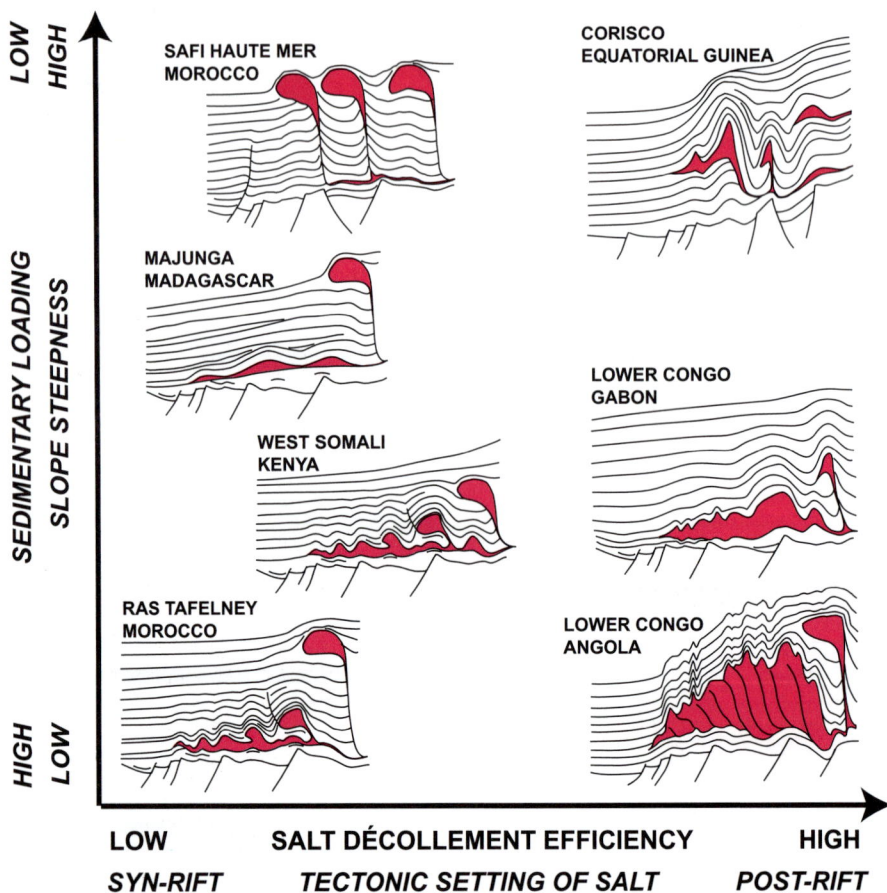

Fig. 15. Speculative comparison of the toe-thrust zone geometries in African salt basins modified from Tari *et al.* (2001). For explanation see text.

contrasts, the differences in map-view are equally important. Our database only allowed direct comparison of two of the West African salt basins, Morocco and Angola, representing syn-rift and post-rift salt settings, respectively (Fig. 16). Other salt basins along the African margin (e.g. Somalia) are much less known due to the lack of adequate and/or accessible reflection seismic and well data coverage.

The Moroccan salt basin on average is between one-third and half as wide (*c.* 50–150 km) as the Angolan salt basin (*c.* 200–300 km). We interpret this difference as a first-order reflection of the origin of these salt basins, i.e. fault-bounded syn-rift setting in Morocco and unconstrained, 'steer's-head' post-rift geometry (White & McKenzie 1988) in Angola. Note that this width difference has been further enhanced by gravity spreading due

to the sedimentary load of the Congo Cone since the Mid-Tertiary.

The Angolan margin has many elongated salt structures which tend to parallel the coastline but still show a minor influence of the underlying basement structure. In contrast, individual salt structures on the Moroccan margin are largely isometric and, with the exception of the Essaouira segment, clearly confined to the east by a basement hinge zone. The outline of the basinward edge of the salt deformational front is affected by a chain of Early–Mid-Cretaceous(?) seamounts in Angola. Furthermore, the edge of the Angolan salt basin always has a bathymetric expression, the Angola escarpment, unlike offshore Morocco where the toe-thrust zone is typically dormant.

Despite these differences, there are some similarities between the two salt basins. Angola is the

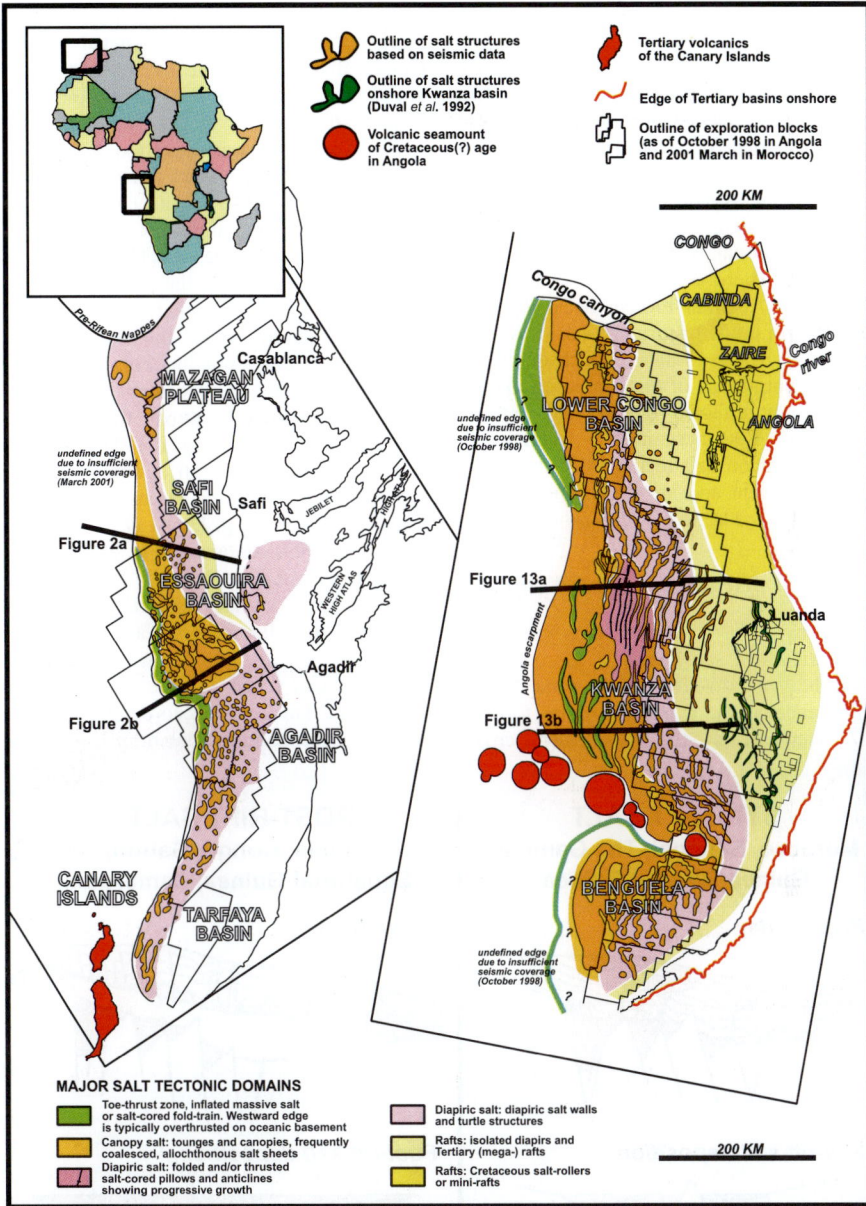

Fig. 16. Comparative salt tectonic maps of Morocco and Angola drawn on the same scale. Note that both maps were rotated away to different degrees from the true geographic North. Simplified salt tectonics on: (**a**) the Atlantic margin of Morocco (based on Hinz *et al.* 1982; Tari *et al.* 2000; and this work); and (**b**) on the Angolan margin (based on Tari *et al.* 1988; Marton *et al.* 2000). Note the location of transects shown in Figures 2 and 13.

locus typicus for rafting (Burollet 1975) and there is a segment offshore of Morocco where analogous structures can be found (Fig. 3). Whereas in Angola the origin of rafts is fairly complex and multi-staged (e.g. Duval *et al.* 1992), in Morocco it appears to be fairly simple (Fig. 3). Just basinward

of the raft zone a few turtle structures are present in the diapir zone, very similar to ones described from the Lower Congo Basin. The salt basin south of Agadir, offshore Morocco, is classified and characterized by simple diapirism driven dominantly by the load of the suprasalt sediments and not by extension via gravity sliding, as in Angola.

There are major differences concerning the style of the toe-thrust zone, as discussed above (Fig. 15). While a toe-thrust zone extends along the entire length of the Angolan salt basin, a corresponding compressional domain is present in Morocco only to the north of the Agadir Basin (Fig. 16). This highlights an additional difference concerning the post-salt deformational history of these African passive margins. There are no major along-strike variations in the Angolan margin because of its relatively uniform uplift history and basinward tilt (cf. Marton et al. 2000; Cramez & Jackson 2000). However, the central part of the Moroccan margin has been strongly affected by the Mid-Tertiary inversion of the Atlas system (e.g. Hafid et al. 2000). The impact on salt tectonics of this compressive Atlasic event appears to diminish rapidly to the south and to the north of the Essaouira Basin.

Comparison of syn-rift v. post-rift salt tectonics

The characteristics of salt basins in Morocco and Angola can be generalized to reflect some fundamental differences in the way these basins formed. Figure 17 summarizes the first-order cross-sectional differences between West African salt basins of syn-rift (Morocco, Senegal, The Gambia, Guinea-Bissau and Guinea) v. those of post-rift origin (Cameroon, Equatorial Guinea, Gabon, Congo, Democratic Republic of Congo and Angola).

The Aptian salt was deposited during a post-rift subsidence in the Southern Atlantic (Marton et al. 2000). Consequently, the salt has a fairly uniform thickness over large areas, providing a continuous and very efficient zone of detachment once gravity sliding/spreading on the passive margins commences. As the décollement connects all the salt-related structures from the extensional updip regime to the corresponding compressional downdip regime, the displacement pattern of any marker immediately above the salt shows a characteristic pattern (Fig. 17). The structural restoration of individual raft structures on the shelf typically indicates finite extension in the order of a few kilometres (e.g. Duval et al. 1992; Eichenseer et al. 1999; Anderson et al. 2000). The extension on any raft contributes to the downdip translation in a monotonic fashion. Further downdip, some of the diapirs also accommodate some extension and therefore the accumulated total extension can reach a value maximum value of 20–30 km. This is in accordance with an independent estimation of downdip displacement in

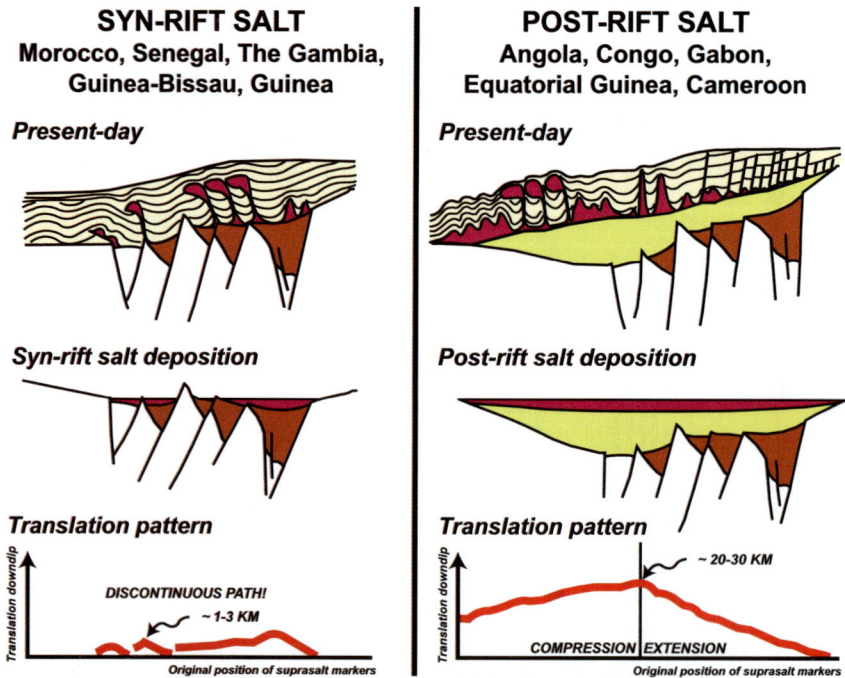

Fig. 17. Cross-sectional evolution of salt structures on an idealized passive margin, in syn-rift v. post-rift settings. See also Figure 18 for the corresponding map-view characteristics. For detailed explanation see text.

the Kwanza Basin (Fig. 14). Once the extension is replaced by compression downdip within the diapir domain, the total displacement begins to decrease monotonically. The overall displacement curve is not symmetric as the basinward edge of the salt is thrust onto oceanic or transitional crust by about 5–15 km (Cramez & Jackson 2000; Marton *et al.* 2000).

In sharp contrast, the syn-rift stage salt typically does not form a basinwide detachment surface and the displacement curve for post-salt markers shows a discontinuous curve (Fig. 17). The salt supply may be very limited for the development of allochthonous salt structures because the salt is separated by tilted blocks. Also, the timing of salt deformation can be very different in individual sub-basins, depending on the areal/temporal distribution of the sedimentary load.

The map-view differences between syn-rift and post-rift salt basins are summarized in Figure 18. The starting point is an idealized rift-basin which opens up to the west and where the rifting propagates towards the north. Continental extension occurs in this setting via west-dipping normal faults, creating half-grabens with a map-view typical of fault growth development (e.g. Schlische 1991). Some of the normal faults may merge together, creating elongated basement highs, or hinge zones, the easternmost being the rift shoulder. Individual syn-rift volcanoes are randomly distributed across the rift.

In the syn-rift scenario, the distribution of salt is patchy, following the outlines of the half-grabens and the associated footwall uplifts. In the southernmost part, where syn-rift subsidence is the most advanced, the salt may be continuous across the entire rift basin. In contrast, in the post-rift scenario, the salt stretches across the entire basin, ignoring the underlying basement structure. The outline of the salt deposition at the northern edge of the basin, however, may be effected by syn-rift structural elements, such as normal and transfer faults, volcanoes and basement highs.

When salt deformation occurs as the result of basinward tilt of the passive margin and/or by sedimentary load, the southernmost part of both syn-rift and post-rift salt basins responds very similarly. The continuous salt detachment is responsible for the formation of linked extensional and compressional salt structures, such as rafts, turtles, diapirs, tongues/sheets/canopies and a toe-thrust zone. The edge of the salt basin may be thrust onto oceanic crust due to the downdip displacement.

Further to the north, however, the syn-rift v. post-rift salt deformation shows a very different picture (Fig. 18). As the syn-rift salt basin is subdivided into separate narrow bands there is no space to develop the whole suite of salt structures.

Instead, the isolated salt patches typically cannot pass the pillow or diapir stage. Deformation in these isolated salt basins could be diachronous as a function of the loading history.

In the post-rift case there is still a variety of extensional and compressional structures. Note that, to the north, at the fault-bounded edge of the salt, the toe-thrust zone does not have a fold train, as opposed to the western edge of the salt basin, where the salt gradually thins to its pinchout. As all the structures are connected via the efficient salt detachment, deformation occurs synchronously basinwide.

Hydrocarbon exploration implications

The classification scheme shown in Figure 15 may have several consequences for exploration in deepwater salt basins around Africa. The toe-thrust zone generally provides very attractive structural targets and, most of the time, the fold train provides the first structures out of the basin where hydrocarbons may have generated. The actual water depth of these structures varies considerably (Tari *et al.* 2001), e.g. the fold belt of Angola (Fig. 13) is located in ultradeep water (>3500 m), beyond currently feasible drilling/development scenarios, as opposed to the Tafelney fold belt of Morocco which is in water depths of 1500–2500 m (Fig. 2b).

Another important exploration aspect concerns the timing of the trap formation. The toe-thrust zone in a syn-rift salt setting appears to create more dormant structures, as opposed to the highly efficient post-rift salt case, where the structural traps tend to be redeformed continuously, which may lead to the loss of early hydrocarbons that must be replenished by later generation.

As an important observation, the salt-cored fold belt may not always be present in the toe-thrust zone, regardless of the syn-rift v. post-rift setting of the salt (e.g. Safi segment of Morocco and Corisco segment of Equatorial Guinea). This is attributed to a sharp, most probably fault-controlled boundary of the original salt. It appears that salt-cored fold trains develop only when the original salt is fairly continuously distributed and its thickness decreases gradually towards the original basinward pinchout.

Regarding the updip, extensional salt domains, the most prolific structural targets to date in Angola, the turtle features, may not be expected in syn-rift salt basins, with a few exceptions (see Fig. 3). Similarly, the widespread raft-play in the Lower Congo Basin does not find a well-developed counterpart in the syn-rift basins of northwestern Africa.

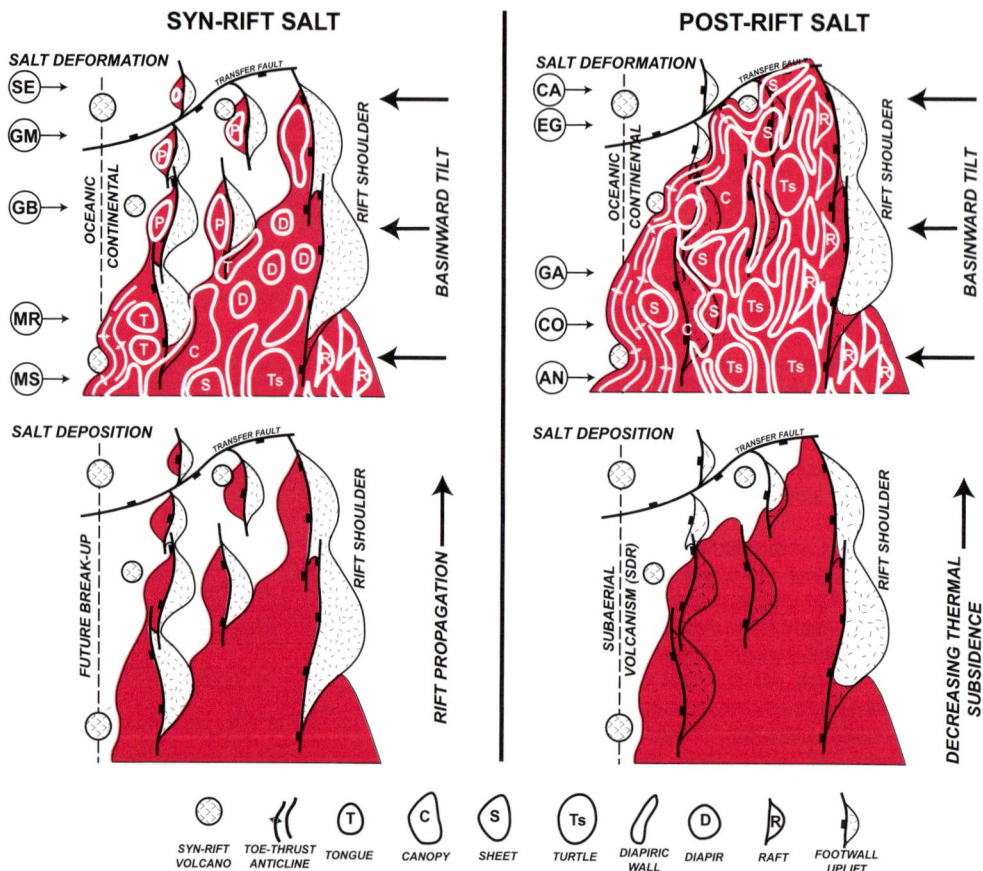

Fig. 18. Map-view evolution of salt structures on an east-facing, north-propagating passive margin, in syn-rift v. post-rift settings. See also Figure 17 for the corresponding cross-sectional characteristics. Note that all the West African salt basins described in this paper can be assigned to a cross-section in this idealized model marked by arrows. Syn-rift salt basins: SE, Senegal; GM, The Gambia; GB, Guinea-Bissau; MR, Morocco-Ras Tafelney; MS, Morocco-Safi. Post-rift salt basins: CA, Cameroon; EG, Equatorial Guinea; GA, Gabon; CO, Congo; AN, Angola. For detailed explanation see text.

Discussion

The speculative and qualitative classification scheme shown in Figure 15 is considered to be very preliminary. Other factors, such as the temporal/spatial distribution and rate of sedimentary loading, the ratio between the initial thickness of the salt and the sedimentary cover etc., should be equally important in formulating more refined classifications in the future. Also note that the salt-cored fold belt is not always present (e.g. Safi segment of Morocco and Corisco segment of Equatorial Guinea), therefore we prefer to generalize all the examples shown in Figure 15 under the term 'toe-thrust zone'.

The variance between West African salt basins is much broader than the examples shown in this

paper might suggest. Among the many important subjects for future research are the geometry of the West African salt basins prior to or immediately after the continental break-up. In the case of the Aptian salt basin of the South Atlantic, Jackson *et al.* (2000) convincingly argued that the salt basin stretching along the Brazilian passive margin was separated from its West African counterpart by a subaerial volcanic ridge. An interesting question arises at this point: can we extrapolate this model for the Triassic–Jurassic salt basins of the Central Atlantic as well? Most of the published plate-tectonic reconstructions (e.g. Klitgord & Schouten 1986) show a single salt basin during the Late Triassic between, for example, Nova Scotia and Morocco. As more modern reflection seismic data become available on both sides of the Atlantic, these questions should be answered in the near future.

Conclusions

In the better-known post-rift salt basins of Equatorial Guinea, Gabon and Angola updip extension is represented by a broad rafted domain. The extension is balanced by downdip contraction in the form of salt tongues, sheets, canopies and the progressive inflation of a massive salt domain at the basinward edge of the salt basin. The efficiency of this gravity sliding/spreading across the whole margin is due to the more or less uniform original distribution of Aptian salt in the post-rift succession forming a continuous detachment level.

In contrast, the typically uneven original distribution of the Late Triassic and Early Jurassic syn-rift salt in Morocco, Mauritania, Senegal, The Gambia and Guinea-Bissau is due to basement highs separating rift half-grabens and creating a different structural pattern. Individual salt structures, such as pillows and diapirs, originated from isolated patches of the autochthonous salt. In the case of syn-rift salt, updip extension may not always be the ultimate driving force for the contractional salt-deformation downdip. Nonetheless, where basinwide syn-rift salt was deposited, unaffected by the underlying basement structure, the resulting salt structures, including allochthonous salt tongues, sheets and canopies, have identical geometries to those of the post-rift salt basins of the South Atlantic. The similarities and differences between West African salt basins have important exploration implications.

We thank A. W. Bally, M. Hafid, M. Zizi and G. Marton for helpful discussions of the geology of eastern Canada, Morocco and Angola. P. Bentham kindly validated a hand-drawn reconstruction of a portion of transect H going through a computer-aided rigorous restoration exercise. M. P. A. Jackson kindly provided two manuscripts prior to publication. A. Danforth, I. Davison, and D. Roberts are thanked for their useful reviews of this paper. We acknowledge Agenzia Generale Italiana Petroli (AGIP), First Exchange and the Total Astrid Marin Group for permission to show selected reflection seismic data.

References

ANDERSON, J. E., CARTWRIGHT, J., DRYSDALL, S. J. & VIVIAN, N. 2000. Controls on turbidite sand deposition during gravity-driven extension of a passive margin: examples from Miocene sediments in Block 4, Angola. *Marine & Petroleum Geology*, **17**, 1165–1203.

BAUMGARTNER, T. R. & VAN ANDEL, T. H. 1971. Diapirs of the continental margin of Angola, Africa. *Geological Society of America Bulletin*, **82**, 793–802.

BUROLLET, P. F. 1975. Tectonique en radeau en Angola.

Bulletin de la Societé Géologique de France, **17**, 503–504.

COFFIN, M. F. & RABINOWITZ, P. D. 1988. *Evolution of the Conjugate East African – Madagascar Margins and the Western Somali Basin*. Geological Society of America Special Papers, **226**.

COWARD, M. P., PURDY, E. G., RIES, A. C. & SMITH, D. G. 1999. The distribution of petroleum reserves in basins of the South Atlantic margins. *In*: CAMERON, N. R., BATE, R. H. & CLURE, V. S. (eds) *The Oil and Gas Habitats of the South Atlantic*. Geological Society, London, Special Publications, **153**, 101–131.

CRAMEZ, C. & JACKSON, M. P. A. 2000. Superposed deformation straddling the continental–oceanic transition in deepwater Angola. *Marine & Petroleum Geology*, **17**, 1095–1109.

DAILLY, P. 2000. Tectonic and stratigraphic development of the Rio Muni Basin, Equatorial Guinea: the role of transform zones in Atlantic Basin evolution. *In*: MOHRIAK, W. & TALWANI, M. (eds) *Atlantic Rifts and Continental Margins*. American Geophysical Union, Geophysical Monograph Series, **115**, 105–128.

DILLON, W. P., POPENOE, P., GROW, J. A., KLITGORD, K. D., SWIFT, B. A., PAULL, C. K. & CASHMAN, K. V. 1983. Growth faulting and diapirism; their relationship and control in the Carolina Trough, eastern North America. *In*: Poag, C. W. (ed.) *Geologic Evolution of the United States Atlantic Margin*. American Association of Petroleum Geologists Memoirs, **34**, 21–46.

EICHENSEER, H. TH., WALGENWITZ, F. R. & BIONDI, P. J. 1999. Stratigraphic control on facies and diagenesis of dolomitized oolitic siliciclastic ramp sequences (Pinda Group, Albian, offshore Angola). *American Association of Petroleum Geologists Bulletin*, **83**, 1729–1758.

HAFID, M. 2000. Triassic–Early Liassic extensional systems and their Tertiary inversion, Essaouira Basin (Morocco). *Marine & Petroleum Geology*, **17**, 409–429.

HAFID, M., AIT SALEM, A. & BALLY, A. W. 2000. The western termination of the Jebilet–High Atlas system (Offshore Essaouira Basin, Morocco). *Marine & Petroleum Geology*, **17**, 431–443.

HART, W. & ALBERTIN, M. 2001. Subsalt trap archetype classification: a diagnostic tool for predicting and prioritizing Gulf of Mexico subsalt traps. *Gulf Coast Section of the Society of Economic Paleontologists and Mineralogists Transactions*, **21**, 619–637.

HEYMAN, M. A. 1988. Tectonic and depositional history of the Moroccan continental margin. *In*: TANKARD, A. & BALKWILL, H. (eds) *Extensional Tectonics and Stratigraphy of the North Atlantic Margin*. American Association of Petroleum Geologists Memoirs, **46**, 323–340.

HINZ, K., DOSTMANN, H. & FRITSCH, J. 1982. The continental margin of Morocco: seismic sequences, structural elements and geological development. *In*: VON RAD, U., HINZ, K., SARTHEIN, M. & SEIBOLD, E. (eds) *Geology of the Northwest African Continental Margin*. Springer-Verlag, Berlin, 34–60.

JACKSON, M. P. A., CRAMEZ, C. & FONCK, J. -M. 2000. Role of subaerial volcanic rocks and mantle plumes in creation of South Atlantic margins: implications for salt tectonics and source rocks. *Marine & Petroleum Geology*, **17**, 477–498.

KARNER, G. D. & DRISCOLL, N. W. 1999. Tectonic and stratigraphic development of the West African and eastern Brazilian Margins: insights from quantitative basin modelling. *In*: CAMERON, N. R., BATE, R. H. & CLURE, V. S. (eds) *The Oil and Gas Habitats of the South Atlantic*. Geological Society, London, Special Publications, **153**, 11–40.

KLITGORD, K. D. & SCHOUTEN, H. 1986. Plate kinematics of the central Atlantic. *In*: VOGT, P. R. & TUCHOLKE, B. E. (eds) *The Western North Atlantic Region*. Geological Society of America, Decade of North American Geology, **M**, 351–378.

LANCELOT, Y., SEIBOLD, E. *et al.* 1977. *Initial Reports of the DSDP*. US Government Printing Office, Washington, **41**, 1259.

LANCELOT, Y. & WINTERER, E. L. 1980. Evolution of the Moroccan oceanic basin and adjacent continental margin: a synthesis. *In*: LANCELOT, Y. & WINTERER, E. L. (eds) *Initial Reports of the DSDP*. US Government Printing Office, Washington, **50**, 801–821.

LETOUZEY, J., COLLETTA, B., VIALLY, R. & CHERMETTE, J. C. 1995. Evolution of salt-related structures in compressional settings. *In*: JACKSON, M. P. A., ROBERTS, D. G. & SNELSON, S. (eds) *Salt Tectonics: A Global Perspective*. American Association of Petroleum Geologists Memoirs, **65**, 41–60.

LEYDEN, R., ASMUS, H., ZEMBRUSCKI, S. & BRYAN, G. 1976. South Atlantic diapiric structures. *American Association of Petroleum Geologists Bulletin*, **60**, 196–212.

LIRO, L. M. & COEN, R. 1995. Salt deformation history and postsalt structural trends, offshore southern Gabon, West Africa. *In*: JACKSON, M. P. A., ROBERTS, D. G. & SNELSON, S. (eds) *Salt Tectonics: A Global Perspective*. American Association of Petroleum Geologists Memoirs, **65**, 323–331.

LUNDIN, E. R. 1992. Thin-skinned extensional tectonics on a salt detachment, northern Kwanza Basin, Angola. *Marine & Petroleum Geology*, **9**, 405–411.

MARTON, G., TARI, G. & LEHMANN, C. 2000. Evolution of the Angolan passive margin, West Africa, with emphasis on post-salt structural styles. *In*: WEBSTER, M. & TALWANI, M. (eds) *Atlantic Rifts and Continental Margins*. American Geophysical Union, Geophysical Monograph Series, **115**, 129–149.

MORABET, A. M., BOUCHTA, R. & JABOUR, H. 1998. An overview of the petroleum systems of Morocco. *In*: MACGREGOR, D. S, MOODY, R. T. J. & CLARK-LOWES, D. D. (eds) *Petroleum Geology of North Africa*. Geological Society, London, Special Publications **132**, 283–296.

PEEL, F. 2001. Emplacement, inflation and folding of an extensive allochthonous salt sheet in the Late Mesozoic (ultra-deepwater Gulf of Mexico). *American Association of Petroleum Geologists Annual Convention, Abstracts*, **10**, 155.

ROWAN, M., TRUDGILL, B. & FIDUK, J. 2000. Deepwater, salt-cored foldbelts: lessons from the Mississippi Fan and Perdido Foldbelts, Northern Gulf of Mexico. *In*: WEBSTER, M. & TALWANI, M. (eds) *Atlantic Rifts and Continental Margins*. American Geophysical Union, Geophysical Monograph Series, **115**, 173–191.

SCHLISCHE, R. W. 1991. Half-graben basin fill models: new constraints on continental extensional basin development. *Basin Research*, **3**, 123–141.

SPENCER, J., TARI, G., JERONIMO, P. & HART, B. 1998. Comparison between offshore Angola and the Gulf of Mexico in terms of salt tectonics. *American Association of Petroleum Geologists Bulletin*, **82**, 1970.

TARI, G., DOMINEY, J., JERONIMO, P. & ALEXANDER, S. 1998. Influence of salt structures on sea-floor topography and sediment dispersal, Lower Congo Basin, offshore Angola. *Geological Society of America, Abstracts with Programs*, **30**, 364.

TARI, G., MOLNAR, J., ASHTON, P. & HEDLEY, R. 2000. Salt tectonics in the Atlantic margin of Morocco. *Leading Edge*, **19**, 1074–1078.

TARI, G. C., ASHTON, P. R. *et al.* 2001. Examples of deepwater salt tectonics from West Africa: are they analogs to the deepwater salt-cored foldbelts of the Gulf of Mexico? *Gulf Coast Section of the Society of Economic Paleontologists and Mineralogists Transactions*, **21**, 251–269.

VENDEVILLE, B. C. & JACKSON, M. P. A. 1992. The rise of diapirs during thin-skinned extension. *Marine & Petroleum Geology*, **9**, 331–353.

VON RAD, U., HINZ, K., SARTHEIN, M. & SEIBOLD, E. 1982. *Geology of the Northwest African Continental Margin*. Springer-Verlag, Berlin.

WHITE, N. & McKENZIE, D. 1988. Formation of the 'steer's head' geometry of sedimentary basins by differential stretching of the crust and mantle. *Geology*, **16**, 250–253.

Syn-rift regional subsidence across the West African continental margin: the role of lower plate ductile extension

G. D. KARNER[1], N. W. DRISCOLL[2] & D. H. N. BARKER[3]

[1]*Lamont-Doherty Earth Observatory, Palisades, New York 10964, USA*
[2]*Scripps Institution of Oceanography, La Jolla, California 92093, USA*
[3]*University of Texas, Institute for Geophysics, Austin, Texas 78759, USA*

Abstract: New ostracode data from the West African margin indicate that the Outer Basin Sediment Wedge (also termed the 'pre-salt wedge' and the 'pre-salt sag basin') is Neocomian to Aptian in age and is contemporaneous with syn-rift deposits developed inboard of the Atlantic hinge zone. Despite the fact that the Outer Basin Sediment Wedge is clearly a syn-rift deposit, it does not exhibit any of the diagnostic characteristics of brittle deformation, such as the existence of normal faults and the faulting and rotation of crustal blocks. Such features are common between the Atlantic and Eastern hinges for the early stages of rifting between West Africa and Brazil, which occurred as a series of extensional phases commencing in the Berriasian and culminating in the Late Aptian. To reconcile the concomitant development of fault-controlled subsidence between the hinges and across the Atlantic hinge zone and sag-basin development seaward of the Atlantic hinge zone requires that: (1) extension seaward of the Atlantic hinge is the result of strain-partitioning between a relatively non-deforming upper crust (i.e. the upper plate) and a ductile-deforming lower crust and lithospheric mantle (i.e. the lower plate) during the second and third rift phases, while (2) between the hinges, early brittle deformation (normal faulting) progresses to ductile deformation in the third rift phase. During the third rift phase, lower plate ductile deformation across the entire region generated regional subsidence both seaward of the Atlantic hinge and between the hinges with little attendant brittle deformation. This extension style produced, directly or indirectly, a sequence of crucial events across the West African margin: (1) the development of the pre-Chela unconformity as lake level dropped in the Early Aptian, exposing the prograding deltas of the Argilles Vertes Formation; (2) the regional development of the Chela unconformity and transgressive lag deposits of the Chela Formation in the Mid-Aptian; (3) the development of regionally extensive, shallow-water, restricted marine conditions across the entire margin (between West Africa and Brazil) immediately prior to evaporite precipitation; and (4) the development of significant post-rift accommodation (deposition of the Late Cretaceous, Paleogene and Neogene formations) in the same region previously characterized by minor syn-rift faulting, repeated dessication cycles (allowing the precipitation of thick evaporites) and negligible erosional truncation of earlier syn-rift units.
Previous workers have suggested that the Loeme evaporites were formed as part of the rapid, early post-rift phase of basin subsidence as the region became inundated by sea water across the Walvis Ridge. In this model, it is difficult to develop the restrictive environments required to deposit the thick (>1 km) evaporites of the Loeme Formation (and the equivalent Ezanga and Ibura evaporites of Gabon and Brazil, respectively) across the entire West African–Brazilian rift system. The existence of shallow-water environments across the entire region is not consistent with water depths determined from the relief of clinoform foresets existing immediately prior to evaporite deposition thus requiring tectonic uplift of the deep-water regions. These evaporites, therefore, appear to be part of the late-stage syn-rift sediment package and the break-up unconformity, if it exists, separates the Loeme evaporites below from the overlying Albian carbonates.
A direct consequence of ductile extension is one of increased heat input accompanying the rift stage in those areas dominated by syn-rift sag-basin development. The distribution and amplitude of the heat pulse is governed by the geometry of the mid-crustal weak zone and the distribution and amplitude of the lower plate extension. Seaward of the Atlantic hinge zone, the maximum heat flow is predicted to be in excess of 200 mW m^{-2}, whereas between the hinge zones, the heat flow is significantly less and ranges between 20 mW/m^2 and 100 mW/m^2. Because sediment temperature is a function of thermal conductivity and thickness of sediment

From: ARTHUR, T. J., MACGREGOR, D. S. & CAMERON, N. R. (eds) *Petroleum Geology of Africa: New Themes and Developing Technologies.* Geological Society, London, Special Publications, **207**, 105–129. 0305-8719/03/$15
© The Geological Society of London 2003.

overburden, the viability of syn-rift sources and prospectivity of the deep-water West African margin will, to a large degree, depend on the delicate interplay between the cooling of the extended lithosphere and subsequent burial of source rocks as a function of time.

Rifting, by definition, deals with extension of the crust and lithospheric mantle. Brittle deformation of the upper crust leads to the generation of horst and graben morphologies. In general, the onset of rifting is recognized by a number of key observations: normal faulting, divergence and rotation of seismic reflectors indicative of differential subsidence and block rotation, and sediment wedge geometries. Most importantly, the wedge geometry is a consequence of stratal onlap rather than truncation. The rift-onset unconformity separates parallel concordant reflectors below from divergent reflectors above (Falvey 1974; Driscoll *et al.* 1995). Such criteria were used by Karner and Driscoll (1999*a*) to map the distribution and timing of extensional deformation of the lithosphere along and across the Gabon–Cabinda continental margin. Spatial partitioning of extension was responsible for the development of two major tectonic hinge zones: an inner/onshore (Eastern) and outer/offshore (Atlantic) hinge zone subparallel to the margin (Fig. 1). The Eastern hinge zone delineates the eastern limit of broadly distributed Early Neocomian brittle deformation that resulted in the formation of deep, anoxic lacustrine systems. Based on gravity trends seaward of the present-day shelf break, individual rift basins of probable Early Neocomian–Early Barremian age tend to exhibit an en-echelon arrangement but with limited along-strike connectivity (Fig. 1).

The Atlantic hinge zone marks the eastern limit of a second phase of extension that began in the Hauterivian (Karner *et al.* 1997; Fig. 2). Analysis of gravity data, supplemented with local seismic grids and well data, indicate that the Atlantic hinge zone is a series of short segment, en-echelon high-standing blocks (Karner & Driscoll 1999*a*; Fig. 1). Landward of the Atlantic hinge zone there are a series of local depocentres with depth to basement ranging from 2 km to 4 km. The depositional packages to the west of the Atlantic hinge are collectively termed the 'Outer Basin Sediment Wedge', 'pre-salt sag basin' or 'pre-salt wedge' (e.g. Henry *et al.* 1995; Karner *et al.* 1997; Marton *et al.* 2000). Material eroded from the Atlantic hinge zone and earlier syn-rift units provided a sediment source to the Outer Basin Sediment Wedge. Sediment delivery was predominantly axial parallel to the basin trend, given that the wedge generally thins by onlap onto the Atlantic hinge zone. According to Karner and Driscoll (1999*a*), a third and final phase of extension began in the Early Barremian and was

ultimately responsible for breaching of the continental lithosphere and the start of sea-floor spreading. The ocean/continent boundary has been interpreted by Karner and Driscoll (1999*a*) as the strong positive/negative gradient in the gravity anomaly (Fig. 1).

The rifting between Africa and Brazil occurred in three distinct phases: (1) Berriasian (Early Neocomian; Fig. 2, rift onset 1); (2) Hauterivian (Fig. 2, rift onset 2); and (3) Early Barremian (Fig. 2, rift onset 3a), with each phase being recorded by an onlap surface followed by an overall regressive package representing the subsequent infilling of rift-induced accommodation (Fig. 2). Each rift phase resulted in the formation of deep, anoxic lacustrine systems. The third of these rift phases (Fig. 2, rift onset 3a) was accompanied by minimal brittle deformation and large regional subsidence across the entire margin. Because of the lack of significant faulting, some workers suggest that the pre-salt wedge is a post-rift unit resulting from the thermal subsidence of the lithosphere (e.g. Marton *et al.* 2000), implying an earlier age of break-up (Early Barremian; Marton *et al.* 2000).

Recent studies by Driscoll and Karner (1998) and Karner and Driscoll (1999*b*) have focused attention on the rapid and regional syn-rift subsidence that characterizes the break-up of the northwest Australian region from Greater India (Fig. 3). The style of deformation varied continually as a function of space and time. Initially, the deformation was characterized by a broadly distributed Late Permian event. During Late Triassic to Mid-Jurassic time, deformation became more localized and formed a series of fault-controlled sub-basins: the Exmouth, Barrow, and Dampier sub-basins. This localized phase of deformation was followed by a substantially more regional deformation event in the Tithonian–Valanginian that generated large post-Valanginian regional subsidence across the Exmouth Plateau with only minor accompanying brittle deformation and erosion. Sea-floor spreading magnetic anomalies provide timing estimates for the cessation of break-up and the onset of sea-floor spreading. The onset of sea-floor spreading propagated from north to south along the northwest Australian margin (Fig. 3); in the Argo Basin sea-floor spreading began during the Callovian (marine magnetic anomaly M25), whereas in the Gascoyne, Cuvier and Perth basins, sea-floor spreading began in the Valanginian (anomaly M10; Exon *et al.* 1982; Boote & Kirk 1989; Exon *et al.* 1994;

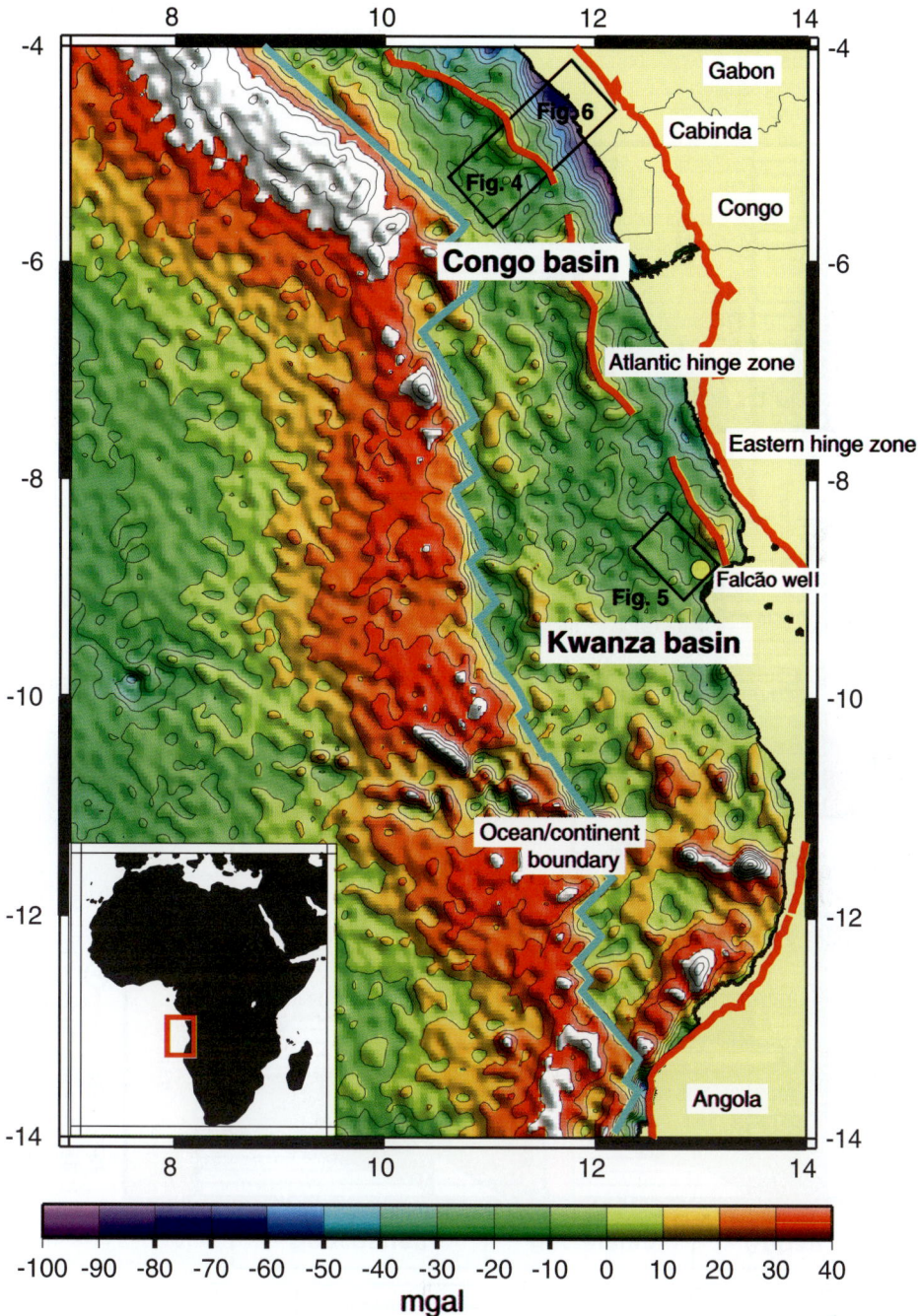

Fig. 1. Location map showing the major tectonic and structural features along the Congo–northern Angola continental margin. The base map shows the offshore crustal Bouguer gravity anomaly of the region. Scale is in mgals with values greater than +40 in white. Seismic reflection profiles are located in the two outlined boxes on the Congo and northern Angolan margins (Kwanza Basin). Two tectonic hinge zones, the Eastern (thin red line) and Atlantic (bold red line), trend subparallel to the margin. The onshore Eastern hinge zone demarcates the eastern limit of Neocomian extension and separates continental margin sediments from Precambrian basement. The Atlantic hinge zone occurs beneath the outer shelf/upper slope transition and consists of a series of en-echelon high-standing blocks. A number of small but in general unconnected sub-basins, identified as localized negative-gravity anomalies, exist on either side of the Atlantic hinge zone. The ocean/continent boundary, approximately delineated by the strong positive/negative gradient in the gravity anomaly, is shown as a bold aqua line. The Falcão-1 well (bold yellow circle), critical because of its definition of the complete Outer Basin Sediment Wedge, is located seaward of the Atlantic hinge zone.

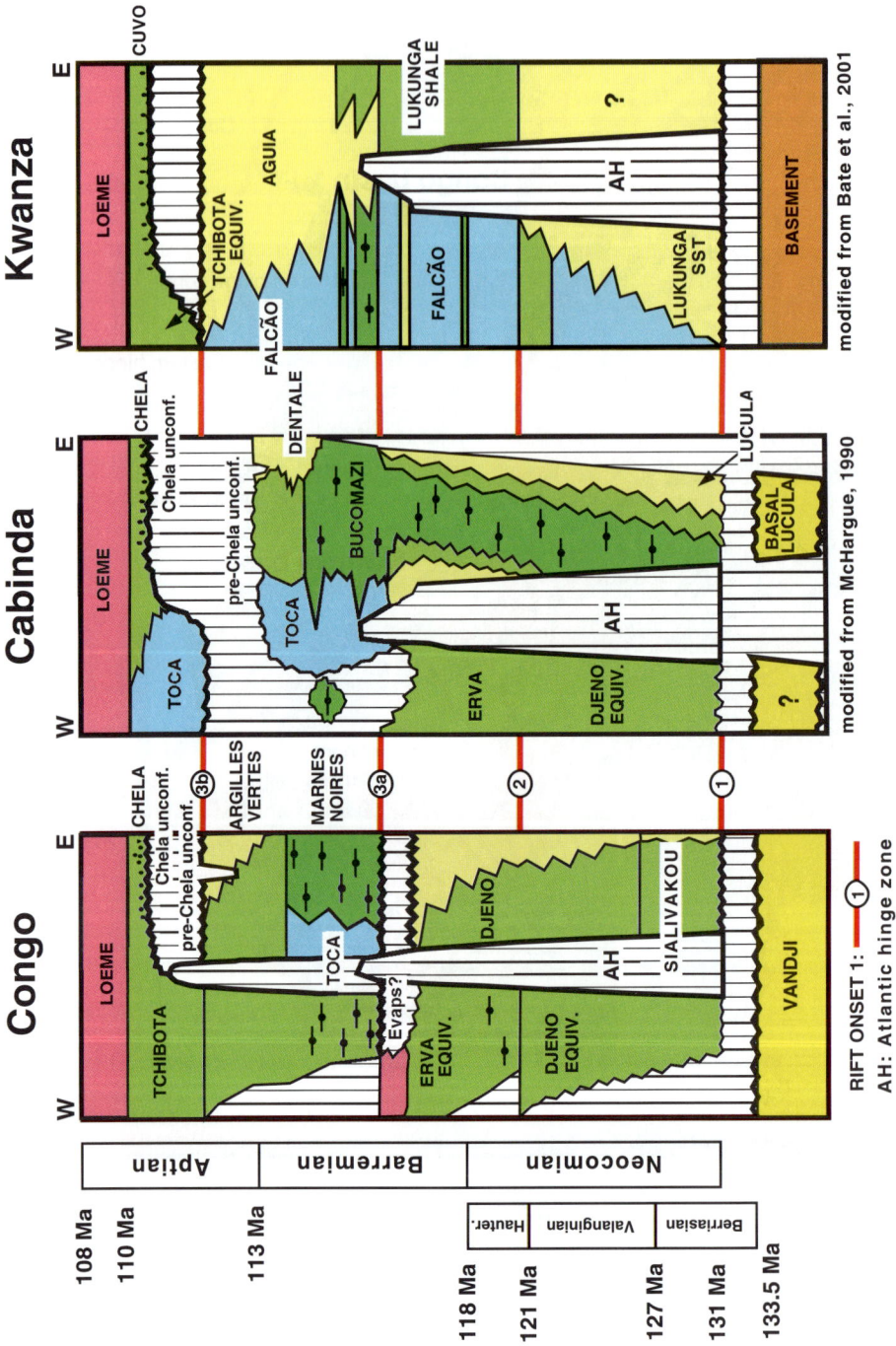

Fig. 2. Generalized chrono-tectonochronostratigraphy of the Congo, Cabinda and Kwanza (northern Angolan) margins, illustrating the link between extensional phases and the stratigraphic development of the margin. Bold red lines mark the onset of rifting for the three main episodes of tectonic deformation. The third and final rift phase, because of its complexity, is subdivided into both a rift onset and the time of generation of the pre-Chela unconformity. Yellow, sandstones; dark green, shales; light green, composite of sandstones and shales; pink, evaporites; blue, limestones.

(a)

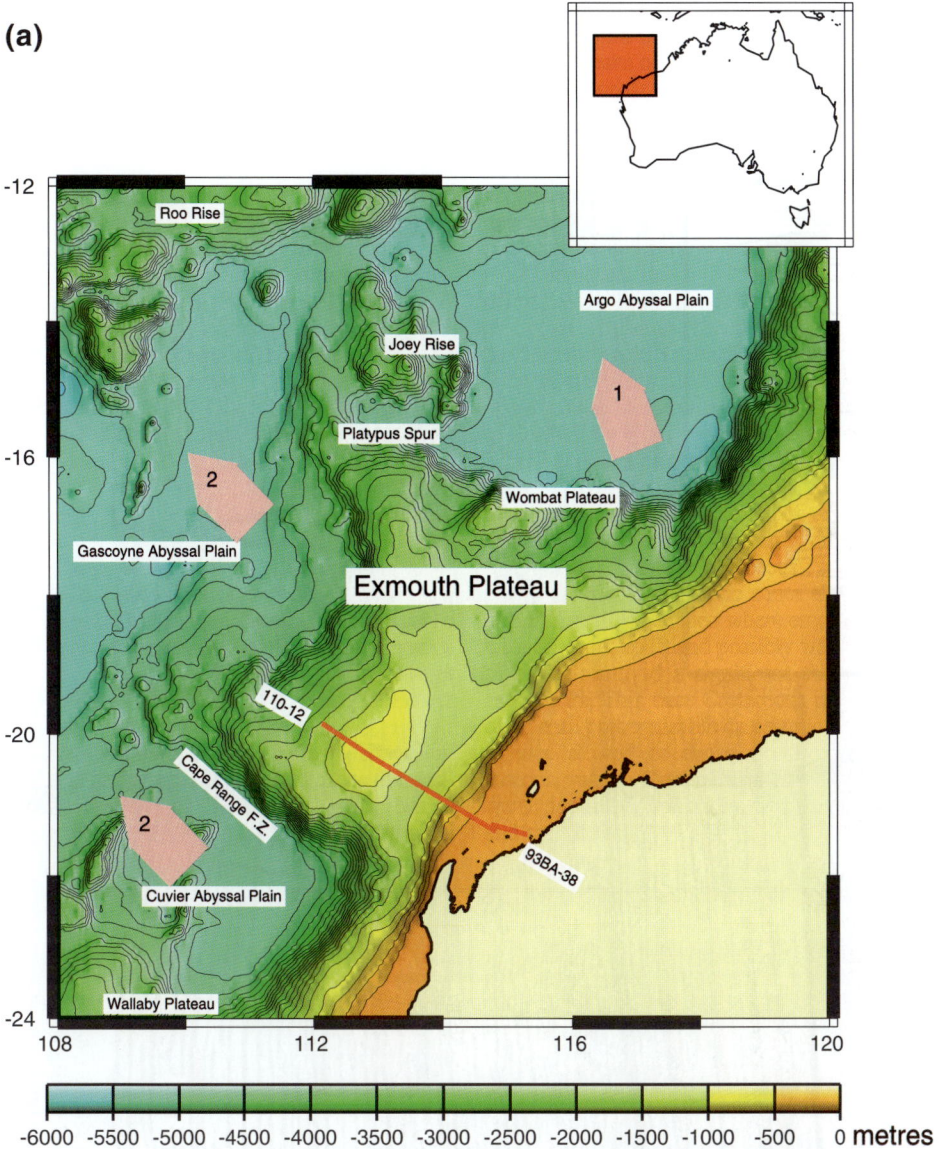

Fig. 3. (a) Bathymetry map of the Exmouth Plateau region, showing the relationship between the plateau and the Argo, Gascoyne and Cuvier Abyssal Plains. Bathymetry contour interval is 200 m. Note the marked difference in margin architecture along the northwest Australian region in response to the north–south propagation of Mesozoic rifting, post-extensional volcanic constructions, and the Mesozoic to Holocene margin progradation. The Argo and Cuvier Abyssal Plains represent narrow margins, whereas the Exmouth Plateau represents a wide margin. The Exmouth, Barrow and Dampier sub-basins have no bathymetric expression and are located beneath the present-day continental shelf. The large transform faults that separate the Exmouth Plateau from the Argo Abyssal Plain to the north and the Cuvier Abyssal Plain to the south are well imaged by the bathymetry data. The onset of sea-floor spreading propagated from north to south along the northwest margin: Callovian in the Argo Abyssal Plain and Valanginian in the Gascoyne and Cuvier abyssal plains. Flow lines and ages are shown by rose-coloured arrows and numbers: 1, Callovian; 2, Valanginian.

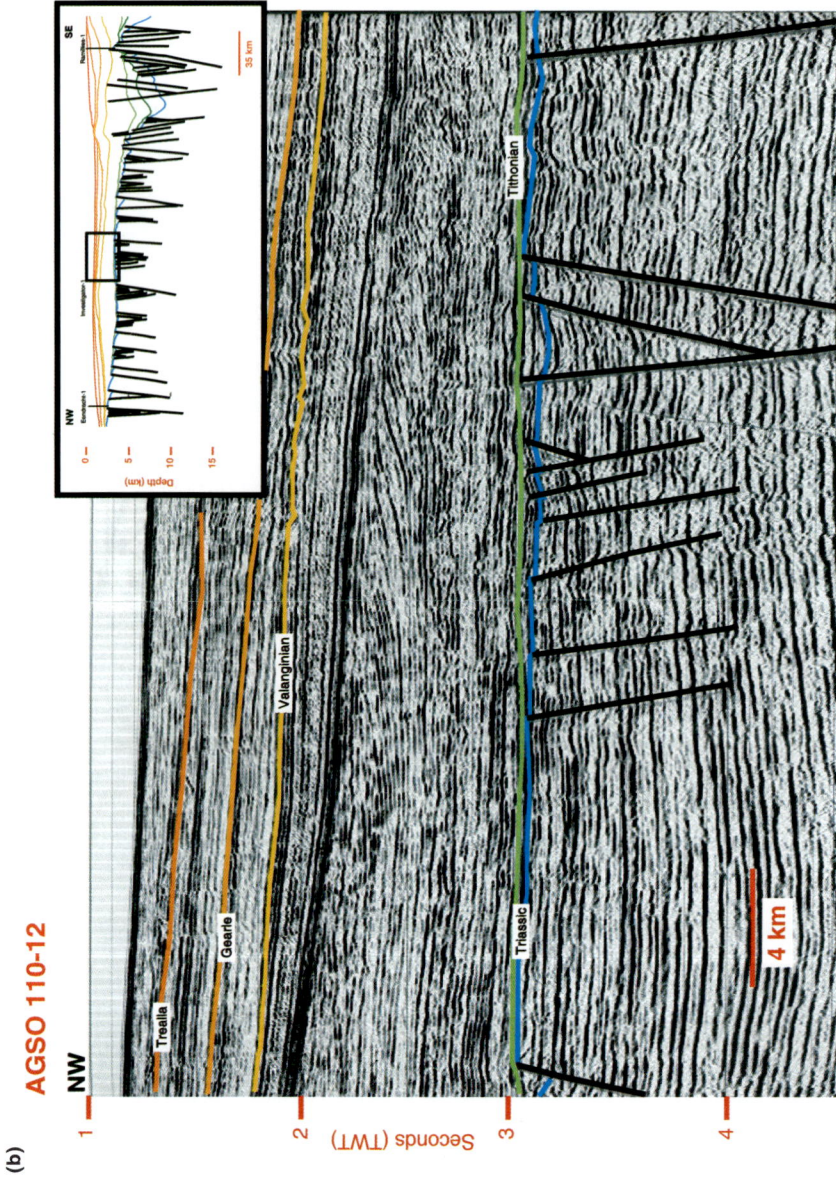

Fig. 3. (b) Interpreted Geoscience Australia seismic reflection profile AGSO110-12, illustrating the clinoforms comprising the Early Cretaceous Barrow delta and the lack of any structuring across the margin (e.g. normal faulting and block rotation) since the Mid-Jurassic. The oblique nature of the clinoforms indicates that space was rapidly filled in the Tithonian, allowing for clinoform progradation (but not aggradation). This stratal pattern suggests that the top of the clinoforms was at, or near, sea level during the time of deposition. Irrespective of the absolute age of break-up on this margin, large post-Valanginian regional subsidence occurred across the Exmouth Plateau with only minor accompanying brittle deformation and erosion.

Mihut & Müller 1998). A marked change in spreading direction from NNW to WNW occurred between the Callovian spreading in the Argo Basin and the Valanginian spreading in the Gascoyne, Cuvier and Perth basins (Fig. 3).

Geoscience Australia seismic reflection profile AGSO110-12 (Fig. 3), which crosses the Alpha Arch and Exmouth Plateau, imaged a Tithonian angular unconformity that records minor truncation of the underlying Mesozoic stratigraphic successions and onlap and downlap of the overlying Early Cretaceous sequences (Fig. 3). This truncation is observed across most of the region, suggesting that large portions of the plateau were emergent or at very shallow-water depths prior to the development of the overlying Barrow delta system. A broad shallow-water environment (200–500 m; Tait 1985) across large portions of the platform is also consistent with facies analysis of the Barrow delta sediments (the Barrow Group), in addition to the mapped northward and westward distribution of the delta foresets (Erskine & Vail 1988). The depositional packages that make up the Barrow Group display a systematic evolution from lower turbidite packages, across which prograded thin pro-delta shales that, in turn, coarsen upward into delta-top sands typical of wave-dominated systems. The clinoform foresets mark the location of increased palaeowater-depth. Palaeowater depth estimates for the Valanginian, together with the present-day bathymetry and thickness of the Late Cretaceous and Tertiary stratigraphic sequences, indicate that significant subsidence has occurred across the northern Exmouth Plateau since the Valanginian.

Pervasive and high-angle normal faulting characterizes the Late Triassic and Mid-Jurassic extension across the Exmouth Plateau (Fig. 3b). However, the amount of brittle extension represented by this faulting is minor. Consequently, after the cessation of these rifting events, any subsidence induced by thermal cooling across the plateau was negligible. While the exact timing for the onset of Early Cretaceous rifting and break-up is questionable (ranging in age from sea-floor spreading magnetic anomaly M7–M10), the fact remains that many kilometres of accommodation were created across the entire plateau and occurred in the absence of brittle deformation of the upper crust and pre-rift sediments. This accommodation space comprises both sediments and unfilled bathymetry, from the Tithonian to the present day (Fig. 3). As shown by the lower section of the Barrow delta clinoforms, deposition was initially rapid since the clinoforms are dominated by progradation. As subsidence outpaced sediment supply (post-Valanginian), the Barrow delta progressively began to aggrade prior to rapidly back-stepping to the

ESE. We conclude that the regional distribution and amplitude of the post-Valanginian subsidence is clearly not consistent with either the distribution or the amount of Tithonian–Valanginian brittle upper crustal extension observed across the margin (Fig. 3; Karner & Driscoll 1999b).

Syn-rift subsidence characterized by regional subsidence with only minor amounts of normal faulting of the crust and overlying sedimentary successions implies that extension can generate sag-type or 'thermal-type' subsidence during rifting. The purpose of this paper is to reinvestigate the syn-rift stratigraphic development of the West African margin in order to define the palaeoenvironment and tectonic implications of the Outer Basin Sediment Wedge and the Loeme evaporites. In particular, we reject the view that the lack of brittle deformation necessarily precludes active lithospheric extension. We will also show that the Loeme Formation was formed during the late transgressive system tract of the final rift phase, in marked contrast to the formation of the low-stand Messinian evaporites in the Mediterranean region.

Late syn-rift stratigraphic packages

New ostracode dating has required us to revise the chrono-tectonostratigraphic correlation chart of Karner et al. (1997) and Karner and Driscoll (1999a), the result of which allows rather precise dating of the rift-onset events along the West African margin (Fig. 2). Crucial ostracode dates used in the construction of Figure 2 are: (1) ostracode zones AS5 and AS6, Middle Djeno; (2) AS8 and AS9, Marnes Noires; (3) AS8 and AS9, Falcão source rock; (4) AS10, Dentale; and (5) AS3–AS6, Erva (Grosdidier et al. 1996; Braccini et al. 1997; Katz & Mello 2000; Bate et al. 2001). We term this a 'chrono-tectonostratigraphic correlation chart' because it classifies the various formations according to tectonic setting and the accommodation space that is created or destroyed by both syn-rift and post-rift tectonics (Fig. 2). Each rifting event results in the rapid generation of accommodation space, the base-bounding surface being either an onlap or a downlap surface. The resulting depositional package is characterized by an overall regressive sequence representing the subsequent infilling of the basin. In general, each rift phase resulted in the formation of deep, anoxic lacustrine systems. The preserved syn-rift stratigraphy of the West African margin is the result of a series of discrete rifting events distributed along and across the margin (Karner & Driscoll 1999a): Berriasian, Hauterivian and Early Barremian. The resulting regressive packages can be grouped as Berriasian–Valanginian, Hauterivian–Early Barremian and Early Barremian–Late Aptian (Fig. 2).

An enigmatic sediment package, the Outer Basin Sediment Wedge, is well developed seaward of the Atlantic hinge and exhibits distinctive basinward thickening and has been described by many workers (e.g. Teisserenc & Villemin 1990; Henry *et al.* 1995; Lomando 1996; Braccini *et al.* 1997; Pasley *et al.* 1998; Marton *et al.* 2000). A reinterpretation of Cabinda pre-salt stratigraphy using industry seismic reflection and well data by Braccini *et al.* (1997) has shown that, at least across the Atlantic hinge, rotated fault blocks are overlain by the Erva Formation, which itself is in part fault-controlled. However, succeeding the Erva Formation are the TOCA carbonates and clastics of the Dentale Formation, both of which are part of a regional and relatively undeformed pre-salt sag basin. Henry *et al.* (1995) describe a similar pre-salt sag-basin wedge seaward of the Atlantic hinge zone on the northern Angolan margin. The pre-salt relationship is crucial because it implies that the sag basin is part of the syn-rift phase of basin development.

Figure 4 is part of a multichannel seismic reflection profile located across the Atlantic hinge zone on the Congo margin (Fig. 1). Landward of the Atlantic hinge zone, syn-rift accommodation appears to be in part controlled by normal faulting of the basement, allowing deposition of the Djeno Formation. This space was produced during the first rift phase (Berriasian–Valanginian). As summarized by Karner *et al.* (1997), the Lower Djeno Formation represents lacustrine turbidite or fan deposits derived from adjacent basement highs. The Upper Djeno Formation represents a progradational clastic wedge, the top of which was subaerially exposed, as evidenced by the existence of palaeosol horizons with iron-stained sands. Toward the Atlantic hinge zone, the Djeno Formation appears to thin by downlap onto the underlying basement. The Early Barremian generation of the 400–650 m of accommodation space required for the deposition of the Marnes Noires Formation, which unconformably overlies the Djeno Formation, occurred without localized brittle deformation. Seaward of the Atlantic hinge zone, rifting is required to produce space for the thick turbidites of the Erva Formation (McHargue 1990) and the formation of the Atlantic hinge zone (Fig. 2). Uplift and erosion of the Atlantic hinge zone, a rift flank to Erva accommodation, was responsible for the local input of sands sourced from the west (Karner *et al.* 1997).

Seaward of the hinge zone there exists a large pre-salt basin: the Outer Basin Sediment Wedge or the pre-salt sag basin (base sag basin; Fig. 4). Similarly, for northern Angola, interpretation of the multichannel seismic profile shown in Figure 5 (location given in Fig. 1) shows a similar feature seaward of the Atlantic hinge, i.e. a depositional package bounded by a sequence boundary (basal onlap) and an upper truncational unconformity. Even though this is a relatively short seismic section, and despite considerable disruption caused by salt tectonics, there is no obvious faulting controlling the generation of accommodation space (Fig. 5). The regional sag is defined by the base of the Loeme evaporites (red reflector) and the base of the Falcão Formation (orange/brown reflector). The internal yellow reflector represents the top of the Aguia Formation. Near-parallel reflectors beneath the Falcão Formation are pre-rift fluvial sands and clays of the Lukunga Formation (Fig. 2). Formation tops for the Falcão-1 well (Fig. 5, bold vertical line) were based on Katz and Mello (2000) and Bate *et al.* (2001). Conversion of two-way travel time to depth was accomplished by using stacking velocities. Most importantly, ostracode-dating of this regional sag package indicates that the entire Outer Basin Sediment Wedge is Neocomian to Early Aptian in age and is thus contemporaneous with syn-rift deposits developed inboard of the Atlantic hinge zone (Bate *et al.* 2001).

While it is clear from age dating that the Outer Basin Sediment Wedge is a syn-rift deposit, it nevertheless does not exhibit any of the diagnostic structural and stratigraphic characteristics of rifting, such as normal fault control and the rotation of crustal blocks (Figs 4 & 5). Such features are common between the Atlantic and Eastern hinges for the early stages of rifting between West Africa and Brazil. The interpretation by Karner and Driscoll (1999*a*), that the Outer Basin Sediment Wedge plus the evaporites are part of the rift phase, remains contentious because many workers believe that the lack of normal faulting and brittle deformation indicates that rifting is complete. As a consequence, many workers attribute the accommodation created for at least the latter units of the Outer Basin Sediment Wedge, the Marnes Noires Formation and the Loeme evaporites to generation by post-rift thermal subsidence (e.g. Henry *et al.* 1995; Marton *et al.* 2000). Based on this approach, the age of break-up is Late Barremian, significantly older than what is usually assumed for the West African margin (e.g. Brice *et al.* 1982; Karner *et al.* 1997) and is certainly not consistent with the rift history of the conjugate Brazilian continental margin (e.g. Feijó 1994).

Regional accommodation developed between the hinge zones during the Early Barremian–Aptian allowed the deposition of the deep-water (400–650 m) lacustrine Marnes Noires and Upper Bucomazi formations, which are separated from the Berriasian–Hauterivian regressive packages by an unconformity (Fig. 2). Seaward of the Atlantic hinge zone at this time, continued regional subsidence augmented the Outer Basin Sediment Wedge.

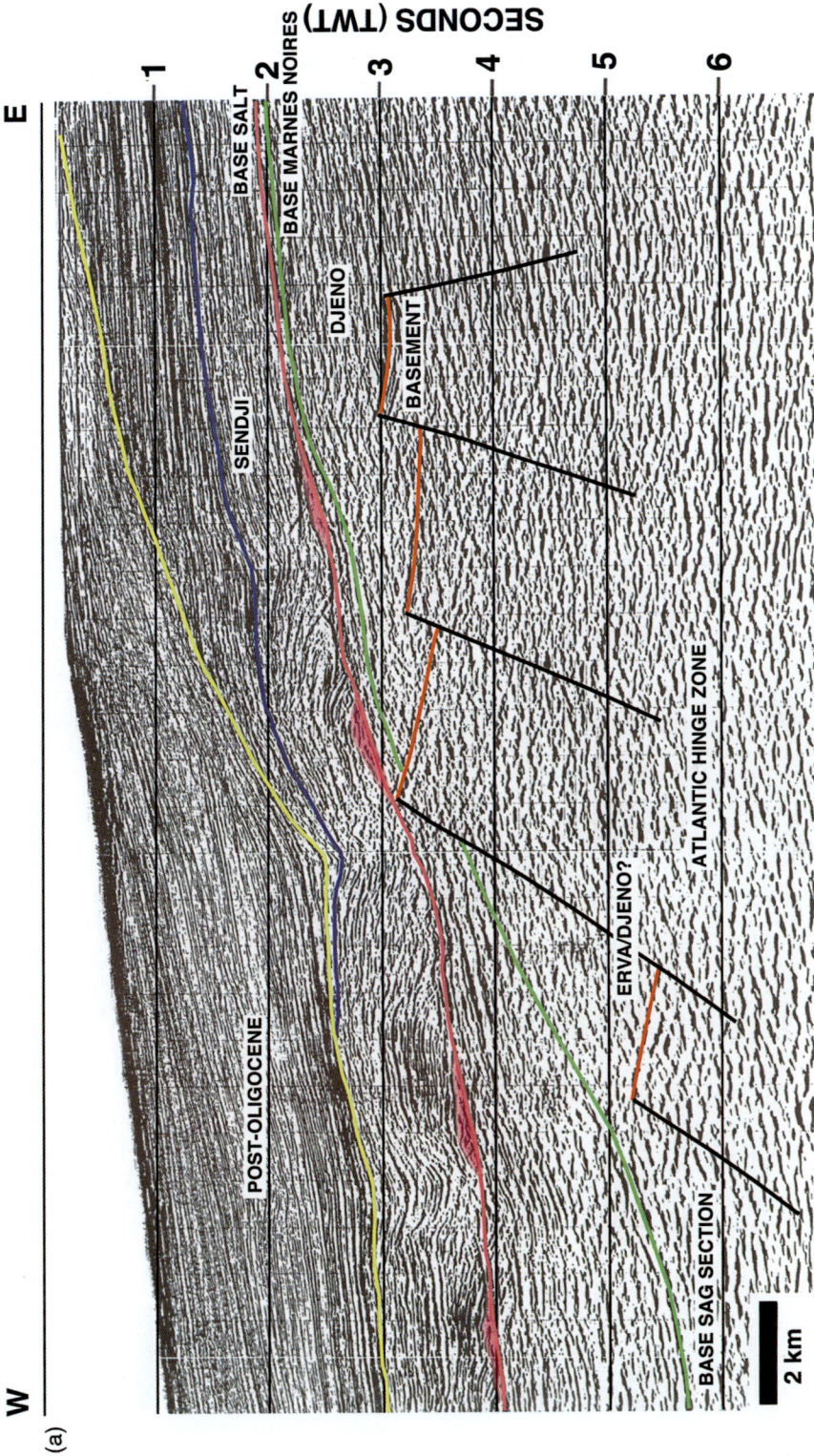

Fig. 4. (a) Seismic reflection profile located across the Atlantic hinge zone of the Congo margin. Seaward of the Atlantic hinge zone, a large pre-salt basin exists termed 'the Outer Basin Sediment Wedge', or 'pre-salt sag basin' (green, base sag basin). Toward the Atlantic hinge zone, the Djeno Formation appears to thin by downlap onto the underlying basement and the top Djeno (green) is truncated. The base Marnes Noires (green) shows no structuring, despite the creation of Early Barremian accommodation. Base salt/Chela Formation (pink) is a strong reflector across the entire margin, as is Top Sendji Formation (blue). The prominent unconformity (yellow) is Oligocene in age and was produced by the vertical truncation of ~2000 m of the shelf, slope and rise over a width of 150–200 km. Seismic data courtesy of TGS-Calibre Geophysical Company.

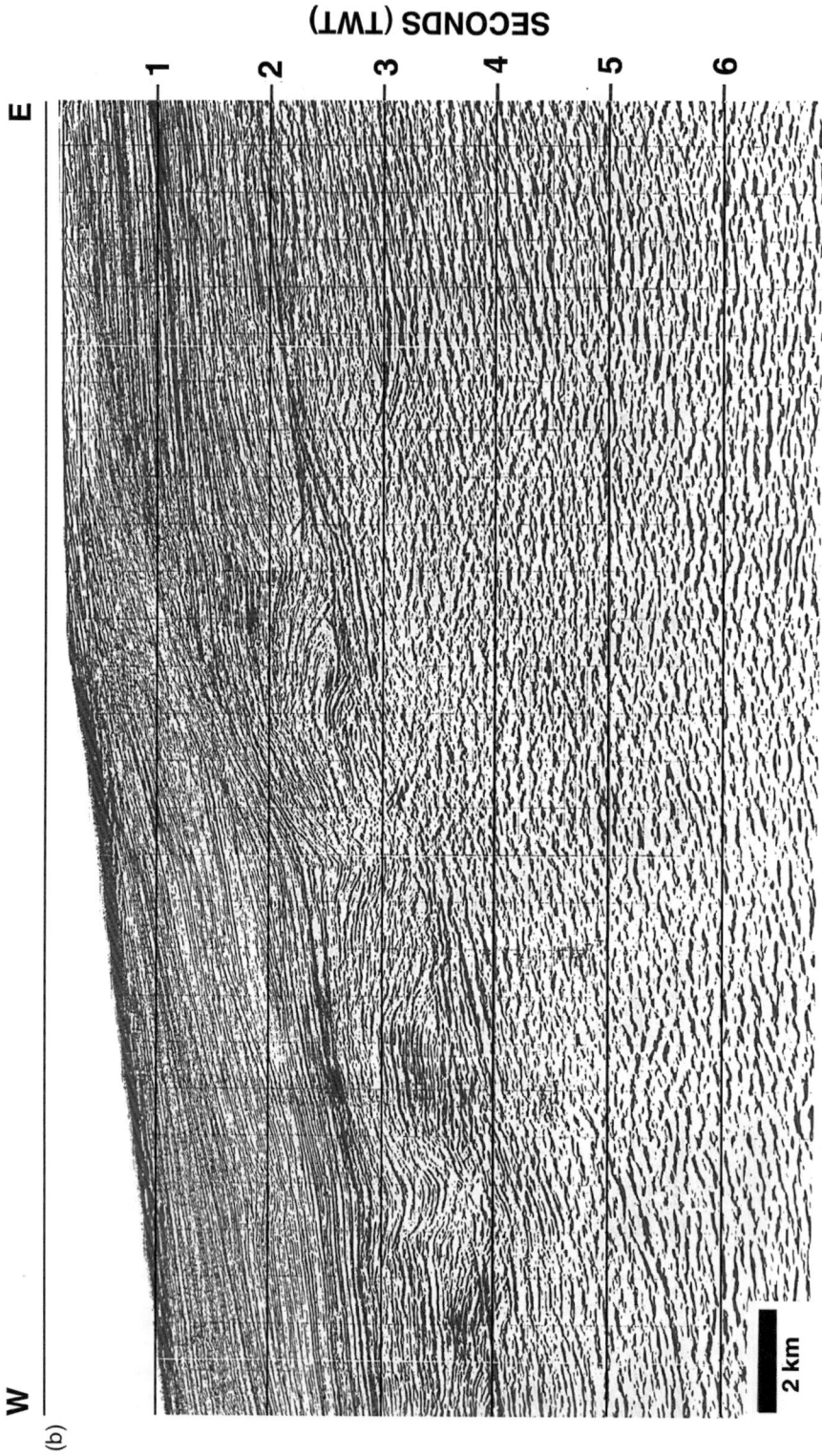

Fig. 4. (b) Uninterpreted seismic reflection profile located across the Atlantic hinge zone of the Congo margin.

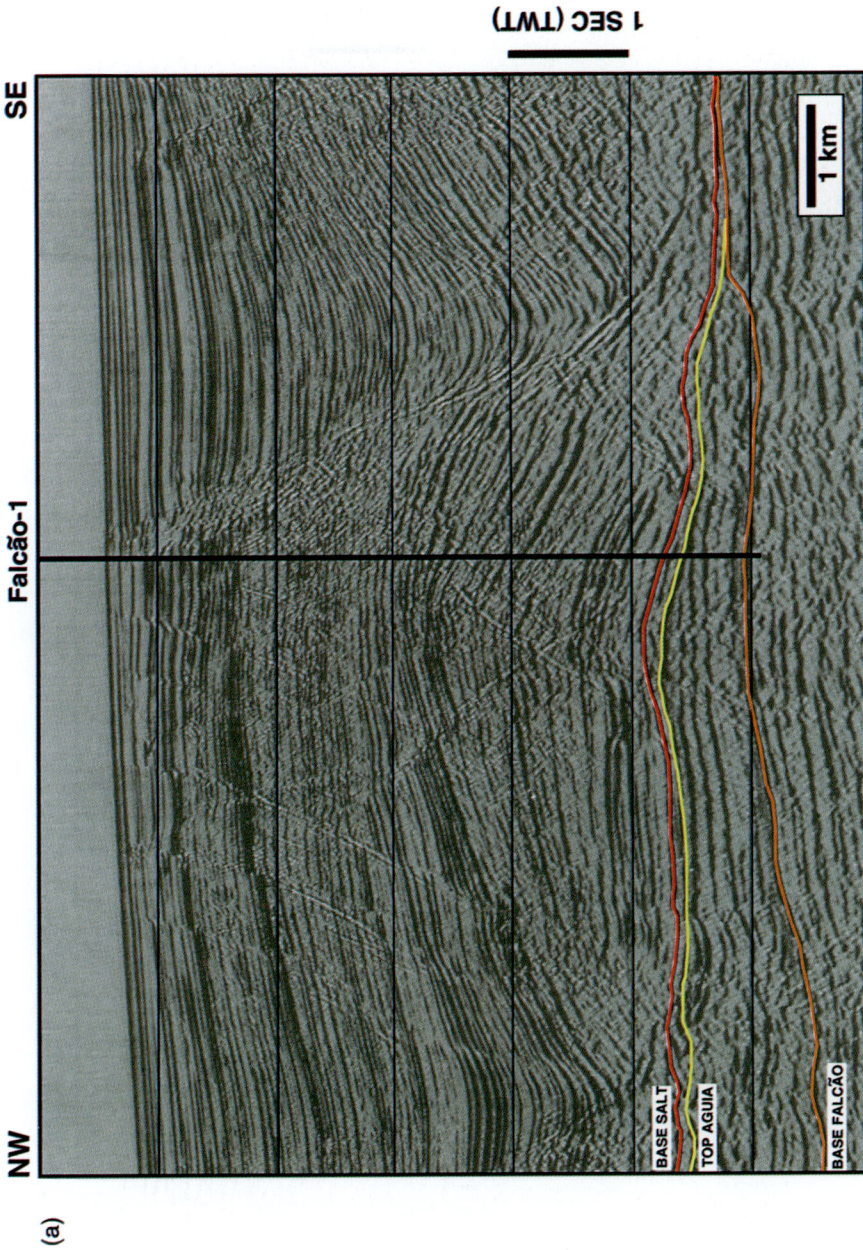

Fig. 5. (a) Seismic reflection profile located seaward of the Atlantic hinge zone on the northern Angolan margin. Well location and drilling depth of Falcão-1 well is shown (bold vertical line), along with a simplified interpretation of the pre-salt sediment package. Formation tops: base salt/top Cuvo (red), top Aguia Formation (yellow) and base Falcão Formation (orange/brown). The base Falcão Formation appears to onlap the underlying parallel pre-rift reflectors, probably sandstones of the Vandji Formation. Overlying the top Aguia is the Chela Formation, which shows truncation towards the east. No obvious brittle deformation structures exist in this section.

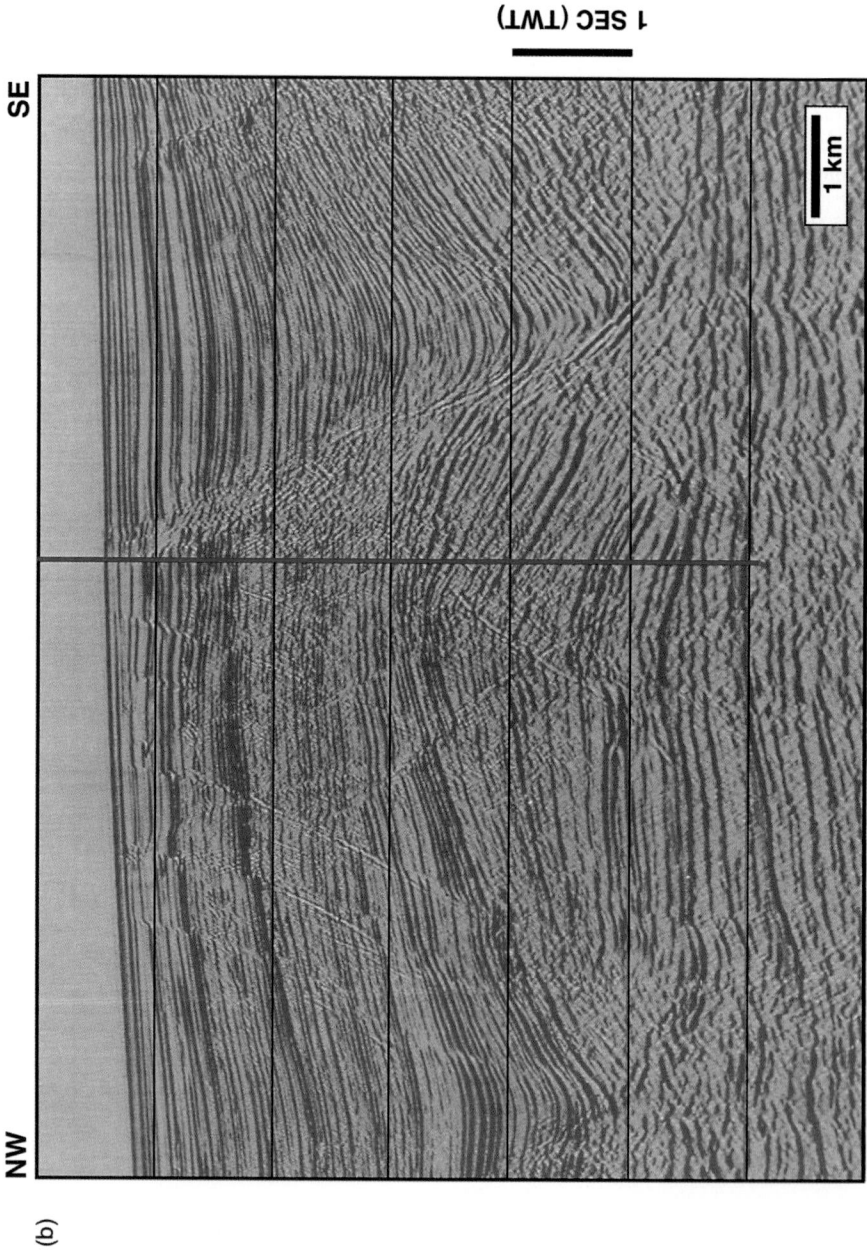

Fig. 5. (b) Uninterpreted seismic reflection profile located seaward of the Atlantic hinge zone on the northern Angolan margin.

Based on the Congo and northern Angola seismic sections, we postulate that the regional subsidence generated across the West African margin consists of two components: (1) Berriasian–Hauterivian regional subsidence seaward of the Atlantic hinge concomitant with fault-induced subsidence between the hinges and continuing to some degree across the Atlantic hinge, and (2) Early Barremian–Aptian regional subsidence across the entire region. The Berriasian–Hauterivian fault-related sediment packages comprise the Sialivakou, Djeno and Lower Bucomazi formations while the sag-related packages consist of the Erva, Dentale and Falcão formations (Fig. 2). The Early Barremian–Aptian sag-related sediment packages comprise the Marnes Noires, Argilles Vertes, Tchibota, Chela, Cuvo, Falcão and Aguia formations (Fig. 2). Note that we are employing the revised nomenclature of the pre-salt wedge package of the Kwanza Basin recently advanced by Bate *et al.* (2001). In this revision, the Lukunga Sandstone Formation is a basal fluvial deposit that is a pre-rift unit to the overlying Falcão Formation. It is equivalent to the basal Lucula and Vandji formations of the Congo and Cabinda margins. The main pre-salt wedge formations consist of the Falcão, Aguia and Lukunga Shale formations. The Falcão Formation is an extensive carbonate platform that is laterally equivalent to the prograding sands of the Aguia Formation (Bate *et al.* 2001).

Figure 6 is crucial for understanding the sequence of events that occurred following the onset of the third and final rift phase (Fig. 2, rift onset 3a). The seismic reflection profile is located between the Eastern and Atlantic hinge zones on the Congo margin (Fig. 1). The top Djeno Formation (or base Marnes Noires; green line) is characterized by either toplap or truncation. The lower boundary of the Argilles Vertes Formation (light brown line) is identified by downlap and the upper boundary (yellow line) by minor erosional truncation and onlap of the Tchibota Formation (Fig. 6). The exposure and incision of the Argilles Vertes clinoforms and the onlapping Tchibota Formation sediments delineate the pre-Chela unconformity (Fig. 6, yellow line) while the lag produced during the ensuing transgression of the margin resulted in the Chela unconformity (Fig. 6, red line). Also shown in Figure 6 is the base of the Chela Formation (red line) and top of the Loeme evaporites (pink line).

In direct response to the onset of the third rift phase, regional subsidence was developed across the entire region, resulting in deep, anoxic freshwater lakes. This anoxia was responsible for the high total organic carbon content of the Marnes Noires, Upper Bucomazi and Falcão source rocks (Fig. 2). By Early Aptian time, the regional accommodation between the hinge zones was partially filled by prograding deltaic systems of the Argilles Vertes, Dentale and Aguia formations (Figs 2 & 6). The clinoform amplitude of the Argilles Vertes deltas suggests maximum lake depths of 500–650 m (Fig. 6). An Early Aptian lake-level fall exposed the topsets of the Argilles Vertes clinoforms; subsequent truncation produced the pre-Chela unconformity (e.g. Teisserenc & Villemin 1990). Downdip, the pre-Chela unconformity is defined by the onlap of the lowstand sequences of the Tchibota Formation onto the Argilles Vertes Formation (Figs 2 & 6, yellow line). The same unconformity exists in the Campos Basin of Brazil, where it is termed 'the pre-Alagoas unconformity' and is followed by gentle regional sagging in the absence of significant brittle deformation (Karner 2000).

The Chela unconformity was developed during the reflooding of the exposed and incised Argilles Vertes Formation and the deposition of the Chela Formation (Figs 2 & 6, red line). The Chela Formation is a transgressive sequence comprising fluvial sandstones and conglomerates (a lag deposit) at the base grading upward, with increasing marine affinity, into lagoonal facies and the evaporites of the Loeme Formation (Figs 2 & 6; Teisserenc & Villemin 1989; Bate *et al.* 2001). As such, the Loeme evaporites were deposited during the late stages of the transgressive system tract. The observation of regional shallow marine and subaerial conditions during the Early Aptian is difficult to explain if the margin was in the early phases of rapid post-rift subsidence. Equally problematic is the need to modify the basin shape and depositional environment, especially seaward of the Atlantic hinge zone, which suggests a change in bathymetry from shallow water in the east to deep water in the west (Fig. 6). However, evaporites were precipitated across the entire margin, from West Africa to Brazil. For this to occur, the deeper-water regions need to shallow (seaward of the Atlantic hinge) while maintaining shallow-water conditions across the upper parts of the Argilles Vertes clinoforms (i.e. between the hinges). We suggest that this shallowing of water depth seaward of the Atlantic hinge and increasing marine affinity of the lakes was part of the late stage of the third rift phase, which commenced in early Aptian time (Fig. 2, rift onset 3b). Such a modification of depositional environment, i.e. via differential uplift across the Atlantic hinge zone, requires a tectonic explanation. Note that, in this analysis, the Chela unconformity (also termed 'the Chela break') is not the break-up unconformity, but was engendered as part of the late rift tectonics immediately prior to salt deposition and the onset of drifting. The break-up

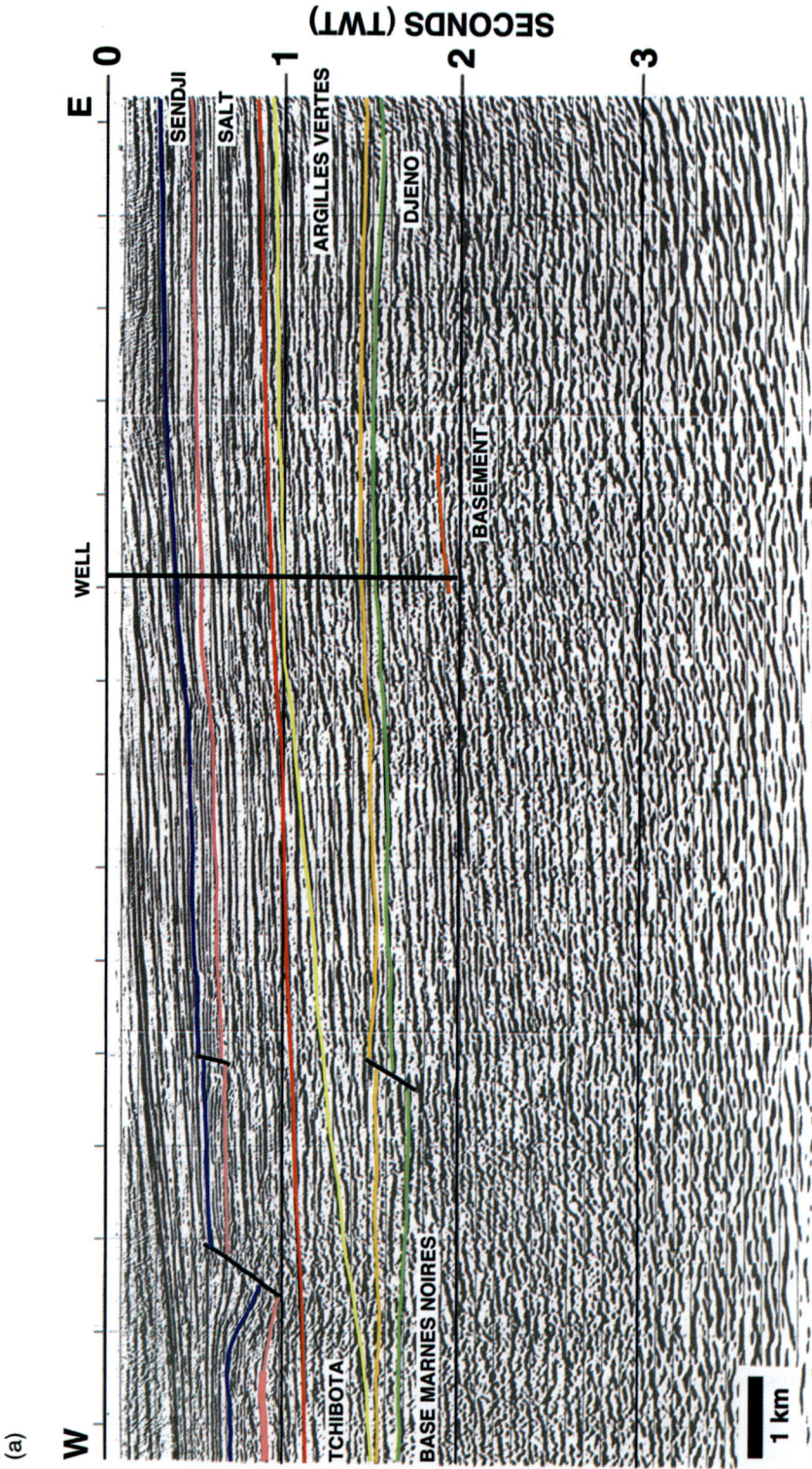

Fig. 6. (a) Seismic reflection profile located between the Eastern and Atlantic hinge zones on the Congo margin. The top of the Djeno Formation (green) is characterized by either toplap or truncation. The Djeno Formation can be separated into two distinct facies units: a lower unit characterized by high-amplitude, laterally continuous reflectors and an upper unit characterized by prograding parallel-oblique. The lower boundary of the Argilles Vertes Formation (light brown) is identified by downlap and the upper boundary by erosional truncation or toplap associated with the pre-Chela unconformity (yellow). Chela unconformity/base Chela Formation (red) shows truncation, produced during the transgression and reworking of the top Tchibota Formation (red). Shown also are top Loeme (salt; pink) and top Sendji (blue) formations. Bold vertical line locates the well control. (Seismic data courtesy of HydroCongo.)

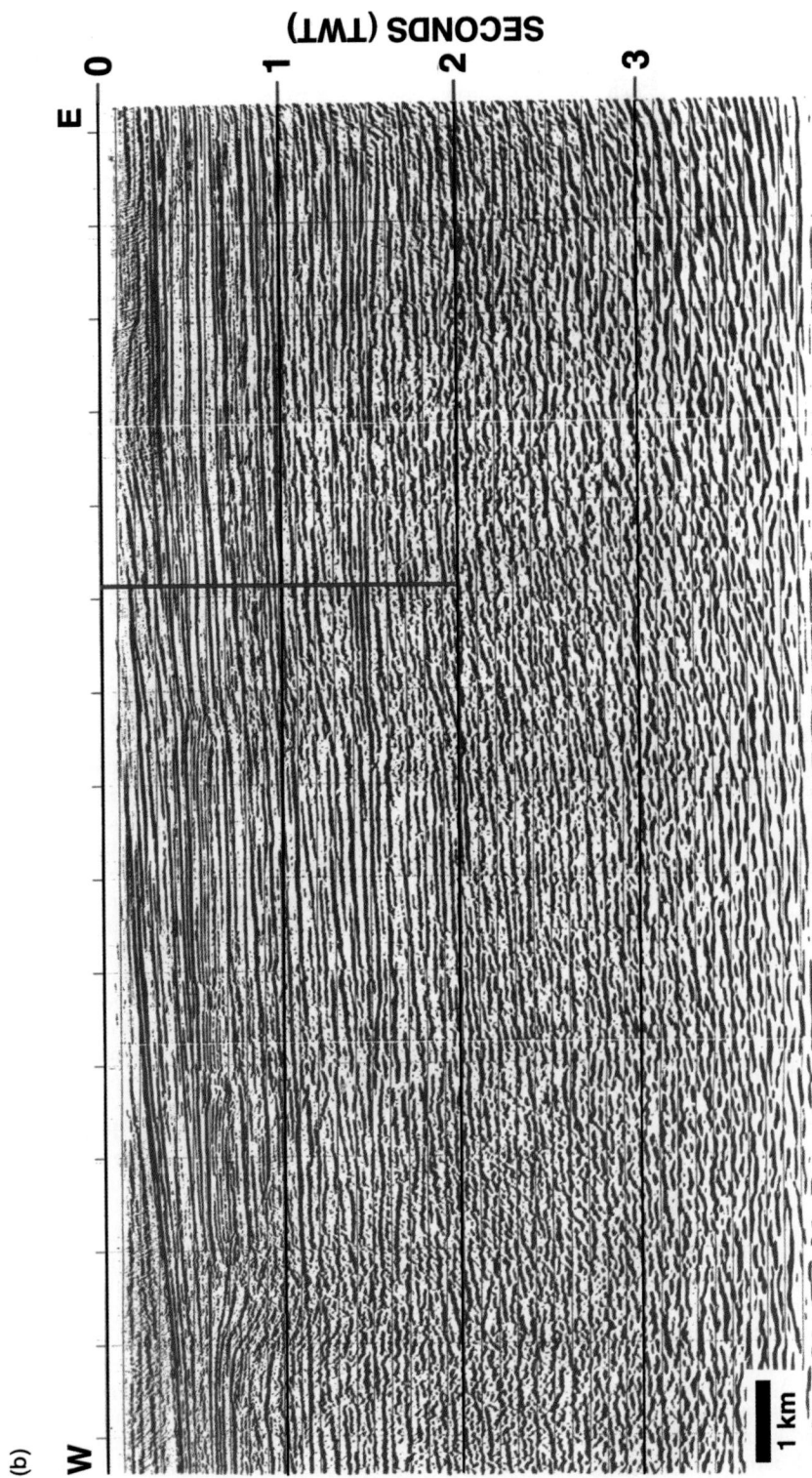

Fig. 6. (b) Uninterpreted seismic reflection profile located between the Eastern and Atlantic hinge zones on the Congo margin.

unconformity, if it exists, is correlative with the top of the Loeme evaporites.

The increasing complexity of ostracode form observed within the Marnes Noires and subsequent formations reflects increasing lake salinity as a function of time (Bate 1999). This change in lake salinity suggests that the lakes were either drying out or were being contaminated by marine waters (or both). Unambiguous marine faunas do not appear until the development of the Chela unconformity and deposition of the Chela and Loeme formations (Teisserenc & Villemin 1990). The Early Aptian lake-level fall was augmented by diminished freshwater input into the lake system because of a drier climate at this time. The high potassium and magnesium salts that characterize the Loeme evaporites (carnallite and sylvite) require a low relative humidity (<35%), suggesting that the palaeoenvironment necessarily needed to be exceedingly dry (Schrieber, pers. comm. 1999). This is consistent with the lack of clastic delivery in areas immediately preceding salt deposition, characterized by the large deltaic systems of the Argilles Vertes and Dentale formations (i.e. indicative of large fluvial systems). Early Aptian eustatic variations (e.g. Haq *et al.* 1988) may also have played a role in modulating the chemistry of the brines and thus the cyclicity preserved within the evaporite sequence. Open marine conditions across the entire region were not established until the Early Albian (Dale *et al.* 1992). After deposition of the evaporite sequences, carbonate platforms dominated the stratigraphic development of both the West African and Brazilian margins. In general, clastic input was not re-established until the Late Cretaceous and Paleogene.

The changing style of extensional deformation

The regional distribution and amplitude of the post-Aptian subsidence is clearly not consistent with the minor amounts of Neocomian–Aptian brittle crustal extension interpreted from seismic sections across the West African margin. Even more problematic, the regional subsidence seaward of the Atlantic hinge is not associated with any obvious rift structures at all (Figs 4 & 5). However, the form, distribution and longevity of the post-rift subsidence are characteristic of the cooling and contraction of extended lithospheric mantle. The magnitude of extension needed to generate the regional post-Aptian subsidence precludes it from being the thermal-subsidence phase engendered by the minor and laterally restrictive brittle extension that occurred during the Berriasian, Hauterivian and Early Barremian. We will now explore various extensional mechanisms in order to ascertain how and why extending lithosphere may sometimes be associated with brittle deformation and rift-induced subsidence while, at other times, only a gentle sag basin is produced.

Syn-rift and post-rift stratigraphy provides first-order information on the amplitude, timing and depth-partitioning of extension. We demonstrate how extension style modifies basin geometry and syn-rift stratigraphy by modelling rifting as a series of discrete pulses with intervening phases of lithospheric cooling and post-rift subsidence. Extension commences by multiple slip across an eastward-dipping border fault (Fig. 7a). The amplitude and wavelength of the footwall uplift is a function of the heave across the fault (5 km), the dip of the fault (30°), and the effective elastic thickness of the lithosphere (T_e, assumed to be a constant value of 20 km in these examples) during rifting. The typical half graben structural and stratigraphic response is generated; rifting results in both the creation and destruction of accommodation. The progressive collapse and rotation of the hangingwall block results in the generation of a series of onlap surfaces and a sediment wedge thickening toward the border fault. In addition, the flexural uplift of the rift flank represents a relative sea-level fall allowing erosional reworking of pre-rift sediments and/or basement.

In contrast, the basin geometry and syn-rift stratigraphy resulting from extension partitioning at mid-crustal levels (i.e. depth-dependent extension) leads to completely different stratal geometries; lower crustal extension thins the crust and isostatically induces regional sagging and an onlapping syn-rift stratigraphy (Fig. 7b). In this case, the extension of the lower plate has a maximum thinning factor of 1.4. It is crucial to note that, despite the extreme differences in basin geometry and stratigraphy produced by brittle and ductile extension, both are equivalent representations of the rifting process, except that one involves ductile deformation of the lower crust and the other brittle deformation of the upper crust (Fig. 7). The standard definition of rifting, i.e. the active period of normal faulting, needs to be broadened to include possible ductile deformation of the lower crust. Extension is thus more appropriately viewed as the process during which the crust is being thinned, albeit by upper (brittle) and/or lower crustal (ductile) thinning.

Any extension within the crust needs to be balanced by an equal amount of extension within the remainder of the lithosphere, although the lateral distributions of upper crustal, lower crustal and lithospheric mantle extension need not be the same (e.g. Kusznir *et al.* 1987; Weissel & Karner 1989). However, in order to have permanent subsidence

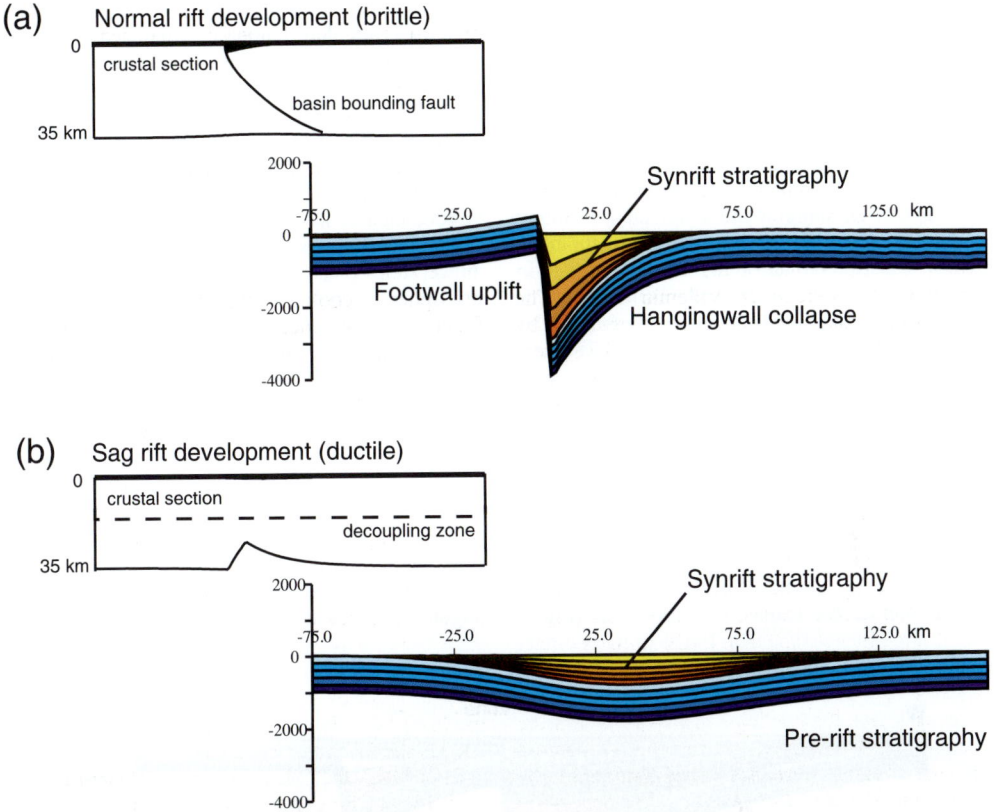

Fig. 7. Modelled stratigraphic and structural response to lithospheric extension. (**a**) Normal rift-basin development when lithospheric extension involves both crustal (brittle) and lithospheric mantle extension. Characteristic features include hanging wall collapse, footwall uplift and an onlapping syn-rift stratigraphy. Assumed effective elastic thickness, 20 km. (**b**) Sag-basin development engendered by a mid-crustal detachment that allows lower crustal and lithospheric mantle deformation to occur in the absence of upper crustal extension. Lower crustal extension thins the lower crust and isostatically induces regional sagging and an onlapping syn-rift stratigraphy. Because only the lower crust is involved in this region, there is a complete lack of brittle deformation structures. Strain-balancing requires significant upper crustal deformation in an adjacent region. Assumed effective elastic thickness, 10 km. The age-range of each model is the same: 20 million years. Both modelled processes are equivalent representations of the rifting process except that one involves ductile deformation of the lower crust and the other brittle deformation of the upper crust.

in an extensional tectonic setting, the crust must be thinned. Since the pre-salt sag basins are not associated with normal faulting, the upper crust cannot be involved in the same region as the sag basins, i.e. the upper crust immediately above the deforming lower plate cannot be involved in the extension process requiring the upper crust to be deformed in an adjacent region. Furthermore, the negative exponential form of the post-salt subsidence requires that the lithosphere mantle also needs to be involved in the extension process, primarily because it is the cooling of the thinned lithosphere mantle that generates post-rift subsidence. The exact form and location of the counterbalancing upper crustal extension is not known, but it needs

to be laterally displaced and presumably exists in the vicinity of the ocean/continent boundary. It is possible that the upper crustal counterbalance may be similar to the extreme extension of the Iberian margin, where the extensional balance through the crust occurs by a combination of thinned and 'rafted' crustal blocks that exposes the continental mantle (e.g. Müntener & Hermann 2001; Manatschal *et al.* 2001; Whitmarsh *et al.* 2001). We use the term 'rafted' in the same sense as the displaced and translated sediment blocks that are disrupted and translated by salt tectonics (e.g. Lundin 1992). The deformed upper continental crust adjacent to either the exposed continental mantle or ocean/continent boundary is likely to be highly intruded and overprinted by volcanism associated with rift-induced decompression melting.

Figure 7 introduces the main concepts that determine the uplift and subsidence of extended continental lithosphere. To summarize, in order to have permanent subsidence, the continental crust needs to be thinned. Extension of the upper crust engenders normal faulting, rift basin subsidence and rift-flank uplift. Thinning of the lower crust results in syn-rift sag basin development but then the question of extensional balancing through the lithosphere needs to be addressed. Thinning of the lithospheric mantle advects heat through the lithosphere, produces a competing thermal uplift to crustal subsidence and, after break-up, cooling of the extended lithosphere engenders slow but long-lived post-rift subsidence (in essence, the development of a post-rift sag basin). We explore these mechanical interactions by modelling the general development of the Neocomian (Djeno and Erva formations) and Barremian (Marne Noires Formation) syn-rift sequences on the West African margin (Fig. 8) in terms of brittle (Fig. 2, rift onsets 1 & 2) and ductile extension components (Fig. 2,

rift onset 3a). For rift phases 1 and 2, we assume that the brittle deformation is dominated by a number of major basin-bounding faults that are responsible for both Djeno and Erva accommodation space being generated (Fig. 8a). Finite rates of rifting result in a series of syn-rift packages. Each syn-rift sediment wedge progressively onlaps the basement towards the west. The extension across each border fault is also responsible for generating the Eastern (rift phase 1) and Atlantic (rift phase 2) hinge zones (Fig. 8). For the third phase of rifting, we have incorporated a general ramp–flat–ramp intracrustal detachment across the West African margin where the western ramp breaches the surface of the crust at the location of what will be the ocean/continent boundary. The second ramp dips down at the locality of the Atlantic hinge and eventually merges with the Moho in the vicinity of the Eastern hinge. The 'flat' component of the detachment is thus located under the region that will be dominated by syn-rift sagging. The modelled Marnes Noires, both between the hinges and seaward of the Atlantic hinge zones, is shown in Figure 8. Notice that we allow the entire rift space

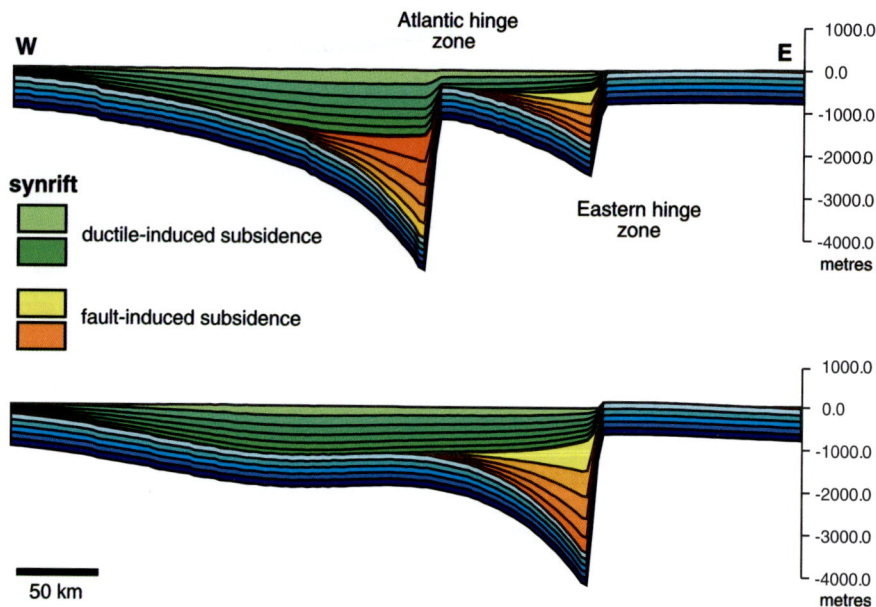

Fig. 8. General kinematic and flexural modelling of the Neocomian (Djeno and Erva formations) and Barremian (Marne Noires Formation) syn-rift sequences on the West African margin in terms of brittle (rift onsets 1 & 2) and ductile extension components (rift onset 3a). Rift phases 1 and 2 generate the Eastern (rift phase 1) and Atlantic (rift phase 2) hinge zones (warm-coloured syn-rift stratigraphy). Third rift phase (green, syn-rift stratigraphy) is controlled by a general ramp–flat–ramp intracrustal detachment across the margin. The 'flat' component of the detachment is located under the region that will be dominated by syn-rift sagging. The modelling is not attempting to match stratigraphic detail but indicates the structural form and sequence geometries that are produced when rifting style changes in time from brittle to ductile. The accommodation generated by lower plate extension is distributed regionally and centred west of the Atlantic hinge zone. Sediment loading in the sag basin induces additional subsidence landward of the Atlantic hinge zone, accentuating the non-faulted and sag nature of the Marnes Noires Formation.

to be infilled, contrary to well-data information for the Marnes Noires (Karner *et al.* 1997). The modelling shown in Figure 8 is not attempting to match stratigraphic detail but indicates the structural form and sequence geometries that are produced when rifting style changes in time from brittle to ductile. The accommodation generated by lower plate extension is, as we would expect, distributed regionally and centred west of the Atlantic hinge zone (Fig. 8). The sediment loading in this region induces additional subsidence landward of the Atlantic hinge zone (Fig. 8), accentuating the non-faulted and sag nature of the Marnes Noires Formation.

While the general basin modelling presented in Figures 7 and 8 shows clearly how a syn-rift sag may form, it is important also to forward-model the change in depositional environment across the West African margin consistent with the chrono-stratigraphy and mapped paleo-water depths of the Marnes Noires, Argilles Vertes, Tchibota, Chela and Loeme formations. The pertinent stratigraphic Aptian sequencing is: (1) development of the pre-

Chela unconformity as lake level was lowered in the Early Aptian, exposing the prograding deltas of the Argilles Vertes Formation; (2) the regional development of the Chela unconformity and transgressive lag deposits of the Chela Formation in the Mid-Aptian; (3) the development of regionally extensive, shallow-water, restricted marine conditions across the entire margin between West Africa and Brazil immediately prior to evaporite precipitation; and (4) the development of significant post-rift accommodation (deposition of the Late Cretaceous, Paleogene and Neogene formations) in the same region previously characterized by minor syn-rift faulting, repeated dessication cycles (allowing the precipitation of thick evaporites) and negligible erosional truncation of earlier syn-rift units. Figure 9 shows the predicted change in palaeo-environment from the deposition of the Argilles Vertes Formation, its exposure and erosion between the hinge zones, transgression and deposition of the Tchibota Formation, followed by the tectonic uplift of the deep-water regions seaward of the Atlantic hinge and the minor subsid-

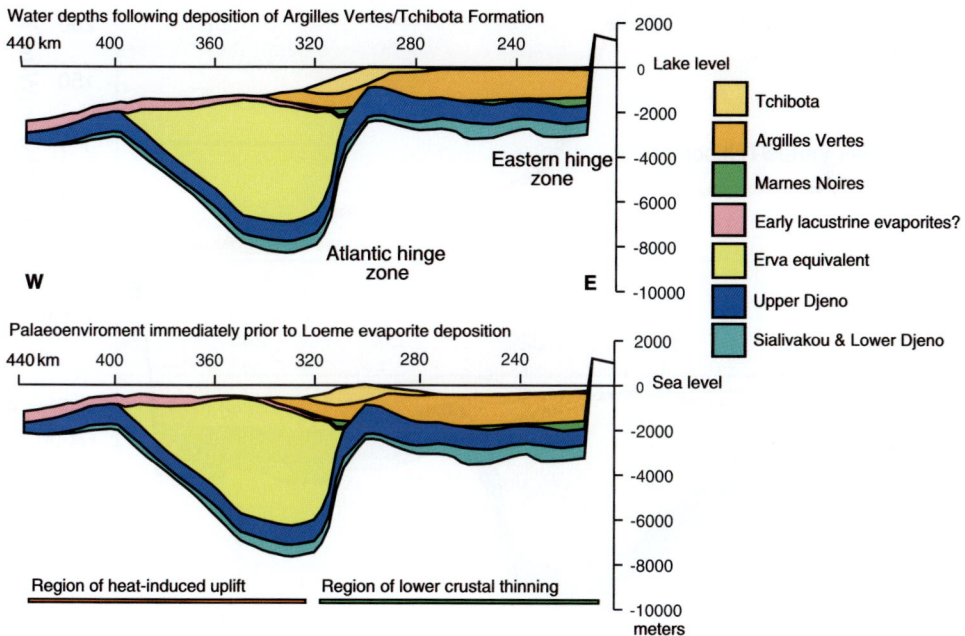

Change in depositional environment during the final rift phase (Barremian-Aptian)

Fig. 9. Kinematic and flexural modelling of the depositional environments across the West African margin that are consistent with the chronostratigraphy of the Marnes Noires, Argilles Vertes, Tchibota, Chela and Loeme formations. The objective is to generate restrictive but shallow marine conditions across the entire West African margin. Following deposition of the Tchibota Formation, rift-induced tectonic uplift of the deep-water regions seaward of the Atlantic hinge and minor subsidence developed between the hinge zones driven by extreme extension of the lower crust and lithospheric mantle. This sequence of events and modification of the depositional environments requires that, during the third phase of rifting, extension became progressively partitioned with depth.

ence developed between the hinge zones. The tectonic uplift and regional rotation of the margin seaward and across the Atlantic hinge is driven by advected heat caused by extreme extension of the lower plate. Resulting water depths are predicted to be approximately constant at 400–500 m across the entire profile, which, following repeated desiccation and flooding, results in the 1000–2000 m-thick Loeme evaporites to be precipitated between West Africa and Brazil (Fig. 9).

A direct consequence of the extensional partitioning described in Figure 9 is the increased input of heat accompanying the third rift stage in those areas dominated by ductile extension, the distribution and amplitude of which is governed by the geometry of the mid-crustal weak zone and the distribution and amplitude of lower plate extension.

As was demonstrated for the Exmouth Plateau of northwest Australia by Driscoll and Karner (1999b), the rapid and large input of transient heat flow engendered by ductile extension is of fundamental importance because of its role in source-

rock maturation. However, such an input of heat would generally not be included in source-maturation studies because sag-basin development is generally presumed to be part of the post-rift. Indeed, it is difficult to drive sufficient generation between the West African hinge zones given maximum sediment thicknesses of only 2–4 km plus the cooling effects of diapiric evaporites (e.g. Mello *et al.* 1995). Figure 10 presents the predicted rift-induced heat flow (i.e. the heat flow at the top of the crust; to determine absolute heat flow, the background heat flow needs to be added to these values), which is consistent with the tectonic modification of the depositional environment described in Figure 9 for two time periods: after deposition of the Argilles Vertes/Tchibota formations and immediately prior to the deposition of the Loeme evaporites. Seaward of the Atlantic hinge zone, the maximum heat flow is predicted to be in excess of 200 mW/m^2 whereas between the hinge zones, the heat flow is significantly less and ranges between 20 mW/m^2 and 100 mW/m^2. The thermal effects of ductile extension are obvious. However, even with

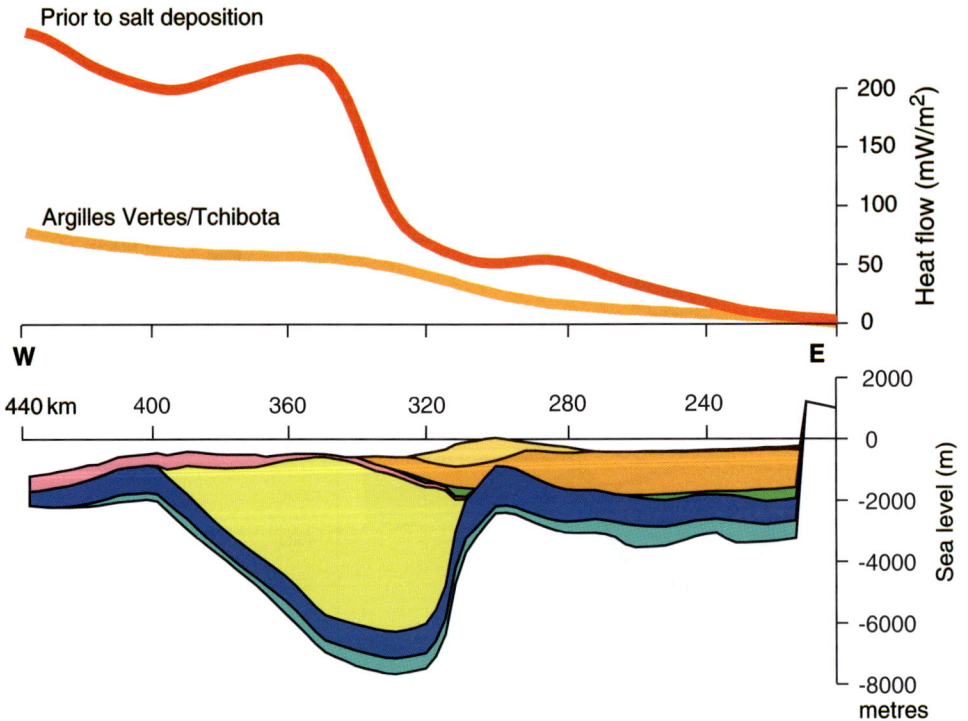

Fig. 10. Predicted heat flow immediately before salt deposition (red) and immediately after Argilles Vertes/Tchibota deposition (light brown) and associated with the rift-induced uplifts, rotations and subsidences shown in Figure 9. Seaward of the Atlantic hinge zone, the maximum heat flow is predicted to be in excess of 200 mW/m^2 whereas, between the hinge zones, the heat flow is significantly less and ranges between 20 mW/m^2 and 100 mW/m^2. The distribution and amplitude of the heat flow is governed by the geometry of the mid-crustal weak zone and the distribution and amplitude of the lower plate extension.

such extreme heat flow, it does not imply that sediment temperatures are necessarily high. Sediment temperature is a function of thermal conductivity and thickness of sediment overburden. For the relatively thin sediment packages overlying the Marnes Noires Formation (and Falcão Formation seaward of the Atlantic hinge zone) during the late rift phase, it is unlikely, even seaward of the Atlantic hinge zone, that source rocks were necessarily rendered overmature at this time. The viability of syn-rift sources and prospectivity of the deep-water West African margin will, to a large degree, depend on the delicate interplay between the cooling of the extended lithosphere and subsequent burial of source rocks as a function of time.

Depositional environment of evaporites: Mediterranean versus West Africa

A common analogue for the South Atlantic Aptian salts has been the drawdown model for the Messinian evaporites of the Mediterranean. 'Drawdown' refers to the kilometre-scale fall in sea level induced by extreme desiccation of the Mediterranean Basin during the Messinian. Evaporites were deposited first at about 6.8 Ma in the highest perched basins in the northern Mediterranean and then, as the sea level continued to fall over about 0.6 million years, other evaporites were produced. Strontium isotope data date the evaporites from the deep Mediterranean at about 5.3–5.5 Ma, suggesting that the whole period of drawdown and evaporation of the Mediterranean basin took place in less than 1.5 million years.

In contrast, the West African evaporites formed after the exposure and reflooding of earlier progradational systems, being deposited only during the late stages of the transgressive system tract. This depositional environment is not consistent with the drawdown model based on the formation of the Mediterranean Messinian evaporites. Despite these differences in depositional environment, many researchers continue to employ the drawdown model to understand the development of the Loeme evaporites. The details of the change in depositional environment from the Argilles Vertes to the Loeme evaporites recorded the vertical tectonics experienced by the West African margin as break-up was approached. Consequently, we will compare and contrast the formation of the Mediterranean and West African evaporites in order to underscore the climate v. tectonic controls on palaeodepositional environment and evaporite precipitation.

Mediterranean evaporites When first proposed, the hypothesis that basin-wide evaporite deposition in the Mediterranean during the Messinian was the result of repeated evaporative drawdown of a pre-existing deep (>1 km) basin was greeted with much scepticism (e.g. Hsü *et al.* 1973). For the model to gain acceptance, it was critical to demonstrate that a deep basin existed prior to, during, and following the Messinian. In addition, it was also necessary to demonstrate that evaporite deposition occurred through evaporation in shallow-water to subaerial conditions. Such evidence came from geophysical and geological data, and geologic studies of now-uplifted basin margins (e.g. Deep Sea Drilling Project [DSDP] Legs 13 and 42A; Ocean Drilling Program [ODP] Legs 107 and 160).

Extension in the western Mediterranean commenced at 32–30 Ma and was primarily controlled by subduction rollback and back-arc basin extension (Rosenbaum *et al.* 2002). The earliest basins began to form in the Late Oligocene in the Gulf of Lion, the Ligurian Sea and Valencia Trough, many going to breakup. In the Early Miocene, back-arc extension propagated to form the Provençal, Algerian and Alboran basins, and in the upper Miocene, extension in the Tyrrhenian Sea commenced (Rosenbaum *et al.* 2002). Thus, the Messinian evaporites represent either late syn-rift or early post-rift stratigraphic units with respect to the western Mediterranean basins. Rapid subsidence allowed the development of significant palaeowater depths (1000–1500 m) by the Messinian (Hsü *et al.* 1973). Support for significant palaeowater depths is consistent with onlapping subsalt sediment wedges at the basin margins. Micropalaeontological and lithofacies data from DSDP/ODP drilling show that the evaporites overlie sediments containing deep-water (>1000 m) benthic fauna. Sediment layers intercalated with the evaporites show bathyal faunal assemblages, but with foraminifera exhibiting dwarf morphologies due to elevated salinities (Hsü *et al.* 1973). The evaporites are overlain by deep-water, open marine hemipelagics, with no evidence for post-evaporite local tectonic-induced subsidence. Over the 1.5 million years development of the Messinian evaporites, accommodation generated by thermal subsidence was minor. Observed reflectors of the pre- and post-salt stratigraphy are parallel or near parallel, in spite of the intervening unconformity, providing further evidence for minimal local tectonic deformation in the region during the Messinian.

Abundant evidence suggests that the evaporites were formed by evaporation in shallow-water environments as opposed to being precipitated from deep-water brines. For example, anhydrite, a highly soluble salt unlikely to precipitate from cold, deep water, is abundant (Schreiber 1988). Strontium isotopic ratios show periods of increased continental runoff contribution. The distribution of bromine records cycles of halite precipitated from sea water and exposure of halites as well as deposition of secondary halite. Most importantly, the presence of shallow-water diatoms and algal stromatolites is definitive (Schreiber 1988) of the shallow-water setting. Volumetric calculations and deposition of recycled halite require multiple cycles of marine inundation and evaporation.

The question, however, remains: is the shallow-water condition and evaporite deposition a consequence of drawdown or regional uplift followed by subsidence (i.e., tectonics)? Abyssal palaeowater depths existing prior to, and intermittently during, the Messinian salt deposition preclude eustatic sea-level fluctuations as a mechanism for the large-amplitude changes in sea level. Nevertheless, eustatic cycles may have played a role in controlling multiple marine incursions into the Mediterranean in the sense that Atlantic Ocean spillage across the Straits of Gibraltar sill, which was responsible for isolating the Mediterranean, would have occurred during highstands. In addition, no evidence exists to support rapid, regional uplift prior to Messinian evaporite deposition or the existence of rapid regional subsidence following the Messinian. In the absence of a tectonic mechanism, we expect that a significant drawdown of sea level by isolation of the Mediterranean should expose shelves and slopes of the basin margins, leading to erosion and deep valley incision. During the subsequent flooding of the basin, these valleys were either filled by Pliocene sediments or partially filled and evident today as submarine canyons. Pliocene infilling of Miocene submarine canyons has long been recognized along the margin of southern France (the Pontian Regression; e.g. Hsü *et al.* 1973). Valley incision at the time of drawdown is consistent with the floodplain silts and channel gravels recovered from DSDP site 133 at the base of the western Sardinian continental rise. It would thus seem reasonable to concur with Hsü *et al.* (1973) that the Messinian drawdown hypothesis is consistent with the observations and is independent of tectonics.

West African evaporites The geological and geophysical observations forced Hsü *et al.* (1973) to propose their rather provocative model of drawdown for the Messinian evaporites of the Mediterranean. Such observations are inconsistent with the evaporite development history of the West African margin and thus we question the applicability of the evaporative drawdown model for this margin.

The regressive Argilles Vertes and Dentale formations, which filled the basin to lake level between the hinge zones, were incised during the drawdown of the lake to form the pre-Chela unconformity. Erosion during the formation of the pre-Chela unconformity, coupled with sediment reworking during the ensuing transgression (generating the regionally developed Chela unconformity), resulted in subaerial exposure, minimal erosion and deposition of the transgressive lag deposits of the Chela Formation. Only during the late stages of this transgressive system tract were the highly soluble Loeme evaporites deposited both between the hinges and seaward of the Atlantic hinge zone.

Following the deposition of the evaporites, post-rift subsidence commenced. In turn, the evaporites are overlain by shallow-water platform carbonates of the Sendji and Madiela formations. These sequences are condensed sections and devoid of a clastic component. In summary, if the drawdown model was applicable to the West African margin, evaporite deposition would necessarily be part of the lowstand system tract (as it is for the Messinian crisis in the Mediterranean). Salt distribution would be restricted to the region seaward of the Atlantic hinge zone. In order to deposit salts both between the Atlantic and Eastern hinges and seaward of the Atlantic hinge zone requires regional uplift to generate regionally shallow-water palaeoenvironments. These environments also need to be suitably restrictive for salt precipitation across the region for a finite (albeit, extremely short) period. We interpret this Early Aptian uplift as part of the final phase of extension that led to the break-up between West Africa and Brazil.

Finally, other workers have suggested that the Aptian evaporites were formed as part of the rapid, early post-rift phase of basin subsidence as the region became inundated by sea water across the Walvis Ridge. However, as noted by Dingle (1996) and Bate (1999), using ostracode data, the Walvis Ridge was not breached until Cenomanian–Turonian times. It was only after this time that South African ostracodes were freely able to migrate northward into the developing South Atlantic. It is thus not possible that the Loeme evaporites are simply the repeated flooding and desiccation of South Atlantic Ocean sea water across the Walvis Ridge sill.

Conclusions

Based on ostracode and seismic reflection data from the West African margin, the Berriasian–Mid-Aptian Outer Basin Sediment Wedge (or the pre-salt wedge and the sag basin) is interpreted to be a syn-rift unit. Furthermore, it was only during the late stages of the rifting process that the necessary restrictive environments and space for the various source rocks that characterize the West African margin were created. Accommodation was generated both between the Eastern and Atlantic hinge zones (e.g. Marnes Noires Formation) and seaward of the Atlantic hinge (e.g. source rock units in the Falcão Formation). Post-Marnes Noires and equivalent formations are unfaulted, implying that the brittle deformation had ceased on this margin at this time. However, the age determinations of the Outer Basin Sediment Wedge suggest that crustal extension had not ceased. For example, the transition from the deep-water restricted lake environment of the Marnes Noires and Upper Bucomazi formations to the regionally extensive, shallow-water, restricted marine environment that allowed evaporite precipitation is not diagnostic of post-rift subsidence but of a competition between lower crustal extension and subsidence with lithospheric mantle thinning and uplift.

Understanding the development of the pre-Chela and Chela unconformities across the West African margin (and their equivalents on the Brazilian margin) have proved crucial in understanding the extension partitioning and modification of accommodation space associated with the third and final rift phase that lead to the breakup of Africa and Brazil. The older unconformity is a sequence boundary, termed 'the pre-Chela unconformity', associated with an Early Aptian relative lake-level fall and it was produced by erosional truncation of the exposed Argilles Vertes clinoforms and onlap of the lowstand sequences of the Tchibota Formation onto the Argilles Vertes Formation. The younger Chela unconformity was developed during the reflooding of the exposed margin, allowing deposition of the Chela Formation. The Chela Formation is a transgressive sequence of basal conglomerates and sands grading upward, with increasing marine affinity, into lagoonal facies and the evaporites of the Loeme Formation. The West African evaporites, in marked contrast to evaporites of the Messinian crisis, formed after the exposure and reflooding of earlier progradational systems, being deposited only during the late stages of the transgressive system tract. This depositional environment is not consistent with the drawdown model (i.e. lowstand systems tract) for the formation of the Mediterranean Messinian evaporites. The requirement for regionally extensive, shallow-water, restricted marine conditions across the entire margin between West Africa and Brazil conducive to thick evaporite precipitation is inconsistent with a post-rift tectonic setting. Evaporite deposition during an overall transgression that produces regionally restrictive depositional environments requires that the Loeme evaporites are part of the late syn-rift stratigraphic succession of the West African margin rather than the earliest units formed by post-rift subsidence. Given this result, the Chela unconformity (or Chela break) cannot be the break-up unconformity because it was developed during the later stages of the third rift

phase immediately prior to salt deposition. The break-up unconformity, if it exists, separates the Loeme evaporites from the overlying carbonates.

The change in water depth from the depositional environment of the Argilles Vertes Formation, its exposure and erosion between the hinge zones, transgression and deposition of the Tchibota Formation, followed by the regional shallowing of the deep-water regions seaward of the Atlantic hinge and the minor subsidence developed between the hinge zones, are all manifestations of the regional uplift of the margin seaward and across the Atlantic hinge in response to advected heat caused by extreme extension of the lower crust and lithospheric mantle (i.e., the lower plate). This sequence of events and modification of depositional environments requires that, during the third phase of rifting, extension became progressively partitioned with depth. It is because of this extensional partitioning, with depth and decoupling of the upper brittle crust from the ductile lower crust, that a 'sag-basin morphology' is formed with little or no attendant brittle deformation.

A direct consequence of this extensional partitioning with depth is the increased input of heat accompanying the rift stage in those areas dominated by ductile extension, the distribution and amplitude of which is governed by the geometry of the mid-crustal weak zone and the distribution and amplitude of lower plate extension. Heat-flow calculations seaward of the Atlantic hinge predict extremely high values at the end of the third rift phase immediately prior to break-up (in excess of 200 mW/m^2). Between the hinge zones, the heat flow is significantly less and ranges between 20 mW/m^2 and 100 mW/m^2. Since sediment temperature is a function of thermal conductivity and thickness of sediment overburden, the viability of syn-rift sources and prospectivity of the deep-water West African margin will, to a large degree, depend on the delicate interplay between the cooling of the extended lithosphere and subsequent burial of source rocks as a function of time.

We are grateful to N. Cameron, R. Bate, W. Brumbaugh, M. Norton and an anonymous reviewer for critically reviewing the manuscript. N. Cameron, R. Bate and W. Brumbaugh are also thanked for their help with interpretation of ostracode and Falcão-1 well-drilling data. Discussions with C. Schreiber concerning the facies and palaeoenvironments of evaporite systems are also appreciated. We acknowledge the support of Western Geophysical during the execution of this project. The Wessel and Smith (1995) GMT software was used in the construction of Figure 1 shown in the paper. This work was supported by National Science Foundation grant OCE 99-12007 (GDK) and OCE 98-09612 (NWD) and TEXACO post-doctoral funding for Barker. Lamont-Doherty Earth Observatory publication #6441.

References

BATE, R. H. 1999. Non-marine ostracod assemblages of the Pre-salt rift basins of West Africa and their role in sequence stratigraphy. *In*: CAMERON, N. R., BATE R. H. & CLURE, V. S. (eds) *The Oil and Gas Habitats of the South Atlantic*, Geological Society, London, Special Publications, **153**, 283–292.

BATE, R. H., CAMERON, N. R. & BRANDÃO, M. G. 2001. The Lower Cretaceous (pre-salt) lithostratigraphy of the Kwanza Basin, Angola. *Newsletters on Stratigraphy*, **38**, 117–127.

BOOTE, D. R. D. & KIRK, R. B. 1989. Depositional wedge cycles on evolving plate margin, western and northwestern Australia. *American Association of Petroleum Geologists Bulletin*, **73**, 216–243.

BRACCINI, E., DENISON, C. N., SCHEEVEL, J. R., JERONIMO, P., ORSOLINI, P. & BARLETTA, V. 1997. A revised chronostratigraphic framework for the pre-salt (Lower Cretaceous) in Cabinda, Angola. *Bulletin des Centres de Recherche Exploration-Production Elf-Aquitaine*, **21**, 125–151.

BRICE, S. E., COCHRAN, M. D., PARDO, G. & EDWARDS, A. D. 1982. Tectonics and sedimentation of the South Atlantic rift sequence: Cabinda, Angola. *In*: WATKINS, J. S. & DRAKE, C. L. (eds) *Studies in Continental Margin Geology*. American Association of Petroleum Geologists Memoirs, **34**, 5–18.

DALE, C. T., LOPES, J. R. & ABOLIO, S. 1992. Takula oil field and the greater Takula area, Cabinda, Angola. *In*: HALBOUTY, M. T. (ed.) *Giant Oil and Gas Fields of the Decade 1978–1988*. American Association of Petroleum Geologists Memoirs, **54**, 197–215.

DINGLE, R. V. 1996. Cretaceous Ostracoda of the SE Atlantic and SW Indian Ocean: A stratigraphical review and atlas. *In*: JARDINE, S., DE KLASZ, I. & DEBENAY, J.-P. (eds) *Géologie de l'Atlantique Sud*. Elf Aquitaine Memoire, **16**, 1–11.

DRISCOLL, N. W. & KARNER, G. D. 1998. Lower crustal extension across the northern Carnarvon basin, Australia: Evidence for an eastward dipping detachment. *Journal of Geophysical Research*, **103**, 4975–4992.

DRISCOLL, N. W., HOGG, J. R., CHRISTIE-BLICK, N. & KARNER, G. D. 1995. Extensional tectonics in the Jeanne d'Arc Basin: Implications for the timing of breakup between Grand Banks and Iberia. *In*: SCRUTTON, R. A., STOKER, M. S., SHIMMIELD, G. B. & TUDHOPE, A. W. (eds.) *The Tectonics, Sedimentation, and Palaeoceanography of the North Atlantic Region*. Geological Society, London, Special Publications, **90**, 1–28.

ERSKINE, R. & VAIL, P. R. 1988. Seismic stratigraphy of the Exmouth Plateau. *In*: BALLY, A. W. (ed.) *Atlas of Seismic Stratigraphy*, vol. 2. American Association of Petroleum Geologists Studies in Geology, **27**, 163–173.

EXON, N. F. & VON RAD, U. 1994. The Mesozoic and Cainozoic sequences of the northwest Australian margin, as revealed by ODP core drilling and related studies. *In*: PURCELL, P. G. & PURCELL, R. R. (eds) *The Sedimentary Basins of Western Australia*. Proceedings of the Petroleum Exploration Society of Australia Symposium, Perth, 181–199.

EXON, N. F., VON RAD, U. & VON STACKELBERG, U. 1982. The geological development of the passive margins of the Exmouth Plateau off northwest Australia. *Marine Geology*, **47**, 131–152.

FALVEY, D. A. 1974. The development of continental margins in plate tectonic theory. *Australian Petroleum Exploration Association Journal*, **14**, 95–106./

FEIJÓ, F. J. 1994. Bacias de Sergipe e Alagoas. *Boletim de Geociências Petrobas*, **8**, 149–161.

GROSDIDIER, E., BRACCINI, E., DUPONT, G. & MORON, J.-M. 1996. Non-marine lower Cretaceous biozonation of the Gabon and Congo basins. *In*: JARDINE, S., DE KLASZ, I. & DEBENAY, J.-P. (eds) *Geologie de l'Afrique Sud. Bulletin des Centres de Recherche Exploration – Production Elf-Aquitaine*, **16**, 67–82.

HAQ, B. U., HARDENBOL, J. & VAIL, P. R. 1988. Mesozoic and Cenozoic chronostratigraphy and eustatic cycles. *In*: WILGUS, C .K., HASTINGS, B. S., KENDALL, C. G. STC., POSAMENTIER, H. W., ROSS, C. A. & VAN WAGONER, J. C. (eds) *Sea-Level Changes: An Integrated Approach*. Society of Economic Paleontologists and Mineralogists Special Publications, **42**, 71–108.

HENRY, S. G., BRUMBAUGH, W. & N. CAMERON, N. 1995. Pre-salt source rock development on Brazil's conjugate margin: West African examples. *1st Latin American Geophysical Conference, Rio de Janeiro, August, Extended Abstracts*, 3.

HSÜ, K. J., CITA, M. B. & RYAN, W. B. F. 1973. The origin of the Mediterranean evaporites. *In*: RYAN. W. B. F., HSÜ, K. J. *et al.* 1973. *Initial Reports of the Deep Sea Drilling Project*. US Government Printing Office, Washington, **13**, 1011–1020.

KARNER, G. D. 2000. Rifts of the Campos and Santos basins, southeast Brazil: Distribution and timing. *In*: MELLO, M. & KATZ, B. (eds) *Petroleum Systems of South Atlantic Margins*. American Association of Petroleum Geologists Memoirs, **73**, 301–315.

KARNER, G. D. & DRISCOLL, N. W. 1999a. Tectonic and stratigraphic development of the West African and eastern Brazilian Margins: Insights from quantitative basin modelling. *In*: CAMERON, N. R., BATE, R. H. & CLURE, V. S. (eds) *The Oil & Gas Habitats of the South Atlantic*. Geological Society, London, Special Publications, **153**, 11–40.

KARNER, G. D. & DRISCOLL, N. W. 1999b. Style, timing, and distribution of tectonic deformation across the Exmouth Plateau, northwest Australia, determined from stratal architecture and quantitative basin modelling. *In*: MACNIOCAILL, C. & RYAN, P. D. (eds) *Continental Tectonics*. Geological Society, London, Special Publications, **164**, 271–311.

KARNER, G. D., DRISCOLL, N. W., McGINNIS, J. P., BRUMBAUGH, W. D. & CAMERON, N. 1997. Tectonic significance of syn-rift sedimentary packages across the Gabon-Cabinda continental margin. *Marine & Petroleum Geology*, **14**, 973–1000.

KATZ, B. J. & MELLO, M. R. 2000. Petroleum systems of South Atlantic marginal basins – An overview. *In*: MELLO, M. & KATZ, B. (eds) *Petroleum Systems of South Atlantic Margins*. American Association of Petroleum Geologists Memoirs, **73**, 1–13.

KUSZNIR, N. J., KARNER, G. D. & EGAN, S. 1987. Geometric, thermal and isostatic consequences of detachments in continental lithosphere extension and basin formation. *In*: BEAUMONT, C. & TANKARD, A. J. (eds) *Sedimentary Basins and Basin-Forming Mechanisms*.

Canadian Society of Petroleum Geologists Memoirs, **12**, 185–203.

LOMANDO, A. J. 1996. Exploration for lacustrine carbonate reservoirs: Insights from West Africa. *American Association of Petroleum Geologists Bulletin*, **80**, 1308–1309.

LUNDIN, E. R. 1992. Thin-skinned extensional tectonics on a salt detachment, northern Kwanza Basin, Angola. *Marine & Petroleum Geology*, **9**, 405–412.

McHARGUE, T. R. 1990. Stratigraphic development of proto-South Atlantic rifting in Cabinda, Angola – A petroliferous lake basin. *In*: KATZ, B. J. (ed.) *Lacustrine Basin Exploration. Case Studies and Modern Analogs*. American Association of Petroleum Geologists Memoirs, **50**, 307–326.

MARTON, L. G., TARI, G. C. & LEHMANN, C. T. 2000. Evolution of the Angolan passive margin, West Africa, with emphasis on post-salt structural styles. *In*: MOHRIAK, W. & TALWANI, M. (eds) *Atlantic Rifts and Continental Margins*. American Geophysical Union, Geophysical Monograph Series, **115**, 129–149.

MANATSCHAL, G., FROITZHEIM, N., RUBENACH, M. & TURIN, B. 2001. The role of detachment faulting in the formation of an ocean-continent transition: Insights from the Iberia Abyssal Plain. *In*: WILSON, R. C. L., WHITMARSH, R., TAYLOR, B. & FROITZHEIM, N. (eds) *Non-Volcanic Rifting of Continental Margins: A Comparison of Evidence from Land and Sea*. Geological Society, London, Special Publications, **187**, 405–428.

MELLO, U. T., KARNER, G. D. & ANDERSON, R. N. 1995. Role of salt in restraining the maturation of subsalt source rocks. *Marine & Petroleum Geology*, **12**, 697–716.

MIHUT, D. & MÜLLER, R. D. 1998. Volcanic margin formation and Mesozoic rift propagators in the Cuvier Abyssal Plain off Western Australia. *Journal of Geophysical Research*, **103**, 27.135–27.149.

MÜNTENER, O. & MERMANN, J. 2001. The role of lower crust and continental upper mantle during formation of non-volcanic passive continental margins: Evidence from the Alps.

PASLEY, M. A., WILSON, E. N., ABREU, V. S., BRANDÃO, M. G. P. & TELLES, A. S. 1998. Lower Cretaceous stratigraphy and source rock distribution in pre-salt basins of the South Atlantic: Comparison of Angola and southern Brazil. *In*: MELLO, M. R. & YILMAZ, P. O. (eds) *American Association of Petroleum Geologists International Conference, November 8–11, 1998, Rio de Janeiro, Brazil, Extended Abstracts*, 822–823.

ROSENBAUM, G., LISTER, G. S. & DUBOZ, C. 2002. Reconstruction of the tectonic evolution of the western Mediterranean since the Oligocene. *In*: ROSENBAUM, G. & LISTER, G. S. *Reconstruction of the evolution of the Alpine-Himalayan orogeny. Journal of the Virtual Explorer*, **8**, 107–130.

SANDWELL, D. T. & SMITH, W. H. F. 1992. Global marine gravity from ERS-1, Geosat and Seasat reveals new tectonic fabric. *American Geophysical Union, EOS Transactions*, **73**, 133.

SCHREIBER, B.C. 1988. Subaqueous evaporite deposition. *In*: SCHREIBER, B. C. (ed.) *Evaporites and Hydrocarbons*. Columbia University Press, New York, 182–255.

TAIT, A. M. 1985. A depositional model for the Dupuy Member and the Barrow Group in the Barrow sub-basin, Northwestern Australia. *Australian Petroleum Exploration Association Journal*, **25**, 282–290.

TEISSERENC, P. & VILLEMIN, J. 1990. Sedimentary basin of Gabon – geology and oil systems. *In*: EDWARDS, J. D. & SANTOGROSSI, P. A. (eds) *Divergent/Passive Margin Basins*. American Association of Petroleum Geologists Memoirs, **48**, 117–199.

WEISSEL, J. K. & KARNER, G. D. 1989. Flexural uplift of rift flanks due to mechanical unloading of the lithosphere during extension. *Journal of Geophysical Research*, **94**, 13.919–13.950.

WESSEL, P. & SMITH, W. H. F. 1995. New version of the generic mapping tools released. *American Geophysical Union, EOS Transactions*, **76**, 329.

WHITMARSH, R.B., MINSHULL, T. A., RUSSELL, S. M., DEAN, S. M., LOUDEN, K. E. & CHIAN, D. 2001. The role of syn-rift magmatism in the rift-to-drift evolution of the west Iberia continental margin: Geophysical observations. *In*: WILSON, R. C. L., WHITMARSH, R., TAYLOR, B. & FROITZHEIM, N. (eds) *Non-Volcanic Rifting of Continental Margins: A Comparison of Evidence from Land and Sea*. Geological Society, London, Special Publications, **187**, 107–124.

Play fairways of the Gulf of Guinea transform margin

D. S. MACGREGOR[1,2], J. ROBINSON[1,3] & G. SPEAR[1]

[1]PGS Reservoir Consultants, Thames House, 17 Marlow Road, Maidenhead, SL6 7AA, UK

[2]Present address: Sasol International, 93-95 Wigmore Street, London, W1U 1HJ, UK
(e-mail: duncan.macgregor@sasol.com)

[3]Present address: Brovig-RDS, Peregrine Rd, Westhill Business Park, Aberdeen, UK

Abstract: The margin between Côte d'Ivoire and the Niger Delta is a region with a common structural history, this being reflected in similarities in the stratigraphic response and play fairways identified across the region. There has been significant exploration on the narrow shelf characterizing the margin, resulting in a series of modest oil and gas discoveries. It is shown in this paper that many of the aspects of the plays in the unexplored deep-water regions of the margin are considerably more favourable to the development of giant fields than those on the shelf. This play-fairway review is based on the integration of existing publications with focused studies of multiclient 3-D seismic data over a number of areas.

Play fairways are classified by seismic sequence and trap type, with an analysis of each undertaken. The most attractive deep-water play types are: (1) anticlinal traps involving late syn-transform (Apto-Albian) and early post-transform (Late Cretaceous) reservoirs, (2) combination traps involving ponded turbidites on the shoreward flanks of these highs, and (3) stratigraphic traps associated with large Late Cretaceous submarine fan complexes. The anticlinal play is associated with the terminations of the St Paul and Romanche fracture zones, with the more recent structuring generally associated with the latter. 3-D imaging and amplitude mapping is critical to prospect delineation, particularly for the combination and stratigraphic plays. Active kitchens are evidenced involving Early and Late Cretaceous source rocks in the Côte d'Ivoire and western Ghana to Nigeria segments of the region, which are consequently upgraded. Considerable volumetric potential is indicated that promises to make the region one of significant new exploration activity in coming years.

Exploration success in West Africa has to date been concentrated along the Angola–Nigeria segment of the margin, with only a few scattered fields found elsewhere. Most recent discoveries have been in deep water, in turbidite reservoirs. Outside the Angola–Nigeria area, there has been relatively little recent exploration, mostly concentrated in shallow water. This paper thus evaluates the potential for extending the trend of deep-water oil occurrences north and west of the area of currently defined success into the region of the Gulf of Guinea transform margin.

This is undertaken by collating a regional play-fairway study between Côte d'Ivoire and western Nigeria, covering both the shallow and deep waters of this narrow margin (Fig. 1). Work is undertaken at two scales: the regional play fairway level, based largely on previous publications, and the sub-basin to prospect scale, as evaluated on a series of PGS 3-D seismic surveys in deep-water areas along the

margin, including Benin, Togo and Côte d'Ivoire (Fig. 1). In areas not covered by 3-D, our evaluation is based on a series of industry papers and publications, which are concentrated in the Ivorian Basin and the Tano Basin of Ghana (e.g. St John 2000; Morrison et al. 1999, 2000). The greatest uncertainty in this analysis lies in the correlation of the seismic megasequences identified because little 2-D seismic data and few well ties are available to the authors to tie the 3-D surveys and a heavy reliance on unconformity ties from published cross-sections is necessary.

All discoveries to date, and all except a handful of the exploration wells, have been located in water depths less than 500 m. Only scout data is available on the few deep-water wells, the most recent of which is the Ocean Energy East Grand Lahou-1 well, in the CI 105 3-D survey area off Côte d'Ivoire, drilled in 1999 and with a probable Late Cretaceous ponded turbidite objective. The other

From: ARTHUR, T. J., MACGREGOR, D. S. & CAMERON, N. R. (eds) *Petroleum Geology of Africa: New Themes and Developing Technologies.* Geological Society, London, Special Publications, **207**, 131–150. 0305-8719/03/$15
© The Geological Society of London 2003.

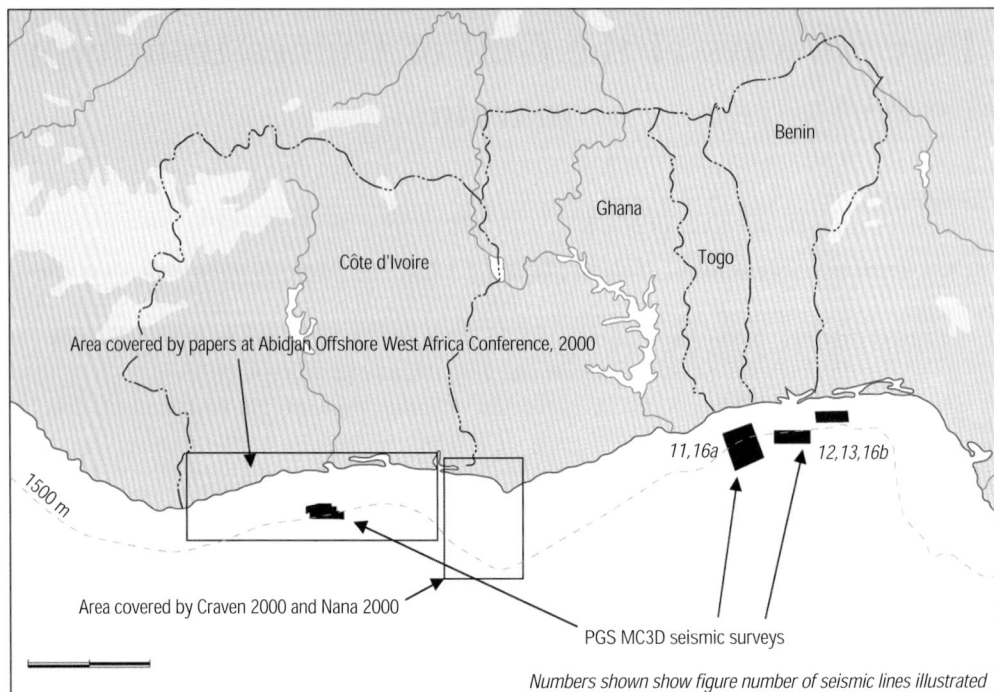

Fig. 1. Location map and database. This paper uses a combination of MC3-D seismic data, predominantly from the east of the area, and published data from African conferences elsewhere.

two deep-water wells are both located in the Cape Three Points area of Ghana, the most recent of which is Hunt Cape Three Points, drilled in 1998, with a probable Late Cretaceous stratigraphic objective. While this chapter was in press, some deepwater wells were drilled in Togo and Benin.

Oil and gas occurrences on the shelf and adjacent onshore extensions of the respective basins are concentrated in Cretaceous reservoirs. These occur along two main trends, namely that through the Ivorian and Tano basins (Fig. 2) and that extending from the Lomé discovery offshore Togo to the Aje Field of western Nigeria. Both areas are associated with seep belts along areas of Late Cretaceous outcrop, particularly in western Nigeria and Benin, where large in-place volumes are reported to occur in a tar belt (Haack *et al.* 2001; B. Frost, pers. comm.; Fig. 3). The tar-belt volumes probably exceed the reserves quoted in any subsurface field along the margin, the largest of which is the Espoir Field of Côte d'Ivoire, with a reported 400 million barrels (MMBO) in place. Mention must also be made of the isolated Saltpond Field of offshore Ghana, which lies in much older (Devonian) reservoirs and on neither of these trends.

Structural development

The structural development of this margin (Figs 4 & 5) is adequately covered in a number of papers (Emery *et al.* 1975; Blarez & Mascle 1988; De Matos 2000) and will only be discussed in condensed form here. The separation of Brazil from the Gulf of Guinea margin took place along transform faults and thus the tectonics observed on the margin are over large areas associated with transtension and transpression. The Gulf of Guinea margin shows two important tectonic differences to the 'passive' South Atlantic margin south of Nigeria, namely the influence of transform tectonics and the absence of salt and therefore of halokinesis.

As detailed by De Matos (2000), three main structural phases are identified, namely pre-transtension (Neocomian–Barremian), syntrans-tension/syntransform (Aptian to latest Albian) and post-transform (Cenomanian–Holocene) The boundaries between these intervals are marked by unconformities on both sides of the Atlantic (Fig. 5). The pre to syn-transtension boundary marks the onset of plate separation along transform faults, and the syn/post-transform boundary the

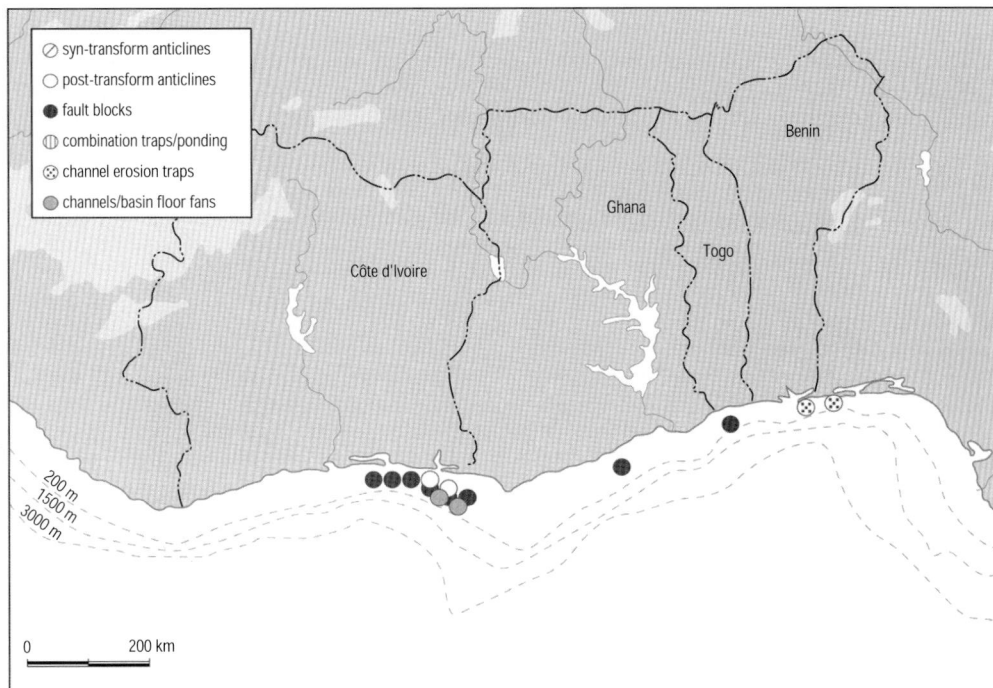

Fig. 2. Distribution of trap types in proven fields on the shelf. The trap-type terminology used is discussed later in this paper.

final separation of continental crust. Our analysis of the 3-D datasets has identified most of these bounding unconformities, with the exception of a clear pick for a pre to syn-transtension unconformity, which, if it exists, must occur at considerable depth.

We have treated these structural phases identified by De Matos as megasequences on which to define the main play fairways (Fig. 5). A description of each follows.

Pre-transform

The underlying Pan-African structural grain across the region runs NNE–SSW, typified by the trend of the Voltaian foreland basin. The Voltain stuctural lineament runs south to Accra and could have influenced the trend of the later Romanche ridge. Sedimentary rocks of Devonian, Carboniferous and Permo-Triassic age are known from the Tano High along to Togo.

Pre-Aptian Cretaceous sediments are known only in the eastern part of the area (Fig. 5), within the Dahomey Embayment (lower section of Ise Formation, Elvsborg & Dalode 1985), in the Keta Basin (Kjemperud *et al.* 1992) and in the Benue Trough. This follows trends on the Brazilian side, where Neocomian sediments are seen only in the

Potiguar Basin (De Matos 2000). These sediments are presumed to have formed in extensional grabens, coeval and similar to those of the South Atlantic, which did not extend into the eastern part of the Gulf of Guinea.

Syn-transform

Oceanic crust was first emplaced in the ?Early Aptian in two areas offshore Côte d'Ivoire and Togo/Benin, bounded by the St Paul and Romanche fracture zones, accompanied by the opening of a number of pull-apart basins (Blarez & Mascle 1988; De Matos 2000). The initial areas of continental v. oceanic crust ('syn-transform') were small, increasing as the two continental plates moved apart. The Aptian saw the onset of sedimentation in the Ivorian Basin and in the Brazilian basins west of the earlier Potiguar rift. This interval is equivalent to that described in other publications as 'syn-rift'; this term is avoided in this paper, as in all the studied 3-D areas and on the Brazilian sides of the Atlantic (De Matos 2000), classic synrift signatures, such as divergent wedging into faults, are rare. However, these features have been reported from areas removed from the transfer zones in Côte d'Ivoire (J. Morrison, pers. comm.

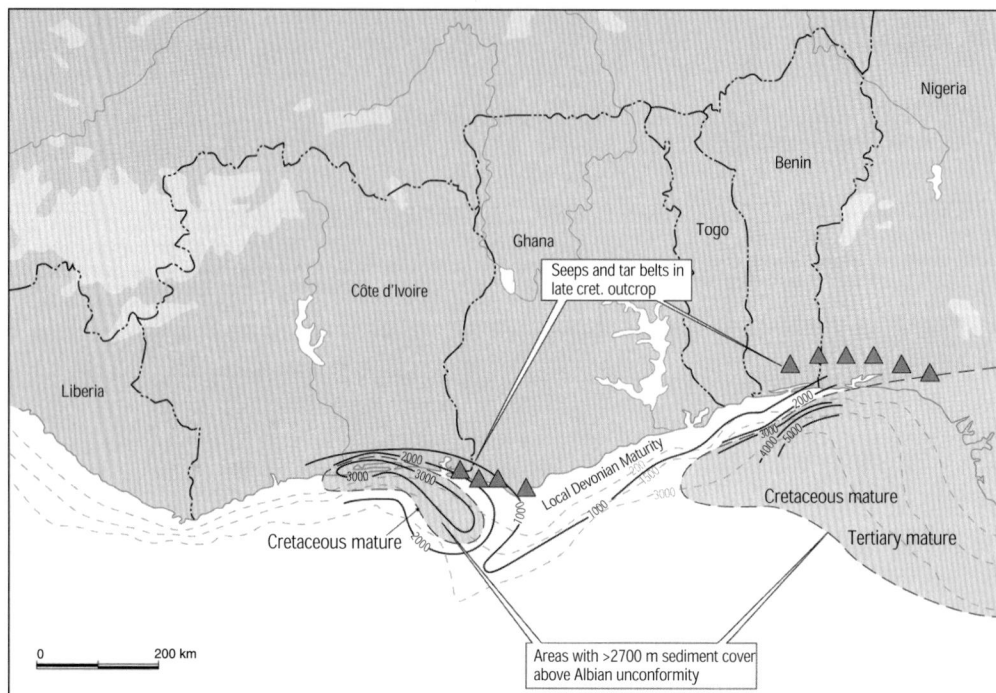

Fig. 3. Distribution of probably hydrocarbon kitchens where source rocks above or below the Late Albian unconformity reach temperatures in excess of 100 °C below overburdens in excess of 2700 m. Timing of generation is believed to be Late Cretaceous onward in the Côte d'Ivoire area and Late Miocene onwards in the eastern kitchen shown.

2001), indicating limited areas of more conventional dip-slip tectonics.

Post-transform

The time of final contact between continental crust of the two plates is documented as Late Albian, being marked in all areas by a major unconformity marking foundering of the current deep-water area and collapse of the shelf margin. In Côte d'Ivoire, a period of transform tectonics seems to accompany the Albian unconformity and leads to the first development of large structures (see below); this seems to have been predominantly a transtensional event (St John 2000). Between the major strike-slip faults, the structural regime may be dominated by oblique extension (J. Morrison, pers. comm. 2001). Following the break-up, marine sediments were deposited on topography that largely mirrored that of present day, with a shelf only tens of kilometres wide and cut by several deep canyons.

The most pronounced event within the post-transform interval is an erosional unconformity and lowstand, presumably of Oligocene age, which appears to affect the whole West African margin (Emery *et al.* 1975; Burke *et al.* 2003). A less significant unconformity and lowstand of Senonian age has been identified in Côte d'Ivoire (Chierchi 1996; Morrison *et al.* 1999; Emery *et al.* 1975) and tentatively also on the Togo and Benin 3-D datasets, though no direct correlation is possible between these areas due to lack of 2-D seismic data. These two unconformities are used for the purposes of this paper to subdivide the post-transform section into early, middle and late phases. Together with the Late Albian unconformity, the three unconformities appear to represent three major lowstands over which submarine fan development was concentrated.

Over most of the margin, the post-transform section is marked by typical passive-margin downwarping. Gravity sliding seems to be a feature of the middle post-transform section in the Togo–Benin area and is also reported from portions of the Brazilian margin. Structuring is locally developed in two areas close to the terminations of the St Paul and Romanche fracture zones (St John 2000; Fig. 4). The St Paul association manifests itself as a number of anticlines in the western Ivorian basin, with uplift that occurred from Albian through to Maastrichtian (St John 2000), as illustrated on the Côte d'Ivoire 3-D dataset (Fig. 6). The Romanche association, as seen on the Togo

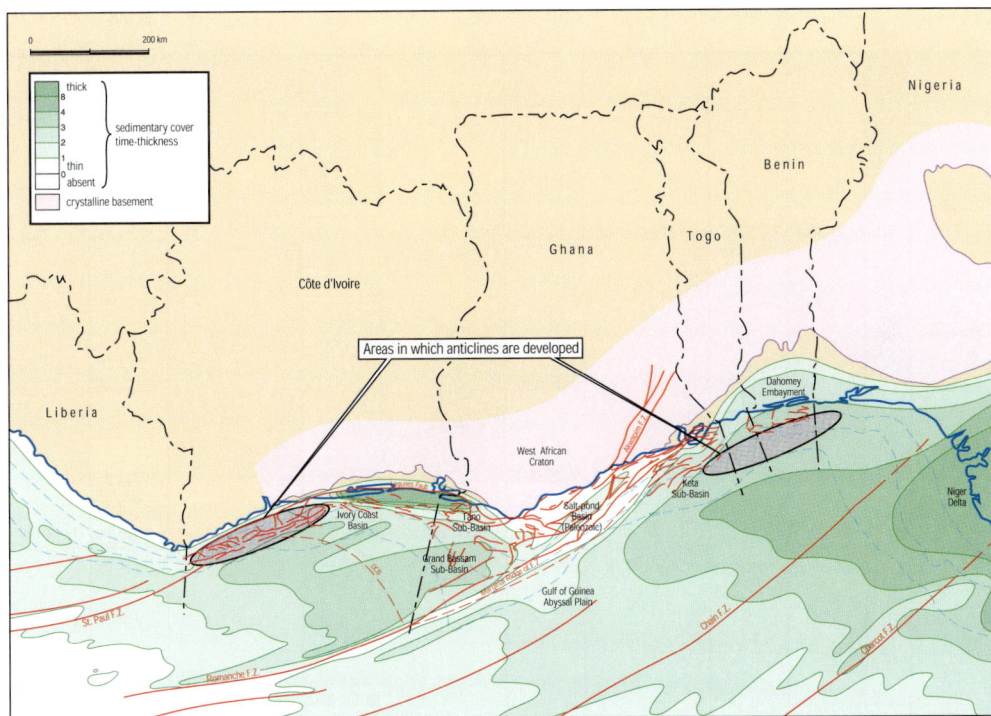

Fig. 4. Tectonic elements illustrated on a sediment cover isopach above Basement. The two main fracture zones, and areas of transpressional structuring associated with these, are shown. Note the major depocentre related to the Niger pro-delta tongue, which has a significant impact on hydrocarbon generation.

and Benin 3-D datasets, seems to comprise multi-phase and generally younger (?Santonian–Holocene) uplifts over earlier half-grabens or pull-apart basins (Fig. 7), and may extend from the Keta Basin into the Niger Delta, where transpressional effects have also been reported. This system may still be active, as evidenced by recent earthquakes on and offshore Ghana and offshore Benin (St John 2000).

Reservoirs

Reservoir potential in the deep water is predicted from a combination of extrapolations from shelfal geology, from seismic facies analysis and by sequence stratigraphic principles, particularly the likelihood of fans being established over major lowstands. Clearly, when extrapolations from shelfal and onshore geology are concerned, facies changes are probable into the offshore. Generally speaking, the equivalent reservoirs in the southern deep-water areas will be more marine in their aspect, with a change probable from fluvial into marine environments in the syntransform section, and from shallow marine to deep marine in the post-transform. In both cases, an improvement in

the quality of the reservoirs would be expected due to increased sediment winnowing.

The stratigraphic levels of proven hydrocarbon-bearing reservoirs in shallow-water wells are illustrated in Figure 5. Practically all the main hydrocarbon levels, including seeps and tar sands, occur in clastic reservoirs between the latest syntransform section (Albian) and the middle post-transform (Maastrichtian). The younger syntransform sections have generally very low net/gross, as typified by the low net/gross Barremian oil-bearing section in the Lomé wells in Togo, and in several wells on the eastern Ghana shelf, with deep erosion of the syntransform section evidenced in these areas, (Kjemperud et al. 1992). A major reduction in sand supply in Côte d'Ivoire is apparent in Maastrichtian times (Morrison et al. 1999). The Tertiary section also appears to be shale-prone in most shelfal wells, although there are indications on some 3-D data, which will be presented later, for significant fan development over the Oligocene unconformity and lowstand. The reduction in sand content up the post-transform section is thought to be due to a number of factors. These include the progressive erosion through time of highs established during the Albian

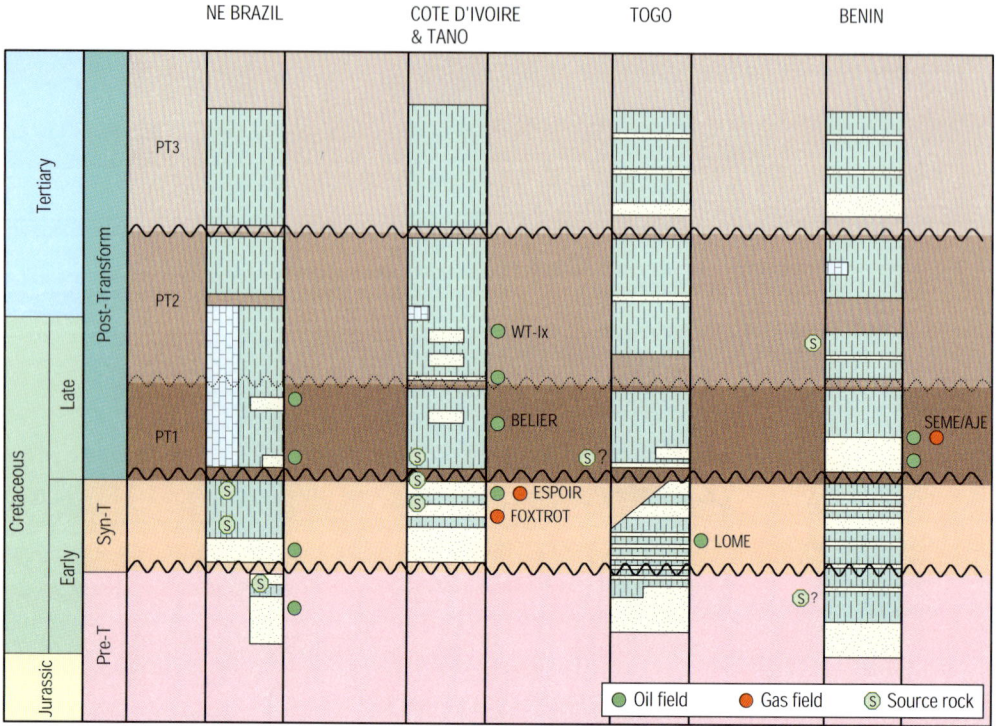

Fig. 5. Chronostratigraphy, reservoirs and source rocks of the Gulf of Guinea margin. The megasequences differentiated in the text are highlighted as are the reservoirs in the main proven fields.

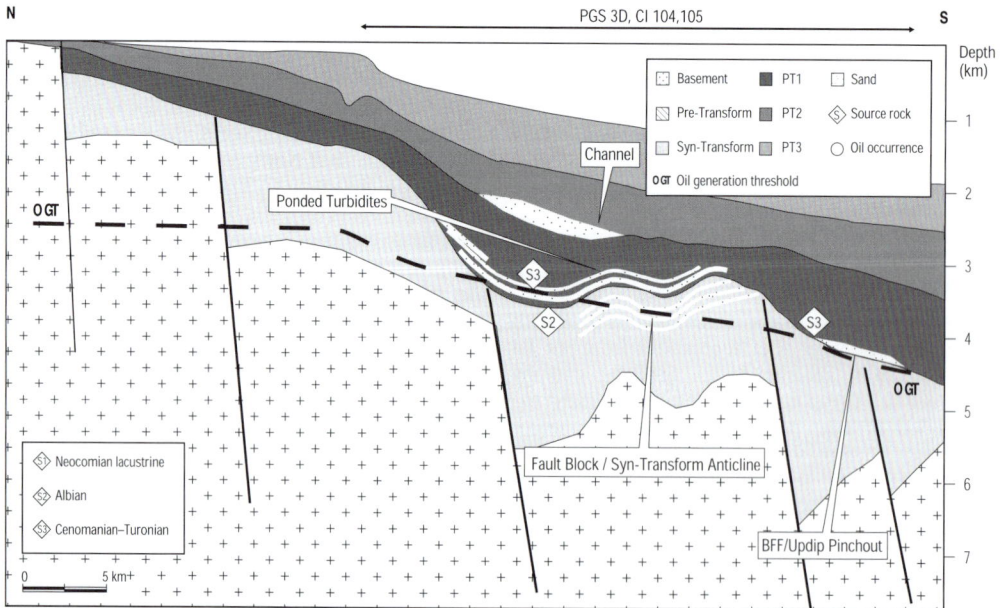

Fig. 6. Schematic cross-section: western Côte d'Ivoire play types. Play-type terminology is presented later in this paper. A critical issue in this region is source-rock maturity.

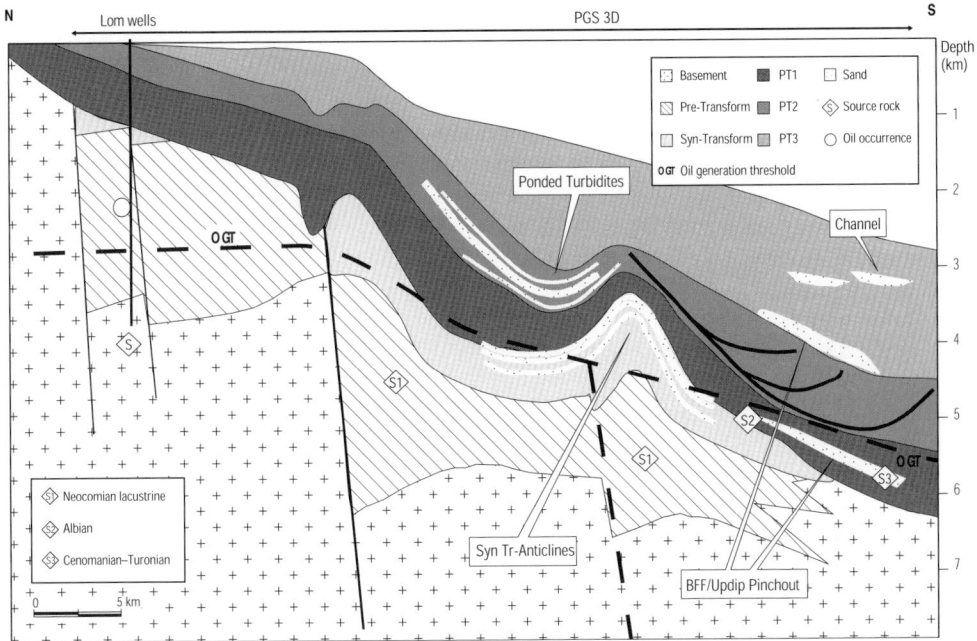

Fig. 7. Schematic cross-section: Togo play types. The dominant feature is a large transpressional anticline, formed by inversion of an older pull-apart basin. The hope for this play is that reservoir potential will improve into the full syntransform section present in such features.

and the probable capture of many of the rivers supplying the margin by the headwaters of the Niger River (see Fig. 8). A major increase in sediment supply via the Niger River commences around the Santonian, which has to be compensated for by a reduction in the supply carried by other rivers.

The two main potential reservoir sections identified are the late post-transform (Albian) sandstones and the early (Cenomanian–Maastrichtian) post-transform marginal marine to turbidite sands. These are typified respectively by the Espoir and Belier fields of Côte d'Ivoire and the latter group also by the Seme Field of Benin, and the Aje Field of Nigeria (Fig. 9).

Albian (late syn-transform/syn-transtension) sandstones

Albian reservoirs are the main hydrocarbon-bearing reservoir in Côte d'Ivoire and it is in this area that most data exists on the interval. They are also known to occur in parts of the Tano and Keta basins of Ghana. The relevant section does not seem to have been penetrated in the Benin or Nigeria offshore wells, but the interval is the equivalent of the upper part of the Ise Formation fluvial sandstones in outcrop to the north. Late syn-rift reservoirs contain reserves in excess of 2000 MMBO

in the general area of central West Africa and its Brazilian conjugate margin (Coward et al. 1999), the interval being a significant producer. particularly in the Sergipe–Alagoas and Reconcavo basins of Brazil. Depositional environments of these reservoirs vary from fluvial, through fluviodeltaic, to lacustrine or marine submarine fan. The marine reservoirs, e.g. the Late Albian sands of Côte d'Ivoire, are of the best petrophysical quality, with porosities up to 25% and permeabilities into the hundreds of mD (millidarcies) though large variations can occur over small areas and many sands are not laterally continuous. Practically all well penetrations are over active fault blocks, where, on seismic data, the section is observed to thin and/or be truncated, and reservoir quality may well improve off-structure. Marine facies seem to be developed only at the top of the syntransform in the available of Côte d'Ivoire penetrations, and erosion on the break-up unconformity, as is frequently observed over the fault blocks that have formed the main drilling objectives, is therefore a major issue. The deeper Mid-Albian sands in Côte d'Ivoire, of fluvial facies, are of considerably poorer quality (Morrison et al. 1999; Craven 2000).

For the Albian to be a viable objective in deep waters, where high flow rates are required to make fields economic, reservoir quality clearly needs to

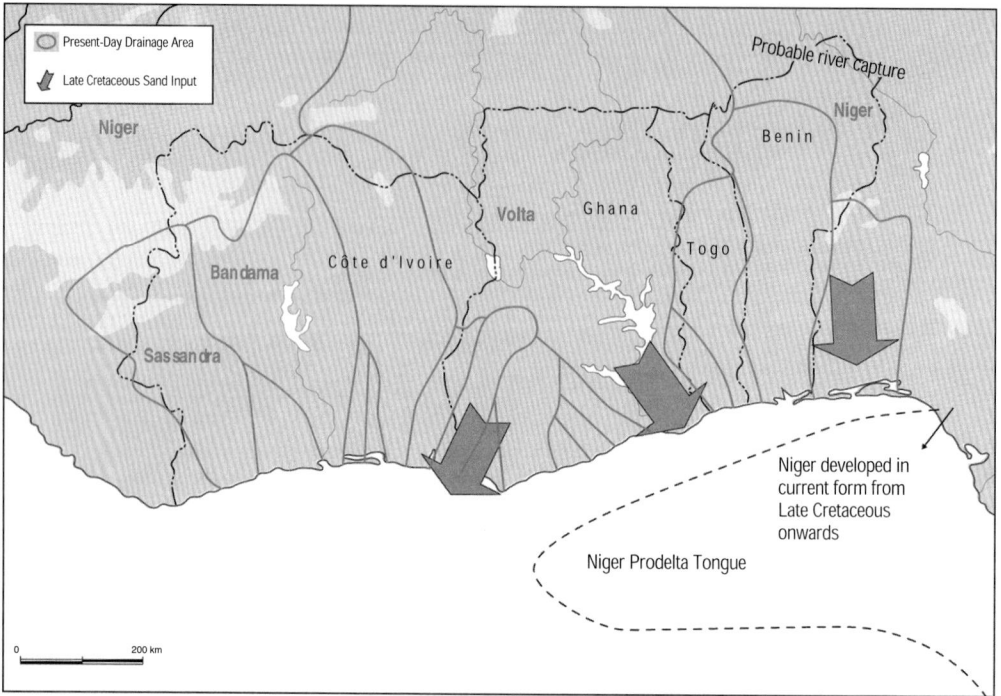

Fig. 8. Main areas of reservoir sand input. A comparison is shown between present-day drainage areas and documented areas of major sediment input in the Cretaceous. It is probable that sediment input was substantially reduced by the capture of the headwaters of the major rivers by the Niger.

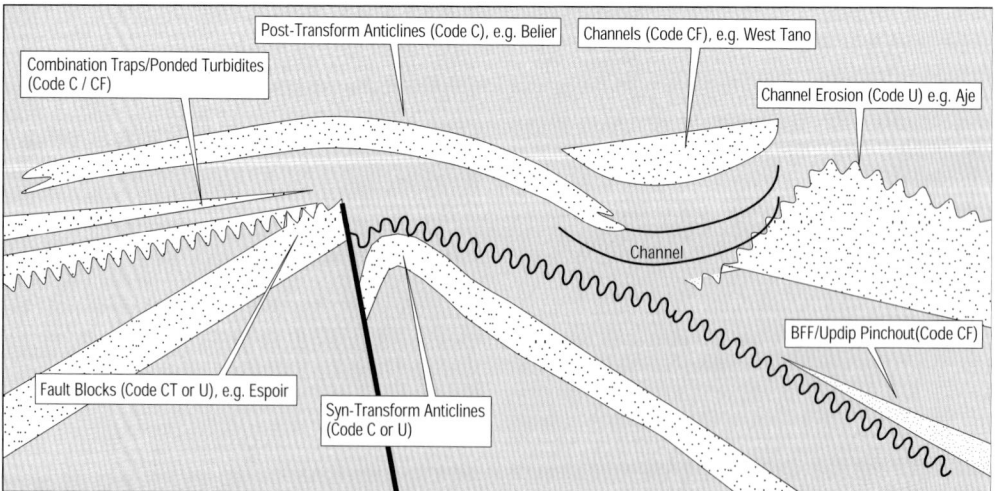

Fig. 9. Schematic classification of play types as applied in this paper, with analogues shown and Milton and Bertram (1992) terminology also applied. Many of the play types in deep water are as yet untested.

improve. It is to be expected that the Albian interval will become more marine in its aspect into the present deep-water area, particularly in the Benin–Togo and Côte d'Ivoire regions, where oceanic conditions will have existed at this time, and this will hopefully lead to an improvement in reservoir quality. Consideration of the seismo-stratigraphic setting in the deep water might allow for the creation of open marine environments, particularly within the deep pull-apart basins formed at this time, which could have formed a focus for turbidite deposition. As these pull-apart basins have often been later inverted to form anticlines, a relationship could conceivably be invoked between structure and areas of improved reservoir thickness and quality. Due to the thickening of the section associated with inverted anticlines and the lesser chances of erosion of the upper sections, such trap types are thought to carry lower reservoir risks than the thinned and frequently eroded sections of the fault block play.

Cenomanian–Maastrichtian (early and middle post-transform) sandstones

By the Late Cretaceous, the steep shelf that has since characterized the margin had developed and thus sedimentary environments should closely parallel that at present day. During the Late Cretaceous, it is thought that a series of north–south-flowing rivers fed the margin, following Dahomeyde (Pan-African) lineaments, as at the present day (Fig. 8). There is no direct manner of reconstructing past drainage patterns, but it is apparent that, prior to capture by the Niger River, these rivers would have drained much larger areas extending much further north. With the exception of the Volta River, drainage areas of each river are of moderate size (Fig. 8) and clearly smaller than the drainage areas which feed the reservoirs of major hydrocarbon provinces such as the Niger and Congo deltas. The recent Ceiba discovery in Equatorial Guinea does now demonstrate that a large delta fed by a large drainage area is not a prerequisite for a significant deep-water hydrocarbon province. The exception, the Volta, flows down the axis of a Precambrian basin, and could therefore be a very long-lived system, leading to reservoir input to deep water at a number of levels.

On the basis of available well and published data, the two areas which appear to have received the greatest supply of sands into shallow water are the Tano area of western Ghana (Tucker 1992; Nana 2000) and the Benin–western Nigeria area (Elvsborg & Dalode 1985). The Tano River of Ghana may well have been larger prior to later capture by tributaries of the Volta (K. Burke, pers. comm.) A major north–south-flowing river is thought to have fed the sands of the Cenomanian–Turonian Abeokuta Formation in Benin. The areas outboard of these sandy deltas would appear the most prospective for the development of turbidite reservoirs. However, as the shelf is generally very steep and narrow along the margin, being cut by numerous canyons, it would certainly have been possible for sands in some areas to feed directly into deep water, without the development of a shallow-water equivalent. As will later be shown, seismic indicators are often the most efficient manner of locating such reservoirs. Evidence for sand input on seismic is particularly well marked on the Côte d'Ivoire CI 105/106 datasets, where large channels are seen of roughly Maastrichtian age, suggesting a further sediment focus in this region. Large fans can be speculated to lie in the deep-water areas south of these blocks.

Petroleum systems

Burke et al. (2003) map out the source-rock fairways of West Africa, highlighting potential within this region specifically within Neocomian source rocks in the east of the region and at Cenomanian–Turonian level along the margin. A proven Neocomian source-rock fairway extends along South Atlantic rifts from Angola to Gabon, with a northern extension proposed as far as Benin, on the basis of palaeotectonic and palaeoclimatic parameters. A Late Albian–Cenomanian–Turonian source-rock fairway, influenced by upwelling and oceanic anoxia, is interpreted to extend the length of West Africa and is the most regionally extensive of any of the major African source-rock trends.

Morrison et al. (2000) document three petroleum systems in the area of greatest well control in the Côte d'Ivoire region, namely:

- Middle Albian terrestrial gas-prone source rocks feeding Albian reservoirs;
- Late Albian marine transgressive oil-prone source rocks feeding Albian reservoirs;
- Cenomanian-Turonian open marine oil-prone source rocks feeding Albian and younger reservoirs.

The quality of these source rocks increases towards the marine environment.

In the eastern part of the study area, at least two further systems are present, namely:

- A small Devonian-sourced system: the Saltpond and Lomé discovery oils are believed, on the basis of stratigraphic association and likely migration paths, to originate from a Devonian (Takoradi Shale) source rock. Recent biomarker analysis of the Saltpond crude has confirmed this (N. Cameron, pers. comm. 2001).

- An Early Cretaceous-sourced system in the eastern part of the region: Haack *et al.* (2001) believe the considerable volume of tar sand in onshore western Nigeria derives from an Early Cretaceous lacustrine source rock. It is also suspected that the deeper oil pools in the Seme and Aje fields derive from this source, as downward migration from the Cenomanian–Turonian source to the Late Albian seems unlikely.

Of these probable petroleum systems, the most extensive and probably the most significant system for deep-water exploration is expected to be the Late Albian and Cenomanian source rocks. These are of wide regional extent and, by analogy to observations in deep-water Angola, are expected to increase in thickness and quality into deep water. Deeper lacustrine source rocks are also indicated to be significant in the east of the region, where the Early Cretaceous section is present within a rifted section. A volumetrically significant deep-water lacustrine system must exist in this area, as it is only in the deep water of the Benin area that sufficient sediment thickness is attained to achieve maturity and to generate the volume of oil seen in the onshore tar sands.

Given the stratigraphic levels of these source rocks (Fig. 5), the best indicator of maturity and the presence of an oily petroleum system can be obtained by maturity mapping at the level of the Late Albian unconformity. Available data from Côte d'Ivoire, Togo and Nigerian wells suggest a uniformly high heat flow along the margin, at around 50 mW/m2 or above, translating into an oil generation threshold ($R_O \approx 0.6\%$) of around 2700 m below sea bed (Morrison 2000; PGS internal analysis of Togo wells). Morrison suggests heat flow may increase into the deep water, as continental crust is thinner, giving shallower generation thresholds. Very shallow observed palaeothresholds on

the Ghanian shelf, which are attributed to late-stage uplift and erosion (Kjemperud *et al.* 1991), are not likely to affect the deep water.

Study of a map on the Late Albian unconformity (Fig. 3) compiled from a number of sources, including some regional maps (Emery *et al.* 1975), regional cross-sections (e.g. De Matos 2000) and our own mapping, reveals two main depocentres and likely kitchen areas. These lie in the Ivorian–Tano Basin areas and along a deep-water trend from the Keta Sub-basin into the Niger Delta. The Keta–Niger depocentre is associated with, and reliant on, a tongue of Neogene sediments thickening toward the Niger Delta and generation is probably active at present day. In the intervening area between the two depocentres, a higher source risk is envisaged, with a reliance on deeper source rocks, such as the Devonian Takoradi shale.

Trap and play types

In this paper, as an aid to the description of play types, a trap-type categorization has been devised (Table 1), which is a modification of that proposed by Milton and Bertram (1992). The categories identified for use in this paper are illustrated schematically in Figure 9, with their characteristics and relationship to the Milton & Bertram classification listed. The trap types proposed cover both the fields drilled to date (Fig. 2) and the deep-water play types discussed later. A spectrum of trap types is proposed, ranging from pure structural (one-seal) traps, including anticlines with conformable and unconformable seals (codes C and U respectively) and fault blocks (code CT or UT), through to stratigraphic (polyseal) traps associated with stratigraphic pinchouts (code CF). To aid further discussions, a differentiation is made between anticlinal traps affecting the syn- and post-transform sections.

Table 1. *Classification of trap types*

Trap type (this paper)	Seismic sequence (reservoir)	Trap type*	Shallow-water field examples	*Critical factors*
Fault blocks	Syn-transform	CT	Espoir (in part), Tano	*Reservoir (erosion and quality)*
Anticlines	Syn-transform	U	None – play not present in shallow water	*Reservoir (quality)*
	Post-transform 1, 2	C	Belier	
Channel erosion	Post-transform 1, 2	U	Aje	*Thief zones in channel*
Ponded turbidites	Post-transform 1, 2	C or CF	La Ceiba, Equatorial Guinea (in part)	*Charge, updip seal for CF traps*
Stratigraphic traps	Post-transform 1, 2	CF	West Tano, Lion (upper pool)	*Updip seal*

*Milton & Bertram (1992) classification

With this classification in mind, the trap types that have proven successful in the shelfal area can be briefly reviewed (Fig. 2). Most working traps proven to date are structural traps, particularly fault-block traps (code CT, e.g. Espoir; Grillot *et al.* 1991), post-transform anticlinal traps associated with inversions (code C, e.g. Belier), or channel-erosion traps bounded by shale-filled channels (code U, e.g. Aje). Some small stratigraphic traps (code CF) are reported from the Cenomanian–Maastrichtian interval in the Ivorian and Tano basins, e.g. a Maastrichtian oil-bearing channel in Dana's West Tano discovery (Craven 2000). This and other examples must rely on updip shale plugging of the channel in order to create a trapping geometry and provide useful analogues for many of the deep-water stratigraphic plays, as will be discussed later. Late Cretaceous channels are also reservoirs where these are draped over Albian highs at the Panthere, Gazelle, Kudu and Ibex fields (J. Morrison, pers. comm. 2001).

Each trap type is illustrated on seismic from the 3-D datasets (Figs 10–16) while examples are also shown on a series of schematic cross-sections along the margin (Figs 6, 7 & 17–19). The distribution of each trap type, both areally and within the geological column, is shown in Figure 20 and Table 2. A discussion of each category follows, commencing with structural traps.

Late syn-transform reservoir/fault-block traps (code CT and UT)

The fault-block play has been the main target of exploration on the shelf, being best represented by the discoveries at Espoir (Grillot *et al.* 1991) and Tano (Fig. 19). In some pools, the Late Albian unconformity is subcropped by the reservoir and provides seal, categorizing the trap type as code UT. The play extends onto the slope into moderate water depths, as represented by the recent Baobab discovery, and is also represented as large highs along the Romanche Fracture Zone, e.g. on the Ghana–Ivory Coast ridge (Fig. 18) and in western Côte d'Ivoire (Fig. 6). The critical factors in all cases are reservoir presence, quality and preservation, particularly the ability to produce hydrocarbons at rates that would be economic in a deep-water setting. Erosion on the outer highs and also on the Ghanian shelf seems to be particularly deep, with sea-bed cores having proven that, in some areas, the Albian crops out on the sea bed.

One of the main areas in which this fairway is identified in deep water is in western Côte d'Ivoire,

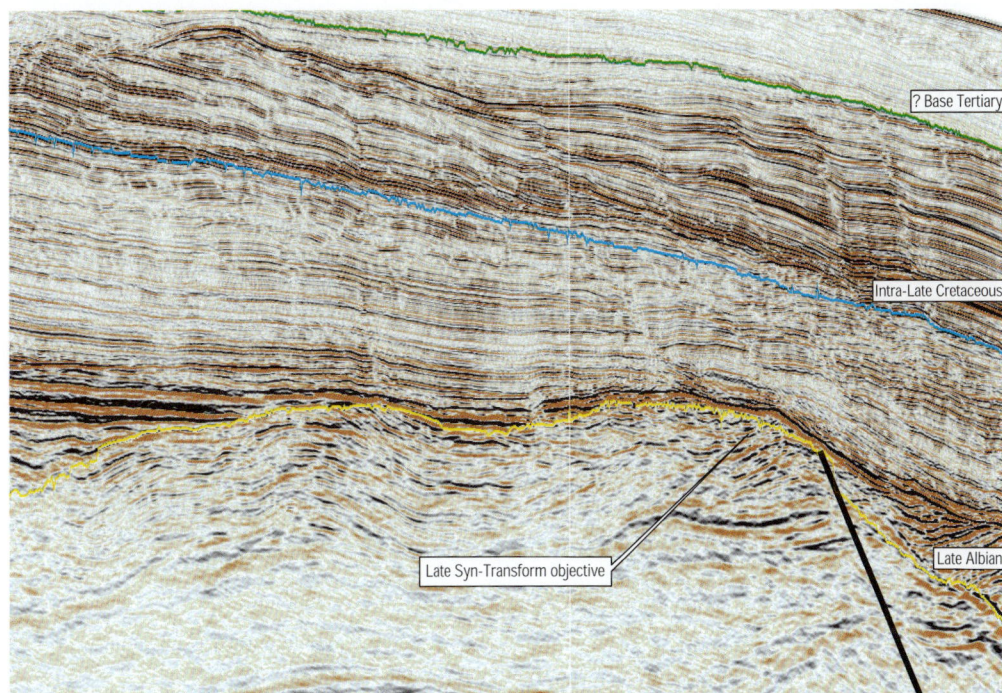

Fig. 10. Seismic example: fault-block trap. Reservoir objectives will be close to the top of the syntransform section and erosion is a concern close to the fault scarp.

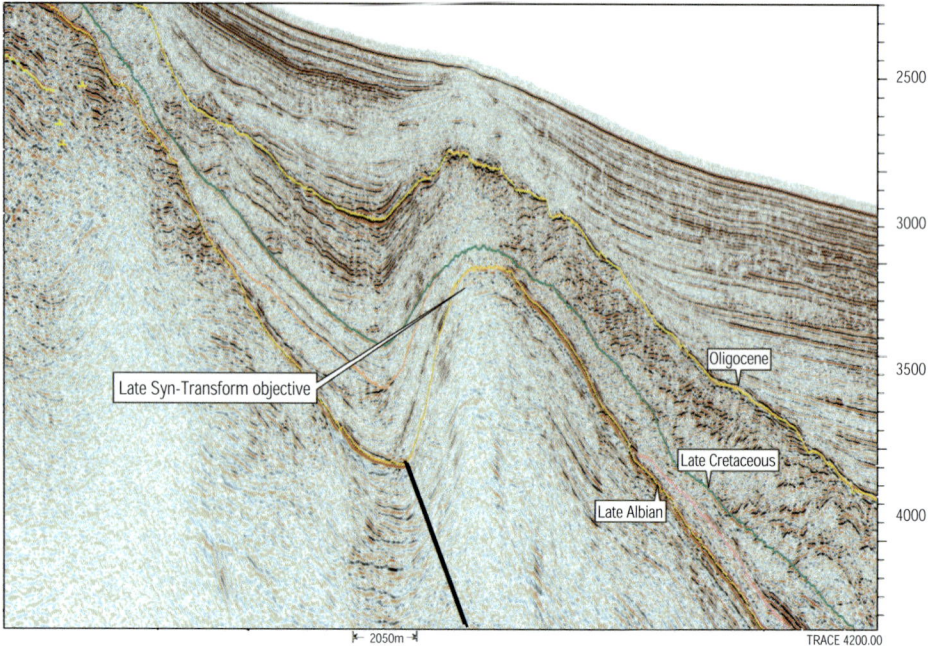

Fig. 11. Seismic example: syntransform anticline. Reservoir objectives will again be in the late syntransform section. No erosion at this level is expected over these structures and the reservoir section may well be thickened.

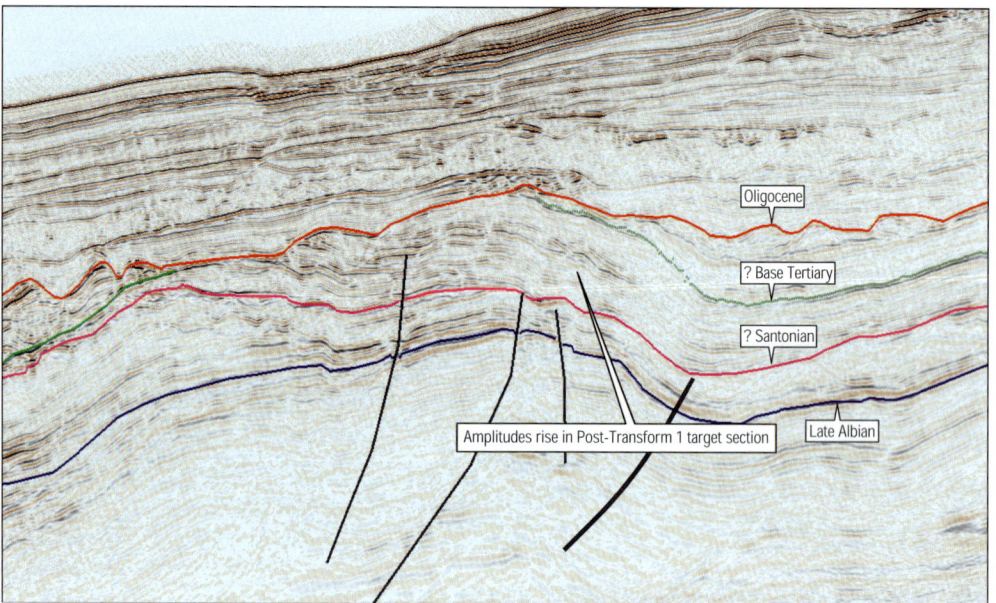

Fig. 12. Seismic example: post-transform anticline. Trap formation here clearly post-dates the deposition of Late Cretaceous reservoirs. Trapping potential higher in the section requires shales overlying the Oligocene unconformity to seal.

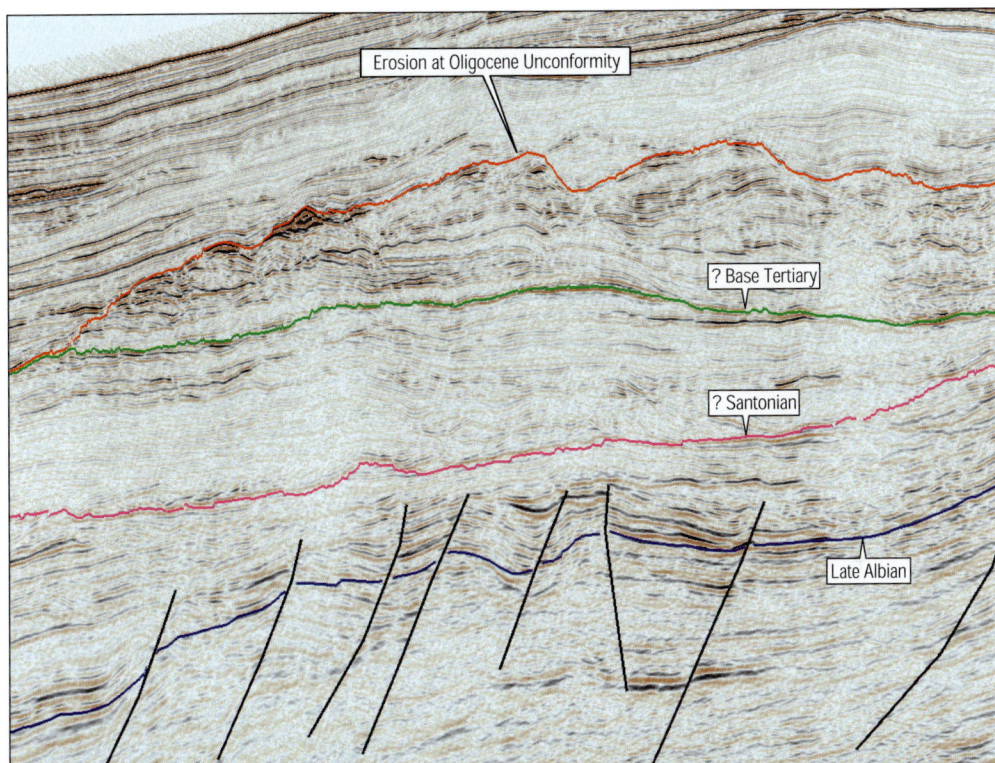

Fig. 13. Seismic example: channel-erosion trap. Closure on a small structure is enhanced by Oligocene erosion. Oligocene–Miocene channel fill is required as seal, with the positive analogue being the Aje Field.

where reservoir quality is poorly controlled due to absence of well penetrations in both shelf and deep water. The highs in this area have frequently been uplifted at a later stage, though the fault-block geometry to the traps is preserved. An example of the play is shown from the PGS 3-D in this area (Fig. 10), while further examples are documented further west by St John (2000).

Late syn-transform reservoir/anticlinal traps (code U)

This play covers the transpressional anticlines described earlier, with a potential reservoir equivalent to that described above for the fault-block play, i.e. the highest quality sands at the top of the syntransform section. The play is identified as being associated with the terminations of oceanic fracture zones in two areas (Fig. 4), centred around Togo, Benin and the Keta Basin of Ghana and at least the western part of the Ivorian Basin. An example is shown from the former area, which is typified by large, relatively young structures (Fig. 11). This feature is interpreted as being formed by inversion from approximately the Santonian onward along an

original NE–SW-trending transtensional fault, bounding an original half-graben or pull-apart basin. This interpretation would suggest that a thick section of syntransform sediments can be expected to be preserved in the original half-graben unconformity and, as no uplift is evident until later in the Cretaceous, a full section including the Late Albian reservoirs should be present. Depositional environment at this point in the basin is expected to be marine, with the half-graben potentially acting as a pond for the accumulation of turbidite sandstones, derived from the north along rivers such as the palaeo-Volta. Charge could either be from Early Cretaceous lacustrine source rocks, formed in the original half-graben/pull-apart geometry, or from Albo–Cenomanian source rocks onlapping the Late Albian unconformity to the south. Other structures close to the termination of the Romanche Fracture Zone are likely to have similar play kinematics.

Early–middle post-transform reservoirs/anticlinal traps (code C)

This play is also associated with the St Paul and Romanche transpressional structures, where struc-

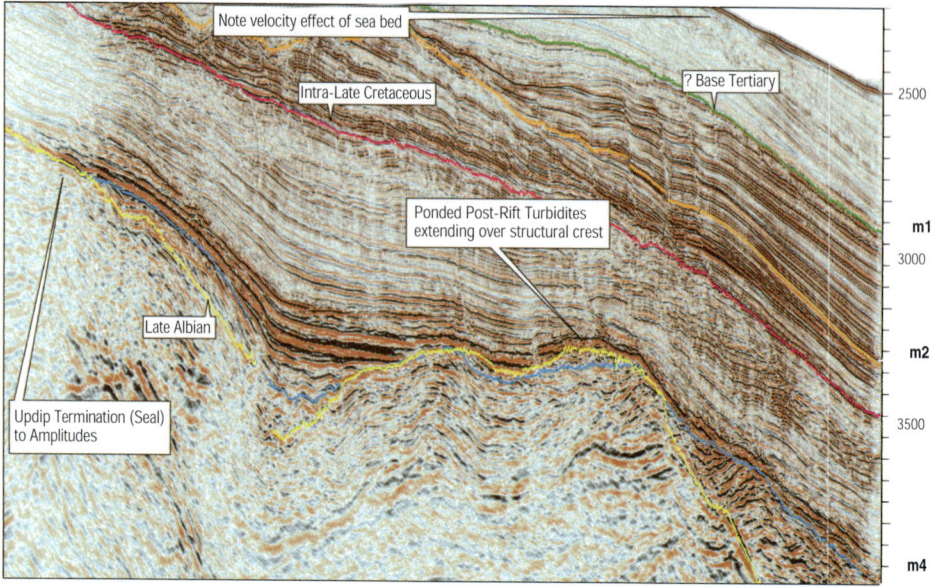

Fig. 14. Seismic example: ponded turbidite trap. Delineation of prospects in this play is heavily reliant on 3-D amplitude mapping. Trapping may be structural, where the sands drape over highs, or stratigraphic, with updip seal as the key risk.

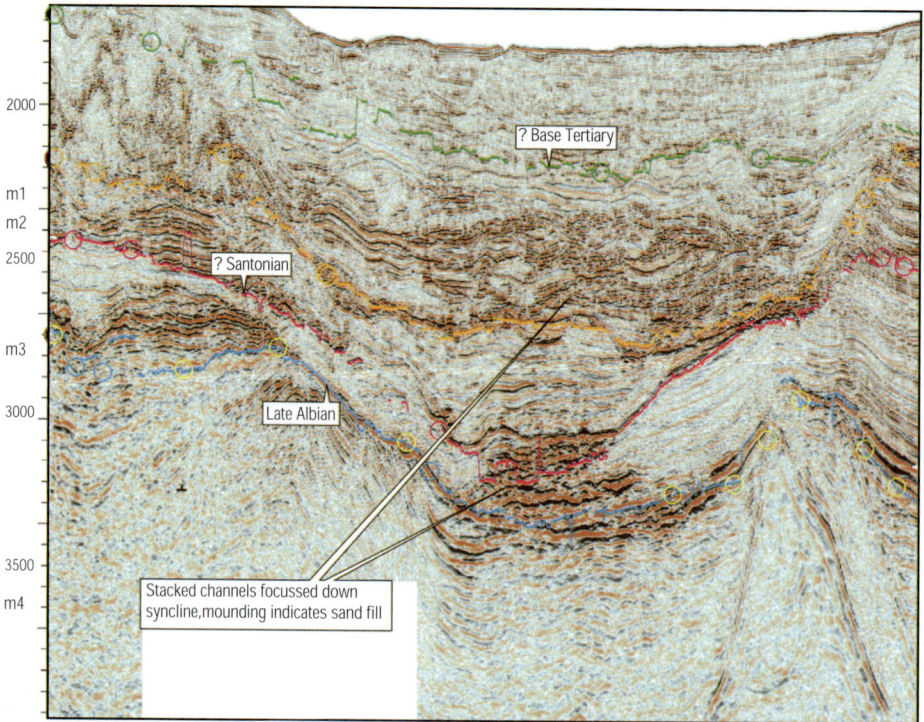

Fig. 15. Seismic example: channel/stratigraphic trap. Discoveries in shelfal wells in this play type bode well for the larger features, such as this seen in the deep water. Amplitude mapping is again key to evaluating this play.

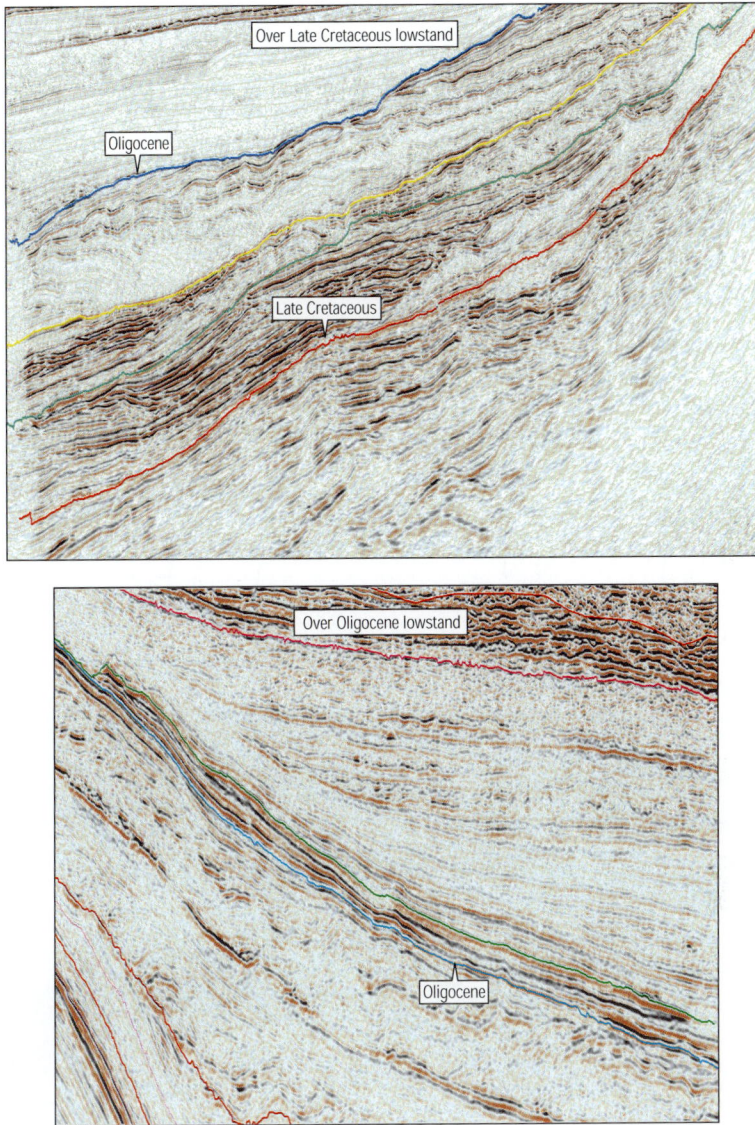

Fig. 16. Seismic examples: basin floor fan traps and stratigraphic traps. These features are found over palaeo-toes or bases of slope over the main sequence-defining unconformities. The steep nature of the shelf increases the likelihood of detached sand bodies and therefore of updip seal.

turing and associated closures extend into the post-transform section. For the St Paul structures, St John (2000) shows that, in some cases, structuring extends as high as the Maastrichtian. A trans-pressional origin can also be proposed for the Lion, Foxtrot, Espoir and Quebec highs, and the trapping geometries for some Late Cretaceous fields, such as Belier (Fig. 19). An alternative model would be an oblique-extension regime with the highs offset by small transfer systems (J. Morrison, pers. comm. 2001)

The Romanche structures are generally younger – over that illustrated in Figure 11, structuring is evident until very recent times. On the example shown on Figure 12, closure is terminated by the downcutting Oligocene unconformity. Closure here at the higher levels would be dependent on the transparent section above the unconformity providing a seal and this would therefore be an example of an unconformity/channel-cut code U trap, similar to the Aje Field on the shelf nearby.

Reservoir quality in the post-transform section is

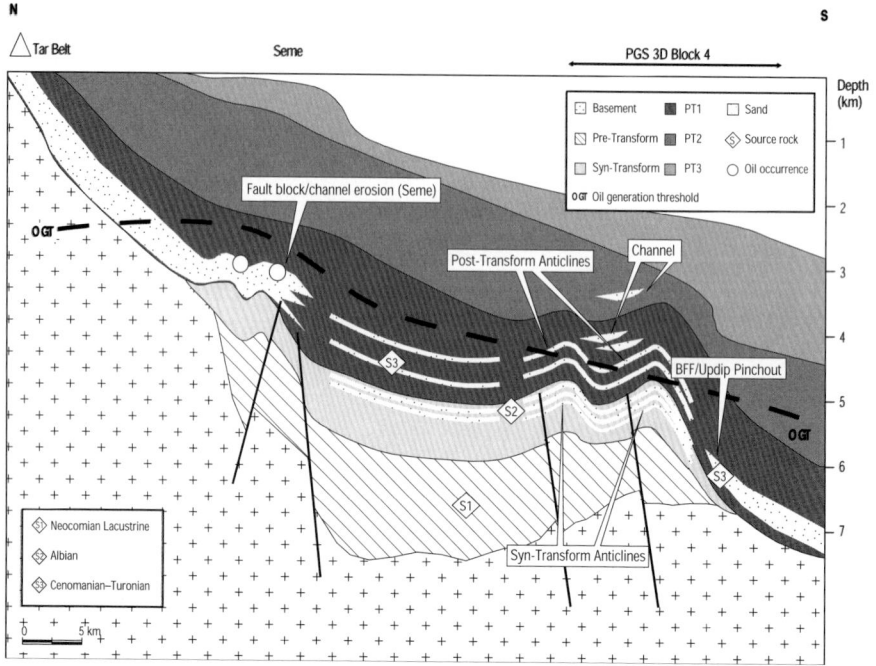

Fig. 17. Schematic cross-section: Benin play types. Key plays in this area are post-transform anticlines and stratigraphic plays. The seep belt in the onshore indicates long-distance lateral migration from an offshore lacustrine kitchen.

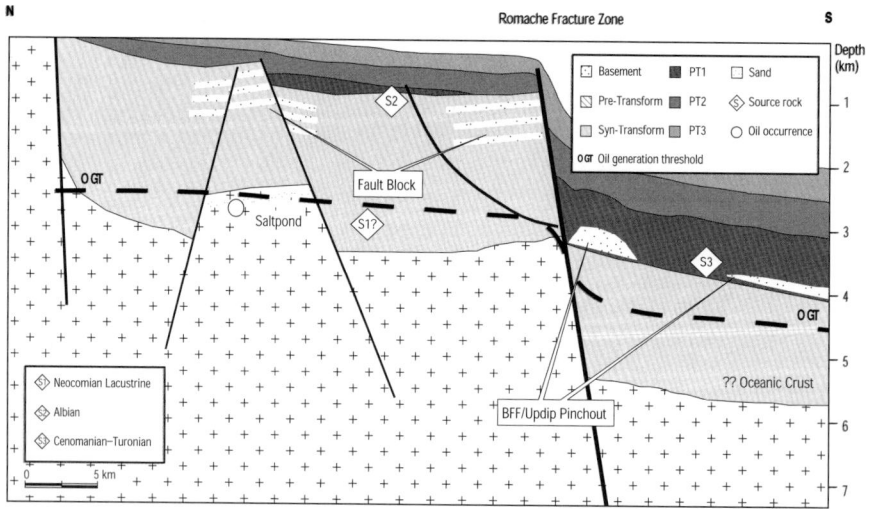

Fig. 18. Schematic cross-section: eastern Ghana play types. Deep-water plays are largely limited to Late Cretaceous basin floor fans.

likely to be substantially improved over the syn-transform, as there are few instances of post-rift/post-transform turbidite sands not having high reservoir quality in West Africa. The critical factor for this play type is reservoir distribution. The potential target sections were deposited in a slope or base-of-slope environment, where sediment supply may have been focused down canyons and

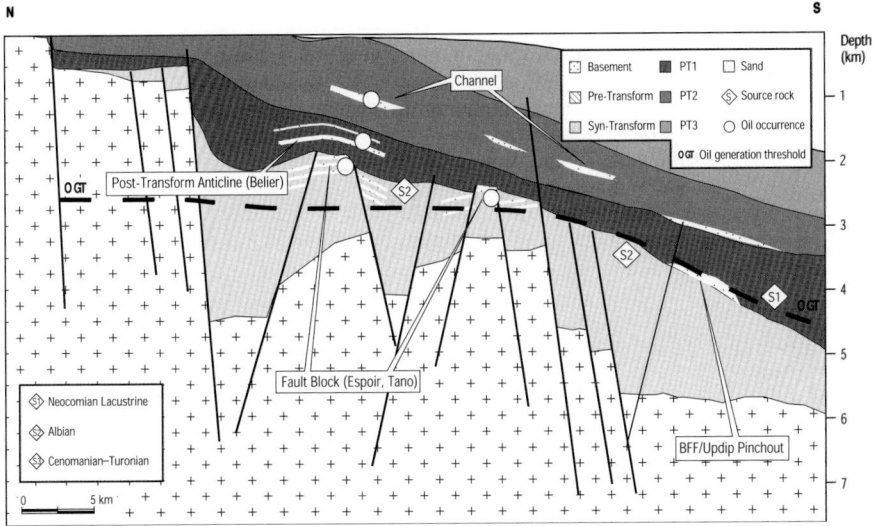

Fig. 19. Schematic cross-section: western Ghana and eastern Côte d'Ivoire play types. Limited deep-water structure implies that the main prospects in this areas are stratigraphic.

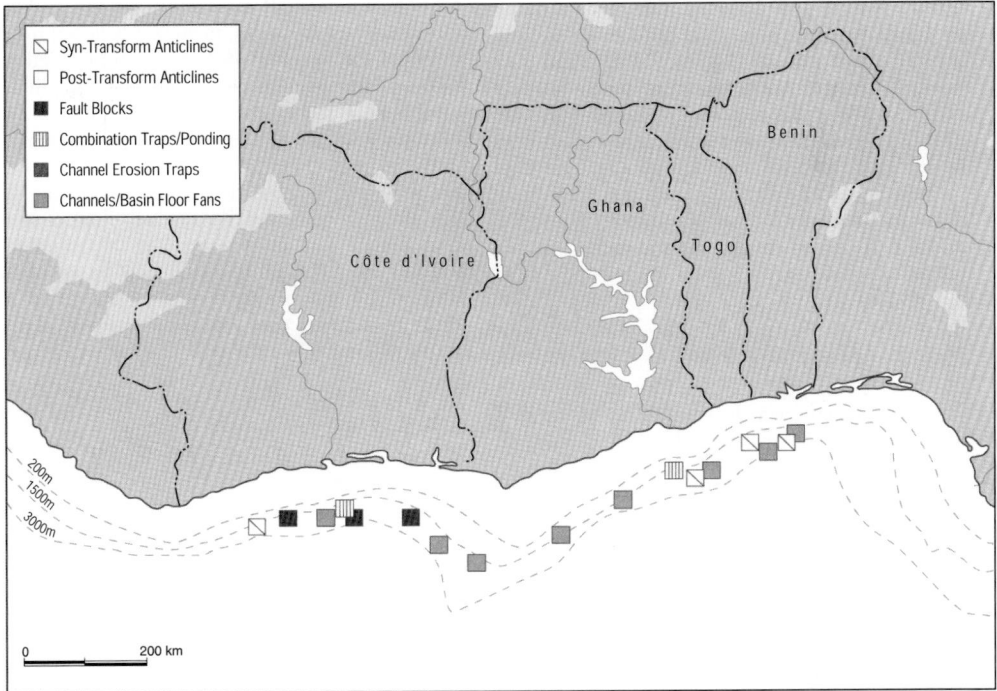

Fig. 20. Distribution of trap types: deep-water prospects. Structural plays are linked to the main fracture zones, while stratigraphic plays occur along the margin. Note that we have little data to evaluate play types in some regions and this map should not be regarded as complete.

Table 2. *Trap-type analogues and critical factors*

	Syn-transform	Post-transform 1	Post-transform 2	Post-transform 3	*Critical factors*
Anticline	*Togo, Benin, Keta Basin, western Côte d'Ivoire*	*Benin, western Cote d'Ivoire*	Benin		*Timing*
Fault block	Shallow–moderate deep-water in all areas				
Channel-erosion trap		Aje/Seme area, Benin, (?) other areas			*Reservoir erosion*
Ponded turbidites		*Western Côte d'Ivoire*	Togo		
Channels and basin floor fans		Most areas, esp. Benin, Tano, palaeo-Volta, western Côte d'Ivoire	*Côte d'Ivoire, Tano,*	Togo, (?)other areas	*Updip seal*
Critical factors	*Reservoir quality*			*Migration path*	

Italics, high-graded plays on volumes and risk criteria (including source).

channels. It is therefore important to reconstruct the sedimentary conditions over each prospect individually in order to assess reservoir risk and, where possible, to relate this back to sedimentary facies on the shelf. It is also key to analyse when the structure developed because this would have influenced the deposition of sands. One of the more favoured structures in this regard is shown in Figure 12, where a thick-layered post-transform section is observed predating the uplift of the structure, which lies basinward to and coeval with the deposition of thick reservoir-quality Late Cretaceous sands on the shelf. This prospect can again be fed by both mature Early Cretaceous lacustrine and Late Cretaceous marine source rocks.

Channel/Unconformity erosion traps (code U)

This category of trap is typified by the Aje Field of western Nigeria, which is sealed on its western side by a deep downcutting shale-filled channel, shown schematically in Figure 9. A further example is the Ceiba Field of Equatorial Guinea (P. Dailly, Triton, pers. comm. 2001). In deep water, potential traps of this type are most commonly associated with the highly erosional Oligocene unconformity. An example is shown in Figure 13, where downcutting of the unconformity has accentuated the closure of a small anticlinal feature, creating a large trapped volume on Palaeogene deep-water sediments. The critical factors for this play type are reservoir quality in the underlying section, which is less sand-prone than many other intervals, and the continuity of the post-unconformity seal.

Ponded turbidites lapping against structures (code C or CF)

Areas of high amplitude are observed in the post-transform section shoreward of and lapping against major transpressional and extensional highs. While some may be limestones, many are believed by the nature of their geometries to represent ponded turbidites. The level of these 'ponded' amplitudes usually post-dates a period of uplift of the high concerned, such that the amplitudes occur immediately above the Late Albian event in western Côte d'Ivoire (Fig. 14), but in younger parts of the section along the younger 'Romanche' trend of structures (Fig. 11). The geometry of the amplitudes controls the type of trap that can be proposed: where the amplitudes drape over the closures of the highs concerned, pure structural trapping can be invoked (code C); otherwise partial stratigraphic trapping must be invoked (code CF). The critical factor in the latter case is updip seal, with the turbidites required to be detached from the feeder area, as is suggested in Figure 14. In order to define this, amplitude mapping on 3-D seismic is deemed to be critical for prospect definition.

Charge risk for this play varies with the stratigraphic level at which the ponding occurs relative to the source rock, with plays in the Late Cretaceous section clearly more favoured, and the maturity of the source in that area.

Early–middle post-transform reservoirs/stratigraphic traps (code CF)

This play covers mounded and channelized features on seismic data for which no structural element to trapping can be invoked, i.e. they must be invoked as pure stratigraphic traps (code CF). In some cases, they may grade into prospects over which an element of drape closure is observed. These features occur at various levels, but predominantly over the three major unconformities/lowstands in the Late Albian, Senonian and Oligocene.

Large mounds can be observed in the Late Cretaceous section overlying the Late Albian unconformity along the margin on 3-D data (Fig. 16a) and occasionally on 2-D seismic data. Amplitude mapping on 3-D datasets is vital to the development of this play. These mounds are extremely large, covering areas often in excess of 80 km^2, and, with thick porous sands expected, the volumetric potential of this play clearly could be very large.

The play is also identified, though clearly less well delineated, on 2-D data at various points along the margin, including the deep-water Tano Basin (Craven 2000) and outboard of the Saltpond area in Ghana (Fig. 18). These mounds are developed at the base of the Late Cretaceous slope and are likely to be intimately associated with mature Cenomanian source rocks. The critical factor for this play is updip seal, for which shale plugging in the feeder channel may have to be invoked; analogues would be the small proven traps of this group identified in Côte d'Ivoire. This requires detailed analysis of amplitudes on 3-D seismic data, firstly to identify whether amplitude lows within the feeder channel could be shale plugs and secondly to delineate amplitude changes along structural contours that could be invoked as oil-water contacts.

Features over the Senonian unconformity are concentrated in the Côte d'Ivoire region, where channels centring around the Maastrichtian are a well-developed play in both the shallower and deep waters. Oil has already been found in stratigraphically sealed channels at West Tano (Craven 2000) and in a shallower pool of the Lion Field. Much larger features are seen in the deep-water focused down synclinal trends such as the Grand Lahou Depression (Fig. 15). Drape is observed over the feature illustrated in Figure 15, strongly suggesting that the mound is filled with sand. The critical factor for this play is clearly updip seal, with the shallow-water discoveries to date establishing the model of shale plugging updip.

The Tertiary section generally appears more seismically transparent and less reservoir-prone than the Cretaceous. However, in at least one area, a well-defined toe-of-slope fan is observed directly over the Oligocene lowstand (Fig. 16b) and this may be indicative of the presence of further features at this level elsewhere on the shelf. The origin of sands at this level is unknown, but they could conceivably originate through canyon erosion of earlier sand-bearing intervals. Again, updip seal on the feeder channel is required for a trap geometry to form. The critical factor for this younger play will be communication with deeper source rocks.

Conclusions

The deep-water area of the northern Gulf of Guinea is indicated to be a frontier area with considerable potential within a series of structural and stratigraphic play types. Key factors that can be used to identify the areas of highest potential along the margin are:

- The presence of depocentres likely to mature Early and Mid-Cretaceous source rocks;
- Association with onshore seep belts;
- Areas of structuring associated with the terminations of the St Paul and Romanche fracture zones;
- Areas of coarse sediment input during the two main Cretaceous lowstands.

The classification of play fairways of the area in this paper is based on a matrix of trap-type v. seismic sequence. This allows a logical evaluation of the risks and uncertainties on each of the plays. This approach, as summarized in Table 2, high grades a number of plays within the Grand Lahou Basin area of Côte d'Ivoire, the South Tano Basin, and offshore Togo and Benin. The key plays are:

- Syn-transform anticlinal traps within the Togo–Benin, Keta Basin (Ghana) and western Côte d'Ivoire areas;
- Post-transform anticlinal traps along the Romanche trend;
- Ponded turbidite combination traps shoreward of and onlapping major extensional and compressional highs;
- Stratigraphically sealed channels and basin floor fans, particularly in the Late Cretaceous section off the main sediment sources discussed in this paper (e.g. Tano Basin, western Côte d'Ivoire, Benin, palaeo-Volta).

Many of the plays identified have a strong stratigraphic component and prospect definition in these relies heavily on 3-D seismic imaging. In this respect, the 3-D areas provide models for areas that are currently not covered in this way.

PGS Exploration are thanked for permitting the use of their multiclient database for this study and for funding of the colour figures.

References

BLAREZ, E. & MASCLE, J. 1988. Shallow structures and evolution of the Ivory Coast and Ghana transform margin. *Marine &Petroleum Geology*, **5**, 54–64

BURKE, K., MACGREGOR, D. S. & CAMERON, N. R. 2003. Africa's petroleum systems: four tectonic aces in the past 600 million years. *In*: ARTHUR, T. J., MACGREGOR, D. S. & CAMERON, N. R. (eds) *Petroleum Geology of Africa: New Themes and Developing Technologies.* Geological Society, London, Special Publications.

CHIERCHI, M. A. 1996. Stratigraphy, palaeoenvironments and geological evolution of the Ivory-Coast–Ghana Basin. *Bulletin Centres des Recherches Exploration-Production Elf Aquitaine*, **16**, 293–303.

COWARD, M., PURDY, E. G., RIES, M. A. C. & SMITH, D. G. 1999. The distribution of petroleum reserves in basins of the South Atlantic margins. *In*: CAMERON, N. R., BATE, R. H. & CLURE, V. S. (eds) *The Oil and Gas Habitats of the South Atlantic.* Geological Society, London, Special Publications, **153**, 101–132.

CRAVEN, J. 2000. Petroleum system of the Ivorian/Tano Basin in the W Tano contract area, offshore west Ghana. *Petroleum Systems and Developing Technologies in African Exploration & Production, Geological Society/Petroleum Exploration Society of Great Britain Conference, May 2000, London, Abstracts Volume.*

DE MATOS, R. M. D. 2000. *Tectonic Evolution of the Equatorial South Atlantic, in Atlantic Rifts and Continental Margins.* American Geophysical Union, Geophysical Monograph Series, **115**, 331–354.

ELVSBORG, A. & DALODE, J. 1985. Benin hydrocarbon potential looks promising. *Oil and Gas Journal*, **83** (6), 126–131.

EMERY, K. O., UCHUPI, E., PHILLIPS, J., BROWN, C. & MASCLE, J. 1975. Continental margin off western Africa: Angola to Sierra Leone. *Bulletin of the American Association of Petroleum Geologists*, **59**, 2209–2265.

GRILLOT, L. R., ANDERTON, P. W., HASELTON, T. M. & DEMERGNE, J. F. 1991., Three-dimensional seismic interpretation, Espoir Field area, offshore Ivory Coast. *In*: BROWN, A. (ed.) *3-D Seismic Interpretation.* American Association of Petroleum Geologists Memoirs, **42**, 214–217.

HAACK, R. C., SUNDARARAMAN, P., DIEDJOMAHON, J. O, XIAO, N. J., GANT, N. J., MAY, E. D. & KELSCH, K. 2000. Niger Delta petroleum systems. *In*: MELLO, M. R. & KATZ, B. J. (eds) *Petroleum Systems of South Atlantic Margins.* American Association Petroleum Geologists Memoirs, **73**, 213–232.

KJEMPERUD, A., AGBESINYALE, W., AGDESTEIN, T., GUSTAFSSON, C. & YUKLER, A. 1991. Tectono-stratigraphy of the Keta Basin, Ghana, with emphasis on late erosional episodes. *Geologie Africaine. Proceedings of the 1st Conference on the Stratigraphy and Palaeogeography of West African Sedimentary Basins, Libreville, 6–8 May 1991, Bulletin des Recherches Exploration-Production Elf Aquitaine*, **13**, 55–69.

MILTON, N. J. & BERTRAM, G. T. 1992. Trap styles – a new classification based on sealing surfaces, *Bulletin of the American Association of Petroleum Geologists*, **76**, 983–999.

MORRISON, J., BURGESS, C., CORNFORD, C. & N'ZALASSE, B. 2000. Hydrocarbon systems of the Abidjan margin, Côte d'Ivoire. *Offshore West Africa, Fourth Annual Conference, Abidjan, 21–23 March 2000.* Pennwell Publishing, Tulsa, Oklahoma.

MORRISON, J., TEA, J., N'ZALASSE, N. & BOBLAI, V.1999. A sequence stratigraphic approach to exploration and re-development in the Abidjan margin, Côte d'Ivoire. *Offshore West Africa. Third Annual Conference, Abidjan, March 2000.* Pennwell Publishing, Tulsa, Oklahoma.

NANA, N. 2000. Prospectivity of offshore Ghana. Presentation at Africa Upstream Conference, Cape Town, September 2000. [www.petro21.com]

ST JOHN, B. 2000. The role of transform faulting in the formation of hydrocarbon traps in the Gulf of Guinea, West Africa. *Offshore West Africa, Fourth Annual Conference, Abidjan, 21–23 March 2000.* Pennwell Publishing, Tulsa, Oklahoma.

TUCKER, J. W. 1992. Aspects of the Tano Basin stratigraphy revealed by recent drilling in Ghana. *Geologie Africaine. Proceedings of the 1st Conference on the Stratigraphy and Palaeogeography of West African Sedimentary Basins, Libreville, 6–8 May 1991, Bulletin des Recherches Exploration-Production Elf Aquitaine*, **13**, 153–159.

Prospectivity in ultradeep water: the case for petroleum generation and migration within the outer parts of the Niger Delta apron

R. MORGAN

Veritas DGC Ltd, Crawley, West Sussex, RH10 9QN, UK (e-mail:
richard_morgan@veritasdgc.com)

Abstract: The interpretation of 2-D and 3-D seismic data acquired over the ultradeep water (1500–4000 m), lower delta-slope region of the Niger Delta indicates the continuation of the principal stratigraphic subdivisions out onto the continental rise. The Niger Delta sediment apron is still over 4 km in thickness in 4000 m of water. The continuation of a thick sediment pile into ultradeep water, coupled with the existence of a proven petroleum system on the higher parts of the slope, extends the potential region for mature Palaeogene and possibly older, oil-source rocks, up to and beyond the limit of current drilling technology. Contrary to previous models, the Miocene–Holocene Agbada Formation does not thin out oceanward relative to the underlying Akata Formation. The continuation of this seismic-event-based division of the Tertiary succession into and throughout the ultradeep-water province indicates an abrupt change in depositional style with the arrival of major submarine fans.

The collapse of the Niger Delta sediment cone above overpressured muds of the Akata Formation is manifested in the ultradeep-water region as thrusting. The style of deformation varies across the area, depending on the level of detachment within the Akata Formation. The thrusting is divided into compartments by transfer zones against which toe-thrusts terminate or are offset. The toe-thrusts can be seen to have antecedents in the framework of rifts that predated break-up. These rift elements are imaged beneath the western delta region and reflect the possible extent of extended continental crust in this area.

The lower slope is traversed by large turbidite channels that extend from the upper parts of the slope to beyond the outermost toe-thrusts. These channels provide a route for sand to be brought into these water depths from the shelf and from mud-diapir-bounded mini-basins higher on the slope.

The existence of underlying, presumed Late Cretaceous rift elements beneath parts of the ultradeep-water areas raise the possibility of additional source rocks within the ultradeep waters of the Niger Delta.

Deep-water exploration in locations such as the Gulf of Mexico, Angola, Brazil and Nigeria has shown that active petroleum systems can extend well beyond the continental shelf, given suitable subsidence and burial histories. Major delta-related sedimentary systems have been critically important to petroleum generation in all of these areas, providing both the reservoirs and the necessary overburden to mature source rocks. These major delta systems have built out over passive margins comprising thinned continental or transitional (part-continental/part-oceanic) crust. The loading effect of the sediment cones caused enormous accommodation space to be created, allowing extremely thick sediment successions to accumulate in deep to ultradeep-water settings. The continuation of thick sedimentary sections out into deep water has allowed the hydrocarbon-generation window for pre-delta source rocks to extend well beyond the shelf edge and onto the continental slope.

The Niger and Benue rivers have delivered vast quantities of sediment to the West African continental margin (Fig. 1) since at least the Eocene (Haack 2000). The resulting sediment apron has built out over the continental shelf and onto oceanic crust, loading and depressing the crust and enabling up to 12 km of sediments to accumulate in the accommodation space created (Doust & Omatsola 1989). The substantial thickness of sediment contained within the lower delta slope (3–8 km) is presently situated in deep to ultradeep water depths (500–4000 m) due to the requirements of isostatic equilibrium (Fig. 2). Recently, 2-D seismic data acquired with a 12 s record length

From: ARTHUR, T. J., MACGREGOR, D. S. & CAMERON, N. R. (eds) *Petroleum Geology of Africa: New Themes and Developing Technologies.* Geological Society, London, Special Publications, **207**, 151–164. 0305-8719/03/$15
© The Geological Society of London 2003.

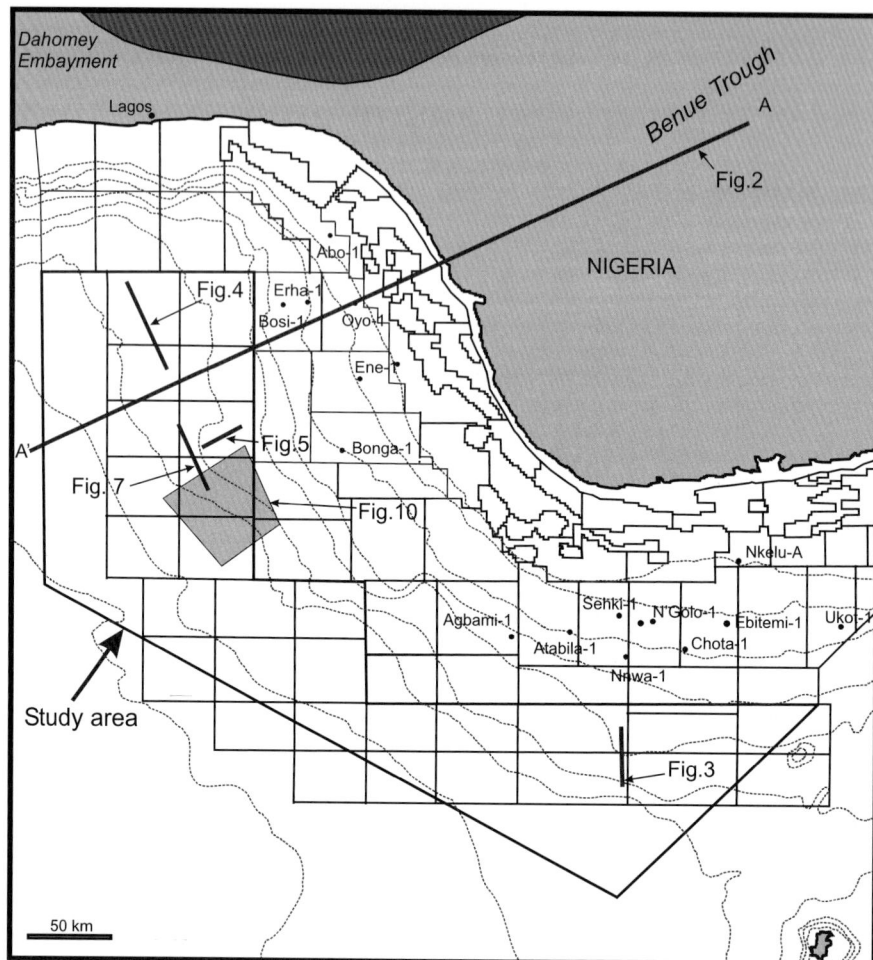

Fig. 1. Location map showing the impact of the Niger Delta sediment apron on coastline and bathymetry. Deep-water well locations are shown and the positions of 2-D and 3-D seismic data used in this study.

have shown these sediments to be partly underlain by an older syn-rift/post-rift succession and these are discussed later.

The sediment pile is deformed by toe-of-slope thrusts that climb from detachments within the overpressured muds of the Akata Formation and ramp up into the overlying mixed clastic sediments of the Agbada Formation (Fig. 3). The outermost thrusts climb to the sea floor and the associated hangingwall anticlines create substantial sea-bed relief. Inboard of these faults, earlier thrusts are buried by sediments and the fault terminations are blind. The hangingwall anticlines above these faults form elongate, laterally continuous closures (Fig. 4).

High-amplitude reflection packets visible in 2-D seismic data show the processes of channel incision, levée-building and channel-stacking, associated with turbidite flows (Fig. 4). These processes have operated throughout the deposition of the Agbada Formation and the lower delta-slope region. The geometry of these depositional systems and their relationship to the structures in the lower slope have become apparent with the availability of 3-D seismic data revealing a dynamic depositional environment across the entire width of the lower slope, from the Oligocene through to the present day.

Stratigraphic framework

The three-fold stratigraphic subdivision of the Niger Delta into the Akata, Agbada and Benin formations, has been described by Knox and Omatsola (1989) as an overall regressive megasequence broken up into a series of offlap cycles. The

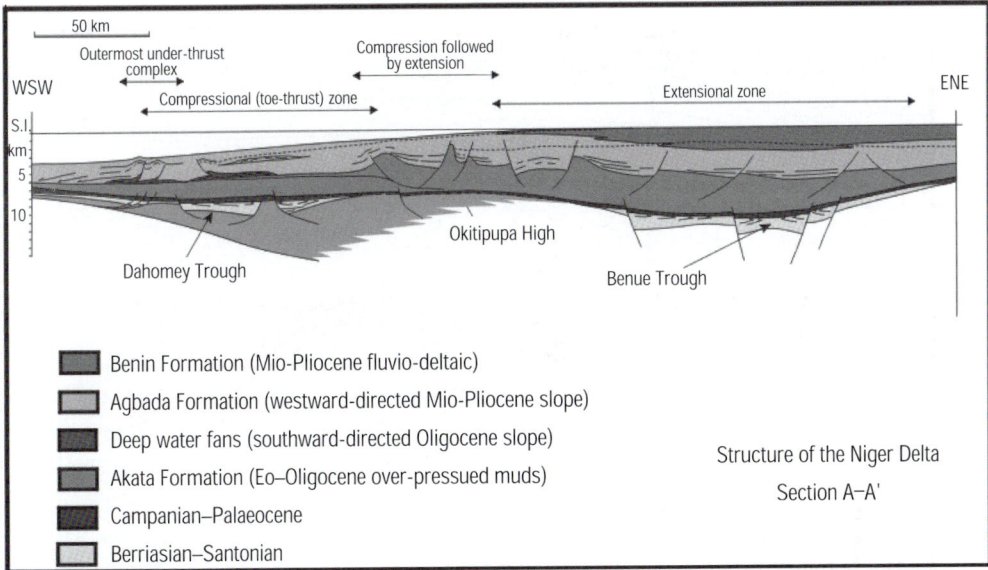

Fig. 2. A diagrammatic, regional scale cross-section taken obliquely through the Niger Delta sediment apron. The three-fold subdivision of the Eocene–Holocene delta sediments displays an overall progradational style. The fault pattern depicts the extensional collapse of the delta cone at its thickest part, detachment within the overpressured Akata muds and toe-thrust/under-thrust style of the western margin of the deep delta. The southwest extension of the Benue Trough and the Dahomey Embayment are projected as Cretaceous rifts beneath the Tertiary sedimentary cover.

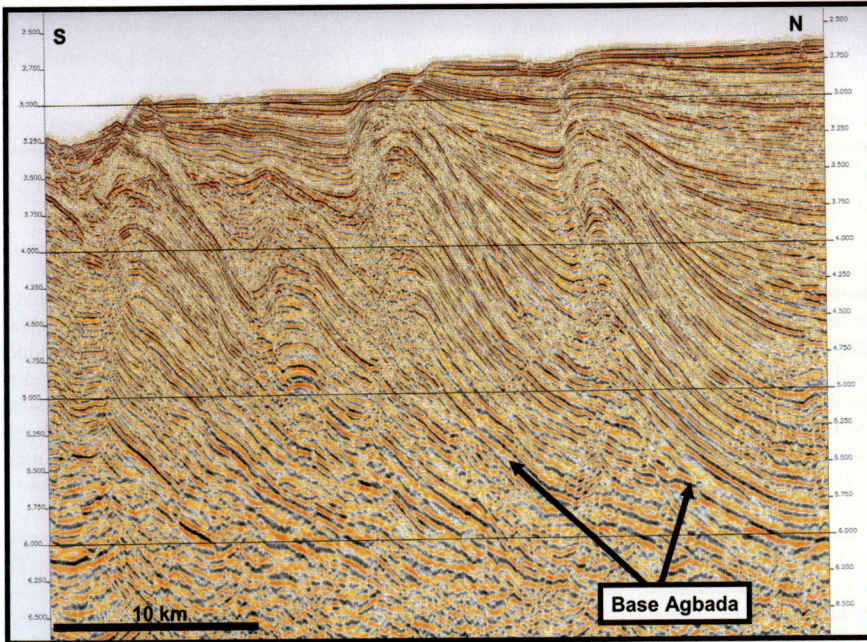

Fig. 3. A seismic profile through the southern toe-thrust belt showing imbrication of thrust sheets within the Agbada Formation and overall oceanward younging of thrust movement. The effects of recent thrusting are apparent in the present sea-floor relief.

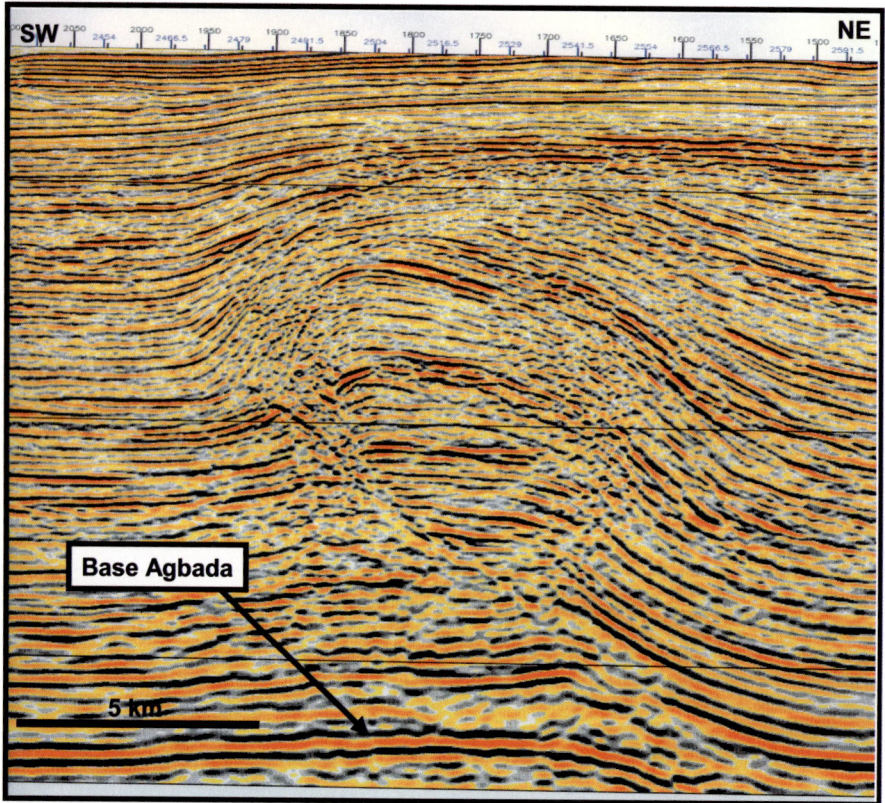

Fig. 4. A discrete thrust ramp with an associated anticline within the Agbada Formation on the western side of the deep delta. Detachment has occurred along the top of the underlying Akata section. Smaller-scale normal faults are visible offsetting the apparent Akata–Agbada contact. This faulting has the same trend as the overlying thrust, NW–SE, and can be correlated along strike with the position of thrust ramps within the overlying Agbada Formation. Both fault and anticline have been buried with the cessation of displacement, while the detachment at uppermost Akata level propagated further to the southwest, carrying the entire section, thrust ramp, anticline and precursor normal faults towards the southwest also.

onshore and shallow-water well data indicates general diachronicity between the formations, the Akata being the distal equivalent of the Agbada, being in turn the marine equivalent of the largely fluvial Benin Formation, over a time span covering the Eocene to Holocene. Within the main body of the delta apron, faulting and mud diapirism have made the characterization of the relationship between the Akata and Agbada formations problematic. However, the continuity of both sequences into the deep and ultradeep water is indicated by seismic-facies analysis, and implies that the Agbada Formation does not thin out in this extremely distal environment, as suggested by Doust & Omatsola (1989) and Cohen and McClay (1996). Instead, the two-fold subdivision of the marine components of the delta apron is maintained throughout the ultradeep-water region and out toward the basin floor. The uppermost Agbada

seismic facies is still over 3 km in thickness in over 4500 m of water and 250 km from the present coastline (Fig. 2).

In the deep-water area to the west of the delta cone the Akata Formation is a 3–4 km-thick, low reflectivity package, above which lies a large southward progradational sediment wedge marking the apparent arrival of the Agbada Formation. So, in the deep- and ultradeep-water areas, the Akata and Agbada formations appear to be divided by a major regional sequence boundary marking an abrupt change in the depositional environment with the appearance of large deep-water fans (Fig. 5).

In the deep water of the northwestern delta area the base of the probable Akata Formation can be seen to onlap an older progradational package and to thin markedly to the north (Fig. 5). The older sedimentary wedge in turn lies above a syn-rift/post-rift succession filling a framework of NE–

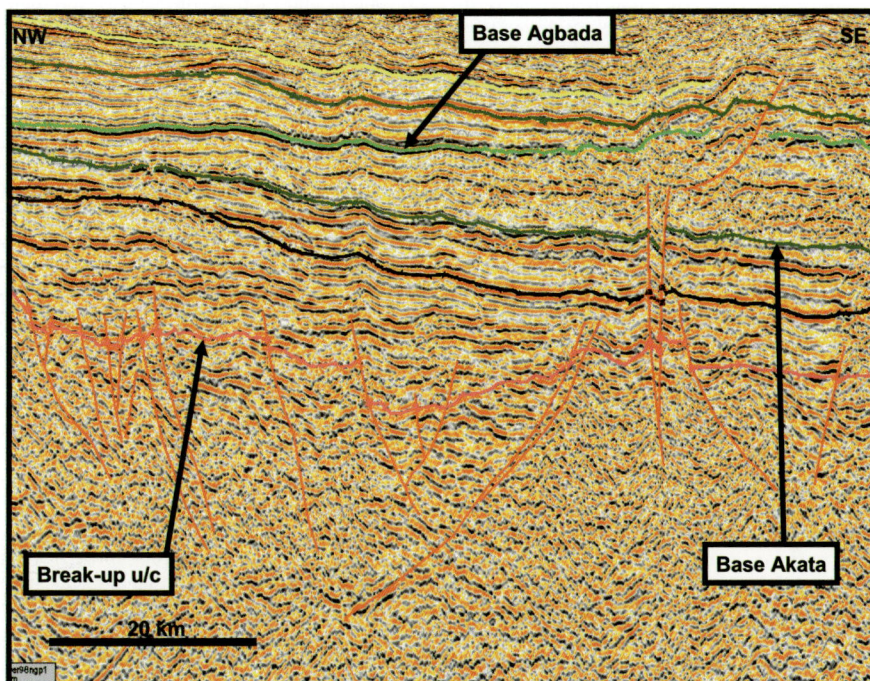

Fig. 5. Rift elements beneath the western deep delta area. Individual rifts typically have half-graben geometry and contain up to 1.5 km of syn-rift fill. An unconformity, angular in places, divides the syn-rift from the post-rift fill. A presumed Late Cretaceous sediment apron overlies the rift-fill succession, indicating significant sediment delivery to this part of the margin during this phase. These sediments are in turn onlapped by the Akata Formation seismic facies, which thins to the north. A second phase of shelf-slope build-out centred on the Dahomey Embayment follows the Akata deposition. This is correlated with the lowermost part of the Agbada Formation.

SW- and WNW–ESE-trending half-graben structures. This pre-Akata Formation sediment wedge is believed to represent a Late Cretaceous to Palaeocene section sitting above a pre-Mid-Aptian syn-rift/post-rift succession preserved in rift elements within the extended continental/transitional margin. An unconformity is clearly imaged between the syn-rift fill of the half-graben form rift elements and the post-rift succession. The unconformity has cut down into the syn-rift fill and records the denudation of footwall crests, indicating some erosion of already thinned crust (Fig. 5). This seismic horizon is the most obvious candidate for the Mid-Aptian break-up unconformity, although why this boundary should display evidence of erosion this far out on the margin is unclear. The rift framework was buried by a passive margin succession including a major sediment apron built out from the Dahomey Embayment region of the onshore Guinea Basin, directly to the north (Fig. 1). It is this apron that is onlapped by the Akata Formation during a presumed period of high stand.

The northward-sourcing of sediment from the Dahomey Embayment was repeated following the deposition of the Akata Formation. The appearance of major deep-water fans marks the end of Akata deposition and, in the northwest of the deep-water area, these can be correlated with the southward build-out of a large sediment apron (see Macgregor *et al.* 2003), possibly of Oligocene–Miocene age (Fig. 5). The main body of sediment representing the Agbada Formation in the deep water onlaps/offlaps this Tertiary post-Akata apron and is sourced from the NE Benue Trough region.

Source rocks and maturity

A current debate regarding the age of the source rocks supplying the Niger Delta sediment cone (Frost 1997; Haack *et al.* 1997; Haack 2000) has focused on the potential of the Akata Formation, presumed to be Eocene–Oligocene in age, and the Late Cretaceous. Late Cretaceous source rocks are recognized along the length of the West African margin (Schiefelbein *et al.* 1999), and rocks of Albian–Turonian age, with some source potential but immature for oil generation, are reported from the Benue Trough onshore (Petters & Ekweozor

1982). Additionally, the Aje Field immediately to the west of the main body of the offshore delta apron and the adjacent Seme Field, offshore Benin, are reported to have been sourced, at least in part, from both Aptian–Albian and Turonian source rocks (Tobias 2000). Other workers (Haack *et al.* 2000) have suggested a Neocomian age for oils in the Benin Basin region, based on lacustrine characteristics.

Any Late Cretaceous sediment preserved beneath the main body of the delta cone is too deeply buried to be imaged by conventional seismic data and is assumed to have long passed into overmaturity. However, the progressive oceanward thinning of the delta cone results in a corresponding reduction in maximum burial depth for the pre-delta section. This brings a notional Late Cretaceous section back into a generally defined hydrocarbon-generation window with increasing water depth and raises the possibility of a Late Cretaceous petroleum system within the deep- to ultra-deep-water province if the source rocks are present.

Cretaceous source rocks

Elements of a pre-delta rift system are clearly imaged west of the delta cone beneath a 3–5 km overburden by the new generation of 2-D seismic data (Fig. 5). These rift elements lie in water depths of 2000 m and more to the west of the main body of the delta and they contain a syn-rift and post-rift section of 2500–3000 m cumulative maximum thickness.

The occurrence of rift elements beneath the delta succession is taken as indicative of the presence of extended continental crust and the distribution of these structures extends the thinned edge of the continental margin to the southwest and oceanward of where previous workers have placed this boundary (Whiteman 1982; Damuth 1994). This area of possible extended continental crust, as defined by the existence of the rift elements, exhibits linear boundaries, particularly along the NE–SW-trending section of the margin, across which there is abrupt transition to apparent oceanic crust (Fig. 6). Here the rift structures terminate along a linear boundary marked by a ridge, outboard of which the event identified as the break-up unconformity landward becomes a high-amplitude horizon broken by tabular fracturing orthogonal to the bounding ridge. The trend and location of the dividing ridge is aligned with the Chain Fracture Zone, one of the major oceanic transform zones in the Gulf of Guinea (Fig. 7). Gravity data further support a correlation between these two structures (Fig. 8), and infer that the NE–SW-trending ridge bounding thinned, rifted crust to the northwest represents a transform margin of the type described by Bird (2001). It is

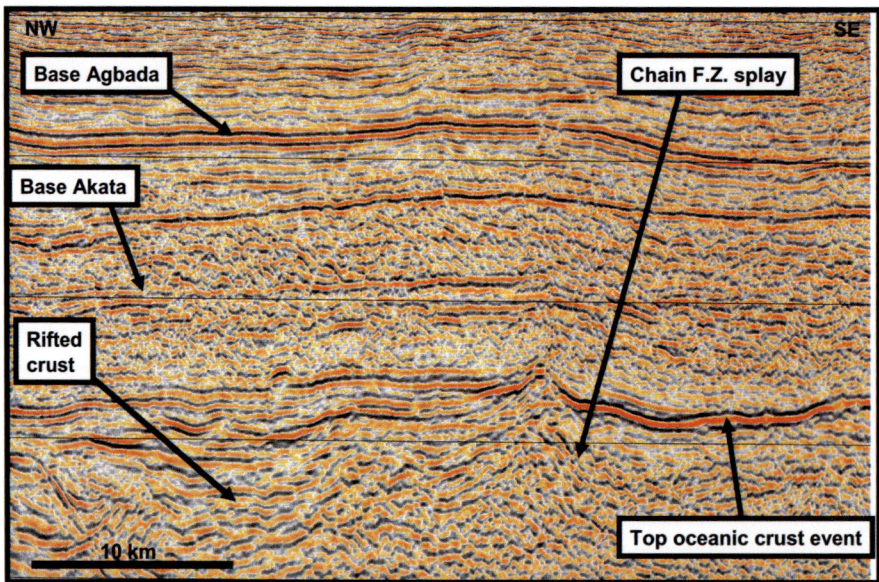

Fig. 6. A NNW–SSE-trending seismic profile across part of the Chain Fracture Zone. The pre-Tertiary section to the northwest exhibits sediment-filled rift elements, presumed to be of the Late Cretaceous, while the area to the southwest exhibits a characteristic bright event that can be followed out to base of the continental rise. This is believed to represent the top of oceanic basement or a volcanic marker directly above oceanic basement.

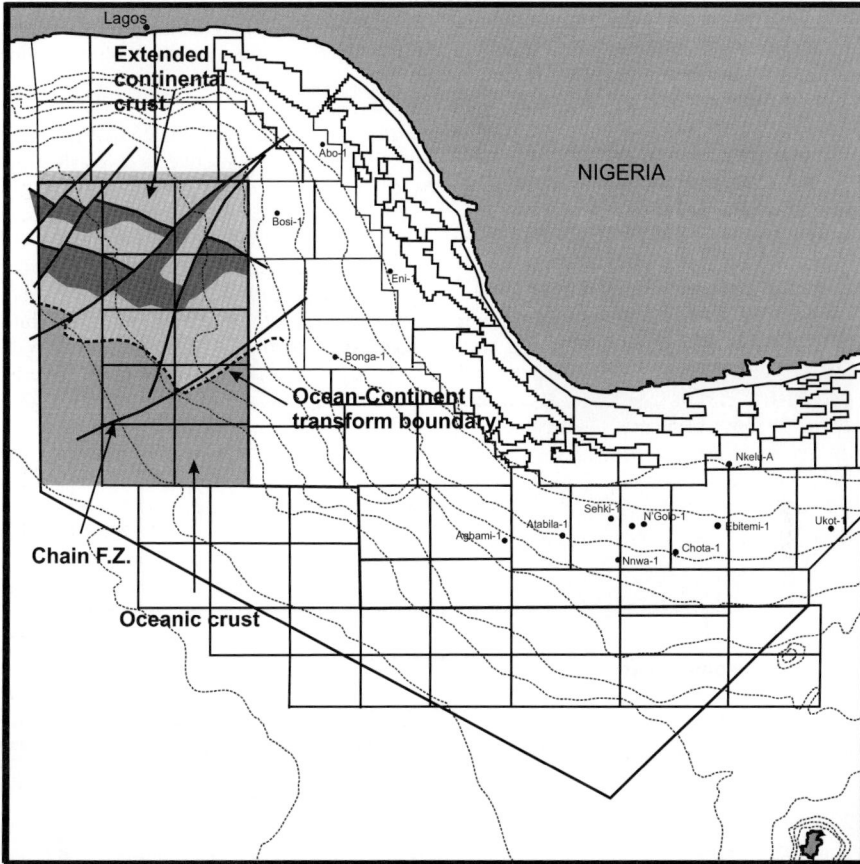

Fig. 7. A model for the crustal structure over the western deep delta area. The distribution of extended/rifted crust has been delineated from the occurrence of pre-delta rift elements. The NW–SE-trending sections of the margin appear to be transitional between continental and oceanic crust while the NE–SW-trending sections represent an abrupt contact between continental and oceanic crustal types as defined on seismic data. These are interpreted as transform boundaries

not known here how far this margin extends landward to the northeast and back toward the main body of the delta.

What is clear from these observations is that the nature of the crust underlying the western side of the Niger Delta toe-thrust zone is heterogeneous and, in some areas, may include rift elements that contain rocks of Early–Mid-Cretaceous age. Elsewhere along the margins of the South Atlantic, namely the Campos, Congo and Benguela basins, lacustrine source rocks preserved within rifts that predate oceanic spreading have been a rich source of hydrocarbons (Schiefelbein *et al.* 1999). The Benue Trough is evidence of the extension of this rift system into the Niger Delta region and the rift elements beneath the western toe-thrust zone may be seen as part of that framework.

The base of the Akata Formation is characterized as a major onlap surface in the northwest of the deep-water Niger Delta, marking the onset of deposition of a thick section of hemipelagic muds above a sediment apron of presumed Upper Cretaceous–Palaeocene age (Fig. 5). The continuation of this event out into ultradeep water has defined an underlying package of sediments comprising the rift–post-rift succession over continental crust and a condensed Upper Cretaceous section over oceanic crust. The age of the onset of sea-floor spreading in the Gulf of Guinea is placed at Mid-Aptian, 118 million years (Hay *et al.* 1998), leaving approximately 60 million years before the Niger Delta began to build out from the re-entrant formed by the Benue Trough during the Palaeogene. Episodes of source-rock deposition have been identified from Deep Sea Drilling Project (DSDP) sites in the Gulf of Guinea and dated as Early Cenomanian, Late Cenomanian–Early Turonian and Maastrichtian (Wagner & Pletsch 1999). Of

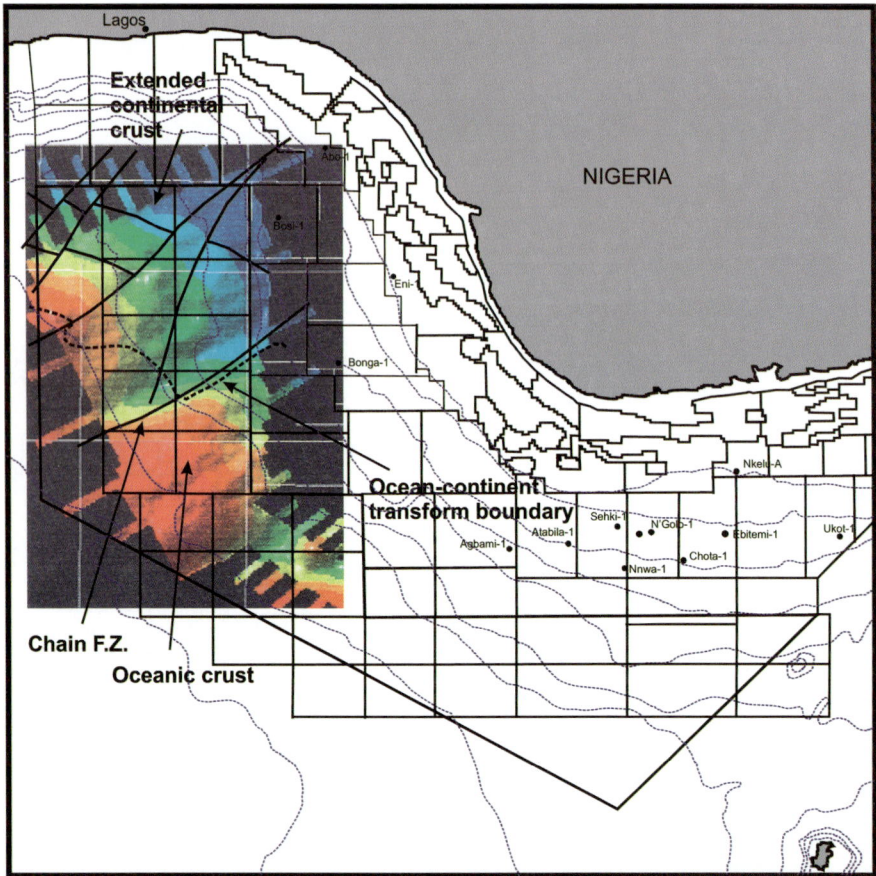

Fig. 8. Bouguer gravity anomaly over the western deep delta area. The juxtaposition of contrasting crustal densities across the Chain Fracture Zone is clearly displayed and demonstrates the link between transfer zones in the pre-delta rift framework and the formation of the continental margin on break-up.

these events, the Cenomanian–Turonian event is recognized as global (Arthur *et al.* 1987) and is expected to be represented generally around the deeper parts of the Gulf of Guinea margin. All the samples from these sites were immature for oil generation (Wagner & Pletsch 1999), a lack of deep burial being implicit in the DSDP, shallow-borehole sampling approach. However, rocks of similar age, deposited in a similar palaeogeo-graphic setting, should be expected beneath the lower slopes of the Niger Delta apron, the delta sediment pile having provided the burial necessary to take the source rocks to maturity.

Tertiary source rocks

The Eocene–Oligocene Akata Formation is con-ventionally viewed as the principal source rock for the hydrocarbons in the Niger Delta region. Over most of the explored parts of the delta, pre-delta

source rocks are very deeply buried and the thick, (upwards of 6 km) pro-delta muds of the Akata Formation are seen as the most likely source of hydrocarbons for most of the known accumulations in the onshore and shallow-water areas (Doust & Omatsola 1989).

The Akata Formation has been described as the distal equivalent of the overlying Agbada Forma-tion (Knox & Omatsola 1989), separated by a diachronous boundary recording the progradation of the delta cone. There is little indication of such a diachronous relationship at this boundary in the deep to ultradeep province, where the contact between these megasequences marks an abrupt change in seismic facies, with the appearance of major fan lobes above parallel-laminated basin-floor deposits. The velocity structure also changes across this contact, with a reduction in seismic velocity of up to 1000 m s^{-1} within the underlying Akata Formation (Fig. 9). The anomalously low velocities at these burial depths imply undercompaction and point to widespread overpressuring. The overpressured zone extends oceanward well beyond the region of large-scale incompetent shale movement (the shale diapir zone located inshore of the ultradeep-water area) and out beneath the zone of toe-of-slope thrusting. The regional extent of overpressuring within the Akata Formation has implications for the maturation of intra-Akata source rocks and the migration of fluids into the overlying Agbada Formation. Geochemical models predict the retardation of late maturity and the onset of gas generation relative to the thermal history in overpressured conditions (Carr 1999). This analysis extends the oil-generation window oceanward out to, and possibly beyond, the toe-of-slope thrust belt in the presence of a raised geothermal gradient over oceanic crust.

Migration pathways

The overpressuring of the Akata Formation inferred by the seismic velocity field suggests a high seal potential in the muds directly beneath the base of the normally pressured Agbada Formation. Late Cretaceous or Akata-sourced hydrocarbons must have passed through this seal on migration into Agbada reservoir sands. The widespread evidence of abnormal pressures within the Akata Formation, in the form of mud diapirism or abnormally low seismic velocities, indicates that this seal has a regional extent.

Faulting within the Akata Formation, and particularly faults passing from the Akata into the overlying Agbada Formation, are expected to have been important in the passage of fluids upward through the section. Detachment faulting is recognized at a number of levels within the Akata Formation and the points at which the faults ramp up into the Agbada Formation represent potential fluid pathways. As the fault ramp into the Agbada Formation is invariably associated with the formation of a hangingwall anticline, the link between the fluid pathway and the trap is implicit. Moreover, the analysis of the development of particular folds above toe-thrusts via the restoration of these structures to a palaeo-sea-bed surface, shows the existence of prethrust structures in the form of subsidiary normal faulting affecting the Akata Formation and the lowermost part of the Agbada Formation (Fig. 10). It follows that the occurrence of these normal faults has been influential in

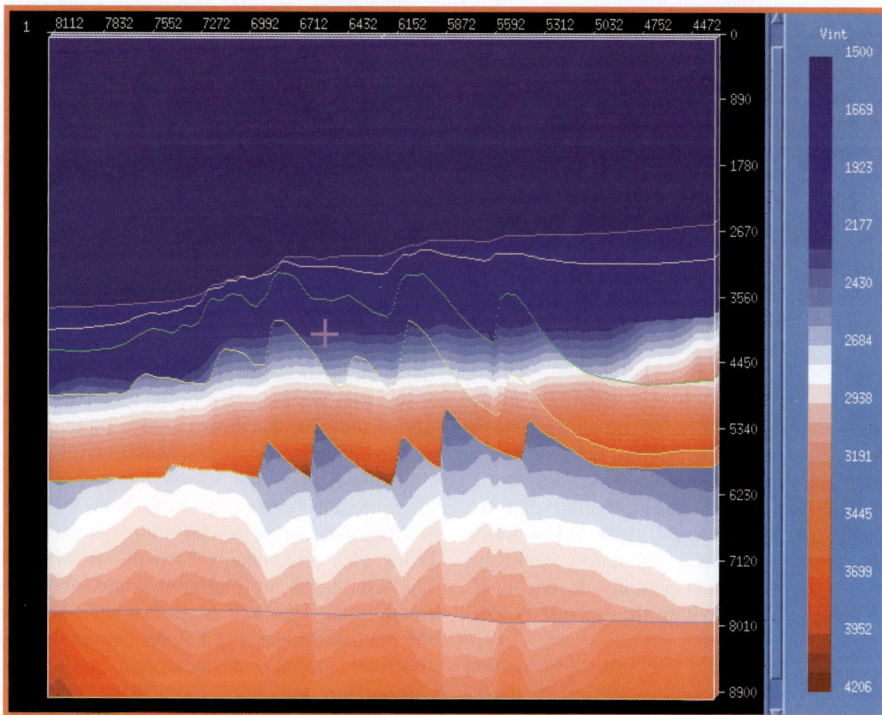

Fig. 9. A 2-D velocity model built from seismic-stacking velocities, demonstrating the abrupt reversal in the velocity gradient at the base of the Agbada section. The sharp reduction in velocity with depth across this boundary is in excess of $1000 \, ms^{-1}$ and indicates undercompaction due to anomalously high pore-fluid pressure with the Akata section.

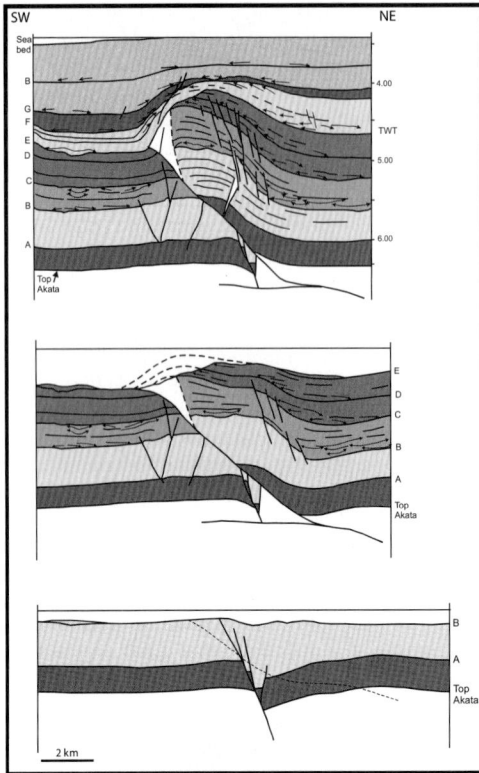

Fig. 10. A sequential section restoration to reconstructed sea-floor relief, showing the growth of a discrete thrust-ramp anticline from the western deep delta area. The thrust ramp probably initiated from the interaction between a detachment within the uppermost Akata Formation and a normal fault zone affecting the Akata and Lower Agbada formations. The thrust fault climbed to sea bed prior to the continued propagation of the detachment at depth. The degrading of the relief created by the periodic growth of the anticline is expressed in the repeated truncation surfaces over the crest of the fold.

determining where the later-formed thrust detachment ramped up into the overlying Agbada Formation. So, the causal relationship between the normal faulting affecting the Akata Formation and the thrust ramps climbing into the Agbada Formation connects potential vertical migration pathways from within the Akata and possibly deeper levels with structural closures within the Agbada Formation.

Sand distribution within the Agbada Formation is probably restricted to within incised or levée-banked valleys (see below), limiting the opportunities for lateral hydrocarbon migration over any distance to the trend of the submarine channel complexes.

Play fairways

The collapse of the main body of the Niger Delta cone in the onshore and inshore regions, and the passage of this displacement oceanward along detachments within the Akata Formation, has driven and continues to drive the growth of toe-of-slope thrusting. These movements, coupled with the involvement of large-scale incompetent muds in the deformation, have caused progressive modification of the depositional slope. With the depositional slope in widespread disequilibrium, sediment has been sequentially reworked downslope and processes associated with the upper parts of the shelf edge, such as channel incision, are equally in evidence towards the base of the slope.

Lauferts (1999) described a three-stage structural/depositional model for the slope, starting with a rim of mini-basins along the shelf edge. These basins record the fill-spill mechanism associated with the growth of faults and the movement of large underlying masses of incompetent mud. The mid-slope setting is dominated by laterally extensive turbidite channels, with terminal fan lobes formed on unconfined slopes, while the lower and base of slope setting comprises extensive fan lobes influenced by toe-of-slope thrusts.

Recent 3-D seismic data imaging the lower and base of slope regions has led to a reconsideration of this division of depositional processes. Discrete, turbidite channel systems involving a combination of incision and levée-building extend across the entire lower slope, with terminal lobe formation occurring out on the basin floor beyond the outermost toe-thrust structures (Fig. 11). These channel systems are typically 1–3 km in width and have formed through repeated use of particular corridors for sediment transfer out to the base of the slope. The channel complexes are flanked by substantial levées and, overall, display low sinuosity, although individual channels within these systems may display high sinuosity. The channel systems both incise and are deflected by hangingwall anticlines above toe-of-slope thrusts where their growth has created bathymetric ridges. Moreover, it is apparent that relay zones and lateral ramps in the toe-of-slope thrust framework have been key in controlling incision points for these systems. The turbidite channel complexes on the western side of the delta apron are clustered around the NE–SW-trending transfer-fault zones that compartmentalize the toe-thrust zone (Fig 12). As the gravity-induced thrusting propagated oceanward, the growth of thrust-ramp anticlines took place in a broadly oceanward-directed sequence. The growth of ramp anticlines upslope from the toe-thrust zone ceased as the active front of fault propagation moved on and the bathymetric ridges formed by the folding

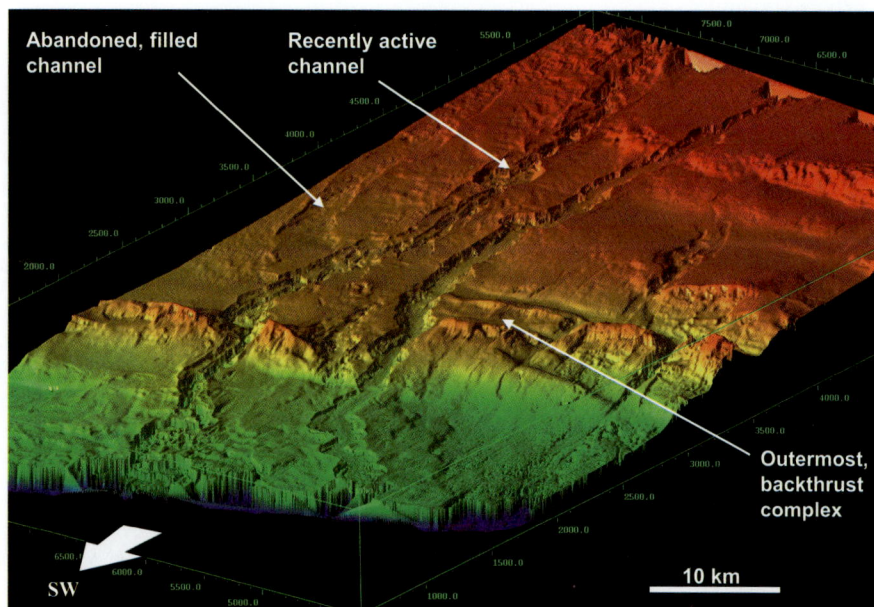

Fig. 11. A perspective relief image covering 3000 km² of the sea bed on the lower slopes of the Niger Delta. The pathways used by successive turbidite flows are visible as partially incised, channel-levée complexes. At least three generations are shown here, completely back-filled, partially filled and recently active. Relatively little deflection of the channel systems occurs over the relief created by NW–SE-trending anticlines above thrust ramps and beyond the outermost thrusts the broader lobes of the distributary channel complex can be seen on the basin floor. The trend of the outer thrust structures is offset mid-way across the image by a relay or transfer zone. The channel systems have exploited this offset in the outer ridge to reach the basin floor.

were buried. However, displacement on the NE–SW-trending transfer zones continued to accommodate different phases of movement in different parts of the toe-thrust zone. So, in the area upslope from the toe-thrust zone, thrusts had ceased movement but transfer zones remained active, accommodating differences in displacement in different compartments. This distribution of fault activity can be observed in the present-day bathymetry of the lower slope, where the most recently active channel systems follow transfer fault zones down to the present toe-thrust front, the transfer fault cutting to sea bed (Fig. 11).

The fill within turbidite channel complexes is expected to be the principal delivery mechanism for sand into the lower delta slope region. These complexes show substantial levée development that contain many episodes of channel formation and overbank splays are uncommon. Individual channel systems also show a degree of internal organization in the form of downslope meander migration and progressive aggradation within the levée containment (Posamentier & Kolla, in press). These channel morphologies imply good sandbody connectivity.

The association between transfer-fault location and turbidite channel systems suggests that particular sediment dispersal pathways have seen repeated use throughout the Neogene while other areas of the lower slope have been bypassed. The continued growth of thrusts through the Neogene to the present day has folded cross-cutting channel systems that had earlier incised the sea-bed relief caused by the formation of hangingwall anticlines. Therefore, it is at the intersection of hangingwall anticline with transfer fault zone that the link between reservoir and structure is most likely.

During periods of sea-level highstand, when lower energy conditions return to the outer shelf, the background deposition of mud in the lower slope environment is likely to have provided extensive sealing potential.

Conclusions

The two-fold stratigraphic subdivision of the Eocene–Holocene delta apron in the form of the Akata and Agbada formation, occurs throughout the lower slope and continues out towards the basin floor. To the northwest of the main part of the delta cone this boundary is marked by the arrival of a major downlapping package, sourced from the

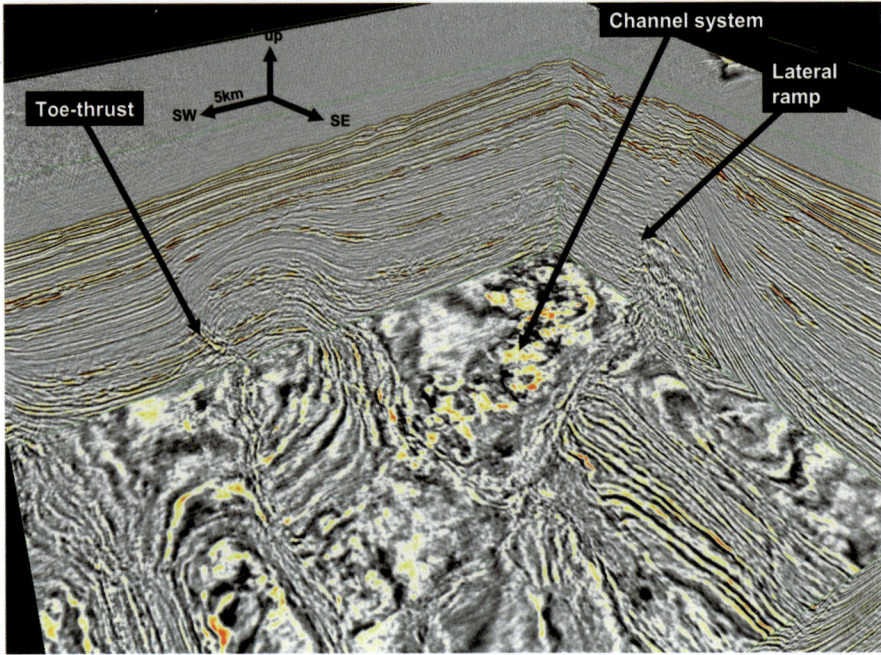

Fig. 12. A chair-cut through a 3-D seismic volume showing the relationship between thrusts and transfer zones in the lower slope. The linear NW–SE-trending toe-thrusts and the associated hangingwall anticlines are abruptly terminated along both sides of the NE–SW-trending transfer fault. This can be seen to have resulted in a near-vertical lateral ramp structure. On the downthrown side of the lateral ramp the undeformed Agbada section contains channel-levée complexes both in the subsurface and at present-day sea bed. These follow the transfer fault out towards the basin floor. The thrust ramp shown has been buried following cessation of movement on the ramp while the transfer fault cuts to present sea bed due to the continued accommodation of differences in displacement across different parts of the toe-thrust zone. Although the ramp anticlines are the most conspicuous structures within the lower slope, it is the transfer zones that have exercised a greater influence on channel location.

Dahomey Embayment area directly to the north, rather than from the Benue Trough region to the northeast. The boundary between the Akata and Agbada formations marks a pronounced change in the depositional environment and can be correlated with a major sequence boundary in the northwest parts of the deep-water area.

Beneath the Agbada–Akata couplet a syn-rift to post-rift to passive-margin section has been identified of presumed Mid-Cretaceous–Palaeocene. The sediment packages contained within rift elements show the delta sediments to be underlain in the northwest of the deep-water area by extended, continental/transitional crust, bounded by a NE–SW-trending transform fault contact with oceanic crust to the SE.

The Akata/Agbada Tertiary delta apron is underlain in the deep-water area on the western side by NE–SW- and WNW–ESE to NW–SE-trending rift elements. These typically take the form of half-grabens, while the NW–SE rift elements are also offset across a number of NE–SE-trending transfer zones. The largest of the transfer zones is broadly aligned with the Chain Oceanic Fracture Zone and defines a rapid transition between extended rifted crust and oceanic crust.

The recognition of a pre-delta rift framework beneath deep-water sediments infers the presence of a Late Cretaceous rift and post rift section. A number of important source-rock depositional episodes of the Late Cretaceous are known in the Gulf of Guinea region and the possibility that one or a number of these episodes are present in the deep-water, pre-delta section should be considered.

The principal mechanism responsible for delivering much of the sediment to the lower slopes of the delta apron during Agbada Formation deposition has been turbidite or related gravity-driven flow. Recently, active channel-levée systems are mirrored in the subsurface and show the reuse of particular corridors for sediment dispersal through the Late Tertiary. In the west of the deep delta, these corridors are clustered around NE–SW-trending transfer-fault systems that compartmentalize the toe-thrust belt. These fault systems, in turn, have their origin in the transform zones that formed at continental break-up in the Aptian.

The link between the turbidite channel systems and the transfer fault zones means that the channels lie largely orthogonal to the trend of the toe-thrust belt. Therefore, the primary location for sand in the lower slope, the channel fill, cross-cuts the trend of the main trap-forming structures, these being hangingwall anticlines above ramps in the toe-thrusts.

The connection between closure, sand fairway and fluid pathway is most clearly made in the earlier formed hangingwall anticlines above toe-of-slope thrusts. This play is most promising where these structures have formed and been subsequently buried and thus remain unbreached in the upper levels. Away from the anticlinal crests abundant opportunity for combination dip/stratigraphic plays exist where stacked channel systems have been tilted on the flanks of folds. However, poorer connectivity to fault-dependent migration pathways may produce additional risk.

Regardless of the predicted confluence of trap, play fairway and source to reservoir pathways, it is the clearly imaged flat-spot/flat-based amplitude anomalies that provide the most compelling evidence for active petroleum systems within the deep- to ultradeep-water province of the Niger Delta.

The author thanks Veritas DGC for permission to show seismic data and Steve Thompson for producing the 3-D seismic images.

References

ARTHUR, M. A., SCHLANGER, S. O. & JENKYNS, H. C. 1987. The Cenomanian–Turonian oceanic anoxic event. II: Palaeogeographic controls on organic matter production and preservation. *In*: BROOKS, J. & FLEET, A. J. (eds) *Marine Petroleum Source Rocks*. Geological Society, London, Special Publications, **26**, 401–420.

BIRD, D. 2001. Shear margins: Continent–ocean transform and fracture zone boundaries. *Leading Edge*, **20**, 150–159.

CARR, A. D. 1999. A vitrinite reflectance kinetic model incorporating overpressure retardation. *Marine & Petroleum Geology*, **16**, 355–377.

COHEN, H. A. & MCCLAY, K. 1996. Sedimentation and shale tectonics of the northwestern Niger Delta front. *Marine & Petroleum Geology*, **13**, 313–328.

DAMUTH, J. 1994. Neogene gravity tectonics and depositional processes on the deep Niger Delta continental margin. *Marine & Petroleum Geology*, **11**, 320–346

DOUST, H. & OMATSOLA, E. 1989. Niger Delta. *In*: EDWARDS, J. D. & SANTAGROSSI, P. A. (eds) *Divergent/passive margin basins*. American Association of Petroleum Geologists Memoirs, **45**, 201–238.

FROST, B. 1997. An alternative Cretaceous age petroleum system model for the Niger Delta. *In*: CAMERON, N. B., BATE, R. H. & CLURE, V. S. (eds) *Oil and Gas Habitats of the South Atlantic, 24–25 February 1997, Abstracts*, Geological Society, London, Special Publications, **153**, 24–26.

HAACK, R. C., SUNDARAMAN, P. & DAHL, J. 1997. Niger Delta petroleum system. *Hedburg Research Symposium, Petroleum Systems of the South Atlantic Margin, Rio de Janeiro, 16–19 November, Extended Abstracts*,

HAACK, R. C., SUNDARARAMAN, P., DIEDJOMAHOR, J. O., XIAO, H., GRANT, N. J., MAY, E. D. & KELSCH, K. 2000. Niger Delta petroleum systems, Nigeria. *In*: MELLO, M. R. & KATZ, B. J. (eds) *Petroleum Systems of South Atlantic Margins*. American Association of Petroleum Geologists Memoirs, **73**, 213–232.

HAY, W. W., DECONTO, R. & WOLD, C. N. 1998. An alternative global Cretaceous palaeogeography. *In*: BARRERA, E. & JOHNSON, C. (eds) *The Evolution of the Cretaceous Ocean/Climate System*. Geological Society of America Special Publications, **332**, 455.

KNOX, G. J. & OMATSOLA, E. M. 1989. Development of the Cenozoic Niger Delta in terms of the 'escalator regression' model and impact on hydrocarbon distribution. *In*: VAN DER LINDEN, W. J. M., CLOETINGH, S. A. P. L., KAASSCHEITER, J. P. K., VAN DER GRAAF, W. J. E., VANDENBERGLIE, J. & VAN DER GUN, J. A. M. (eds) *Proceedings of the Royal Geological and Mining Society of The Netherlands Symposium, Coastal Lowlands, Geology and Geotechnology, The Hague, 1987*. Kluwer Academic, Dordrecht, 181–202.

LAUFERTS, H. 1999. Reservoir depositional facies in the deep offshore Niger Delta slope basin, Nigeria – examples from 3D seismic. *NAPE 17th Annual International Conference and Exhibitions, 15–19 November, Lagos, Nigeria*.

MACGREGOR, D. S., ROBINSON, J. & SPEAR, G. 2003. Play fairways of the Gulf of Guinea transform margin. *In*: ARTHUR, T. J., MACGREGOR, D. S. & CAMERON, N. R. (eds) *Petroleum Geology of Africa: New Themes and Developing Technologies*. Geological Society of London, Special Papers, **207**, 131–150.

POSAMENTIER, H. W. & KOLLA, V. (In press) Seismic geomorphology and stratigraphy of depositional elements in deep-water settings. *American Association of Petroleum Geologists Bulletin*.

PETTERS, S. W. & EKWEOZOR, C. M. 1982. Petroleum geology of the Benue Trough and southeastern Chad Basin, Nigeria. *American Association of Petroleum Geologists Bulletin*, **66**, 1141–1149.

SCHIEFELBEIN, C. F., ZUMBERGE, J. E., CAMERON, N. R. & BROWN, S. W. 1999. Petroleum systems in the South Atlantic margins. *In*: CAMERON. N. R., BATE, R H. & CLURE, V. S. (eds) *The Oil and Gas Habitats of the South Atlantic*. Geological Society, London, Special Publications, **153**, 169–179.

TOBIAS, R. 2000. Exploration potential of offshore Benin. *Oil and Gas Offshore West Africa, Houston, 26–28 January, Strategic Research Institute, Abstracts*.

WAGNER, T. & PLETSCH, T. 1999. Tectono-sedimentary controls on Cretaceous black shale deposition along the opening of the Equatorial Atlantic Gateway (ODP 159) *In*: CAMERON, N. R., BATE, R. H. & CLURE, V. S. (eds)

The Oil and Gas Habitats of the South Atlantic. Geological Society, London, Special Publications, **153**,
241–265.

WHITEMAN, A. 1982. *Nigeria – It's Petroleum Geology,
Resources and Potential.* Graham and Trotman, London, 394.

Frasnian organic-rich shales in North Africa: regional distribution and depositional model

S. LÜNING[1,2], K. ADAMSON[3] & J. CRAIG[4]

[1]*Royal Holloway University of London, Department of Geology, Queens Building, Egham, Surrey, TW20 0EX, UK*

[2]*Present address: University of Bremen, Geosciences, P.O. Box 330 440, 28334 Bremen, Germany (e-mail: Sebastian.Luning@gmx.net)*

[3]*Badley, Ashton & Associates Ltd, Winceby House, Winceby, Horncastle, Lincolnshire, LN9 6PB*

[4]*ENI-LASMO, London Technical Exchange, Bowater House East, 68 Knightsbridge, London, SW1X 7BN*

Abstract: During the Frasnian, organic-rich shales were deposited across much of North African, most notably in parts of Morocco, Algeria, southern Tunisia, western Libya and the Western Desert of Egypt. They are estimated to be the origin of about 10% of all Palaeozoic-sourced hydrocarbons in North Africa. The depositional, palaeoecological and geochemical characteristics of this black shale unit can be best studied in the eastern Algerian Berkine (i.e. western Ghadames) Basin where the thickest and organically richest 'hot shales' occur. In wire-line logs, the Frasnian hot shales are marked by high gamma-ray values, often in excess of 300–400 API, which, according to gamma-ray spectrometry, almost exclusively originate from an elevated uranium content. Comparison with total organic carbon (TOC) data shows that the gamma-ray curve can be used as a proxy for the TOC content of the Frasnian shales, with 150 API correlating approximately with TOCs of about 3% in eastern Algeria.

The hot shale unit usually consists of high-frequency, high-amplitude, metre-scale gamma-ray cycles; however, especially in the thicker hot shale units, the lower frequency envelope curve of the high-frequency gamma-ray cycles has a gradual, bell-shaped form. The gradual increase and subsequent decrease in organic richness over time may be interpreted as evidence for a gradual rise and subsequent fall of the oxygen minimum zone (OMZ), with invasion of oxygen-depleted waters onto the North African shelf. The rise of the OMZ may have been triggered by the Early Frasnian transgression, which has been described in detail from Morocco, where it is now well-dated by conodonts and is associated with characteristic black shales and carbonates. Additional high-resolution biostratigraphic data are still needed in order to better correlate the Frasnian hot shales of Algeria, Tunisia and western Libya with other Late Devonian dysaerobic/anaerobic facies in Morocco, western Egypt, Europe, South and North America.

Of the world's original petroleum reserves, 8% are estimated to have been generated by Late Devonian–Tournaisian source rocks (Klemme & Ulmishek 1991), which are dominantly marine (type II) (Arthur & Sageman 1994). Most of these source rocks are black shales which are widely present on the North and South American, Russian and North African cratons (e.g. Ormiston & Oglesby 1995; Duval *et al.* 1998). In North Africa, the Frasnian 'hot shales' (i.e. organically rich shales) are estimated to be the origin of about 10% of all Palaeozoic-sourced hydrocarbons (Fig. 1) (Boote *et al.* 1998) and were deposited during a major, second-order transgressional phase (Wendt & Belka 1991). These organic-rich shales, together with the basal Silurian shales (Lüning *et al.* 2000), are a major source rock in the Ghadames (Berkine) and Illizi basins (Fig. 2), with total organic carbon (TOC) values of up to 14% (Daniels & Emme 1995). Their distribution is less well defined elsewhere but they appear to have extended at least as far east as the Western Desert of Egypt (Boote *et al.* 1998) and also occur in parts of the western Algerian Ahnet, Sbaa and Reggane basins.

For this contribution, all accessible published and unpublished data of the Frasnian hot shales in North Africa are reviewed, with the aim to better

From: Arthur, T. J., MacGregor, D. S. & Cameron, N. R. (eds) *Petroleum Geology of Africa: New Themes and Developing Technologies.* Geological Society, London, Special Publications, **207**, 165–184. 0305-8719/03/$15

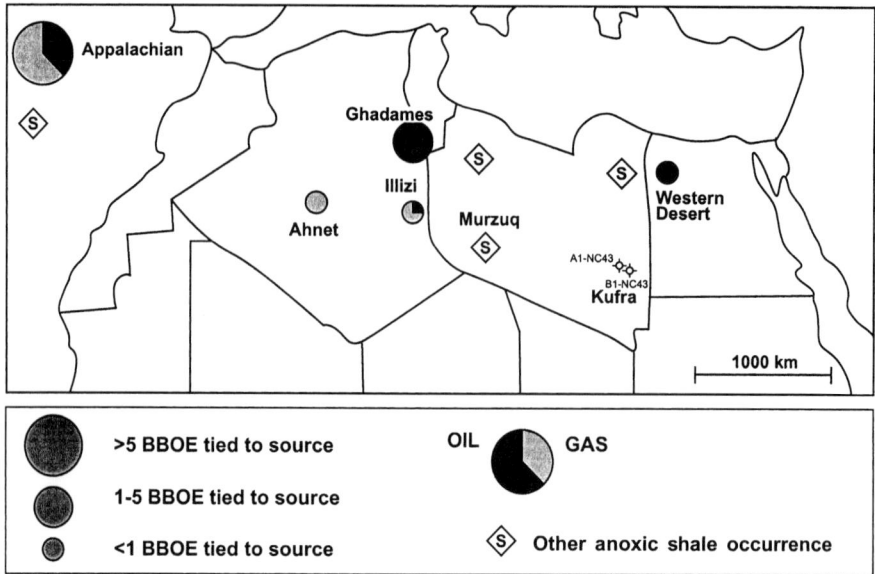

Fig. 1. Oil and gas occurrences in North Africa inferred to have been sourced by Late Devonian shales (modified after Macgregor 1996). Palaeolocation of Appalachian area and hydrocarbon occurrences sourced by Frasnian shales shown for comparison. BBOE, billion barrels of oil equivalent.

understand the depositional mechanisms and palaeoecological conditions which led to their formation. Improved models are necessary for more reliable regional Frasnian source quality predictions, especially in light of the fact that many exploration wells in the Ghadames Basin terminate in reservoir horizons well above the Frasnian hot shales.

Materials and methods

The database for this study originates from the LASMO archive and consists of some 235 wireline logs from petroleum exploration wells drilled in Algeria, Libya and Egypt by various companies (see details below), an extensive organic geochemical dataset for eastern Algeria (Geomark Research Inc., Houston, 1996, unpublished report: Ghadames and Illizi geochemical study) and palynological data (Futyan, Jawzi & Associates, Robertson Research International and Sonatrach, unpublished reports), complemented by published data. Most of the material used in the study is proprietary. Presentation of the data and interpretations here was made possible only by generous permissions granted by the owners of the data (see Acknowledgements).

Wireline logs

Wireline logs from more than 200 Algerian, 30 Libyan and 5 Egyptian petroleum exploration wells have been studied and the gamma-ray amplitudes, thicknesses and depth intervals of the hot shale unit

recorded. Well log parameters studied include lithology, total gamma ray, spectroscopy gamma ray, sonic and resistivity.

Hot shales have a typical response on wireline-logs, especially the gamma-ray, resistivity and sonic curves. Whereas the gamma-ray and resistivity values increase significantly, the sonic values decrease (e.g. Figs 3 & 4) (e.g. Meyer & Nederlof 1984; Passey *et al.* 1990; Stocks & Lawrence 1990). Increases in gamma-ray values often indicate elevated amounts of TOC where non-detrital uranium is associated with organic matter (Stocks & Lawrence 1990). A gamma-ray spectrometer can be used to measure the abundance of the three main naturally occurring radioactive elements: potassium, uranium and thorium.

In marine shales, potassium and thorium abundances are closely related because both are present mainly in the detrital clay fraction. Uranium also occurs in the detrital clay fraction, but, unlike thorium, it is also carried partly in solution as uranyl carbonate complexes (Wignall & Myers 1988; Postma & ten Veen 1999). Under reducing conditions it may be precipitated to enrich the sediment in 'authigenic' (non-detrital) uranium. The uranium is fixed at the sediment–water interface under reducing conditions and in the presence of a sorbent, which is usually organic matter or phosphate (Wignall & Myers 1988). Hence, sediments enriched in authigenic uranium tend to be

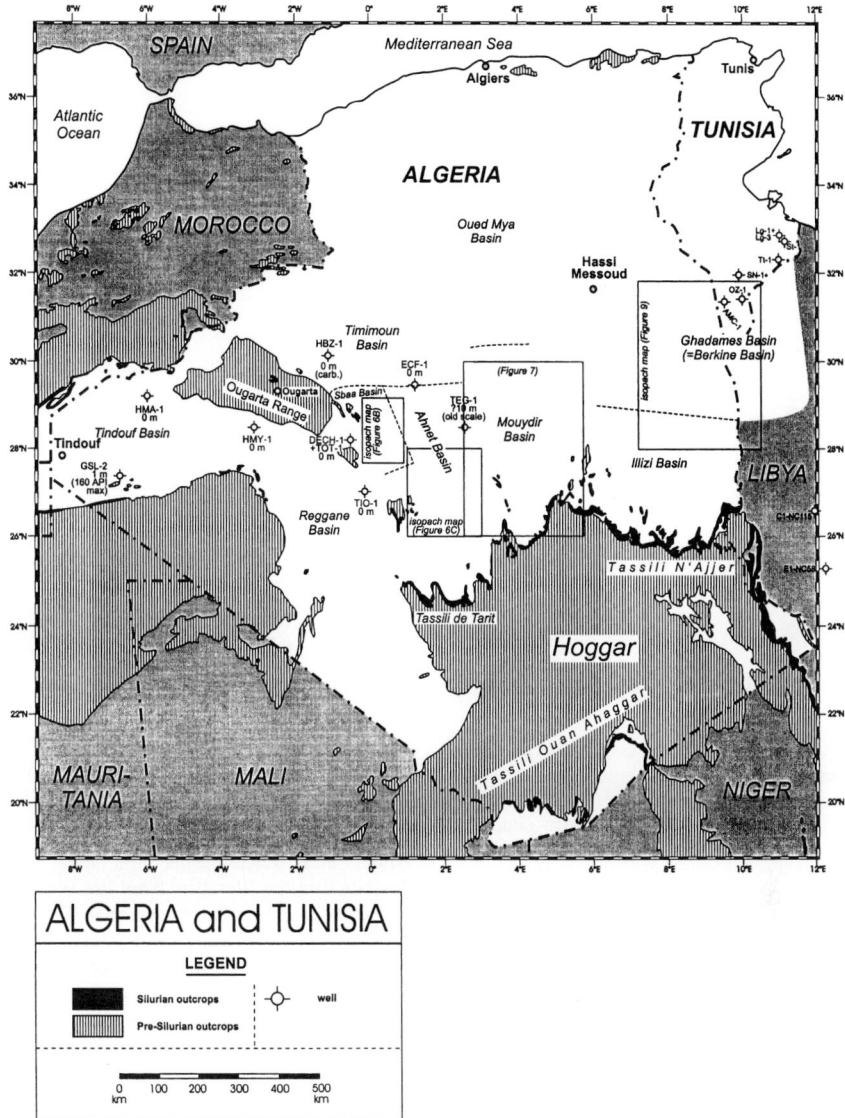

Fig. 2. Palaeozoic sedimentary basins in Algeria and thickness of Frasnian hot shales in areas not covered by maps in Figures 6, 7 and 9.

deposited under anoxic conditions that allow both large amounts of organic matter to accumulate and uranium to be fixed (Wignall & Myers 1988).

In this contribution, we define shales as 'hot' if the gamma values exceed 150 API. This value seems to correlate with TOCs of about 3% (for maturities around the oil window; lower TOCs for shales with higher maturities), as documented in data from eastern Algeria (e.g. Figs 3 & 4), which allows the use of the gamma ray as a proxy for the organic content in this case and to construct

continuous vertical TOC profiles. A similar method has been previously used by Schmoker (1980, 1981) and Fertl and Chilingarian (1990) in the eastern USA. The thickness of the hot shale is measured here from the level where the gamma-ray curve first exceeds 150 API up to the level where values drop permanently below 150 API (e.g. Fig. 3). Short intervals with gamma values <150 API are included in the hot shale unit; however, longer ones of more than 5–10 m are subtracted.

The arbitrary value of 150 API (~3% TOC)

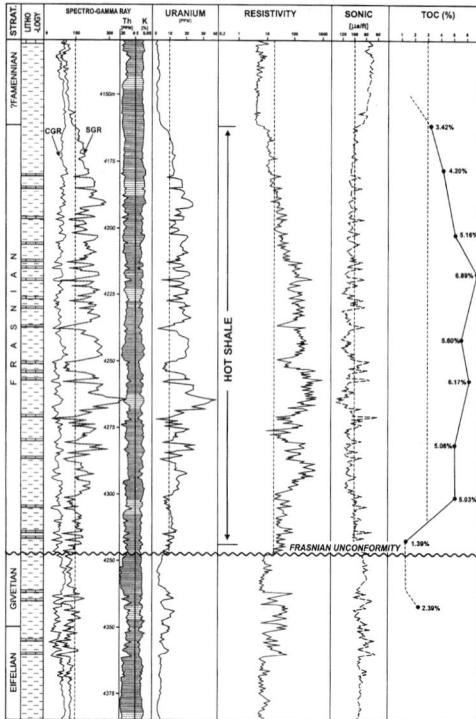

Fig. 3. Frasnian hot shale in well BKE-1 (eastern Algeria, location map in Fig. 9). Upper part of Frasnian–Famennian succession in Figure 12. CGR, computed gamma-ray (Th+K); SGR, spectro-gamma-ray (Th+K+U).

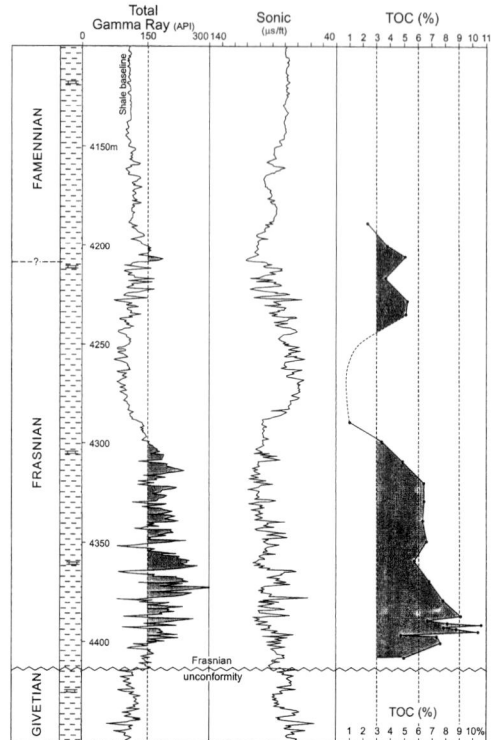

Fig. 4. Frasnian and ?Frasnian–Famennian hot shales in well BK-2 (eastern Algeria, location map in Fig. 9).

allows a clear differentiation between the organic-rich, Frasnian high-gamma shales and less organic-rich shales. Obviously, the thickness of shales with source-rock qualities (>1% TOC) is greater than that of the hot shale interval.

In practice, the comparison of absolute gamma values from different wells is not straightforward. The absolute gamma-ray values recorded vary depending on the setting of the gamma meter, e.g. varying distances between the logging tool and the borehole wall. To allow a direct comparison, it was necessary to recalibrate some of the gamma curves. In order to define a standard, the organically lean shales which make up most of the Frasnian shale succession have been assumed to have values of around 100 API ('shale baseline'), as is the case in most of the Algerian well logs analysed. In cases of mismatch, the gamma scale was modified so that the shale baseline corresponds to about 100 API.

Distribution of Frasnian organic-rich strata in North Africa

The Frasnian hot shales have been positively identified in well logs in the Ghadames (Berkine)/Illizi,

Ahnet and Sbaa basins and they have been described in the literature from the Western Desert of Egypt and the western Algerian Reggane Basin (Figs 1 & 2). Frasnian–Famennian organic-rich strata have also been reported from Morocco. Late Devonian organic-rich strata are markedly absent where areas were dominated by shallower water or terrestrial facies. An overview of the general Mid- to Late Devonian facies conditions in northwest Africa can be found in Wendt (1991).

Morocco

In the eastern Anti-Atlas of Morocco, a facies termed 'Kellwasser facies' occurs in several horizons of the Frasnian and Famennian which consists of black bituminous limestones and shales that were deposited on pelagic platforms and adjacent shallow basins (Wendt & Belka 1991). Conodonts are common to extremely abundant in these horizons, which allowed biostratigraphic dating of the organic-rich horizons with great precision (Bensaïd *et al.* 1985; Wendt & Belka 1991; Belka *et al.* 1999; see Ellwood *et al.* 1999, for cyclostratigraphy in this unit). Wendt and Belka (1991) were able to differentiate a lower organic-rich unit of the earliest Frasnian (lower *asymmetricus* Zone) from an upper organic-rich unit of the Late Frasnian–Early Famennian (Fig. 5). They also associated deposition of both organic-rich units with prominent sea-level rises. The original organic content of the 'Kellwasser deposits' in the eastern Anti-Atlas is complicated to assess, according to Wendt and Belka (1991), because of oxidization of organic matter in surface samples through weathering and present-day high levels of thermal maturity.

As will be discussed in greater detail below, the lower organic-rich unit in Morocco may well correspond to the main Frasnian hot shale in Algeria. Deposition of this anoxic/dysoxic unit clearly predates the 'Upper' and 'Lower' Kellwasser Limestone' horizons (e.g. Joachimski & Buggisch 1993; Wendt & Belka 1991) (Fig. 5) and the extinctional 'Kellwasser Event' in Europe around the Frasnian–Famennian boundary (e.g. Buggisch 1991), which has been classified as one of the seven strongest Phanerozoic faunal turnovers (e.g. Walliser 1996).

Western Algeria

Ahnet Basin The thickness of the Frasnian hot shales in the Ahnet Basin varies between 0 m and about 80 m (Fig. 6c). The hot shales are absent in the southern part of the Ahnet Basin and reach maximum values in the central and eastern parts. These thicknesses are significantly less than those published in Logan and Duddy (1998) for the

Ahnet Basin (Fig. 6a), indicating that those authors used a lower gamma-ray cut-off value to define their 'hot shales'.

While the hot shales occur close to the base of the Frasnian succession in the northwestern part of the study area, they are underlain by 75–200 m of organically lean (?Frasnian) shales in most of the wells in the eastern part of the basin (e.g. HMN-3, HMNE-1, SMH-1, GMD-2, BG-3, BZN-1), including nearly all of the wells with hot shale isopachs greater than 50 m. High-resolution biostratigraphic data is, unfortunately, not available for the Frasnian hot shales in the subsurface of the Ahnet Basin, so that the onset of dysoxia/anoxia in this basin cannot be dated accurately within the Frasnian stage.

At outcrop, Mid-Devonian carbonates are overlain by black shales with a thin bituminous carbonate horizon rich in styliolines (conical, calcareous mollusc tests) separating the two units (B. Kaufmann & J. Wendt, pers. comm. 1999). The styliolinite contains abundant conodonts which, in many localities, yield a basal Frasnian age (*pristina* to *transitans* Zone, former lower *asymmetricus* Zone). Conodonts in limestone interbeds in grey shales above the black shales seem to be of the Mid-Frasnian age (lower *gigas* Zone), pointing to an Early or Mid-Frasnian age for the black shales cropping out around the Ahnet Basin (all data B. Kaufmann & J. Wendt, pers. comm. 1999; see also Wendt *et al.* 1997).

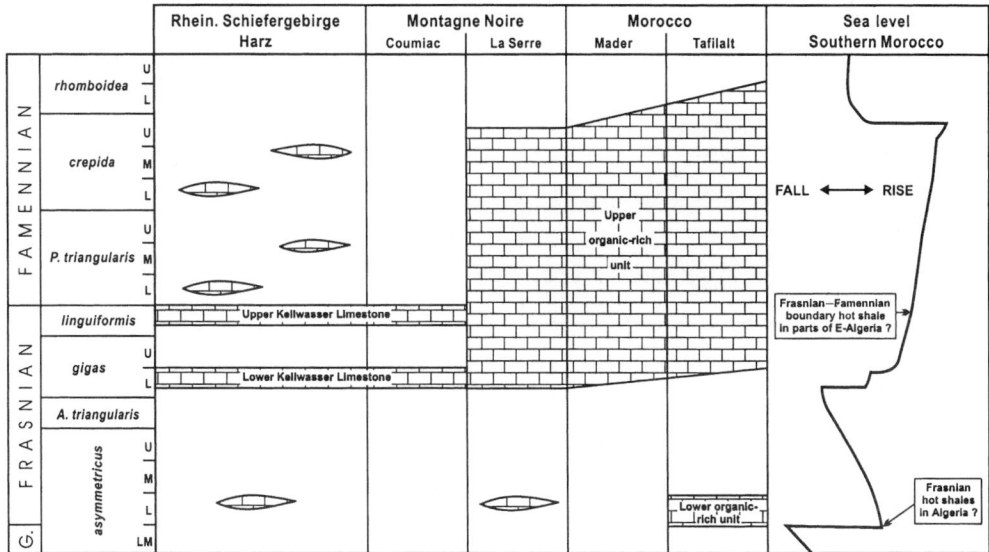

Fig. 5. Occurrence, time range and terminology of Kellwasser lithologies in Europe and North Africa, and regional sea-level curve for southern Morocco (modified after Wendt & Belka 1991). The correlation of the Algerian hot shales with these organic-rich units is currently unclear and requires biostratigraphic conodont data.

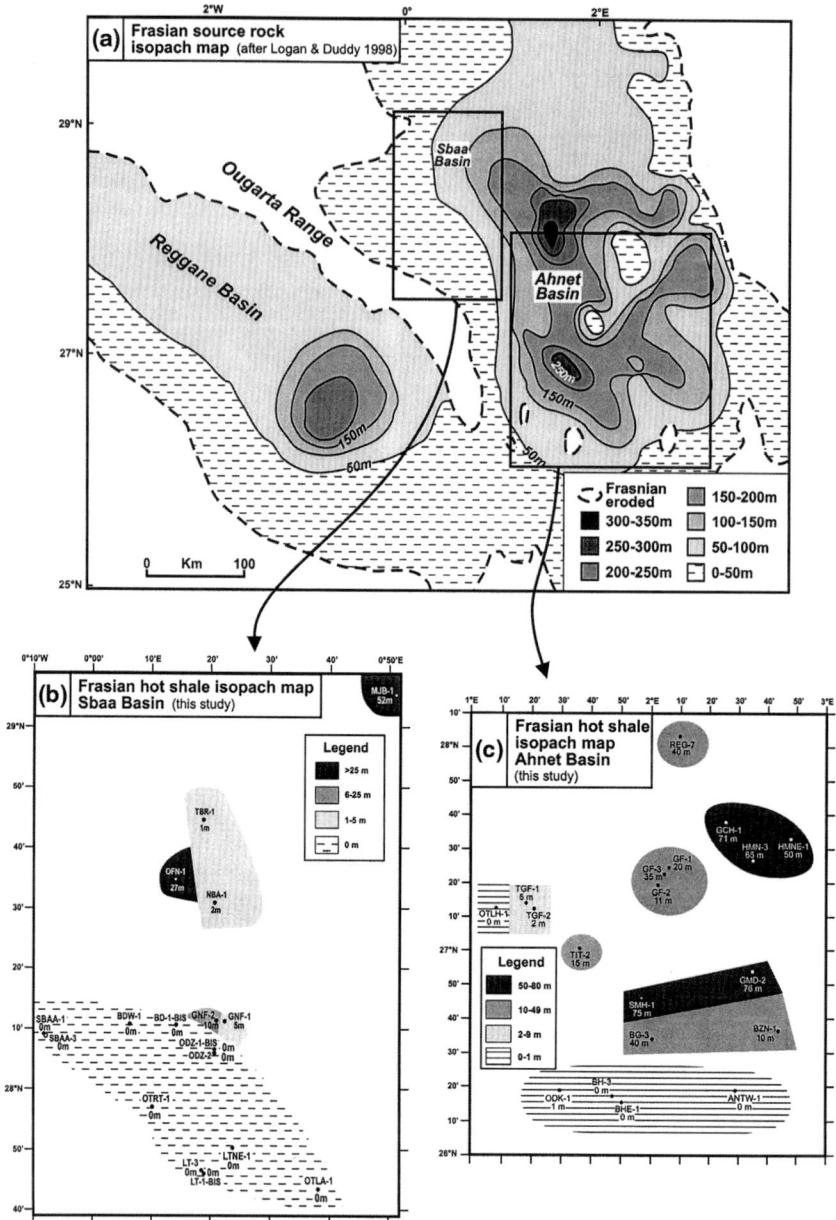

Fig. 6. Frasnian source-rock and hot shale isopach maps of western Algeria. (**a**) Frasnian source rock isopach map for the Ahnet and Reggane basins (after Logan and Duddy 1998). Locations of the newly compiled Frasnian isopach maps for (**b**) the Sbaa Basin and (**c**) the Ahnet Basin are marked. Note that thicknesses in Logan and Duddy's map significantly exceed those determined in our study, indicating that those authors used a gamma-ray cut-off value lower than the one we applied (150 API).

Sbaa Basin Frasnian hot shales are absent from the southern part of the Sbaa Basin (Fig. 6b). North of this organically lean belt, they vary in thickness between 1 m and about 50 m. In general, the organic content in the Frasnian shales of the Sbaa Basin rarely exceeds 1.2%, with the highest values reported from wells OFN-1 (2.03%) and GNF-1 (2.08%) (Fig. 6b). Present-day thermal maturities at base Frasnian level range between R_o 0.6% and R_o 1.2%. According to Boote *et al.* (1998), the Late

Devonian shales in the Sbaa Basin have sourced hydrocarbons trapped in 'Strunian' and Tournaisian sandstone reservoirs.

Peak generation from the Frasnian organic-rich shales in the Sbaa and Ahnet basins is estimated to have occurred immediately before and during the Late Hercynian deformational event (Boote *et al.* 1998). A 'heat spike' of unknown origin at about 200 Ma possibly brought the Frasnian shales of the Ahnet and Reggane basins briefly into the gas window in the Late Triassic–Early Jurassic (Logan & Duddy 1998).

Reggane Basin Frasnian hot shales also occur in the Reggane Basin but appear to be markedly thinner than in the Ahnet Basin (Logan & Duddy 1998) (Fig. 6a).

Timimoun Basin Little data is available for the Frasnian shales of the Timimoun Basin. The hot shales are clearly absent in wells HBZ-1 and ECF-1 (Fig. 2). Frasnian shales with present-day thermal maturities equivalent to the top of the oil window and with TOC values of 2.2% have been described from the Djoua Saddle which separates the Timimoun Basin from the Ahnet Basin (Anon. 1992).

Central Algeria (Oued Mya and Mouydir basins)

The Late Devonian strata have been eroded beneath the Hercynian unconformity across the whole Oued Mya Basin and large parts of the Mouydir Basin, and the subcrop of the Late Devonian succession is now restricted to the southern part of the Mouydir Basin (Figs 2 & 7). The stratigraphy of the Late Devonian in exploration wells in the area is poorly constrained and the exact age of samples analysed for TOC is mostly uncertain. Shales of possibly Frasnian age reach TOC values of up to 3.5% (Fig. 7).

The ?Frasnian organic-rich shales in the Mouydir Basin are believed to have reached their present state of maturity before the Hercynian orogeny and were already dominantly in the gas/condensate generation window before the Triassic reservoirs and the overlying Triassic–Early Jurassic seals were deposited (Boote *et al.* 1998).

Ghadames (Berkine)/Illizi Basin (eastern Algeria, Tunisia)

The Frasnian organic-rich shales of the Ghadames (Berkine)/Illizi Basin are termed 'Argile Frasnienne Radioactive' in eastern Algeria and, in southern Tunisia, are part of 'term III' (Frasnian) of the Aouinet Ouenine Formation (e.g. Aissaoui *et al.* 1996) (Fig. 8).

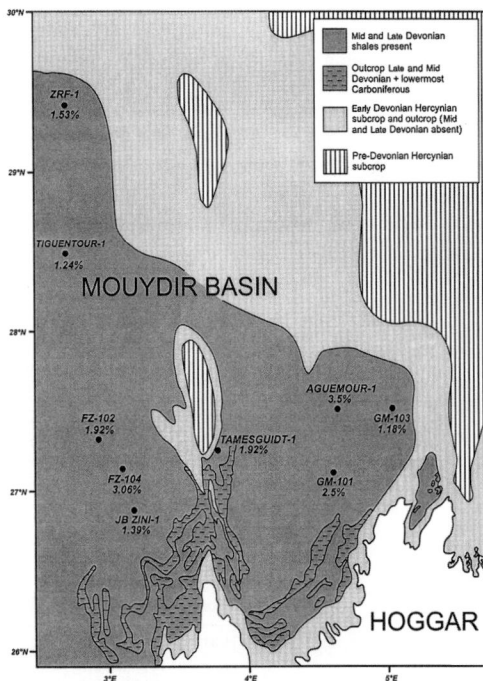

Fig. 7. Devonian subcrop beneath the Hercynian unconformity, Devonian outcrops and maximum TOC values reported from the Upper Devonian (?Frasnian) succession in the central Algerian Mouydir Basin (location of basin illustrated in Fig. 2) (from LASMO internal report).

Isopach and log characteristics The Frasnian hot shales form a laterally almost continuous unit in the Ghadames (Berkine)/Illizi Basin in eastern Algeria (Fig. 9) and are characterized by total gamma-ray values of commonly more than 300–400 API. They range in thickness from 0 m to more than 200 m (Fig. 9) and are typically thicker in the northern part of the basin than in the southern part. Similar trends are shown on a map by Daniels and Emme (1995: fig. 5), who contoured the organic richness of the Frasnian shales for eastern Algeria. The Frasnian hot shales in Algeria occur almost directly above (Figs 3 & 4) or up to 100 m above (Figs 10 & 11) the Frasnian unconformity and often form an important seismic marker. Changes in the thickness of the Frasnian hot shale were probably controlled by the Frasnian palaeorelief (see also Daniels & Emme 1995; Boote *et al.* 1998) associated with the existence of a Devonian palaeohigh in the south. Siluro-Devonian compressional tectonic activity is well documented in the region (e.g. Adamson 1999; Davidson *et al.* 2000) and may have played an important role in forming the transgressed palaeorelief.

The hot shale unit consists of high-frequency,

Stratigraphic period and epoch		E Algeria Shelmani pers. comm. (1999)	Murzuq Basin, SW Libya Seidl and Röhlich (1984)	Bellini and Massa (1980)	Belhaj (1996)	Kufra Basin, SE Libya Bellini *et al.* (1991)	Western Desert NW Egypt Keeley (1989)	Age (Ma) Harland *et al.* (1990)
Carboniferous	Visean	Series A	Mar'ar Formation	Mar'ar Formation	M'rar Formation	Dalma Formation	Desouqy Formation	349.5
	Tournaisian	Series B	Ashkidah Formation	Tahara Formation	Tahara Formation			362.5
Devonian	Famennian	Reservoir F2						
	Frasnian	Shale Series / Argile Frasnienne Radioactive	Tarut Formation / Dabdab Formation / Quttah Formation (Awaynat Wanin Group)	Awaynat Wanin IV Fm. / Awaynat Wanin III Fm. (Awaynat Wanin Group)	Aouinet Ouenine Formation (C / B)	Binem Formation	Zeitoun Formation	367 / 377.5
	Givetian	F3	Idri Formation	Awaynat Wanin II Fm.				
	Eifelian		B'ir al Qasr Formation	Awaynat Wanin I Fm.	A			381
	Emsian	Reservoir F4 F5	Ouan Kasa Formation	Ouan Kasa Formation	Ouan Kasa Formation			385
	Siegenian	Unit C3 C2 C1	Tadrart Formation	Tadrart Formation	Tadrart Formation	Tadrart Formation		390.5
	Gedinnian							396.5
		Reserv. F6 / Unit B A M	Akakus Formation	Akakus Formation	Akakus Formation	Akakus Formation	Basur Fm.	408.5
Silurian		Argile Gothlandien	Tanezzuft Formation	Tanezzuft Formation	Tanezzuft Formation	Tanezzuft Fm.	Kohla Fm.	

Fig. 8. Common stratigraphic subdivisions of the Palaeozoic units within eastern Algeria and the Murzuq Basin after Shelmani (pers. comm. 1999), Bellini and Massa (1980) and Seidl and Röhlich (1984). Note that the Gedinnian and Siegenian stage names are used here to aid comparison and correlation of data from previous studies that have used these terminologies, rather than the newly adopted Lochkovian and Pragian stage names.

high-amplitude, metre-scale gamma-ray cycles; however, especially in the thicker hot shale units, the envelope curve of the high-frequency gamma-ray cycles has a bell-shaped form with gradual increase and subsequent decrease of values (Figs 3 & 4). Gamma-ray spectroscopy reveals that the high gamma-ray values and strong cyclicity in the Frasnian hot shale is largely due to changes in the uranium content of the shales, with only small changes in the thorium and potassium content (Fig. 3). Increases in shale gamma-ray values related to uranium often reflect increases in the organic matter content (Stocks & Lawrence 1990). A comparison of TOC content data and the uranium gamma-ray curve shows that this relationship is also valid for the Frasnian shales of Algeria (Figs 3, 4, 10 & 11). The 150 API gamma-ray cut-off value correlates with TOC values of around 3%. This differs markedly from the 200 API/3% correlation found for the lowermost Silurian hot shales of the same region (Lüning *et al.* 2000). High gamma-ray values in the Frasnian hot shales correlate directly with high resistivity values and low sonic values, a relationship which is typical of organic-rich shales (Figs 3, 4, 10 & 11).

A second organic-rich interval is developed in parts of the northern Ghadames Basin, but is significantly thinner and has lower gamma and TOC values than the main Frasnian hot shale (Fig. 4, at 4200 m; Fig. 12, at 4070 m). This interval also exhibits a positive correlation between high total gamma values, high uranium content based on gamma spectroscopy, high resistivity, low sonic and high TOC content, which again allows the gamma-ray curve to be used as a proxy for the organic matter content. The stratigraphic age of this unit is currently unclear because of the low-resolution of existing palynomorph biostratigraphic data, but it is generally assumed to lie close to the Frasnian–Famennian boundary (based on regional lithological evidence).

High TOC values of locally more than 6% (e.g. Fig. 11) also occur in deltaic shales of a 'Strunian' unit (age around the Devonian–Carboniferous boundary), consisting of interbedded sandstones–siltstones and black shales (Figs 10, 11 & 12). The gamma-ray values, however, rarely exceed 150 API and gamma spectroscopy indicates that most of these gamma peaks are more related to a high thorium content rather than to uranium con-

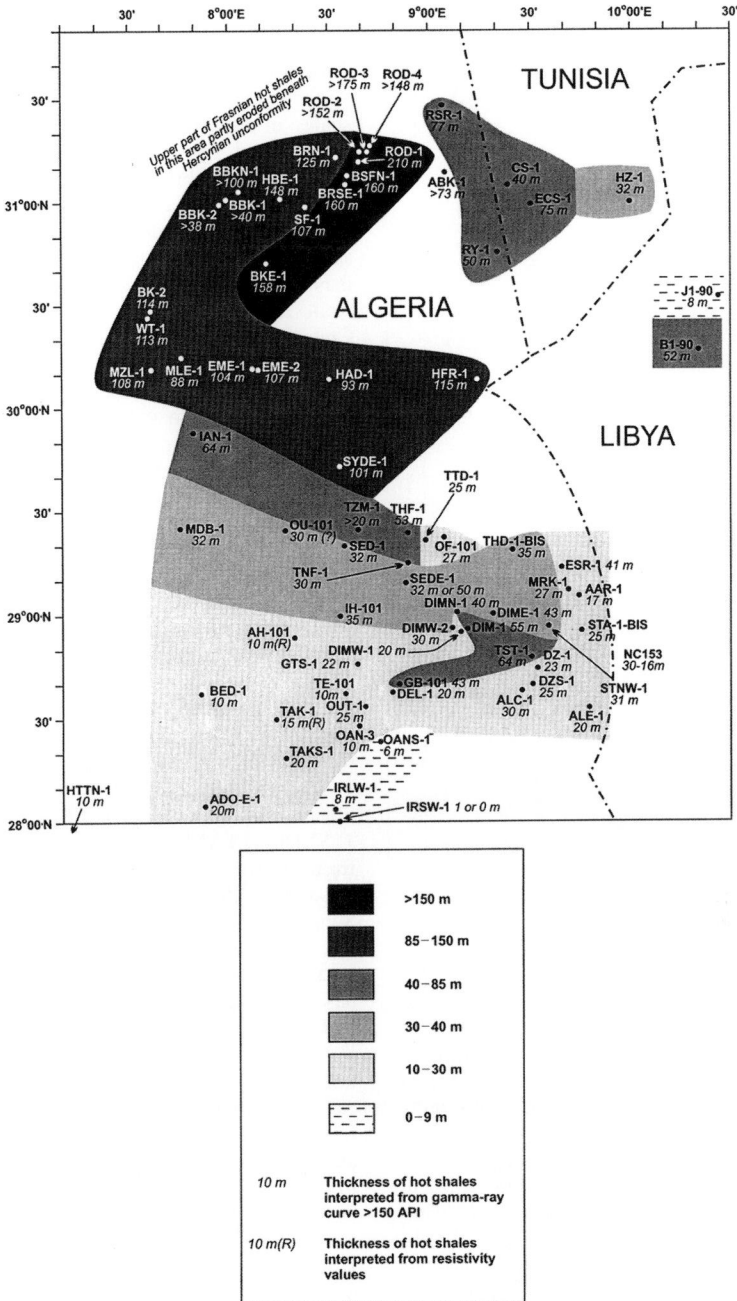

Fig. 9. Frasnian hot shale isopach map of the Algerian Berkine/Ghadames/Illizi Basin (shales >150 API). The thickest hot shales occur in the northern part of the basin, while the reduced thicknesses in the southern part are thought to be related to the presence of a structural high (Moldahar High), uplifted during Devonian times, as evidenced on seismic lines from the area.

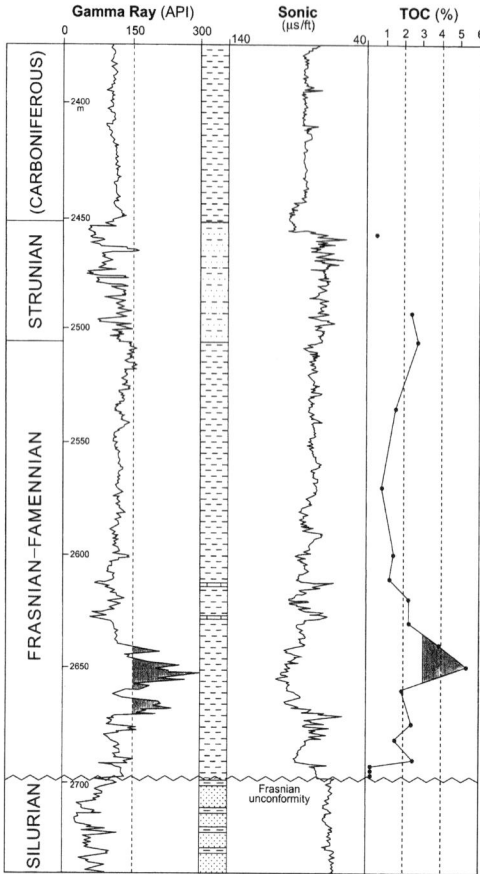

Fig. 10. Upper Devonian–Lower Carboniferous succession in well SED-E-1 (eastern Algeria, location map in Fig. 9).

tent (Fig. 12). The gamma-TOC relationship observed in the Frasnian black shales does not appear to be valid in the sandier 'Strunian' unit.

Regional trends in organic richness and maturity
The Frasnian hot shale is an important petroleum source rock in eastern Algeria because of its high organic matter content (up to 14%) and its, locally, great thickness (>200 m in the northern Ghadames Basin). Boote *et al.* (1998) reported TOC values of up to 8–14% of oil-prone type I–II kerogen from the Algerian part of the Ghadames Basin, with source quality improving to the north. The northward thickening of the Frasnian hot shale in the Ghadames (Berkine)/Illizi area (Fig. 9) is also reflected in a parallel northward increase of the maximum TOC values, from 5% in the south to a maximum of 10% in the north (see MacGregor 1998: fig. 6, based on Daniels & Emme 1995).

High TOC values for the Frasnian black shales of up to 15% have been also described from southern Tunisia (Inoubli *et al.* 1992; Aissaoui *et al.* 1996).

A contoured maturity map for the Frasnian hot shales in eastern Algeria has been published by Daniels and Emme (1995: fig. 7). The thermal maturity of the Frasnian shales in the Ghadames (Berkine)/Illizi area is described as generally mirroring that of the deeper Silurian shales but at slightly less elevated levels. It has been estimated that the main phase of hydrocarbon generation from the Frasnian hot shales started in Mid-Cretaceous times (Boote *et al.* 1998; Daniels & Emme 1995; Ghenima 1995). In the Illizi Basin, organic carbon content ranges from less than 2% in the southeast to 4–6% in the north and west (Boote *et al.* 1998: p. 46). The kerogen is predominantly oil-prone (although a mixed kerogen facies becomes more dominant to the southeast), and present-day maturities increase northwards from 1.1 R_o vitrinite reflectance in the central part of the basin to 1.3 R_o in the northeastern depocentre (Boote *et al.* 1998). It is assumed that generation and expulsion took place in the Late Cretaceous to Mid-Tertiary (Boote *et al.* 1998).

Biostratigraphic age Palynological data from six petroleum exploration wells in the Algerian part of the Ghadames (Berkine) Basin have been evaluated in order to determine the biostratigraphic age of the Frasnian hot shales (ESR-1, HBE-1, EME-1, MLE-1, SYDE-1, TAKS-1; see Fig. 9 for locations; data from unpublished reports by Futyan, Jawzi & Associates, Robertson Research International and Sonatrach). Unfortunately, the Frasnian palynomorph biostratigraphic schemes are of only low resolution, so that comparisons with the conodont-based, well-dated, organic-rich strata in Morocco and Europe are complicated.

A Frasnian age for the organic-rich, high gamma interval in the Ghadames (Berkine) Basin is indicated by variable biostratigraphic evidence which, for example, in well HBE-1, includes an age-characteristic assemblage comprising of *Polyedryxium pharaonis*, *Rhabdosporites parvulus*, *Geminospora lemurata*, *Grandispora riegelii* (all indicating an age not younger than Frasnian) and *Verrucosisporites bulliferus* (indicating an age not older than Frasnian) found in cutting samples. *Gorgonisphaeridium* spp. are very abundant and reach an abundance peak around the hot shale interval in all of the wells studied, a microfloral occurrence which is known to be typical for the Frasnian of the region. In some wells, the hot shale interval contains species which seem to be characteristic of the Early Frasnian and may indicate that the main hot shale unit was deposited during Early Frasnian times only. These tentatively Early Frasnian spec-

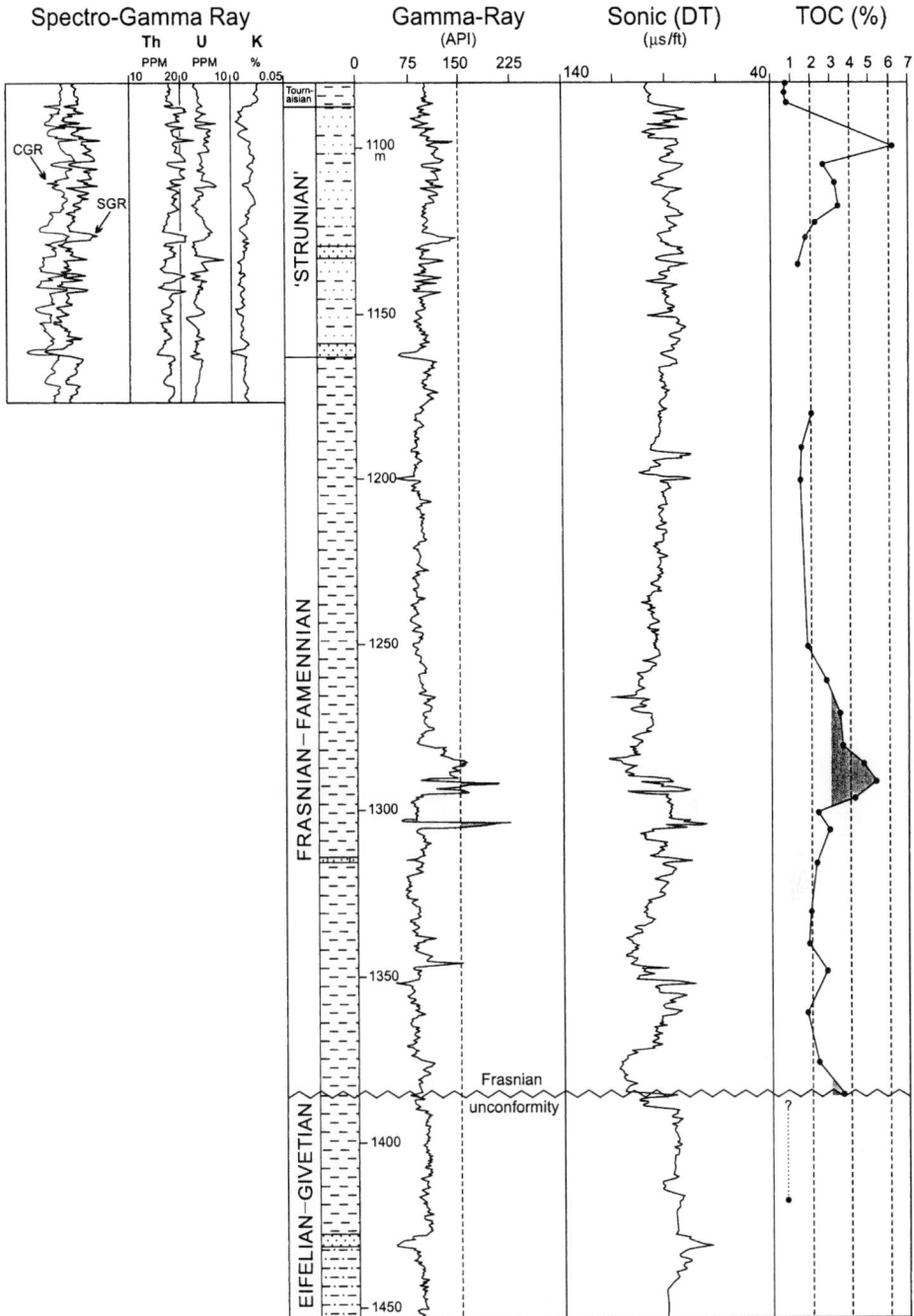

Fig. 11. Upper Devonian–Lower Carboniferous succession in well ADO-E-1 (eastern Algeria, location map in Fig. 9).

ies include *Deltotosoma intonsum* (ESR-1, MLE-1), *Fungochitina* cf. *pilosa* (HBE-1, TAKS-1), and *Duvernaysphaera tessela/angelae* (TAKS-1). In well EME-1 *Verrucosisporites bullatus* and *Lopho-*

zonotriletes media have been reported from the hot shale interval, which have been previously described from the Frasnian of Boulonnais (Taugourdeau-Lantz 1971) where they seem to

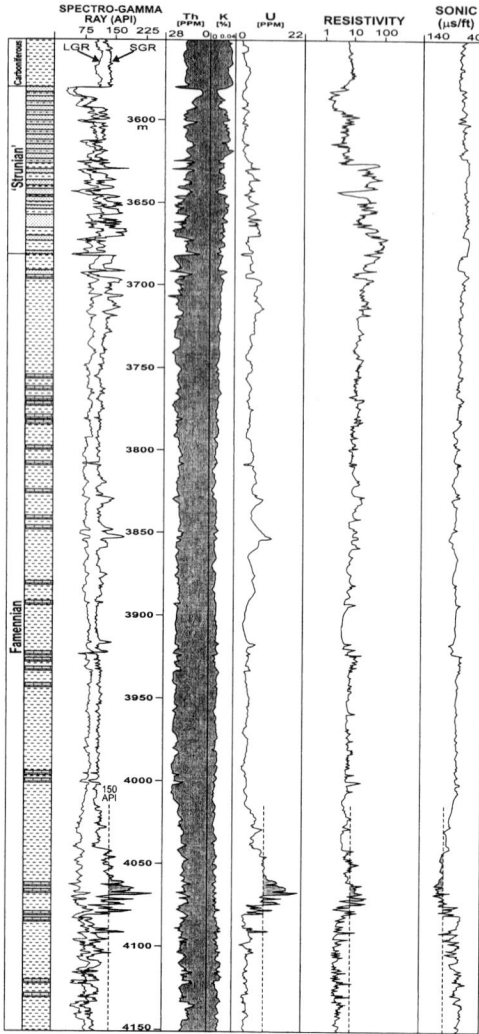

Fig. 12. Upper part of Frasnian–Famennian succession in well BKE-1 (eastern Algeria, location map in Fig. 9). A hot shale with high uranium-related gamma-ray counts occurs in the lower part of the ?Famennian shale succession (weak biostratigraphic control). Fluctuations in the total gamma-ray in the Strunian interval are mainly caused by changes in the thorium concentration.

occur in the conodont *triangularis* Zone (Mid-Frasnian) (Fig. 5).

While the exact age of the Frasnian hot shales in eastern Algeria currently remains unclear, it may be speculated that they are age-equivalent to the lower organic-rich unit in Morocco described by Wendt and Belka (1991), which has been dated using conodonts as the earliest Frasnian lower *asymmetricus* Zone (Fig. 5). A conodont study of Algerian subsurface samples from petroleum

exploration wells is needed in order to allow better intra- and intercontinental correlations of the Frasnian hot shale unit in Algeria.

Depositional environment as evidenced by paly-nology The palynomorph assemblages of the Frasnian hot shale in eastern Algeria indicate marine outer shelf conditions, based on evidence of abundant acritarchs and common chitinozoans (ESR-1), the presence of the deeper-water indicator *Umbellasphaeridium* spp. (MLE-1), scarcity of miospores (MLE-1, TAKS-1), and relatively low-diversity acritarch/algal assemblages (SYDE-1). In some wells, low diversities of the miospore assemblages and moderate diversities of the acritarch assemblages, despite high numbers of recovered specimens (wells MLE-1 and TAKS-1), suggest restricted environments and poorly oxygenated water masses. Such conditions may also be interpreted based on Tasmanaceae blooms reported by Combaz (1966) from the basal Frasnian hot shales of the 'Sahara'.

There seems to be evidence for an intrashale unconformity with a hiatus possibly including parts of the Late Frasnian and Early Famennian in all of the wells. Indicators include abrupt changes in preservation and lack of direct palynological evidence for the presence of Late Frasnian and Early Famennian strata.

The mud particles of the Frasnian organic-rich unit in the Ghadames (Berkine)/Illizi Basin are interpreted to have been fed by a major deltaic system to the south which has been described by Vos (1981).

Ghadames Basin (northwest Libya)

The Frasnian strata in the Libyan part of the Ghadames Basin are termed 'Aouinet Ouenine III Formation' (Massa & Moreau-Benoit 1976) or are attributed to the upper B/lower C units of the Aouinet Ouenine Formation (Belhaj 1996: p. 74) (Fig. 8). Although part of the same basinal system, the Libyan part of the Ghadames (Berkine) Basin is described here separately because the review is based on published data only and a direct comparison with the results from eastern Algeria is therefore complicated.

Within the Frasnian succession, a highly radioactive carbonate–black shale/marl interval occurs throughout much of the Libyan Ghadames Basin, which was named 'Cues' or '*Tornoceras* Limestone' (Massa & Moreau-Benoit 1976; Bracaccia *et al.* 1991) (Fig. 13). According to palynological biostratigraphy, this radioactive unit has probably a Late Frasnian age (Elzaroug & Lashhab 1998).

Because wireline logs from the Libyan part of the Ghadames Basin were not available for this study, an isopach map for the Frasnian radioactive interval cannot be presented here. The only isopach data available for this radioactive interval (gamma-ray values greater than 150 API) includes two wells of concession NC90 (Fig. 9) with 8 m and 52 m, and an unidentified well illustrated by Bracaccia *et al.* (1991) (Fig. 13) where the 'hot' interval attains a thickness of about 10 m. In concession NC153 the thickness of this horizon varies between 3 m and 16 m (absent in the southern part).

Frasnian palaeoenvironments in the Libyan part of the Ghadames Basin range between fully marine outer shelf and marginal marine (Massa & Moreau-Benoit 1976; Belhaj 1996; Elzaroug & Lashhab 1998), as indicated by parameters such as the palynoflora and the amount of sandstone interbeds. Pyritic, organic rich strata is notably associated predominantly with a fully marine facies (Belhadj 1996). A detailed facies and conodont-based biostratigraphic analysis of the carbonates in the *Tornoceras* Limestone is needed in order to better understand the facies and age relationship between the shale- and carbonate-dominated organic-rich

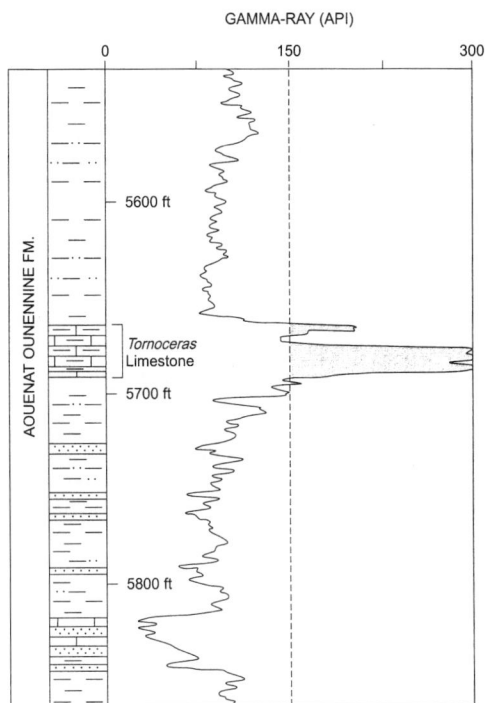

Fig. 13. Gamma-ray log of the Aouenat Ounenine Formation from the Libyan part of the Ghadames Basin, including the strongly radioactive *Tornoceras* Limestone (after Bracaccia *et al.* 1991). An API scale was not included in the figure of Bracaccia *et al.* (1991) and has been added here based on a shale baseline of about 100 API.

Frasnian units in the Algerian and Libyan parts of the Ghadames (Berkine) Basin, respectively.

Murzuq Basin (SW Libya)

Lithologies and biostratigraphy The Mid- to Late Devonian succession that crops out on the northern margin of the Murzuq Basin consists of shales, siltstones and sandstones. Brachiopod assemblages in the uppermost part of the Idri Formation (*Cyrtospirifer* cf. *verneuili* and *Cupularostrum arenosum*; Parizek *et al.* 1984) and in the overlying Quttah Formation (*Cyrtospirifer* cf. *verneuili*, *Cupularostrum arenosum*, *Cupularostrum opulentum*, *Cassidirostrum* cf. *pedderi*, *Neoglobithyris tmisanensi* and *Tarutiglossa platyfaba*; Parizek *et al.* 1984; Seidl & Röhlich 1984) indicate the presence of Early Frasnian and Frasnian strata, respectively (Fig. 8). Sediments of the Idri and Quttah formations are interpreted to have been formed in coastal-deltaic environments (Seidl and Röhlich 1984).

The overlying Dabdab Formation (Fig. 8) comprises a thin succession of ferruginous oolites, shales and siltstones. The unit contains artrypid brachiopods which indicate a pre-Famennian and, within the stratigraphic context, Frasnian age for this unit (Seidl & Röhlich 1984). The Dabdab Formation marks a major change toward a transgressive regime in the basin, with strata often onlapping tectonic unconformity surfaces. While the underlying Devonian succession comprises a number of shallowing upward sandstone-rich cycles, the Dabdab and overlying Famennian Tarut formations (Fig. 8) are shale-dominated open marine series with only minor amounts of sandstones.

Dark grey to black organic-rich shales appear to be absent in the Frasnian–Famennian succession at outcrop. However, black shales of undifferentiated Mid- to Late Devonian age have been described from many wells in the Murzuq Basin. Unfortunately, palynological data from the subsurface are scarce and it may only be speculated that parts of these black shales are of Frasnian age because of the presence of Frasnian strata at outcrop, the major regional Frasnian transgression and elevated gamma-ray values (see below).

Log characteristics. The Mid- to Late Devonian succession in the subsurface is termed the 'Aouenat Ounenine' (or 'Aweinat Wanin') Formation and is characterized on wireline logs by distinct, sea-level-driven, coarsening-upwards cycles and sharp flooding surfaces (Fig. 14) (see Adamson 1999 for details). Gamma-ray values of the shales partly exceed 150 API but rarely reach values greater than 170 API. The elevated gamma-ray values occur mainly in the middle and upper parts of the undifferentiated Mid- to Late Devonian succession, but

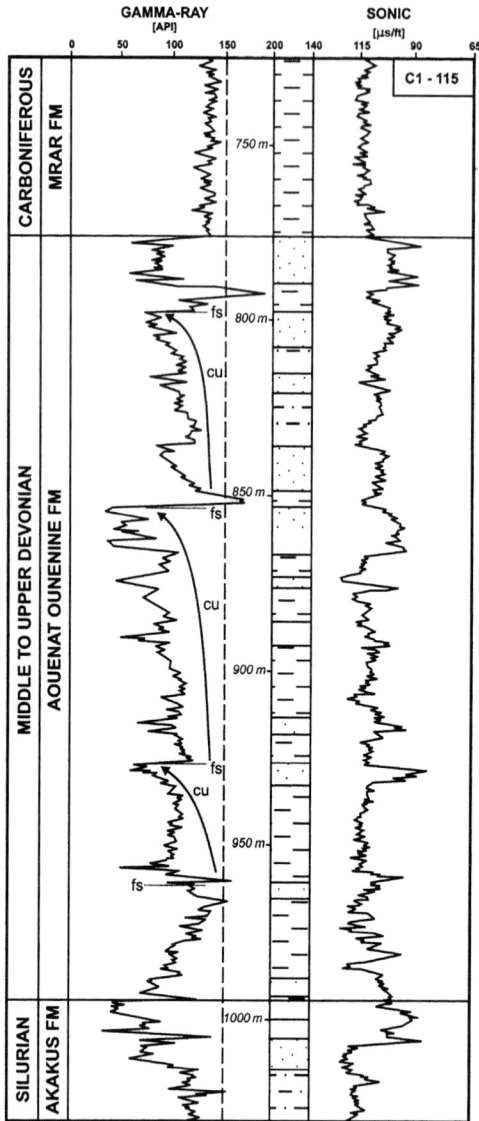

Fig. 14. Wireline log of well C1–115 in the Murzuq Basin (location in Fig. 2). Major coarsening-upward (cu) cycles, separated by flooding surfaces (fs), are developed. The gamma-ray values of shales deposited during maximum sea level are elevated, which may indicate that the anoxic/dysoxic water masses of the relatively deeper environments in the Ghadames (Berkine)/Illizi basins to the northwest (eastern Algeria) also reached the Murzuq Basin during these times.

in some wells also occur in the lower part of the unit. The cumulative thickness of the Late Devonian (?Frasnian) hot shales in the Murzuq Basin ranges between 0 m and 5 m only.

Petroleum source rock potential. The relatively low gamma-ray values, thinness, discontinuous distribution and widespread thermal immaturity all seem to restrict the importance of the Late Devonian (?Frasnian) 'hot shales' in the Murzuq Basin as a hydrocarbon source. Nevertheless, Meister *et al.* (1991) reported TOC values from Devonian shales ranging from 0.1% to 3.8%, with the highest value recorded in well E1-NC58 (R_O around 0.5–0.6), where the high-gamma-ray interval >150 API is about 2 m thick.

Regional context. The typically low organic content of the Mid- to Late Devonian shales in the Murzuq Basin is probably due to the relatively updip position of the Murzuq Basin on the North African shelf during the Late Devonian. While outer shelf conditions prevailed during deposition of the prolific, organic-rich Frasnian hot shales in eastern Algerian southwest Libya was dominated by coastal and deltaic conditions (Idri and Quttah formations) and possibly by a more open marine facies during deposition of the Frasnian Dabdab and Famennian Tarut formations.

Anoxic/dysoxic water masses of the deeper shelf apparently only reached the Murzuq Basin area during periods of high sea level. Due to the lack of biostratigraphic data, it can only be speculated that the thin high-gamma horizons in the Murzuq Basin are the updip expressions of the thick Frasnian hot shale-depositional event in eastern Algeria, with occurrences pulsed by sea-level changes.

Kufra Basin and Cyrenaica Platform (SE and NE Libya)

The Mid- to Late Devonian Binem Formation in the Kufra Basin consists of sandstones and interbedded siltstones and silty shales (Bellini & Massa 1980; Turner 1980). The depositional environment of the Binem Formation has been interpreted by Bellini *et al.* (1991: p. 2161) as 'very shallow and marginal marine' and by Turner (1980, 1991) as tidal, partly subtidal. Detailed palynological investigations in the two exploration wells A1-NC43 and B1-NC43 have proven the presence of Frasnian strata in the basin (Grigagni *et al.* 1991). According to Grigagni *et al.* (1991) the Frasnian palynoflora is characterized by a bloom of *Lophozonotriletes* spp. This may indicate that the Early Frasnian is absent because, in the Libyan part of the Ghadames Basin, such great abundances of this genus are restricted to the Late Frasnian (Massa & Moreau-Benoit 1976).

There is no TOC data available for the Frasnian shales in the Kufra Basin but high TOC values of up to 2.71% have been reported by Agip and Forum Exploration (unpublished reports) from the

Late Famennian–'Strunian' interval in wells A1-NC43 and B1-NC43 (Fig. 1). The organic matter in this latter interval is dominantly type-IV kerogen and has poor hydrocarbon-generating potential (Forum Exploration, unpublished report).

Devonian (including Late Frasnian) palynomorph assemblages have been described by Paris *et al.* (1985) from wells on the Cyrenaica Platform in northeastern Libya, but without lithological or petrophysical log details.

Western Desert (northwest Egypt)

The Devonian (Gedinnian–Strunian) clastic sediments in the Western Desert of Egypt (Fig. 15) are grouped into the Zeitoun Formation and are characterized by a general coarsening-upward cycle (Keeley 1989). Shales are reported as volumetrically the most important lithotypes, being locally rich in pyrite and sapropel (Keeley 1989;

Hantar 1990). The organically richest Frasnian shales in the Western Desert (Ghazalat) Basin are reported from the northwestern part of this basin by Futyan, Jawzi and Associates (1990, unpublished). Sapropelic laminated shales in the Zeitoun Formation have been described by McGarva (1986). In the Foram-1 well at the southwestern margin of the Western Desert Basin, gamma-ray values of grey to dark grey Late Devonian shales reach values of about 170 API (original scale modified according to shale baseline) and the maximum TOC value is 6.27 wt%.

The Late Devonian shales in the Western Desert of Egypt have been described as potential petroleum source rocks by Keeley & Massoud (1998) and Macgregor (1996: fig. 8). Due to the lack of biostratigraphic data, it remains unclear how the organic-rich Devonian shales in the Western Desert correlate with the Frasnian hot shales of Algeria.

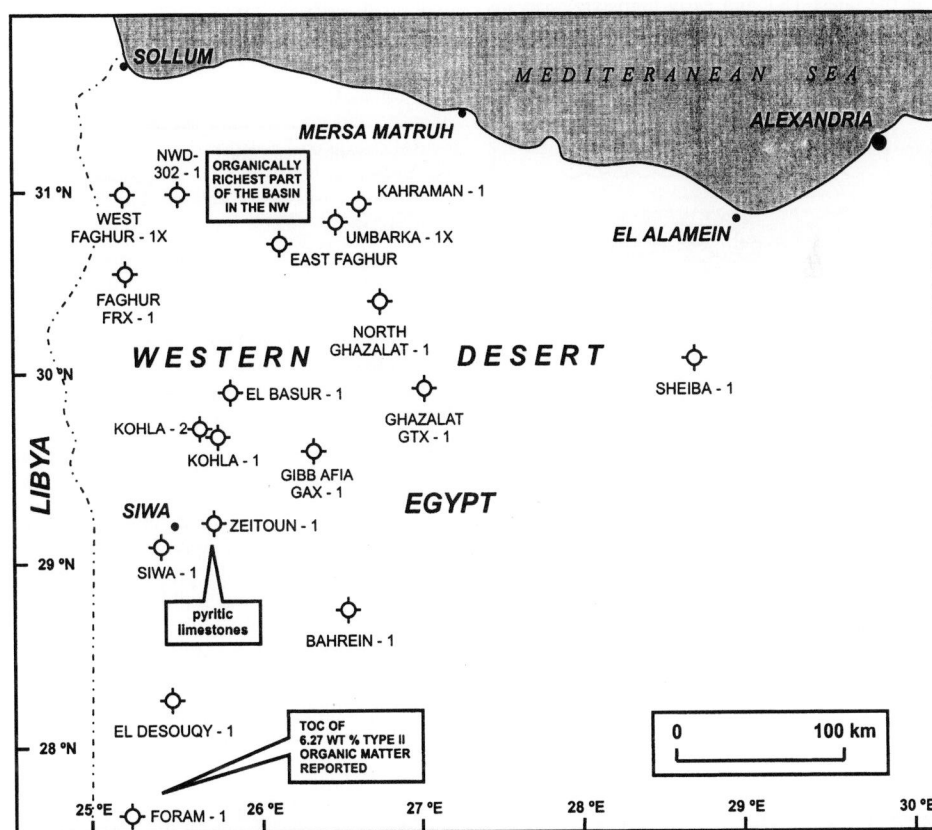

Fig. 15. Location map of wells in the Western Desert of Egypt and characteristics of Upper Devonian black shale unit (base map after Keeley 1989).

Black shale depositional model

The mechanisms of Frasnian black shale deposition in North Africa may be best understood by analysing the detailed datasets available for the Frasnian in the Algerian Berkine (i.e. western Ghadames) Basin. Here, the Frasnian shales reach their maximum thickness, which may be evidence for maximum stratigraphic completeness, and the anoxia/dysoxia was most intense.

The Frasnian hot shales were deposited during a regional transgression and in many areas they directly overlie the Frasnian unconformity (e.g. wells BKE-1 and BK-2; Figs 3 & 4). The onset of Frasnian shale sedimentation is probably associated with the earliest Frasnian eustatic sea-level rise, as recorded in Morocco by Wendt and Belka (1991) (Fig. 5). The palynofacies of the Frasnian hot shales in eastern Algeria indicates outer shelf conditions, with shallower-water deposits of the transgressive systems tract (TST) apparently absent. In agreement with Fekirine and Abdallah (1998), the hot shales are therefore interpreted as having been deposited around the maximum flooding surface during the latest TST and early highstand systems tract (HST). Non-deposition during TST times may

have been related to the rapidity of the Frasnian flooding event (see also sea-level curve of Wendt & Belka 1991; Fig. 5).

The bell-shaped form of the gamma-ray curve and vertical TOC distribution of the ?Early Frasnian hot shales indicate that anoxia/dysoxia developed and faded gradually rather than abruptly (e.g. Figs 3 & 4). This development may be best explained by flooding of oxygen-depleted water masses onto the North African shelf, triggered by a rise in sea level and an associated rise in the oxygen minimum zone (OMZ) (Fig. 16). It is now well established that the Late Devonian was a time of widespread anoxic bottom waters (McGhee 1996). After the onset of Frasnian shale deposition, sea level and the OMZ were still rising (late TST). Maximum oxygen deficiency and associated maximum organic richness developed slightly later, around the peak sea level when, also, the OMZ reached its highest position. Eventually, sea level and OMZ began to fall gradually, as reflected in the gradual decrease in gamma-ray and TOC values.

A similar depositional/ecological mechanism with a fluctuating OMZ was proposed for the (probably younger) organic-rich strata at the Frasn-

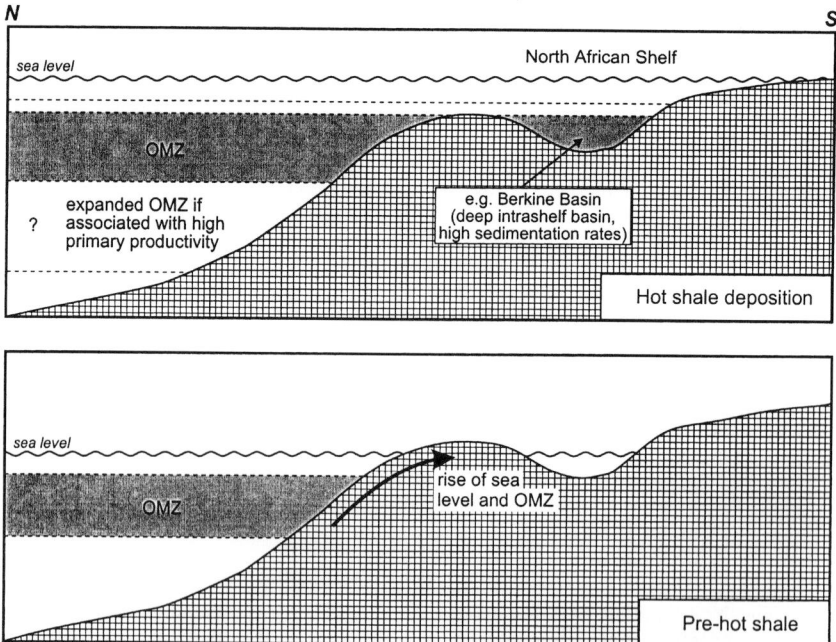

Fig. 16. Model of palaeoceanographic changes associated with the ?Early Frasnian sea-level rise in North Africa. It is assumed that the rise in sea level triggered a rise in the oxygen minimum zone (OMZ), flooding the shelf with oxygen-poor water masses. Due to the lack of detailed faunal and floral data, it remains unclear whether a high primary productivity setting existed, which may have also led to an overall expanded OMZ. Maximum hot shale thicknesses are expected to have occurred in deep intrashelf basins which were characterized by high sedimentation rates and strong oxygen deficiency.

ian–Famennian boundary (Wilde & Berry 1984, 1986; Walliser 1986, 1996; Walliser *et al.* 1988; Schindler 1990). A high primary productivity origin for the Frasnian hot shales in North Africa does not seem likely because palaeoclimate modelling by Ormiston and Oglesby (1995) (in contrast to earlier results by Parrish *et al.* 1983) suggest that no major upwelling took place along the North African coast during the Late Devonian.

The high-frequency cyclicity superimposed on the bell-shaped gamma-ray curve and repetitive occurrence of carbonate interbeds indicate that the overall anoxia was frequently disrupted and the OMZ may have oscillated vertically. Possible causes of these disruptions include high-frequency sea-level changes, Milankovitch effects, changes in seasonal effects and others.

A second hot shale occurs in parts of the Ghadames (Berkine) Basin (e.g. in wells BKE-1 and BK-2; Figs 12 & 4) and has been provisionally interpreted as having an earliest Famennian or Frasnian–Famennian boundary age, based on lithological and limited palynological evidence. This organic-rich unit, like the underlying main Frasnian hot shale, exhibits a 'bell-shaped' gradual increase and subsequent decrease in gamma values and in TOC. It could be speculated that this unit represents in part the upper organic-rich unit described by Wendt and Belka (1991) from Morocco (Fig. 5), and that it is associated with a renewed, though short-lived, rise in sea level and the OMZ. It may be during this time that the Frasnian–Fammenian extinction event took place (e.g. McGhee 1996; Walliser 1996).

The apparent differences in facies, gamma-ray character and vertical TOC distribution indicate that a different black shale depositional model may apply for the deltaic Strunian unit in eastern Algeria (e.g. in wells ADO-E-1, SED-E-1; Figs 10 & 11). The organic richness of these shale deposits may be associated with the 'Hangenberg Event', a large mass-extinction event close to the Devonian–Carboniferous boundary which occurs at the base of a globally extensive black shale (e.g. Walliser 1996; Caplan & Bustin 1999).

Conclusions

Frasnian black shales form an important petroleum source rock in North Africa and account for about 10% of all Palaeozoic-sourced hydrocarbons in North Africa. In well logs these ('hot') shales are characterized by high gamma-ray values, often in excess of 300–400 API, which, according to gamma-ray spectrometry, almost exclusively originate from an elevated uranium content. Comparison with geochemical data shows that the gamma-ray curve can be used as a proxy for the TOC con-

tent of the Frasnian shales. Based on 235 well logs and associated geochemical data from Algeria, Libya and Egypt, the vertical and lateral distribution of the organic richness of the shales has been studied across North Africa. Thickness and source quality of the Frasnian hot shales vary significantly across the region, with the organically richest and thickest hot shale package developed in eastern Algeria. The palaeorelief in combination with the sea-level-related rise and fall of the oxygen minimum zone (OMZ) seem to have been the key controlling depositional mechanisms. More high-resolution biostratigraphic data is needed to improve correlations of this anoxic event within the region and with other parts of the world, e.g. North America (e.g. Schmoker 1980, 1981).

The study was sponsored by LASMO plc. We thank the following national oil companies, oil companies and consultancies for kindly granting permission to use data in this contribution: Agip, Anadarko, BHP, ETAP, Geomark, Korea National Oil Corporation, LASMO, Maersk, NOC, Repsol, and Sonatrach. We are grateful for valuable comments to J. Wendt, B. Kaufmann (both University of Tübingen), Richard Hedley (LASMO), B. Thusu (AGOCO and University College London), B. Schöne, O. H. Walliser (both University of Göttingen), M. Shelmani (ALEPCO, Algeria), J. Schieber (University of Texas at Arlington), J. Warme, A. Y. Huc and J. Macquaker (University Manchester). We acknowledge technical support provided by W. Whittingham (LASMO technical library), Instant Library, Wave Graphics and A. Marzouk (Tanta University, Egypt). We thank the two anonymous reviewers for their useful suggestions which helped to further improve the manuscript.

References

ADAMSON, K. 1999. *Evolution of the Murzuq Basin, Southwest Libya, and Surrounding Region During the Devonian.* Ph.D. thesis, University of Wales, Aberystwyth.

AISSAOUI, N., ACHECHE, M. H., BEN YACOUB, J., M'RABET, A., NEHDI, T., PEZZINO, F. & SMAOUI, J. 1996. Silurian Acacus new play in southern Tunisia. *Proceedings of the 5th Tunisian Petroleum Exploration Conference, Tunis, 15–18 October 1996. Enterprise Tunisienne d'Activites Petrolières (ETAP) Memoirs,* **10**, 1–14.

ANON. 1992. The Frasnian: Hydrocarbon potential source rock in the Central Algerian Sahara – Modeling of generation and migration eras. *29th International Geological Congress, Kyoto, Japan, August–September 1992, Abstracts,* **3**, 814.

ARTHUR, M.A. & SAGEMAN, B. B. 1994. Marine black shales: Depositional mechanisms and environments of ancient deposits. *Annual Reviews of Earth and Planetary Science,* **22**, 499–551.

BELHAJ, F. 1996. Palaeozoic and Mesozoic stratigraphy of eastern Ghadamis and western Sirt Basins. *In:* SALEM, M. J., MOUZUGHI, A. J. & HAMMUDA, O. S.

(eds) *The Geology of Sirt Basin*, vol. 1. Elsevier, Amsterdam, 57–96.

BELKA, Z., KLUG, C., KAUFMANN, B., KORN, D., DÖRING, S., FEIST, R. & WENDT, J. 1999. Devonian conodont and ammonoid succession of the eastern Tafilalt (Ouidane Chebbi section), Anti-Atlas, Morocco. *Acta Geologica Polonica*, **49**, 1–23.

BELLINI, E. & MASSA, D. 1980. A stratigraphic contribution to the Palaeozoic of the southern basins of Libya. *In:* SALEM, M. J. & BUSREWIL, M. T. (eds) *The Geology of Libya*, vol. 1. Academic Press, London, 3–56.

BELLINI, E., GIORI, I., ASHURI, O. & BENELLI, F. 1991. Geology of Al Kufrah Basin, Libya. *In:* SALEM, M. J., SBETA, A. M. & BAKBAK, M. R. (eds) *The Geology of Libya*, vol. 6. Elsevier, Amsterdam, 2155–2184.

BENSAÏD, M., BULTYNCK, P., SARTENAER, P. WALLISER, O. H. & ZIEGLER, W. 1985. The Givetian–Frasnian boundary in pre-Sahara Morocco. *Courier Forschunginstitut Senckenberg*, **75**, 287–300.

BOOTE, D. R. D., CLARK-LOWES, D. D. & TRAUT, M. W. 1998. Palaeozoic petroleum systems of North Africa. *In:* MACGREGOR, D. S., MOODY, R. T. J. & CLARK-LOWES, D. D. (eds) *Petroleum Geology of North Africa*. Geological Society, London, Special Publications, **132**, 7–68.

BRACACCIA, V., CARCANO, C. & DRERA, K. 1991. Sedimentology of the Silurian–Devonian Series in the southeastern part of the Ghadamis Basin. *In:* SALEM, M. J. & BELAID, M. N. (eds) *The Geology of Libya*, vol. 5. Elsevier, Amsterdam, 1727–1744.

BUGGISCH, W. 1991. The global Frasnian–Famennian 'Kellwasser event'. *Geologische Rundschau*, **80**, 49–72.

CAPLAN, M. L. & BUSTIN, R. M. 1999. Devonian–Carboniferous Hangenberg mass extinction event, widespread organic-rich mudrock and anoxia: causes and consequences. *Palaeogeography, Palaeoclimatolology, Palaeoecology*, **148**, 187–207.

COMBAZ, A. 1966. Remarques sur les niveaux a Tasmanacées du Paléozoïque Saharien: *Palaeobotanist*, **15**, 29–34.

DANIELS, R. P. & EMME, J. J. 1995. Petroleum system model, Eastern Algeria, from source rock to trap: when, where and how. *Proceedings of the Seminar on Source Rocks and Hydrocarbon Habitat in Tunisia, Tunis, 15–18 November 1995. Enterprise Tunisienne d'Activites Petrolières (ETAP) Memoirs*, **9**, 101–124.

DAVIDSON, L., BESWETHERICK, S. *et al.* 2000. The structure, stratigraphy and petroleum geology of the Murzuq Basin, southwest Libya. *In:* SOLA, M. A. & WORSLEY, D. (eds) *Geological Exploration in Murzuq Basin*. Elsevier, Amsterdam, 295–320.

DUVAL, B. C., CRAMEZ, C. & VAIL, P. R. 1998. Stratigraphic cycles and major marine source rocks. In: de Graciansky, P.-C., Hardenbol, J., Jaquin, T. & VAIL, P. R. (eds) *Mesozoic and Cenozoic Sequence Stratigraphy of European Basins*. Society of Economic Paleontologists and Mineralogists Special Publications, **60**, 43–51.

ELLWOOD, B. B., CRICK, R. E. & EL HASSANI, A. 1999. The magneto-susceptibility event and cyclostratigraphy (MSEC) method used in geological correlation of Devonian rocks from Anti-Atlas Morocco. *American Association of Petroleum Geologist Bulletin*, **83**, 1119–1134.

ELZAROUG, R. & LASHHAB, M. I. 1998. Palynostratigraphy and palynofacies of subsurface Devonian (Middle–Upper) strata of Al Wafd Field. *Geological Conference on Exploration in Murzuq Basin, 20–22 Sep 1998, Sabha, Libya, Abstracts*, 38.

FEKIRINE, B. & ABDALLAH, H. 1998. Palaeozoic lithofacies correlatives and sequence stratigraphy of the Saharan Platform, Algeria. *In:* MACGREGOR, D. S., MOODY, R. T. J. & CLARK-LOWES, D. D. (eds) *Petroleum Geology of North Africa*. Geological Society, London, Special Publications, **132**, 97–108.

FERTL, W. H. & CHILINGARIAN, G. V. 1990. Hydrocarbon resource evaluation in the Woodford shale using well logs: *Journal of Petroleum Science and Engineering*, **4**, 347–357.

GHENIMA, R. 1995. Hydrocarbon generation and migration in the Ghedames Basin – application to the filling history of the El Borma oil field. *Proceedings of the Seminar on Source Rocks and Hydrocarbon Habitat in Tunisia. Tunis, 15–18 November 1995. Enterprise Tunisienne d'Activites Petrolières (ETAP) Memoirs*, **9**, 3–15.

GRIGAGNI, D., LANZONI, E. & ELATRASH, H. 1991. Palaeozoic and Mesozoic subsurface palynostratigraphy in the Al Kufrah Basin, Libya. *In:* SALEM, M. J., HAMMUDA, O. S. & ELIAGOUBI, B. A. (eds) *The Geology of Libya*, vol. 4. Elsevier, Amsterdam, 1159–1227.

HANTAR, G. 1990. North Western Desert. *In:* R. SAID (ed.) *The Geology of Egypt*. Balkema, Rotterdam, 293–319.

HARLAND, W. B., ARMSTRONG, R. L., COX, A. V., CRAIG, L. E., SMITH, A. G. & SMITH, D. G. 1990. *A Geologic Time Scale 1989*. Cambridge University Press, Cambridge.

INOUBLI, H., ACHECHE, M. H., SAIDI, M., & BELAYOUNI, H. 1992. Geochimie des roches meres de petrole dans l'extreme Sud Tunisien. *Proceedings of the 3rd Enterprise Tunisienne d'Activites Petrolières (ETAP) Conference, Tunis, May 1992*, 235–248.

JOACHIMSKI, M. M. & BUGGISCH, W. 1993. Anoxic events in the Late Frasnian – Causes of the Frasnian–Famennian faunal crisis? *Geology*, **21**, 675–678.

KEELEY, M. L. 1989. The Palaeozoic of the Western Desert of Egypt: *Basin Research*, **2**, 35–48.

KEELEY, M. L. & MASSOUD, M. S. 1998. Tectonic controls on the petroleum geology of NE Africa. *In:* MACGREGOR, D. S., MOODY, R. T. J. & CLARK-LOWES, D. D. (eds) *Petroleum Geology of North Africa*. Geological Society, London, Special Publications, **132**, 265–281.

KLEMME, H. D. & ULMISHEK, G. F. 1991. Effective petroleum source rocks of the world: Stratigraphic distribution and controlling depositional factors. *American Association of Petroleum Geologist Bulletin*, **75**, 1809–1851.

LOGAN, P. & DUDDY, I. 1998. An investigation of the thermal history of the Ahnet and Reggane basins, Central Algeria, and the consequences for hydrocarbon generation and accumulation. *In:* MACGREGOR, D. S., MOODY, R. T. J. & CLARK-LOWES, D. D. (eds) *Pet-*

roleum Geology of North Africa. Geological Society, London Special Publication, **132**, 131–155.

LÜNING, S., CRAIG, J. LOYDELL, D. K., ŠTORCH, P. & FITCHES, B. 2000. Lower Silurian 'hot shales' in North Africa and Arabia: regional distribution and depositional model. *Earth Science Reviews*, **49**: 121–200.

MACGREGOR, D. S. 1996. The hydrocarbon systems of North Africa. *Marine & Petroleum Geology*, **13**, 329–340.

MACGREGOR, D. S. 1998. Giant fields, petroleum systems and exploration maturity of Algeria. *In*: MACGREGOR, D. S., MOODY, R. T. J. & CLARK-LOWES, D. D. (eds) *Petroleum Geology of North Africa*. Geological Society, London, Special Publications, **132**, 79–96.

MASSA, D. & MOREAU-BENOIT, A. 1976. Essai de synthèse stratigraphique et palynologique du Système Dévonien en Libye occidentale. *Revue de l'Institut Français du Pétrole*, **31**, 287–332.

MCGARVA, M. M. 1986. Sedimentology, petrology and diagenesis of the Zeitoun Formation (Devonian) of the Western Desert, Egypt. *Egyptian General Petroleum Company (EGPC), Proceedings of the 8th Exploration Conference, Cairo*, 127–134.

MCGHEE, G. R. 1996. *The Late Devonian Mass Extinction. Critical Moments in Palaeobiology and Earth History Series*, Columbia University Press, New York.

MEISTER, E. M., ORITZ, E. F., PIEROBON, E. S. T., ARRUDA, A. A. & OLIVEIRA, M. A. M. 1991. The origin and migration fairways of petroleum in the Murzuq Basin, Libya: An alternative exploration model. *In*: SALEM, M. J., BUSREWIL, M. T. & BEN ASHOUR, A. M. (eds) *The Geology of Libya*, vol. 7. Elsevier, Amsterdam, 2725–2741.

MEYER, B. L. & NEDERLOF, M. H. 1984. Identification of source rocks on wireline logs by density/resistivity and sonic transit time/resistivity crossplots. *American Association of Petroleum Geologist Bulletin*, **68**, 121–129.

ORMISTON, A. R. & OGLESBY, R. J. 1995. Effect of Late Devonian paleoclimate on source rock quality and location. *In*: HUC, A.-Y. (ed) *Paleogeography, Paleoclimate, and Source Rocks*. American Association of Petroleum Geologists Studies in Geology, **40**, 105–132.

PARIS, F., RICHARDSON, J. B., RIEGEL, W. STREEL, M. & VANGUESTAINE, M. 1985. Devonian (Emsian–Famennian) palynomorphs. *In*: THUSU, B. & OWENS, B. (eds) *Palynostratigraphy of North-East Libya. Journal of Micropalaeontology*, **4**, 49–82.

PARIZEK, A., KLEN, L. & RÖHLICH, P. 1984. *Explanatory Booklet for the Geological Map of Libya 1:250.000, Sheet Idri, NG33–1*. Industrial Research Centre, Tripoli, 108 pp.

PARRISH, J. T., ZIEGLER, A. M. & HUMPHREYVILLE, R. E. 1983. Upwelling in the Palaeozoic era. *In*: THIEDE, J. & SUESS, E. (eds) *Coastal Upwelling: Its Sedimentary Record*. NATO Conference Series IV, 10B, Plenum Press, New York, 553–578.

PASSEY, Q. R., CREANEY, S, KULLA, J. B., MORETTI, F. J. & STROUD, J. D. 1990. A practical model for organic richness from porosity and resistivity logs. *American Association of Petroleum Geologist Bulletin*, **74**, 1777–1794.

POSTMA, G. & TEN VEEN, J. H. 1999. Astronomically and tectonically linked variations in gamma-ray intensity in Late Miocene hemipelagic successions of the Eastern Mediterranean Basin: *Sedimentary Geology*, **128**, 1–12.

SCHINDLER, E. 1990. *Die Kellwasser-Krise (hohe Frasne-Stufe, Ober-Devon)*. Göttinger Arbeiten Geologie Paläontologie, Göttingen, **46**.

SCHMOKER, J. W. 1980. Organic content of Devonian shale in Western Appalachian Basin. *American Association of Petroleum Geologist Bulletin*, **64**, 2156–2165.

SCHMOKER, J. W. 1981. Determination of organic-matter content of Appalachian Devonian shales from gamma-ray logs. *American Association of Petroleum Geologist Bulletin*, **65**, 1285–1298.

SEIDL, K. & RÖHLICH, P. 1984. *Explanatory Booklet for the Geological Map of Libya 1:250.000, Sheet Sabhà, NG33–2*. Industrial Research Centre, Tripoli.

STOCKS, A. E. & LAWRENCE, S. R. 1990. Identification of source rocks from wireline logs. *In*: Hurst, A., Lovell, M. A. & MORTON, A. C. (eds) *Geological Applications of Wireline Logs*. Geological Society, London, Special Publications, **48**, 241–252.

TAUGOURDEAU-LANTZ, J. 1971. Les spores du Frasnien d'une région privilégiée, le Boulonnais. *Memoires de la Société géologique de France nouvelle. Serie Paléobotanique*, **114**.

TURNER, B. R. 1980. Palaeozoic sedimentology of the southeastern part of Al Kufrah Basin, Libya: a model for oil exploration. *In*: SALEM, M. J. & BUSREWIL, M. T. (eds) *The Geology of Libya*, vol. 2. Academic Press, London, 351–374.

TURNER, B. R. 1991. Palaeozoic deltaic sedimentation in the southeastern part of Al Kufrah Basin, Libya. *In*: SALEM, M. J. & BELAID, M. N. (eds) *The Geology of Libya*, vol. 5. Elsevier, Amsterdam, 1713–1726.

VOS, R. G. 1981. Deltaic sedimentation in the Devonian of western Libya: *Sedimentary Geology*, **29**, 67–88.

WALLISER, O. H. 1986. Towards a more critical approach to bio-events. *In*: WALLISER, O. H. (ed.) *Global Bio-Events*. Lecture Notes in Earth Sciences, Springer-Verlag, Berlin, **8**, 5–16.

WALLISER, O. H. 1996. Global events in the Devonian and Carboniferous. *In*: WALLISER, O. H. (ed.) *Global Events and Event Stratigraphy in the Phanerozoic*. Springer-Verlag, Berlin, 225–250.

WALLISER, O. H., LOTTMANN, J. & SCHINDLER, E. 1988. Global events in the Devonian of the Kellerwald and Harz Mountains. *Courier Forschunginstitut Senckenberg*, **102**, 190–193.

WENDT, J. 1991. Depositional and structural evolution of the Middle and Late Devonian on the northwestern margin of the Sahara Craton (Morocco, Algeria, Libya). *In*: SALEM, M. J., SBETA, A. M. & BAKBAK, M. R. (eds) *The Geology of Libya*, vol. 6. Elsevier, Amsterdam, 2195–2210.

WENDT, J. & BELKA, Z. 1991. Age and depositional environment of Upper Devonian (Early Frasnian to Early Famennian) black shales and limestones (Kellwasser Facies) in the Eastern Anti-Atlas, Morocco. *Facies*, **25**, 51–90.

WENDT, J., BELKA, Z., KAUFMANN, B., KOSTREWA, R. & HAYER, J. 1997. The world's most spectacular carbonate mud mounds (Middle Devonian, Algerian Sahara). *Journal of Sedimentary Research*, **67**, 424–436.

WIGNALL, P. B. & MYERS, K. J. 1988. Interpreting benthic oxygen levels in mudrocks: a new approach. *Geology,* **16**, 452–455.

WILDE, P. & BERRY, W. B. N. 1984. Destabilization of the oceanic density structure and its significance to marine „extinction' events: *Palaeogeography, Palaeoclimatology, Palaeoecology,* **48**, 143–162.

WILDE, P. & BERRY, W. B. N. 1986. The role of oceanographic factors in the generation of global bio-events. *In*: WALLISER, O. H. (ed.) *Global Bio-Events*. Lecture Notes in Earth Sciences, Springer-Verlag, Berlin, **8**, 75– 91.

Aptian source rocks in some South African Cretaceous basins

D. VAN DER SPUY

Petroleum Agency SA, PO Box 1174, Parow, 7499, South Africa
(e-mail: plu@petroleumagencysa.com)

Abstract: Source rocks of the Aptian play an important role in petroleum systems operating off the west coast of Africa. Further south, Aptian source rocks are a proven component of a petroleum system described in the Bredasdorp Basin, a sub-basin of the greater Outeniqua Basin. It has been shown that these Aptian source rocks have supplied oil to the Cretaceous sands of the producing Oribi oilfield and its satellite fields. Aptian source rocks have also been intersected by numerous exploration and scientific boreholes in other sub-basins of the greater Outeniqua Basin, and off the west coast of South Africa in the Orange Basin. Data from these wells suggest the regional development of good-quality Aptian source rocks in the deeper parts of the Orange and Southern Outeniqua basins, possibly on a basin-wide scale.

In the Orange Basin, no Aptian source rocks were intersected in the boreholes drilled in the northern part of the South African sector. However, further north, in the Namibian sector of the basin, over 100 m of good-quality Aptian source rock was intersected in the Kudu wells. To the south, at about 31°S, organic carbon rich intervals a few tens of metres thick have been intersected in a number of boreholes. Here, the kerogen is largely gas-prone, with some thin intervals of shale capable of producing oil. Intersections of the Aptian in boreholes south of 32°S demonstrate how the source quality can be expected to improve from a proximal to more distal position. A thin, high total organic carbon (TOC), gas-prone interval in the most proximal position improves in quality and thickness to become up to 140 m thick, capable of generating wet gas and small amounts of oil. The Deep Sea Drilling Project (DPDS) 361 borehole, located to the southwest in the Cape Basin, intersected layered dark grey and black anoxic Aptian shales with very high TOC, with alternating marine and terrigenous influence. The hydrogen indices of over 500 mg HC/g TOC from some layers are clearly indicative of oil potential. There is thus a strong case for the regional development of a good-quality source rock within the Early Aptian succession in the deeper parts of the Orange Basin. Burial history studies show that Aptian sediments should be in the oil window in large areas to the west of the basin depocentre.

The Early Aptian source rock in the Bredasdorp Basin has been well described. Here it can be over 200 m thick and occurs over a large areal extent. The organic material is largely Type II with a Type I component. Early Aptian source rocks have also been intersected in the Pletmos Basin further to the east, where the organic-rich interval is over 80 m thick in places. There is strong seismic evidence that this source interval should be well developed in the greater Southern Outeniqua Basin to the south of these two inboard basins. Burial history studies in this large basin show that the Early Aptian interval is sufficiently mature over large areas to have generated and expelled oil.

Source rocks of the Aptian play a major role in some of the petroleum systems of the oil-rich northern and central west coast of Africa. Off the South African coast, Aptian source rocks of up to 200 m thick are a proven component of the petroleum system described in the Bredasdorp Basin by Davies (1997a). The Bredasdorp Basin is the westernmost sub-basin of the greater Outeniqua Basin off the south coast of South Africa (Fig. 1) and it has been shown that these Early Aptian source rocks have supplied oil to the Early and Mid-Cretaceous sands of the producing oilfields in the central area of this sub-basin (Davies 1997a).

Organic-rich intervals of the Aptian have also been intersected by numerous exploration and scientific boreholes in the other sub-basins of the greater Outeniqua Basin and off the west coast of South Africa in the Orange Basin (Bolli et al. 1978; Broad & Mills 1993; Muntingh 1993).

The Early Cretaceous saw the early stages of break-up of southwestern Gondwana (Ben-Avraham et al. 1993). The incipient sedimentary basins forming at this time in the Afro-Antarctic

From: ARTHUR, T. J., MACGREGOR, D. S. & CAMERON, N. R. (eds) *Petroleum Geology of Africa: New Themes and Developing Technologies.* Geological Society, London, Special Publications, **207**, 185–202. 0305-8719/03/$15
© The Geological Society of London 2003.

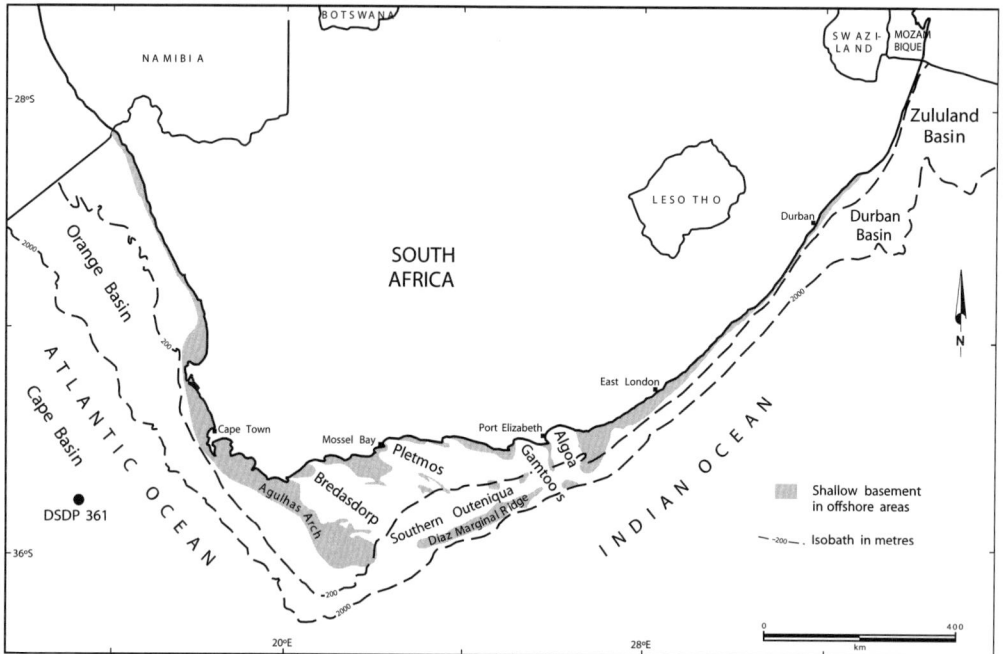

Fig. 1. Sedimentary basins offshore South Africa. The Outeniqua Basin to the south of the country comprises the Bredasdorp, Pletmos, Gamtoos, Algoa and the Southern Outeniqua sub-basins.

Domain (Matthews *et al.* 2001) were separated by major physical barriers and oceanic circulation was inefficient, slow and haline driven. These conditions resulted in the deposition of organic-rich sedimentary sequences throughout the southern Atlantic over a long time span, as demonstrated by the results of the Ocean Drilling Program (ODP) and Deep Sea Drilling Project (DSDP) around eastern Antarctica, the Falkland Plateau and southern Africa (O'Connell & Wise 1990).

Results from DSDP 361 indicate that these conditions certainly existed around South Africa (Jacquin & de Graciansky 1988). The Outeniqua Basin was bounded along most of its southern margin by the Diaz Marginal Ridge (Ben-Avraham *et al.* 1993), while its sub-basins also appear to have been cut off to some extent. For example, the Bredasdorp Basin was bounded to the south by a structural high for most of the Early Cretaceous (Davies 1997*a*).

The Orange Basin, to the west of South Africa, was certainly restricted from open oceanic circulation at this time. To the north, it is bounded by the Walvis Ridge and was cut off from the southern open ocean by the presence of the passing Falkland Plateau (Jungslager 1999*a*). The seaward-dipping reflectors apparent on seismic data in the early part of the section have been interpreted as subaerially deposited continental flood basalts. (Jungslager 1999*b*; Lawrence *et al.* 1999). The margin is thus of volcanic type, characterized by a platform-like continental margin bordered by a narrow deeper ocean. As Coster *et al.* (1989) point out, this scenario would have presented ideal conditions for the formation of anoxic source rocks during marine incursion. As is the case offshore Namibia, predrift seaward-dipping reflector related reservoirs could capture hydrocarbons migrating from drift source rocks (Lawrence *et al.* 1999).

The Late Barremian to Early Aptian saw the first true marine sediments deposited across the Orange Basin. The Aptian interval consists of a regionally distributed marine condensed section within a third-order sequence, resting on a second-order unconformity (Brown *et al.* 1995). The transgression would have formed a relatively shallow sea over the continental margin, resulting in the ideal anoxic source-rock-forming conditions as discussed by Coster *et al.* (1989).

This paper examines the evidence for the regional development of potential petroleum source rocks within the immediately post-Neocomian succession in the distal Orange and Southern Outeniqua basins (see Fig. 2). Borehole data are used to postulate the likely distribution and quality of a possible source interval in the untested areas of the

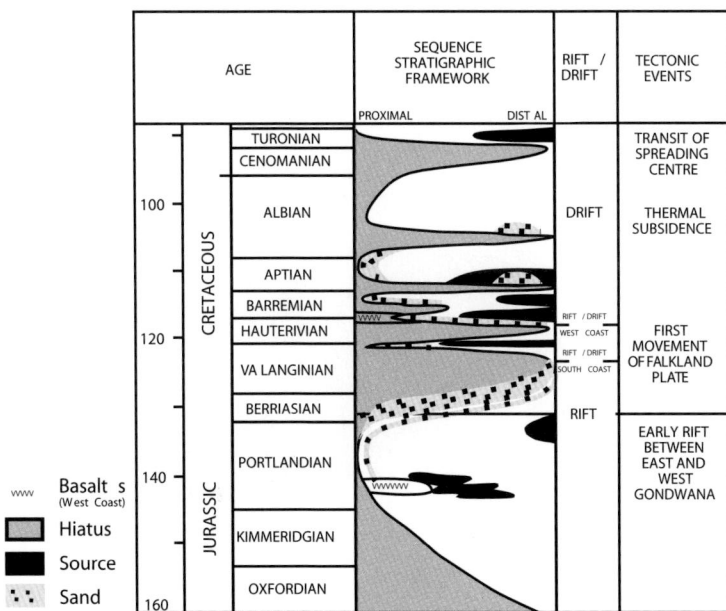

Fig. 2. Simplified chronostratigraphy of the offshore Mesozoic basins.

two basins, while burial history and maturity models elucidate the interval's potential as a source for oil or gas.

First, there is a short overview of the proven and well-described Early Aptian source rocks in the Bredasdorp Basin (e.g. Burden 1992; Davies 1997a), as the character of this source interval could provide some clues as to what can be expected in the other areas where it may be developed. Sources of organic material would have been similar to those in the Bredasdorp Basin for both the Southern Outeniqua and Orange basins. The paper then goes on to examine the evidence for the likely development of a good-quality Aptian source rock in the more distal Orange and Southern Outeniqua basins.

Aptian source rocks in the Bredasdorp Basin

Figure 1 shows the location of the offshore basins developed around the coast of South Africa. Off the south coast is the greater Outeniqua Basin, which is made up of four so-called 'inboard sub-basins' and an elongate basin to the south of these, known locally as the Southern Outeniqua Basin (there is no area referred to as the Northern Outeniqua Basin). The Bredasdorp Basin is the westernmost of the inboard basins. It is NW-SE-trending, and bounded to the west and east by the Agulhas and Infanta basement arches, respectively. It covers

an area of around 18 000 km^2 (Broad & Mills 1993). The basin was formed during the break-up of Gondwana and the subsequent drifting apart of the African and South American plates. It is thought to follow a structural grain inherited from the underlying Cape Supergroup, later modified by dextral motion along the Aghulhas-Falkland Fracture Zone during continental separation (Fouché et al. 1992).

Figure 3 shows depth contours within the Bredasdorp Basin to the Early Aptian type 1 unconformity (Brown et al. 1995), which is generally directly overlain by the source interval. The surface presently has a topographic expression of over 3000 m. Over 200 exploration boreholes have been drilled in the basin and Figure 2 shows the distribution of the Early Aptian source interval based on this well control. Generally, oil-prone kerogen is found in the central area of the basin, in a distribution parallel to the basin's long axis. The central oil-prone zone is surrounded by gas-prone source rocks. The area of non-deposition near the centre of the basin has been interpreted as a structural high, and the thinning of sands over this area in younger parts of the succession supports this interpretation (Davies 1997a). Comparison of the present topography of the surface with the distribution of oil- and gas-prone source rocks suggests that the depositional axis of the basin has migrated northward since Aptian times, as the thickest oil prone source is south of the present deepest area.

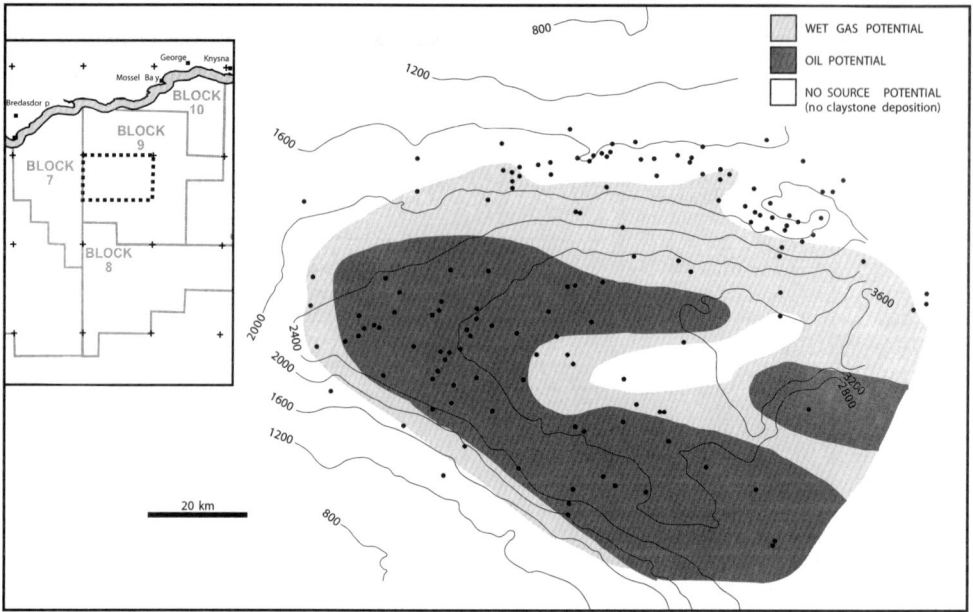

Fig. 3. Depth contours to the Early Aptian unconformity in the Bredasdorp Basin. The ornamentation indicates the source quality of the Aptian source rocks overlying this unconformity.

The plot of hydrogen index (HI) against Rock-Eval Tmax (Fig. 4), clearly shows that the organic material is a mixture of land-derived, higher plant matter and marine material. The terrestrial macro-phyte component accounts for the gas-prone facies of the source rock, as shown in Figure 3. This is borne out by optical examination of the organic material, which commonly shows a mixture of

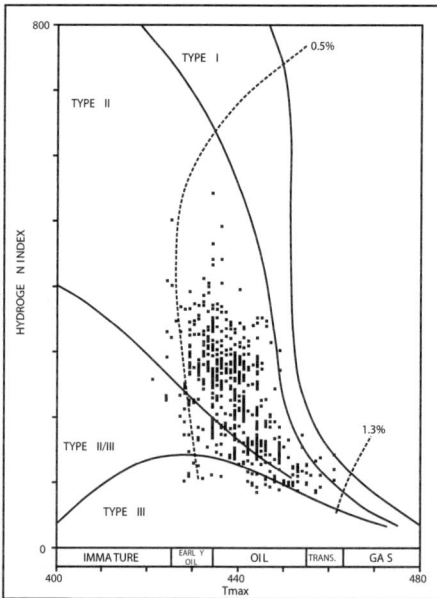

Characteristics of Bredasdorp
Aptian source rocks

Oil prone Type II
80 – 120 m thick (220 m)
2000 km²
S2   8 – 12 kg/tonne

Wet gas Type II/III
50 – 100 m thick
2500 km²
S2   4 – 7 kg/tonne

Fig. 4. HI plotted against Tmax for Aptian source-rock samples from the Bredasdorp Basin. (After Davies 1997*a*).

inert and gas-prone higher plant material, pollen grains and cuticles, along with marine phytoplankton, acritarchs and amorphous kerogen.

The oil-prone source occurs mainly in the south of the basin and covers some 2000 km². It is typically 80–120 m thick, with Rock-Eval S2 values of 8–12 kg hydrocarbons (HC)/t rock at its present maturity, which ranges from the early oil window to the gas window (Burden 1992). Even though fairly mature, there are still HIs above 400 (Fig. 4). Optical examination shows there is a strong correlation between increasing oil potential as measured by Rock-Eval indices and the observed percentage of amorphous kerogen. However, even where amorphous kerogen content is very high, small, inert fragments of higher plant material usually comprise some part of the kerogen.

The gas-prone facies covers about 2500 km² and is typically 50–100 m thick. Rock-Eval S2 values commonly range between 4 and 7 kg HC/t rock (Burden 1992). Interpretation guidelines following Peters and Cassa (1994) indicate that this is a good wet-gas-prone source. Cornford (1998) considers such values as marginal with regard to the ability to saturate the source rock, so that any oil generated will, in any case, not be expelled. Optical examination of samples from this area shows a marked reduction in the proportion of amorphous organic matter, and both these factors (quantity and quality) probably contribute to the gas-prone nature.

The Early Aptian source in the Bredasdorp Basin is in close proximity to the submarine-fan channel complexes and younger Albian sands developed within the basin. The source rock has been proven as the source for oil discovered and currently in production from Aptian and Albian reservoirs. Following Magoon and Dow (1994, p.13), this system, which falls within a group of systems named Outeniqua (!) by Davies (1997a), could be named the Early Aptian-Albian(!) as the largest portion of the reservoir is in the Albian.

The geological model for source deposition in the basin is thought to be that of a silled anoxic basin (Davies 1997a). After the formation of the Early Aptian unconformity, a thick claystone was deposited during the ensuing transgression and culminating flood. Whatever the mechanism of formation, it is clear that organic material sourced from both the surrounding continent and from within the ocean itself was preserved during the deposition of this Early Aptian claystone. The pattern of distribution of the kerogen types, and the resulting source quality occurring within this claystone as discussed above, suggests that an important (detrimental) control on source quality was proximity to a source of terrigenous macrophytic material.

Aptian source rocks in the Orange Basin

Geological framework

The Orange Basin is situated off the southwest coast of Africa (Figs 1 & 5). A description of the Orange Basin and its petroleum geological development is dealt with in detail by Jungslager (1999a) and the references therein. The basin is extremely large, covering some 130 000 km², and for the purposes of this paper, is defined by the limits of the post-Hauterivian sediment pile, as shown in Figure 5 (Gerrard & Smith 1982). The basin is classified as a passive-margin basin, with sediment sourced from the continent to the east being carried into the basin by westward-flowing rivers, the most important of which is the Orange River. The sediment pile exceeds 7 km in some areas, with the thickest succession concentrated in the north-central part of the basin, directly opposite the present-day mouth of the Orange River.

Figure 6 (from Jungslager 1999a) shows a generalized cross-section through the basin at position A-A' from Figure 5. The pre-Barremian syn-rift succession comprises small, isolated half-grabens to the east, with a large central wedge to the west. These grabens are thought to be filled with volcanics and continental clastics (Jungslager 1999b). This sequence is overlain by transitional Barremian-Aptian sediments, which consist of alternating fluvial and marine rocks deposited during various transgressive and regressive cycles. These cycles culminated in a major flooding event during the

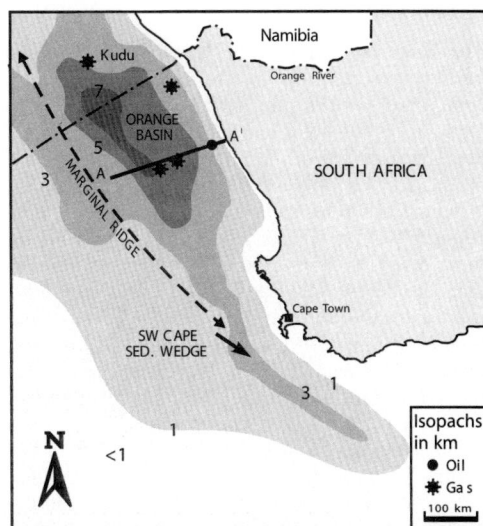

Fig. 5. Post-rift sediment thickness defines the extent of the Orange Basin off the South African west coast. (From Jungslager 1999a).

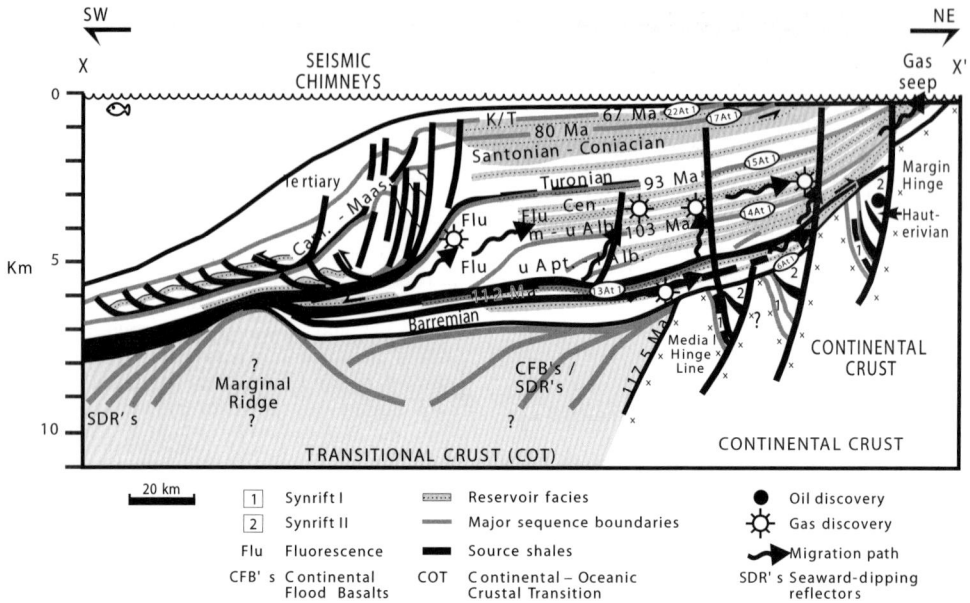

Fig. 6. A schematic cross-section along line A-A' of the previous figure, showing the generalized geology of the Orange Basin. (From Jungslager 1999*a*).

Early Aptian. After this, the sediments in the basin are characterized by a fully developed drift succession, consisting of prograding clastics, typified by a lack of structure on the shelf in the east, and growth-faulting and toe-thrusts in the west (Jungslager 1999*a*).

Aptian source rocks

The Kudu boreholes in the Namibian part of the basin intersected up to 150 m of good-quality source rocks within the Aptian succession. Thirty exploration boreholes and four appraisal boreholes have been drilled in the South African part of the basin. The majority of these wells are in fairly proximal positions, and mostly intersect sediments deposited on the palaeoshelf, while those drilled in more distal positions generally are too shallow to intersect the Aptian–Barremian succession. However, as can be seen from Figure 7, twelve of the boreholes do intersect sediments of the Aptian and older. The discussion below examines these intersections and their implications for the development of source rocks in the deeper basin.

Boreholes A-E1, A-O1, A-F1 and A-D1 in the north all intersected sediments of Aptian age and older. However, all are extremely proximal and the Aptian sediments are continental to transitional in nature. No intervals of organic enrichment have been identified within these sediments.

Figure 8 shows annotated gamma-ray logs across part of the Aptian interval, for boreholes A-A1, K-A2 and K-D1, in the northern and central parts of the basin. Their positions are shown on Figure 7. All other intersections of Aptian sediments are to the south of these wells.

The first well, A-A1, is the most proximal of the three, and intersected transitional to inner shelf Aptian sediments. Up to 30 m of organic rich claystones are interbedded with sands and silts over a 100 m-thick interval between 2730 m and 2830 m, dated as probably Late Aptian. (McMillan & Marot 1997). The total organic carbon (TOC) levels are fairly high, reaching up to 3.4%. The Early Aptian-Barremian was not intersected by this well, according to McMillan's dating (McMillan & Marot 1997). The Rock-Eval analyses show an average S2 value of 2 kg HC/t rock for this organically enriched interval. Based on the HIs of less than 200, the material is gas-prone.

The Aptian interval intersected by K-A2 is interpreted as having been deposited in an upper slope to outer shelf environment (McMillan & Marot 1997). Figure 8 shows the gamma-ray log from this borehole across the relevant depth interval from 5185 m to 5248 m, of which 55 m is claystone. While similar in quality to the organic-rich interval intersected in A-A1, here the interval is more than 20% thicker and far more concentrated, with little in the way of sand and silt interbeds. The interval is deeply buried and is presently post-mature. Present TOC contents reach 2.5% while S2 values range

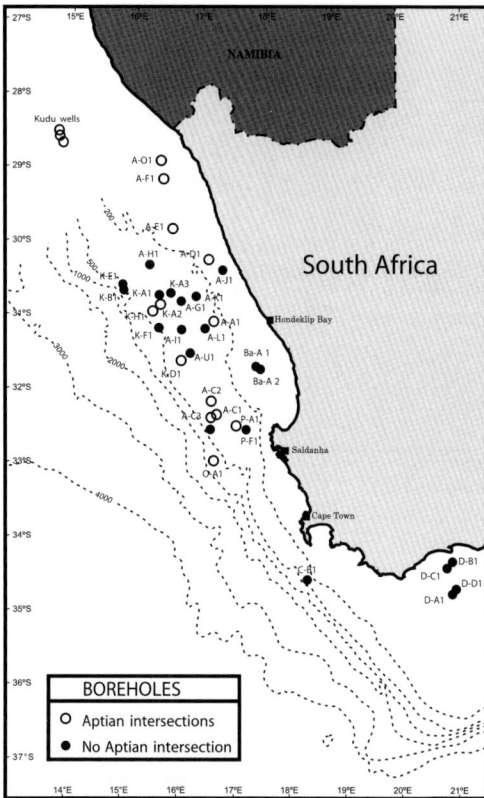

Fig. 7. Map showing the position of exploration and scientific boreholes known to intersect sediments of Aptian age.

from 1.5 to 2 kg HC/t rock and HIs are low. Optical examination shows a majority of vitrinite, although one sample shows up to 40% exinite, which is very well preserved. Assuming this is an exhausted Type II kerogen, original TOC contents could have been around 8% (Cornford 1994).

Further to the south, and in a slightly more distal position, K-D1 intersected about 300 m of sediment dated as Late Aptian. (McMillan & Marot 1997). The 112 Ma-old Early Aptian unconformity (Haq *et al.* 1988; Brown *et al.* 1995) is 4380 m below surface at this position and no analyses of the shale were done due to the obvious overmaturity. However, from Figure 8, which shows the gamma-ray log for this borehole across this unconformity, it can be seen that the log character shows a substantial increase in gamma response over background, as observed in the previous two boreholes. This increased gamma response, commonly observed in organic-rich rocks, suggests that an organic-rich interval of about 40 m thick directly overlies the unconformity.

While the above data by no means represent the

development of a substantial source rock capable of liquid generation on this part of the Aptian palaeoshelf, it is significant that organic enrichment occurs consistently within this interval. Comparison with results from the northern wells suggests that a certain degree of marine influence is required for this to be so. The difference in log character between A-A1 and K-D1 indicates a trend of increasing condensation of the organic-rich interval. Optical and Rock-Eval data demonstrate the presence and preservation of organic material on the shelf at this time.

Within the South African part of the basin, boreholes drilled to the south of 32°S provide far more encouragement in the argument for the regional development of a good-quality Early Aptian source-rock interval on a basin-wide scale. The proximal to distal transition occurs over a shorter distance in the southern sector of the basin and it is inferred that this may be due to less influence by terrestrial input from sediment-laden rivers (Fig. 5).

Five of the seven boreholes drilled in this part of the basin intersected Early Aptian and Barremian sediments, and all of these demonstrate organic enrichment within the interval.

Borehole P-A1 is one of the most proximally positioned wells in this southern sector of the basin (Fig. 9). An early interpretation based on foraminifera suggests that a thin sliver of Aptian sediment overlies the unconformity, which is compounded with a Hauterivian hiatus (McMillan & Marot 1997). Geochemical data show that the sediment directly overlying the unconformity is organic-rich and Figure 9 shows the high response of the gamma-ray log through the Aptian section. A 3 m-thick interval of high organic enrichment occurs directly overlying the unconformity. TOC content reaches 4% in this interval, where chemical and optical analyses demonstrate it to be of poor gas-prone (Type III kerogen) quality and immature.

Further out into the basin is a group of relatively closely spaced boreholes, i.e. A-C1 to 3, and A-N1.

In A-C1, two organic-rich intervals occur directly above the Early Aptian unconformity, within sediments dated as Aptian. These are between 2953 m and 2984 m, and 2998 m and 3027 m. Both of these intervals are gas-prone, with chemical indices indicating fairly low generative potential.

In borehole A-C2, an 89 m-thick interval of organically enriched claystones occurs above the Early Aptian uncomformity. Figure 9 shows the gamma-ray log over this interval. TOC percentages range from 2% to 3 %, and Rock-Eval S2 values are around 1.8 kg HC/t rock on average. The maturity at this level is equivalent to a vitrinite reflectance of around 1.2% and, although the ana-

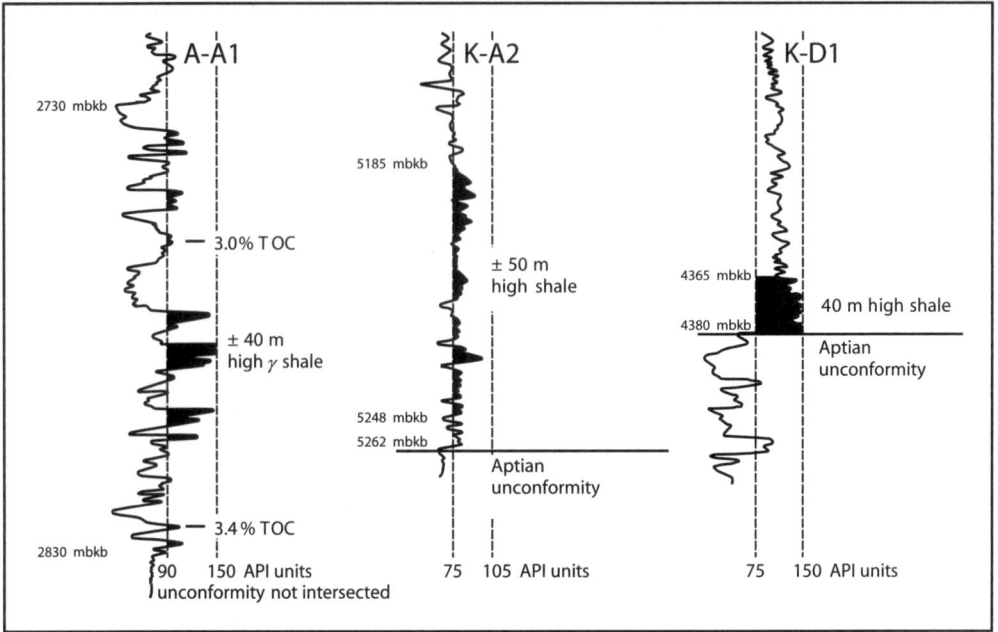

Fig. 8. Annotated gamma-ray logs through the Early Aptian section, from exploration boreholes A-A1, K-A2 and K-D1. For positions of boreholes, see Figure 7.

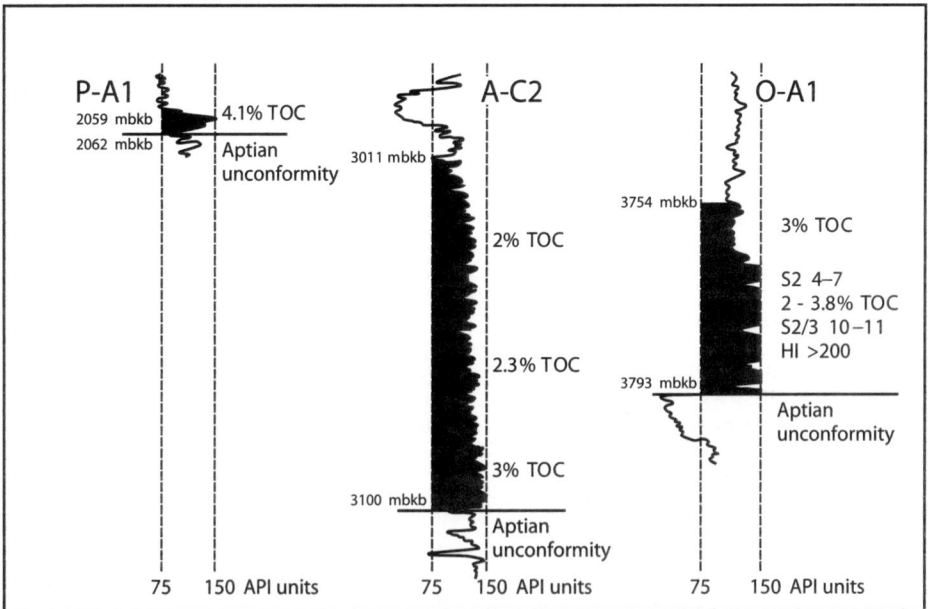

Fig. 9. Annotated gamma-ray logs through the Early Aptian section, from exploration boreholes P-A1, A-C2 and O-A1. For positions of boreholes, see Figure 7.

lytical results indicate poor-quality, largely inertinitic kerogen, some potential for wetter gas may exist in some thin layers where the gamma response increases.

In A-C3, a similar interval, 63 m thick, occurs directly above the Aptian unconformity. Here, average S2 values are higher than those in A-C1 and up to 30% of the organic material is exinitic, made up of a mixture of macrophyte and marine palynomorph material, implying both terrestrial and marine sources of organic matter. This is similar to the nature of the organic material in the wet-gas-prone source in the Bredasdorp Basin. The interval in this borehole is interpreted as being capable of wet-gas generation.

A further intersection of the Aptian occurs in O-A1, to the south (Fig. 7), which is regarded as the most distally positioned of all boreholes drilled within the South African part of the basin. At this position, part of the Aptian has been removed by erosion. The typical high gamma-ray log response (Fig. 9) together with the measured TOC control points, suggests that at least 33 m of organically enriched claystone remains. The TOC percentages range between 2% and 3%, with one sample close to 4%. S2 values of up to 7 kg HC/t rock were recorded and many of the HIs are close to 200 kg HC/t rock. While reflectance data in this borehole are questionable, the Tmax data over the interval in question are consistently in the higher part of the 440–450 °C range. Simply extrapolating back (Fig. 10) suggests a far higher original HI for these samples. Following Cornford's graphical corrections (Cornford 1998), most of the original TOC values would have been over 4%. (See Table 1 for a listing of the above data.)

These few intersections show consistent development of organic enrichment in claystones overlying the Early Aptian unconformity, in boreholes up to 100 km apart, with a trend of improvement in quality as the depositional environment becomes more distal. Figure 10 shows this trend graphically. The HIs of high TOC-bearing intervals within the Aptian are plotted against Tmax for three of the above wells. Kerogen quality clearly improves basinward.

Figure 11 is a highly schematic representation of the interpreted environments of deposition at the various borehole positions during the Aptian, based on foraminiferal studies of McMillan (McMillan & Marot 1997). Also shown are the sand-rich areas resulting from the sediment brought into the basin by the two major river systems in operation during the basin's history. The major input has always been from the Orange River, which has deposited a large amount of terrigenous sediment in the central and northern part of the basin. The northern wells discussed earlier are well up on the Early

Table 1. *Rock-Eval and TOC data for boreholes A-A1, P-A1, A-C2 and O-A1*

Borehole	Depth	Tmax	s1	s2	s3	TOC
A-A1	2755	472	0.11	1.33	0.24	3.01
A-A1	2830	485	0.07	1.24	0.69	3.39
P-A1	2060	437	0.59	2.28	0.99	4.13
A-C2	3024	444	1.49	1.40	1.91	1.86
A-C2	3029	471	0.48	0.40	1.58	1.75
A-C2	3033	422	2.61	2.52	1.87	2.30
A-C2	3042	444	0.53	1.29	1.72	1.88
A-C2	3047	461	0.24	1.03	0.62	1.69
A-C2	3051	449	0.59	1.20	1.53	1.66
A-C2	3060	449	0.77	1.25	0.94	1.73
A-C2	3067	459	0.34	1.18	0.50	2.26
A-C2	3069	433	1.57	1.18	1.31	2.19
A-C2	3072	446	0.52	1.09	1.65	1.73
A-C2	3078	437	1.66	1.48	1.16	2.47
A-C2	3081	449	0.92	1.41	1.09	2.07
A-C2	3087	435	1.92	1.40	1.40	2.56
A-C2	3087	455	0.54	0.73	0.64	2.93
A-C2	3090	449	0.71	1.33	1.72	2.04
A-C2	3096	443	1.87	1.40	1.19	2.53
A-C2	3102	454	0.58	1.10	1.52	2.05
O-A1	3756	453	1.95	4.42	0.52	2.94
O-A1	3763	450	0.73	2.83	0.59	1.94
O-A1	3765	448	0.70	2.74	0.69	1.94
O-A1	3769	449	1.07	3.87	0.62	2.23
O-A1	3771	446	1.15	3.99	0.65	2.31
O-A1	3774	448	1.21	3.64	0.69	2.28
O-A1	3778	447	1.42	4.34	0.47	2.47
O-A1	3780.5	450	3.88	7.64	0.29	3.81
O-A1	3781	448	1.43	4.46	0.48	2.44
O-A1	3783	447	1.46	4.14	0.75	2.54
O-A1	3787	448	1.60	4.68	0.54	2.56
O-A1	3790	445	1.85	4.84	0.55	2.78
O-A1	3792	446	1.78	4.85	0.59	2.74

Aptian palaeoshelf, or fairly high up on the slope, and directly within the major zone of influence of the Orange River. Sediment deposited in this large area would have been very rich in terrestrial material brought into the basin by the river system, accounting for the terrigenous and thus gas-prone nature of the organic material found in the Aptian claystones. Nonetheless, conditions in the Early Aptian, even within zones of the river's influence, were conducive to the preservation of organic material. This is evidenced by the organic enrichment observed within the Aptian section.

The influence of the proto-Olifants River system in the south appears to have been less dramatic. The boreholes in the southern part of the basin are more remote from the influence of terrestrially derived material brought into the basin by the westward-flowing river system and thus show better source quality closer to shore than the boreholes

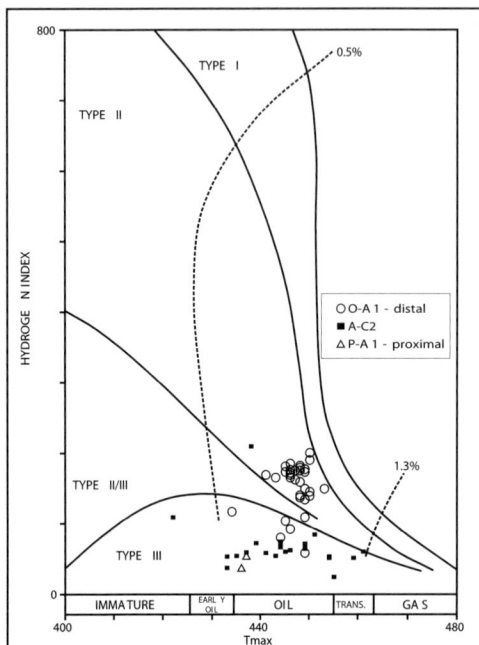

Fig. 10. HI plotted against T_{max} for samples from P-A1, A-C2 and O-A1. Source quality improves with the increasingly distal nature of the depositional environment.

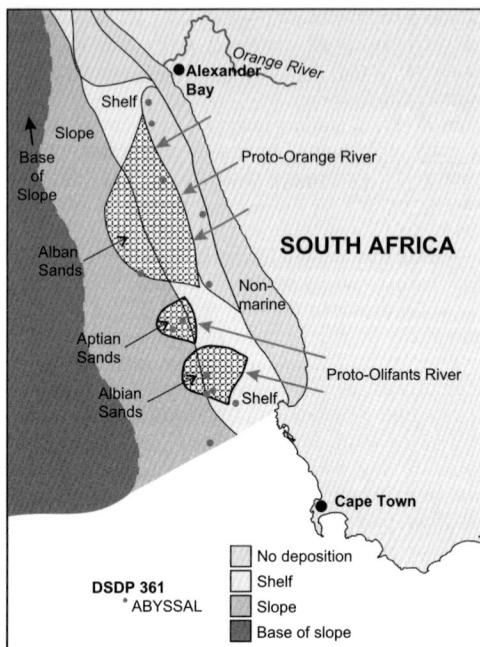

Fig. 11. Palaeogeography of the Orange Basin during the Early Aptian-Cenomanian. Note the influx of coarse sedimentary material brought into the central part of the basin by the Orange River. (After McMillan 1997).

in the north. There is also a clearly developed trend in the south of source-quality improvement from proximal to distal, as shown by Figure 10. This makes it not unrealistic to extrapolate this pattern to the northern areas and thus prognose the development of improved source quality further offshore.

Further data points supporting the regional development of a good-quality Early Aptian source off the southern west coast of Africa include the Kudu wells in Namibia and the DSDP 361 borehole drilled in the southern Cape Basin.

The Kudu wells are more distal, in terms of environment of deposition, than any of the other wells drilled within the basin. The Kudu 9A – 1, 2 and 3 wells all intersected a thick (about 150 m) interval of organic-rich, dark grey shales within the Early Aptian (Davies & van der Spuy 1990). The organic material is predominantly of marine origin and is regarded as being originally oil-prone. Benson (1990) describes it as 'consisting predominantly of clumps of amorphous organic matter, with ragged margins formed by decomposition and complete bacterial reworking of organic debris in organic muds under anaerobic conditions'. McMillan (1990) records a dominance of planktonic species, with very few benthonic forms, implying anaerobic conditions on the sea floor at the time.

Benson does note the presence of small amounts of 'P-wafers', indicating some terrestrial influence, although Davies and van der Spuy (1990) record these as less than 10–5%. This is very similar to the character of the best-quality source rocks described earlier from the Bredasdorp Basin.

Present organic carbon contents (post-maturity) average around 2%, while Rock-Eval indices indicate dry-gas proneness. However, maceral analyses show extremely high proportions of amorphous organic material, suggesting that the interval was oil-prone when immature (Davies & van der Spuy 1990). Bray et al. (1998) have calculated higher pre-maturity organic carbon contents of up to 8%, using the method of Cornford (1994).

A further encouraging data point in the southern part of the basin is provided by the scientific borehole DSDP 361, drilled in the Cape Basin (Fig. 1). This borehole is positioned well to the south, out in the open basin, away from the continuous influence of terrestrial material brought into the basin by river systems but still within reach of occasional turbidite flows. This resulted in the intercalation of turbiditic sediments rich in terrestrial organic matter and hemipelagic intervals richer in marine organic matter (Jacquin & de Graciansky 1988). The borehole position is seaward of magnetic anomaly M4 and the basal sediments have been dated as Aptian (Bolli *et al.* 1978; Jacquin & de Graciansky 1988). Only 27% of the core drilled through the Aptian interval was recovered (Bolli *et al.* 1978). This interval has been described in numerous papers, including Bolli *et al.* (1978), Jacquin & de Graciansky (1988), Zimmerman, Boersma and McCoy (1987) and Stow (1987), among others. It consists of highly organic-rich sapropelic claystone (Zimmerman, Boersma & McCoy 1987), with TOC contents ranging between 3% and 15%, interbedded with coarser sediments interpreted as turbidites. Figure 12 shows a modified van Krevelen plot for samples from the Aptian cores, after Jacquin and de Graciansky (1988). Material from the coarse turbiditic lithologies is terrestrial in origin, and plots within the Type III field. Organic material from the darker pelagic claystones plots along the Type I and II trends, with some HIs close to 600. Rock-Eval S2 values are as high as 73 kg HC/t rock. While it is clear that a portion of the organic material is of terrigenous origin, brought into this environment by turbidites (see Stow 1987), pelagic sedimentation has contributed a marine, oil-prone component to the kerogens in these sediments. Optical analysis of samples with higher HIs shows a kerogen dominated by amorphous material, with a smaller amount of terrestrially derived higher plant material (Fig. 12). Again, there is a similarity with the best source as described in the Bredasdorp Basin. Taking the

unrecovered core interval into account, it is possible that over 200 m of oil- and gas-prone potential source rock could be developed at this position.

Every borehole drilled in this extremely large basin that has encountered Early Aptian sediments deposited in a predominantly marine environment has demonstrated organic enrichment within these sediments. This is true of boreholes many hundreds of kilometres apart. The above evidence makes it clear that organic enrichment is consistently developed in claystones within the Early Aptian. Trends observed from the few data points available give a strong indication of improvement in thickness and source quality developed as one moves basinward, away from the palaeoshelf and the influence of the Orange River.

Jungslager (1999*a*) introduced the concept of the 'Aptian Anoxic Basin'. His figure 11 clearly shows how the area to the south of the Walvis /Rio Grande Ridge was cut off from major open oceanic circulation during the Early Aptian, causing restricted conditions within the early southern proto-Atlantic (compare this with fig. 6 of Zimmerman, Boersma & McCoy 1987). Both these papers suggest a restricted basin environment for the formation of this potential source rock, an explanation that appears very likely. Whatever the mechanism, it is clear that conditions were conducive for the preservation of organic material within sediments deposited in the Early Aptian within this basin. By analogy with source quality in the Bredasdorp

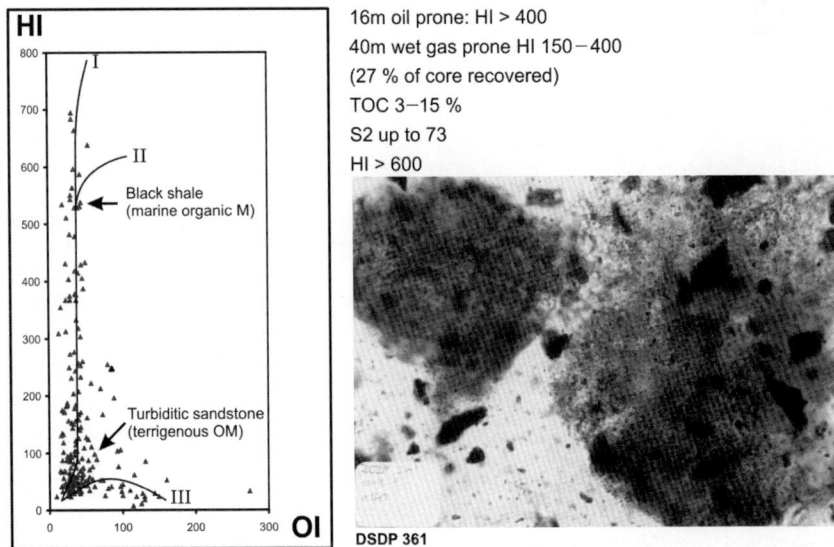

Fig. 12. Modified van Krevelen diagram showing samples from the Aptian cores of DSDP 361. Note the range in source quality corresponding to differences in the source of organic material. The photomicrograph shows the dominantly amorphous character of the kerogen with a small structured component. (Sources: Jacquin & de Graciansky 1988 and Petroleum Agency SA data files).

Basin, Figure 13 shows the probable distribution of good wet-gas- and oil-prone source rocks in the Orange Basin.

Maturity

Except for the Kudu DSDP 361 and O-A1 positions, the good-quality source rocks postulated by this paper have not been intersected by boreholes. However, a large set of maturity data, including vitrinite reflectance, Rock-Eval Tmax, spore colouration and some spore-fluorescence data, is available for the exploration boreholes drilled on the shelf.

A number of wells show anomalously high maturity values in the shallower sections, possibly indicative of some heating event during the Tertiary. This is analogous to the situation further north, as described by Bray *et al.* (1998) offshore Namibia. As these authors mention, such a heating event could have an important influence on the timing of the onset of various maturity levels, and heating events such as this have been included in modelling work done in the past. Modelling has demonstrated that the most significant effect of including a Tertiary heating event is to shorten the residence time of the source rock in the oil and wet-gas windows. Where encountered, the Early Aptian is usually fairly deeply buried and has been

well into the gas window since the Late Cretaceous or Early Palaeogene.

Figures 14a and 14b show BasinMod® burial-history diagrams for the O-A1 position in the south and a pseudowell in a deep-water position to the west of borehole K-F1. This model was calibrated on maturity and present-day temperature data from K-F1, and shows the Aptian to be presently within the oil window at this position. Models such as these have formed the basis of Figure 15, which shows the present-day maturity levels of the Early

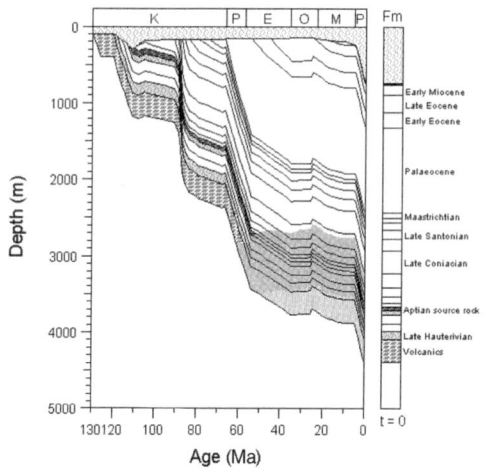

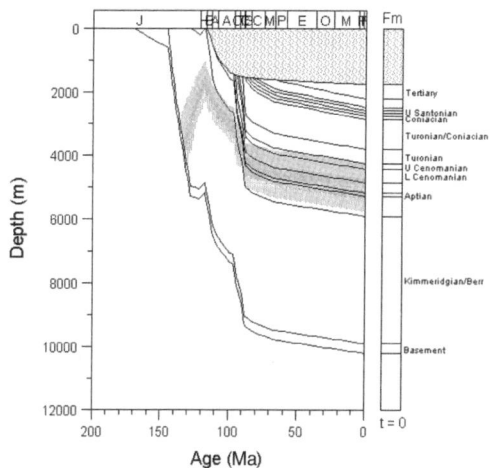

Fig. 14. Burial-history diagrams for boreholes O-A1 and K-F1. (see Fig. 7.) The results of these and other models were used to construct Figure 15.

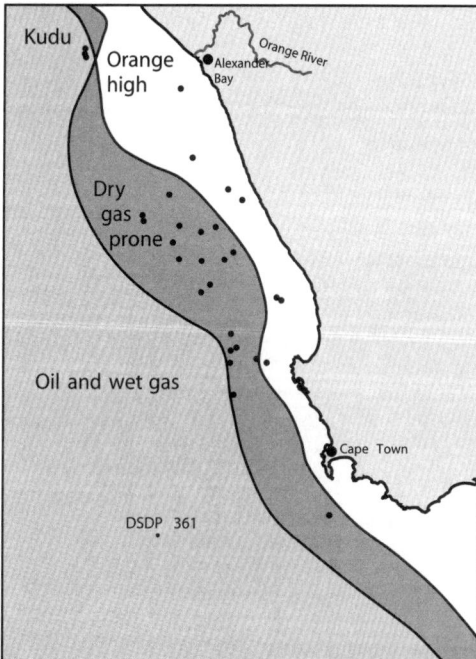

Fig. 13. The possible extent and quality of Aptian source rocks in the Orange Basin.

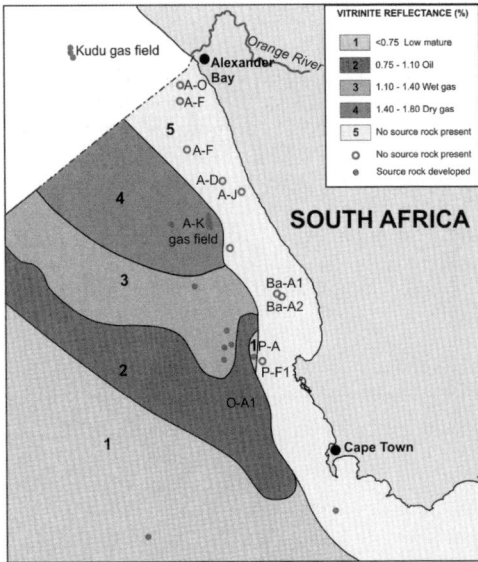

Fig. 15. Present-day maturity levels of the known and postulated Early Aptian source-rock interval in the South African portion of the Orange Basin.

Aptian in the Orange Basin. Where the overburden is thickest, in the centre of the basin, the Early Aptian is in the dry- and wet-gas window. Further to the west and in the south, where overburden is thinner, large areas are currently in the oil window.

Petroleum systems

Early Aptian source rocks form a component of two known petroleum systems in the Orange Basin that have been formally named by Jungslager (1999b). Further possible petroleum systems involving Early Aptian source rocks of better generative potential, as prognosed in this paper, are described in Jungslager (1999b). A short summary of Jungslager's work is presented below.

Barremian/Early Aptian–Barremian(!) petroleum system This system was proven by the Kudu wells off Namibia. Gas is stratigraphically trapped in aeolian sands of the Barremian.

Barremian/Early Aptian–Late Albian/Cenomanian (!) petroleum system This known petroleum system, with gas and condensate stratigraphically trapped in Albian and Cenomanian sands, has been described on the basis of a number of discoveries, the most significant of which are the A-K1 well and others within the Ibhubesi Field. The presence of oleanane in condensate samples from A-K1 suggests some terrestrial component to the kerogen

(Moldowan *et al.* 1994), while C30 steranes confirm the largely marine nature of the source (Waples & Machihara 1991; Peters & Moldowan 1993). The very small amount of oleanane is more consistent with an Early Cretaceous rather than a Late Cretaceous or Tertiary source (Moldowan *et al.* 1994), although there is always the possibility of a younger source contributing some component to the condensate during migration. Optical and chemical characterization of the Aptian source in the Orange Basin shows it to be suboxic marine shale with terrestrial input and, geologically, it is the obvious source component of this system.

Other plays Jungslager (1999a) has also described three more speculative plays that would rely on a source rock of sufficient quality to generate and expel oil. The first of these involves an Early Aptian source rock in conjunction with deep-water turbidite sandstones, ponded against the marginal ridge (see Fig. 6).

The second is a Barremian and Early Aptian sourced oil and gas play, with traps in the Late Cretaceous growth-fault and toe-thrust belt. This system would operate in the northern part of the basin. It is thought possible that the toe-thrust structures contain deep-sea fan turbidite systems.

The third play would involve Barremian and Early Aptian source rocks supplying oil to traps in the Late Cretaceous and Tertiary toe-thrust belts in the central part of the basin. Once again, sands are brought into these plays through the action of turbidites.

Aptian source rocks in the Southern Outeniqua Basin

Geological setting

Figure 16 shows the position of the Southern Outeniqua Basin off the southern coast of South Africa. It is a NE-SW-trending basin off the south coast of South Africa, covering about 20 000 km². Together with the four smaller inboard basins, it makes up the greater Outeniqua Basin, formed during the break-up of Gondwana. The water depth over the basin ranges from 300 m to about 2000 m. The basin is bordered to the south by the Diaz Marginal Fracture Ridge. The formation of this ridge is thought by some to be linked to the movement of the Falklands Plate to the west (Francheteau & Pichon 1972; Roux 1997), although Ben-Avraham *et al.* (1993) point out that this may not be the case.

Source rocks

No boreholes have been drilled in this frontier area, so that source-rock occurrences within the South-

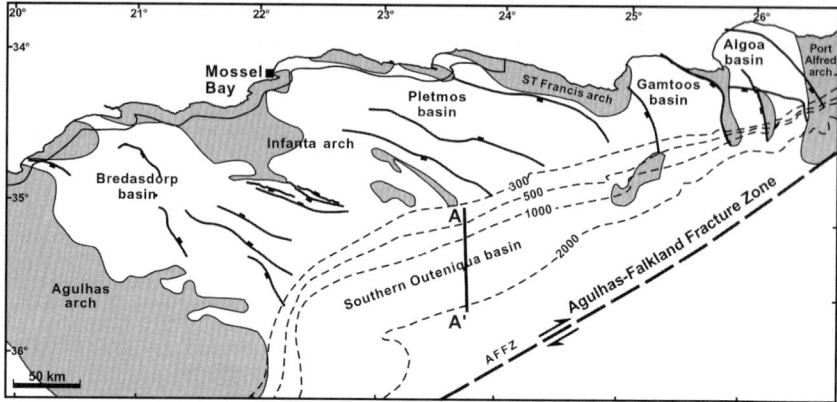

Fig. 16. Map of the Outeniqua Basin off the southern coast of South Africa. The map shows the four 'inboard' basins and the Southern Outeniqua Basin to the south of them.

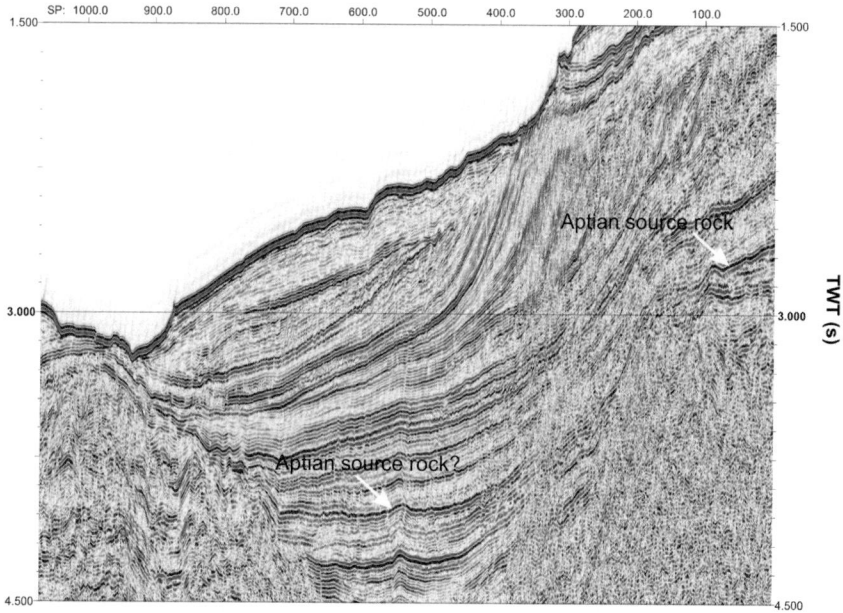

Fig. 17. A seismic section from points A to A' on Figure 16. Note the typical character of the Early Aptian source rocks in the Bredasdorp Basin and the similar character of the Early Aptian interval in the Southern Outeniqua Basin. The line covers about 25 km.

ern Outeniqua are speculative and unproven. However, many wells have been drilled in the southern parts of the four sub-basins to the north, as well a number on the Maurice Ewing Bank of the Falkland Plateau, which, according to most reconstructions, was situated to the south and east of the Southern Outeniqua Basin during the Early Aptian (e.g. Ben-Avraham *et al.* 1997; Zimmerman *et al.* 1987).

One of the characteristics of the Early Aptian source rock in the Bredasdorp Basin, described at the beginning of this paper, is a strong seismic signature (Davies 1997*a*). Figure 17 is a seismic section running from the Bredasdorp Basin in the north, across the Southern Outeniqua Basin to the south. The figure shows how the strong seismic signature of the Aptian source rock (in this case a strong trough signature, shown in black) can be traced down into the Southern Outeniqua Basin, where it shows the characteristic tramline effect.

Based on this distinctive seismic character, it is possible to map the probable occurrence of source rock similar to that known in the Bredasdorp Basin, across the Southern Outeniqua Basin. Figure 18 shows the known distribution of the Aptian source within the inboard basins and the extrapolated extent of the possible source interval across the Southern Outeniqua Basin, based on mapping of this distinctive seismic character (Davies 1997a). By analogy to the Bredasdorp Basin, the distribution of possible oil- and wet-gas-prone source rocks is shown. Gas-prone source rocks can be expected where terrestrial influence is greatest, while oil-prone organic material is expected in the deeper parts of the basin, and toward the seaward side. Indications from peripheral wells and intersections within the depocentres of the more proximal basins suggest that this interval should be oil-prone and may be very thick (over 200 m), similar to the source-rock development in the Bredasdorp Basin.

Evidence from DSDP sites on the Maurice Ewing Bank provides further encouragement for the development of organic-rich Aptian sediments within the Southern Outeniqua Basin. Numerous reconstructions position the Malvinas/Falklands Plate, and particularly the Maurice Ewing Bank, in close proximity to the south and east of the Southern Outeniqua Basin during the Early Aptian (e.g. Zimmerman, Boersma & McCoy 1987; ODSN 1999).

The Early Aptian succession intersected in borehole DSDP 511 is described as black sapropelic shales interbedded with laminated grey marls and comprises a thick interval of black shale containing Type II marine organic material, with HIs of between 300 and 600, and TOC contents of up to 5% (Jacquin & de Graciansky 1988). Benthic forams and metazoans are absent. Jacquin and de Graciansky interpret this as indicative of an anoxic sea floor supplied with varying amounts of organic material as surface productivity waxed and waned. The depocentre of this 'silled, stratified basin' would have been the present-day Southern Outeniqua Basin.

Organic-rich claystones are also described from the Aptian in borehole DSDP 330. Only a small interval of Aptian was recovered (Jacquin & de Graciansky 1988), but TOC contents are up to 3% and the kerogen is described as amorphous (Comer & Littlejohn 1976).

Maturity

While seismic cover over this basin is fairly widely spaced, reasonable quality depth-maps for six major unconformity surfaces have been produced. This has enabled a regional, albeit simplified, approach to burial and maturity studies (van der Spuy 2000).

Platte River's BasinMod® (V7.06) software was used to construct over 40 burial models to constrain maturity levels within the basin. Pseudowell positions are numbered on Figure 19 and range from relatively quiet areas in the central basin, where the stratigraphic succession is most com-

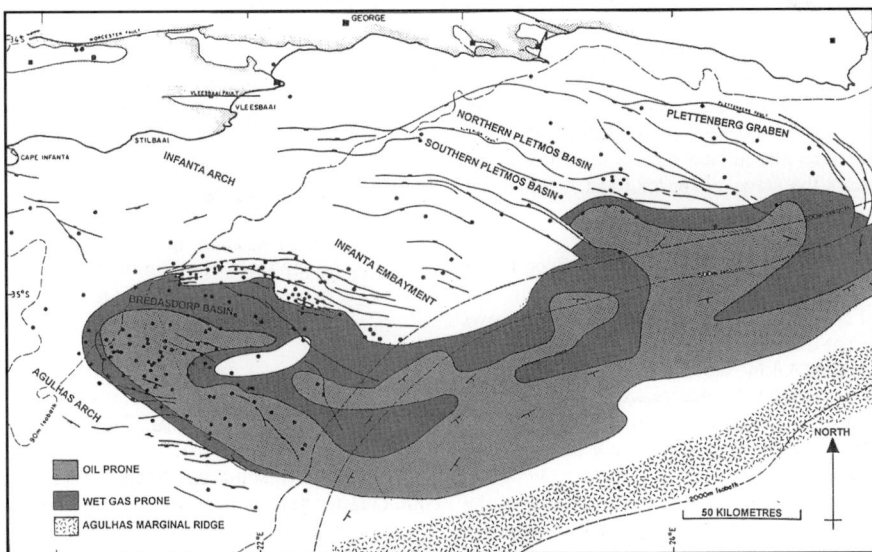

Fig. 18. The likely distribution of oil- and wet-gas-prone Early Aptian source rocks in the Southern Outeniqua Basin, extrapolated on the basis of seismic character. (From Davies 1997a).

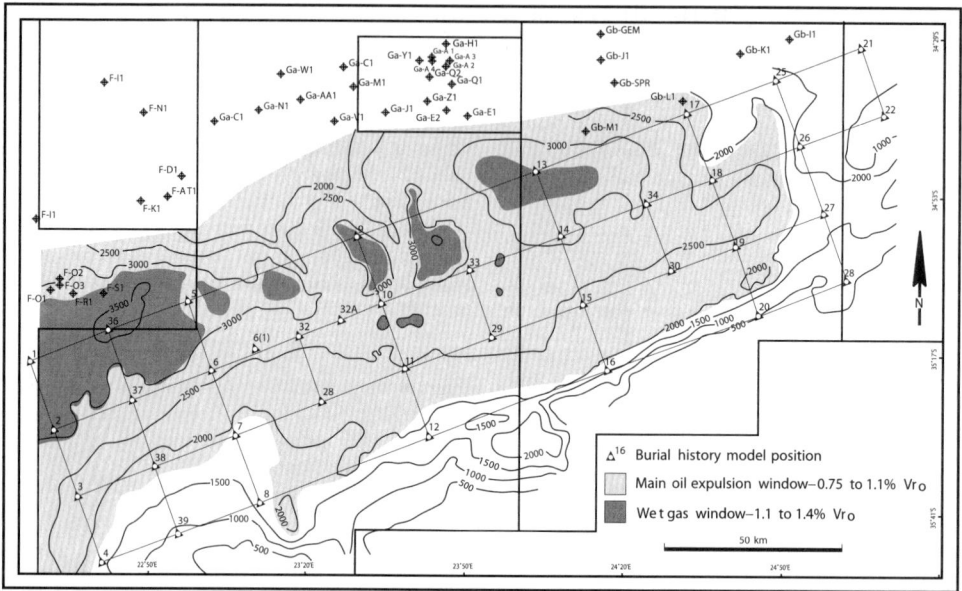

Fig. 19. Contour map showing the depth to the Early Aptian unconformity in the Southern Outeniqua Basin. The annotation shows the present-day maturity levels at the unconformity extrapolated from burial-history models constructed at each of the points on the numbered grid.

plete, to positions against the flanks of the Diaz Fracture Ridge. Here, partial erosion of the syn-rift and some of the later sediments has been a major feature of the burial history. Some areas of the basin have thick Tertiary successions while, in other parts of the basin, the entire Tertiary has been removed by relatively recent erosion.

Thermal histories for each model were calibrated using data from boreholes in the peripheral basins. The heat-flow history was reconstructed using the method of Jarvis and McKenzie (1980), where an exponential decline in heat flow is modelled as the basin enters the stage of thermal sag. Present-day heat flows used in the models ranged from 40 mW/m^2 to 48 mW/m^2.

Figure 19 shows an isopach of sea floor to the Early Aptian unconformity directly underlying the potential source rock. The maturity overlay shows that present-day burial of between 2000 m and about 3500 m below sea floor is required for the potential source to be within the oil window. These depths become somewhat shallower to the west, in line with the higher heat flows used in models in this part of the basin.

The figure shows that a very large area of the Southern Outeniqua Basin is currently mature for oil- and wet-gas generation from a potential source rock developed in sediments overlying the Early Aptian unconformity.

Petroleum systems

Roux (1997) has identified a number of possible sand-rich seismic facies occurring within the Southern Outeniqua Basin. The sandstones in these plays, described below, could form the reservoir half of the source-reservoir couplets of a petroleum system reliant on an Early Aptian source. The relevant plays are:

- Shallow marine sands on tilted fault blocks, occurring around the northern rim of the basin and across the entire Infanta Embayment.
- Slope and basin-floor fan complexes within the Barremian-Albian. Drilling has proven the existence of these to the south, downdip of the palaeoshelf edge.
- Deep marine fans and channels of the Valanginian to Albian, within the central basin.

Conclusions

Evidence from exploration and scientific boreholes shows consistently developed zones of organic enrichment within the Early Aptian succession of the Orange Basin. A trend of improvement in source quality from proximal to distal depositional environments is demonstrated by these data and it is postulated that oil-prone source rocks are developed over large areas in the distal part of the basin. Burial history studies show that both oil and wet gas could be generated by this potential source rock.

The seismic signature of the Aptian succession within the undrilled Southern Outeniqua Basin is similar to that of the proven Aptian source rocks in the Bredasdorp Basin. Mapping of this seismic character suggests that a potential source rock of Aptian age is present across a large part of this basin. Burial-history studies show that the majority of the prognosed source rock is currently in the oil window.

Petroleum Agency SA is thanked for permission to publish. Comments from S. Lawrence, C. Cornford and E. Jungslager helped improve this paper. I. McMillan is thanked for Figure 11. The University of Stellenbosch is thanked for permission to publish Figures 4 and 18.

References

BEN-AVRAHAM, Z., HARTNADY, C. J. H. & KITCHIN, K. A. 1997. Structure and tectonics of the Agulhas-Falkland fracture zone. *Tectonophysics*, **282**, 83–98.

BEN-AVRAHAM, Z., HARTNADY, C. J. H. & MALAN, J. A. 1993. Early tectonic extension between the Agulhas Bank and the Falkland Plateau due to the rotation of the Lafonia microplate. *Earth and Planetary Science Letters*, **117**, 43–58.

BENSON, J. M. 1990. Palynofacies characteristics and palynological source rock assessment of the Cretacous sediments of the northern Orange Basin. (Kudu 9A-2 and 9A-3 boreholes). *Communications of the Geological Survey of Namibia*, **6**, 31–39.

BOLLI, H. M., RYAN, W. B. F. ET AL. 1978. *Initial Reports of the Deep Sea Drilling Project*. US Government Printing Office, Washington, **40**, 29–82.

BRAY, R., LAWRENCE, S. & SWART, R. 1998. Source rock, maturity data indicate potential off Namibia. *Oil and Gas Journal*, **96** (32), 84–89.

BROAD, D. S. & MILLS, S. R. 1993. South Africa offers exploratory potential in variety of basins. *Oil and Gas Journal*, **91** (49), 38–44.

BROWN, L. F., BENSON, J. M. ET AL. 1995. *Sequence Stratigraphy in Offshore South African Divergent Basins, An Atlas On Exploration for Cretaceous Lowstand Traps by SOEKOR (Pty) Ltd*. American Association of Petroleum Geologists Studies in Geology, **41**.

BURDEN, P. L. A. 1992. Soekor, partners explore possibilities in Bredasdorp Basin off South Africa. *Oil and Gas Journal*, **90** (51), 109–112.

COMER, J. B. & LITTLEJOHN, R. 1976. Content, composition and thermal history of organic matter in Mesozoic sediments, Falkland Plateau. *Initial Reports of the Deep Sea Drilling Project*, US Government Printing Office, Washington, **36**, 941–944.

CORNFORD, C. 1994. Mandal-Ekofisk(!) petroleum system in the central North Sea. *In*: MAGOON, L. & DOW, W. (eds) *The Petroleum System -From Source to Trap*. American Association of Petroleum Geologists Memoirs, **60**, 537–571.

CORNFORD, C. 1998. Source rocks and hydrocarbons of the North Sea. *In*: GLENNIE, K. W. (ed.) *Petroleum Geology of the North Sea*. Blackwell Science, Oxford, 376–462.

COSTER, P. W., LAWRENCE, S. R. & FORTES, G. 1989. Mozambique: A new geological framework for hydrocarbons exploration. *Journal of Petroleum Geology*, **12**, 205–230.

DAVIES, C, P. N. 1997a. *Hydrocarbon Evolution of the Bredasdorp Basin, Offshore South Africa: From Source to Reservoir*. Ph.D. thesis, University of Stellenbosch.

DAVIES, C .P. N. 1997b. Hydrocarbon families in the Bredasdorp Basin, offshore South Africa. *In*: MPANJU, F. K., KILEMBE, E. A. & KAGYA, M. L. N. (eds) *Proceedings of the 4th American Association of Petroleum Geologists Conference on Petroleum Geochemistry and Exploration in the Afro-Asian Region, 2–6 June 1996, Arusha, Tanzania*. Afro-Asian Association of Petroleum Geochemists, **69–76**.

DAVIES, C. P. N. & VAN DER SPUY, D. 1990. Chemical and optical investigations into the hydrocarbon source potential and thermal maturity of the Kudu 9A-2 and 9A-3 boreholes. *Communications of the Geological Survey of Namibia*, **6**, 49–58.

FOUCHÉ, J., BATE, K. J. & VAN DER MERWE, R. 1992. Plate tectonic setting of the Mesozoic basins, southern offshore, South Africa: a review. *In*: DE WIT, M. J. & RANSOME, I. G. D. (eds) *Inversion tectonics of the Cape Fold Belt, Karoo and Cretaceous Basins of Southern Africa*. A. A. Balkema, Rotterdam, 33–59.

FRANCHETEAU, J. & LE PICHON, X. 1972. Marginal fracture zones as structural framework of continental margins in South Atlantic Ocean. *American Association of Petroleum Geologists Bulletin*, **56**, 991–1007.

GERRARD, I. & SMITH, G. C. 1982. Post-Palaeozoic succession and structure of the southwestern African continental margin. *In*: WATKINS, J. S. & DRAKE, C. L. (eds) *Studies in Continental Margin Geology*. American Association of Petroleum Geologists Memoirs **34**, 49–74.

HAQ, B. U., HARDENBOL, J. & VAIL, P. R. 1988. Mesozoic and Cenozoic chronostratigraphy and eustatic cycles of sea-level change. *In*: WILGUS, C. K., HASTINGS, B. S., POSAMENTIER, H., VAN WAGONER, J., ROSS, C. A. & KENDALL, C. G. ST C. (eds) *Sea-level Changes: An Integrated Approach*. Society of Economic Paleontologists and Mineralogists Special Publications, **42**, 71–108.

JACQUIN, T. & DE GRACIANSKY, P. CH. 1988. Cyclic fluctuations of anoxia during Cretaceous time in the South Atlantic Ocean. *Marine & Petroleum Geology*, **5**, 359–369.

JARVIS, G. T. & MCKENZIE, D. P. 1980. Sedimentary basin formation with finite extension rates. *Earth and Planetary Science Letters*, **48**, 42–52.

JUNGSLAGER, E. H. A. 1999a. Petroleum habitats of the Atlantic margin of South Africa. *In*: CAMERON, N. R., BATE, R. H. & CLURE, V. S. (eds) *The Oil and Gas Habitats of the South Atlantic*. Geological Society, London, Special Publications, **153**, 153–168.

JUNGSLAGER, E. H. A. 1999b. New geological insights gained from geophysical imaging along South Africa's western margin. *Proceedings of the 6th Biannual Confernce and exhibition, South African Geophysical*

Society, Cape Town, 28 September – 1 October 1999, Extended Abstracts, Paper 14.4.

JUNGSLAGER, E. H. A. 1999c. Geological aspects of petroleum systems and related exploration plays in South Africa's Orange Basin. *Southern Africa and the Falklands Meeting, Petroleum Exploration Society of Great Britain, 29 March 1999, London, Extended Abstracts.*

LAWRENCE, S. R, MUNDAY, S. & SWART, R. 1999. New concepts on Namibian margin evolution and bearing on petroleum prospectivity of offshore Namibia. *Southern Africa and the Falklands Meeting, Petroleum Exploration Society of Great Britain, 29 March 1999, London, Extended Abstracts.*

Magoon, L. B. & Dow, W. G. 1994. The petroleum system. *In*: MAGOON, L. B. & Dow, W. G. (eds) *The Petroleum System – From Source to Trap.* American Association of Petroleum Geologists Memoirs, **60**, 3–24.

MATTHEWS, A., LAWRENCE, S. R., MAMAD, A. V. & FORTES, G. 2001. Mozambique Basin may have bright future under new geological interpretations. *Oil and Gas Journal,* **99(2)**, 70–75.

McMILLAN, I. K. 1990. Foraminiferal biostratigraphy of the Barremian to Miocene rocks of the Kudu 9A-1, 9A-2 and 9A-3 boreholes. *Communications of the Geological Survey of Namibia,* **6**, 23–29.

McMILLAN, I. K. & MAROT, J. E. B. 1997. *Tabulated Interpretative Age Information. Blocks 1 to 5.* Petroleum Agency SA. [Unpublished report]

MOLDOWAN, J. M., DAHL, J., HUIZINGA, B. J., FAGO, F. J., HICKEY, L. J., PEAKMAN, T. M. & TAYLOR, D. W. 1994. The molecular fossil record of oleonane and its relation to angiosperms. *Science,* **265**, 768–771.

MUNTINGH, A. 1993. Geology, prospects in Orange Basin offshore western South Africa. *Oil and Gas Journal,* **91(4)**, 105–109.

O'CONNELL, S. & WISE, S. W. 1990. Development of Mesozoic organic-rich sedimentary facies across southwestern Gondwanaland margins and basins. *American Association of Petroleum Geologists Annual Convention, June 3–6, San Francisco, Technical Program with Abstracts. American Association of Petroleum Geologists Bulletin,* **74**, 732.

ODSN. 1999. ODSN Plate tectonic reconstruction service. World Wide Web Address: http://www.odsn.de/ odsn/services/paleomap/paleomap.html.

PETERS, K. E. & CASSA, M. R. 1994. Applied source rock geochemistry. *In*: MAGOON, L. & Dow, W. (eds) *The Petroleum System – From Source to Trap.* American Association of Petroleum Geologists Memoirs, **60**, 93–120.

PETERS, K. E. & MOLDOWAN, J. M. 1993. *The Biomarker Guide.* Prentice-Hall, New Jersey.

ROUX, J. 1997. Potential outlined in Southern Outeniqua basin off South Africa. *Oil and Gas Journal,* **95** (29), 87–91.

STOW, D. A. V. 1987. South Atlantic organic-rich sediments: facies processes and environments of deposition. *In*: BROOKS, J. & FLEET, A. J. (eds) *Marine Petroleum Source Rocks.* Geological Society, London, Special Publications, **26**, 287–300.

VAN DER SPUY, D. 2000. Potential source rocks and modelled maturity levels in the Southern Outeniqua Basin, offshore South Africa. *Journal of African Earth Sciences, Geological Society of South Africa,* **31** (1a), 83–84.

WAPLES, D. W. & MACHIHARA, T. 1991. *Biomarkers for Geologists: A Practical Guide to the Application of Steranes and Triterpanes in Petroleum Geology.* American Association of Petroleum Geologists Methods in Exploration Series, **9**.

ZIMMERMAN, H. B., BOERSMA, A. & McCoy, F. W. 1987. Carbonaceous sediments and palaeoenvironment of the Cretaceous South Atlantic Ocean. *In*: BROOKS, J. & FLEET, A. J. (eds) *Marine Petroleum Source Rocks.* Geological Society, London, Special Publications, **26**, 287–300.

Geochemical evaluation of East Sirte Basin (Libya) petroleum systems and oil provenance

R. BURWOOD[1,2], J. REDFERN[1] & M. J. COPE[1]

[1]*North Africa Research Group, Oxford Brookes University, Headington, Oxford, OX3 0BP, UK*

[2]*Present e-mail for correspondence: rochemere@waitrose.com*

Abstract: With cumulative reserves exceeding 23 gigabarrels oil recoverable (GBOR), the East Sirte Basin is a prolific oil province hosting supergiants such as the Amal, Augila–Nafoora and Sarir fields. Production from Precambrian–Oligocene reservoirs yields low sulphur and often highly waxy oils.

The Late Mesozoic–Cenozoic Agedabia and older Hameimat, Maragh and Sarir troughs provide the main structural features of the habitat and control hydrocarbon prospectivity. Paleogene subsidence has facilitated the generative process with Mesozoic basin-fill sediments hosting source rocks for productive petroleum system(s). Traditionally the marine Upper Cretaceous Sirte Shale Formation source was thought to provide the dominant charge. Application of geochemical inversion procedures to oil data, however, indicates a greater diversity in oil provenance. Delineation of eight end-member generic oil families indicates a number of complex contributory petroleum systems, mixed-system hybrid oils also being evident. Non-marine (lacustrine) source inputs are also in evidence, enhanced waxiness differentiating petroleums of such provenance. Systematic screening of the stratigraphic section has additionally identified source potential in Nubian (Triassic and Lower Cretaceous), Rachmat–Tagrifet (Upper Cretaceous), Harash (Paleocene) and Eocene formations.

Assignment of oil provenance has been achieved via multivariate oil data analysis and application of a carbon isotope-based source kerogen-oil correlation procedure. End-member petroleum systems have been definitively identified involving the Sirte Shale Formation, Rachmat–Tagrifet Formations and Nubian (Triassic) as the contributory sources. The remaining major systems rely upon Pre-Upper Cretaceous lacustrine sediments specific to the Hameimat and Sarir troughs. Whereas numerous archetypal Sirte Shale Formation oils were recognized (e.g. Messla, Hamid, Sarir-L etc.), reserves for many of the giant fields, including Amal, Augila–Nafoora and Sarir-C, rely on hybrid system charging.

These results confirm that the prospectivity of the Sirte Basin is not exclusively dependent upon the Sirte Shale Formation, with other petroleum systems in operation, often involving hybrid-sourcing.

An extensive geochemical evaluation of the East Sirte Basin hydrocarbon habitat has been undertaken, providing contemporary support to prospectivity assessment within what otherwise can be regarded as a mature petroleum province. Hosting a number of giant fields, including Amal, Augila–Nafoora, Gialo, Messla and Sarir, the eastern sub-basin accounts for some 23 of the 45 gigabarrels oil recoverable (GBOR) discovered to date (Fig. 1). Where aided by use of modern exploration techniques, it is believed that significant potential for realization of further economic discoveries remains. The diversity of the habitat as presently known is reflected in the overview of likely effective reservoirs and source rocks as illustrated in Figure 2.

In the present study an understanding of the operative hydrocarbon habitat has been approached from a petroleum system vantage (Magoon & Dow 1994), employing contemporary geochemical data acquisition practices and interpretative concepts such as 'geochemical inversion' (Bissada *et al.* 1992). Here the petroleum system concept can be used as an effective model to account for discovered hydrocarbon accumulations and, by implication, be applied to evaluation of unexplored or under-explored areas.

For synthesis of such an approach, key objectives to be addressed include:

- crude-oil characterization, leading to the recognition of diversity/multiplicity in generic families, hence petroleum system(s);

From: ARTHUR, T. J., MACGREGOR, D. S. & CAMERON, N. R. (eds) *Petroleum Geology of Africa: New Themes and Developing Technologies.* Geological Society, London, Special Publications, **207**, 203–240. 0305-8719/03/$15
© The Geological Society of London, 2003.

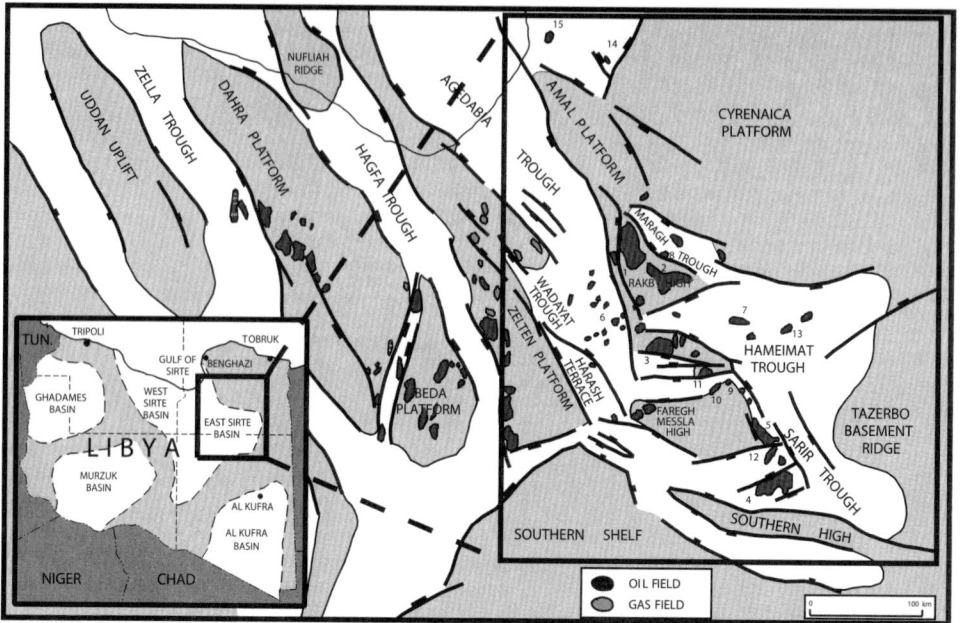

Fig. 1. East Sirte Basin study area identifying major petroleum accumulations, main tectonic elements and depositional centres. Key oilfields include, 1, Amal; 2, Augila–Nafoora; 3, Gialo; 4, Sarir-C; 5, Messla; 6, Intisar/Shatirah area; 7, Bu Attifel; 8, As Sarah–Jakhira; 9, Hamid; 10, Magid; 11, Masrab; 12, Sarir-L; 13, Remel; 14, Antelat and 15, B1-NC152.

- source-rock candidature, characterization and quantification of petroleum potential;
- source-oil assignments, confirming hydro-carbon provenance and the existence, plus any diversity, in operative petroleum system(s).

Application of this knowledge then provides informed and tailored input to basin-modelling procedures leading to:

- evaluation of the circumstances and timing of oil generation and emplacement history, mixed-source aggregate fluids and any post-emplacement/in-reservoir alteration processes;
- quantification of basinal generation and charge volumetrics as input to prospect evaluation and risk analysis.

In the body of this contribution, new data are presented primarily in support of the three former concerns. In part this permits realization of the fourth consideration, revealing that the East Sirte Basin is a complex hydrocarbon habitat where mixed-aggregate oils derived from hybrid petroleum systems are often encountered. This naturally has implications as to the diversity of play types that could be anticipated. The application of thoroughly controlled modelling then provides realistic basin assessment in the formulation of an exploration strategy.

Regional and petroleum geological overview

Reviews by Parsons *et al.* (1980) and Macgregor (1996) reference the Sirte Basin habitat in the context of the wider North African scene. El Alami *et al.* (1989), Ghori and Mohammed (1996) and Ambrose (2000) have provided more local insights.

For the East Sirte Basin a tectonostratigraphic overview is summarized in Figure 2. The basin shows remnant evidence of Early Palaeozoic continental sag with successive deposition of Cambro-Ordovician non-marine clastics (Amal Formation) through Silurian to Permo-Carboniferous aged sediments. With the Amal Formation clastics representative of the oldest sedimentary rocks observed, much of the intervening section was stripped by the Late Permo-Carboniferous (Hercynian) uplift; the resulting broad inversion feature is commonly termed the Sirte Arch.

Subsequent Triassic and later Jurassic–Early Cretaceous rifting led to the progressive fragmentation of the arch. Substantial graben-localized non-marine lacustrine deposition of fine to coarse clastics then took place in the Maragh, Hameimat and Sarir troughs.

The dominant basin architecture was formed by Late Cretaceous rifting, with subsidence occurring

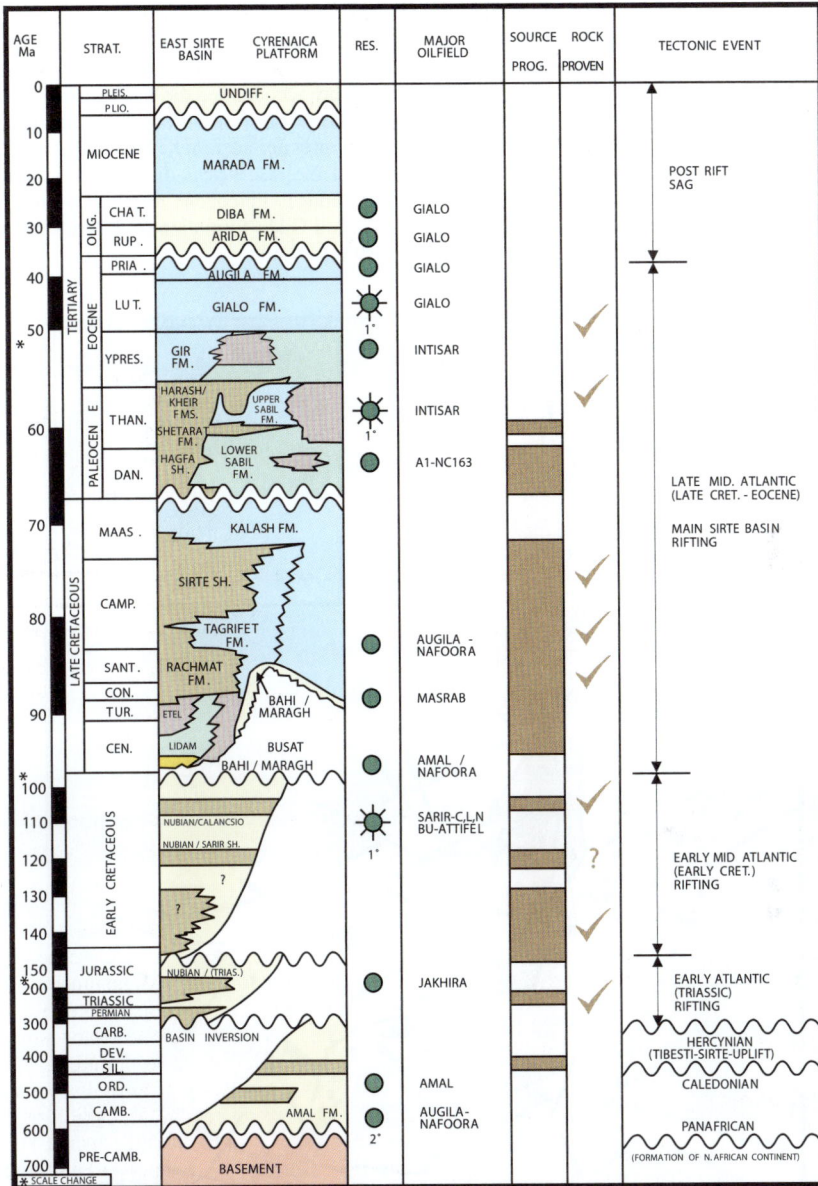

(After Masera 1988 Unpublished)

Fig. 2. Stratigraphic correlation chart for the East Sirte Basin with a tectonic overview and identifying proven reservoir intervals, producing fields and prognosed plus confirmed candidate source rocks.

throughout the area at this time and continuing into the Mid-Paleogene. As the main structural feature of the basin, the Agedabia Trough experienced the greatest subsidence and attendant marine sedimentary fill. Subsequent thermal cooling saw onset of a later Paleogene sag phase of subsidence, with the basin depocentre migrating northward into the present-day Gulf of Sirte. Neogene emergence tops

the stratigraphic succession with the deposition of a relatively thin veneer of predominantly continental sediments.

Structurally the study area is bounded by major basement highs, including the Cyrenaica Platform (northeast), Tazerbo Basement Ridge (east) and Southern Shelf, with the Zelten Platform providing separation from the West Sirte Basin (cf. Fig. 1).

Whereas Paleogene subsidence has facilitated the generative process by providing the required burial, Mesozoic basin-fill sediments, notably the Upper Cretaceous, provide source opportunities for the operative petroleum system(s).

The existing proven 23 GBOR are distributed throughout the whole Phanerozoic section between a diverse range of reservoir horizons (Fig. 3). Multiple pay zones are a feature of the Amal–Nafoora area, with some 10% of the cumulative reserves hosted in Early Palaeozoic (Amal Formation) sandstone traps. Mesozoic non-marine lacustrine clastics of the Nubian (Triassic) and Nubian (Sarir), however, account for the largest reserves (~50%), with the primary reservoir, Sarir Sandstone, hosting the premier accumulations of the area. The balance of the reserves is distributed between Upper Cretaceous to Paleogene reservoirs, with the Upper Sabil and Gialo formation limestones acting as primary reservoirs.

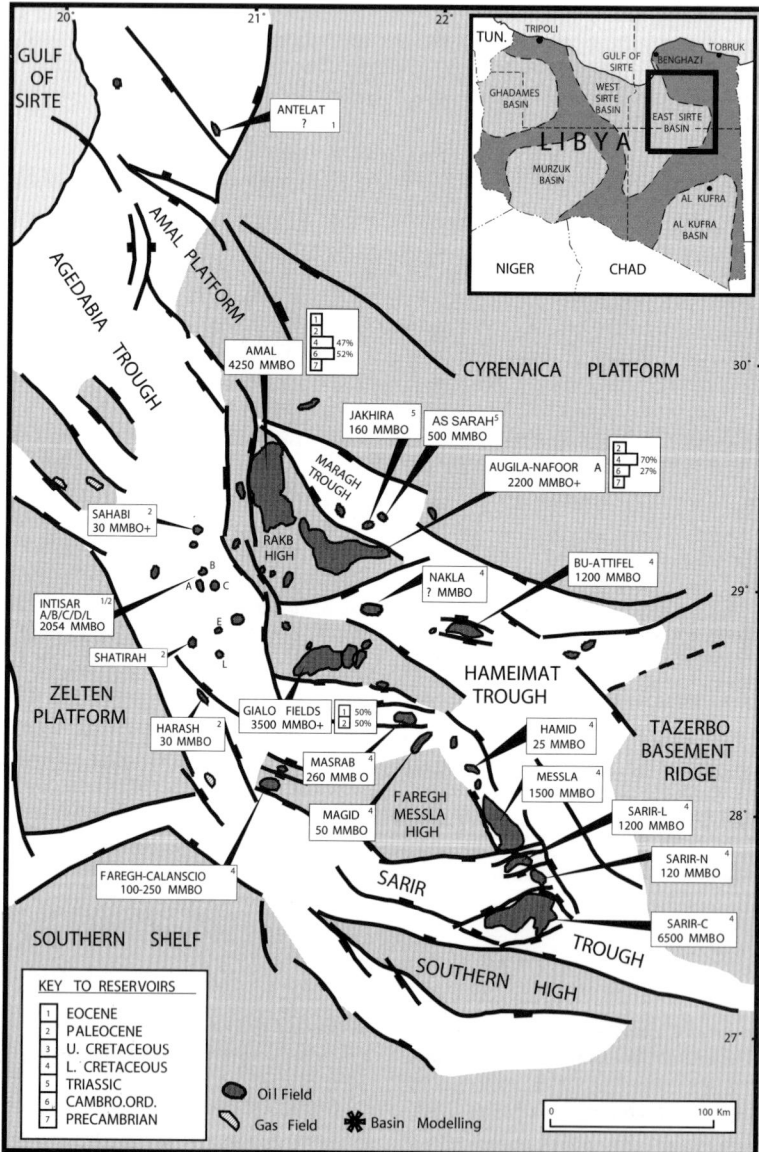

Fig. 3. Summary petroleum reserves as million barrels oil (MMBO) recoverable and reservoir information for a selection of East Sirte Basin fields.

Several Mesozoic to earliest Paleogene candidate source formations have been prognosed for the area (Petroconsultants Ltd 1981; Fig. 2) and so-called 'Nubian' lacustrine siliciclastics of Pre-Late Cretaceous age (commonly termed the Pre-Upper Cretaceous or PUC) deposited as components of:

- Nubian (Triassic), and
- Lower Cretaceous Sarir and Calanscio formations, (collectively termed 'the variegated shales').

Additionally, marine deposits of varying shale to argillaceous lime mudstone lithology have been proposed, including:

- Upper Cretaceous Sirte Shale Formation; Rachmat Formation; Tagrifet Formation and Etel Formation; and
- Paleogene Sheterat Formation and Hagfa Formation.

The distribution of prolific source rocks is not ubiquitous, all candidates being either locally developed within their respective depocentres (e.g. Triassic shales currently identified only within the Maragh Trough) or of highly variable richness (e.g. Sirte Shale Formation; El Alami *et al.* 1989). A regional and vertical intermittence in source-rock development hence provides a constraint on basin-wide prospectivity.

Relaxation of higher heat flows operative during the Mesozoic rifting events (Maragh through Agedabia trough-forming episodes) has resulted in average contemporary geothermal gradients, previously higher values having decayed to *c.*30 °C/km (Gumati & Schamel 1988; Ghori & Mohammed 1996). Present-day heat flows were computed to fall in the range 1.14–1.20 HFU (heat flow units), according to basinal situation (this work). As a consequence of rapid subsidence climaxing in the Paleogene, active generation ensued which, for the various Upper Cretaceous candidate rocks, peaked during the Neogene. Optimal hydrocarbon generation and expulsion in the basin thus occurred relatively late (<40~25 Ma BP).

Database

A comprehensive crude oil database (Fig. 4), comprising stable isotope and quantified biomarker analyses, was assembled for 60 authenticated petroleums to an analytical specification set by GeoMark Research Inc. Three additional key datasets were abstracted from a 1993 proprietary analytical study by Geomark Research Inc. Multivariate statistical evaluation of the data was achieved using Pirouette™ software for principal component analyses (PCA) and hierarchical cluster analyses (HCA). Data for type oils, representative of the

generic families and sub-families recognized, are listed in Table 1.

In excess of 50 km of section from 47 wells have been variously screened for source-rock potential, using conventional Rock-Eval and Pyrolysis-gas chromatography (GeoFina HM) procedures, and incorporated into this study (Fig. 4). Where appropriate, candidate source detailing analyses, including kerogen carbon isotope and kinetic parameter measurements, were performed. For presentation of data, primary source character is attributed to those sediments with S2 $\geq$5 kg/ton, S2 $\geq$3 kg/ton to $\leq$5 kg/ton being ascribed secondary source status. Petroleum potential (PP) is expressed in terms of PP units ($\times 10^6$ m$^3_{(oil\ equiv.)}$/km$^3_{(rock)}$) with the corresponding barrels/acre-foot equivalent also given in the respective tabulations.

The gas–oil production ratio (GOPR) is a normalized parameter (0 to 1) derived from GeoFina HM measurements. For values above 0.35, the kerogen assemblage is adjudged to be of increasing, to ultimately exclusive, gas-proneness. Elsewhere, parameters and petroleum-system nomenclature employed in this paper are defined in the Appendix.

East Sirte Basin oils inversion analysis

The existence of oil in the East Sirte Basin demonstrates the presence of at least one petroleum system. It follows that a regional oil study can be an efficient way of identifying, evaluating and verifying the uniqueness, or multiplicity, of such system(s). Determination of the diversity of effective source units within the basin can be inferred from the number of generically distinct oil families. Such an approach for the present study has been aided by access to a 60-specimen crude oil dataset embracing comprehensive biomarker and stable isotope information (Table 1).

Initial interpretation has led to conclusions relating to thermal maturation considerations, a preliminary grouping of the oils in terms of contributory source depositional environments and evidence for mixed-charge phenomena. Subsequent application of multivariate statistical analysis techniques to selected parameters then facilitated the definitive segregation into generic oil families. For purposes of graphic clarity, many of the text figures show only those type-oils chosen to be representative of the generic families and sub-families deduced from the inversion procedures subsequently applied.

In terms of bulk compositional data the oils showed limited diversity, with American Institute of Petroleum gravities (°API) in the conventional range of 30–39°. Several lower-gravity oils were observed from shallower reservoirs (K1–12 and Amal–Mesdar pool) but significant biodegradative

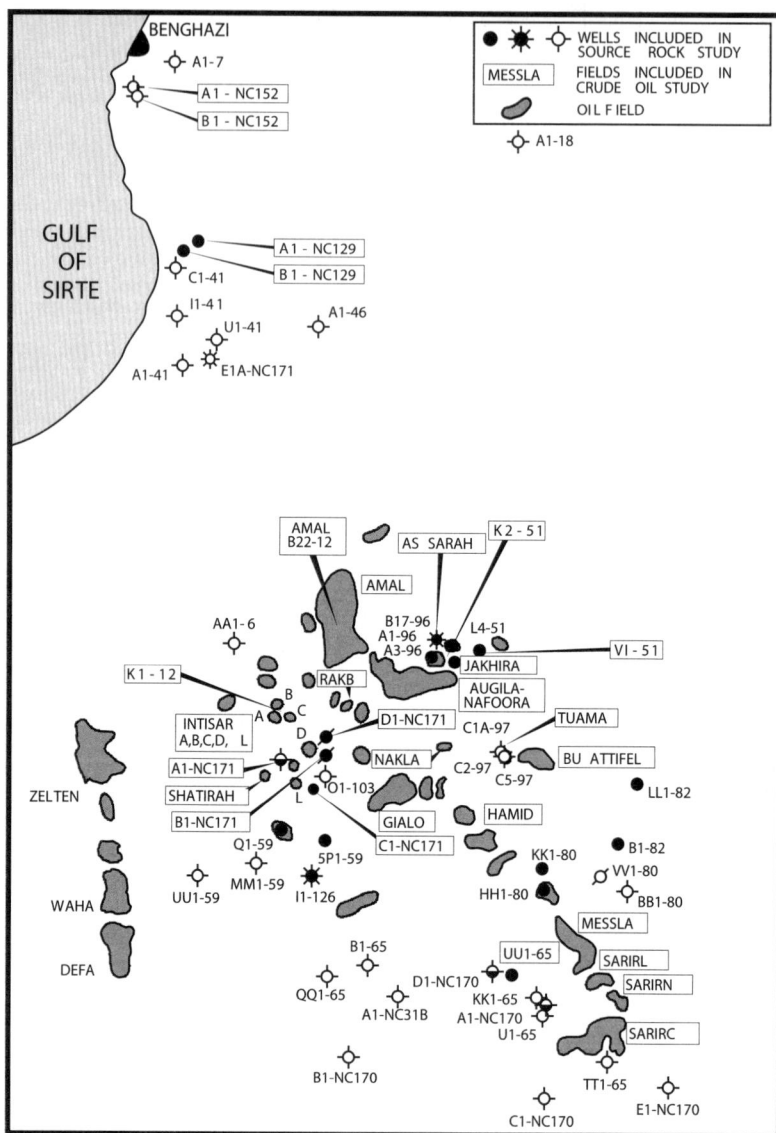

Fig. 4. East Sirte Basin geochemical data acquisition programme: location of crude oils and well sections employed in the study.

processes do not appear to have been operative. Elsewhere, the Intisar oils were of higher gravity (+40° API) but were not extra-mature or condensate-like, as in the case of the E1A-NC171 fluid. Sulphur contents were generally low (≤0.6% and often <0.2%), with higher values again observed only in the case of the shallow-reservoired oils as above and for the younger Augila–Nafoora pools.

The main compositional departure among these oils was the high waxiness observed for the As

Sarah, Jakhira, Nakla and Bu Attifel fluids, plus the Sarir-C and some of the deeper-reservoired Amal and Augila–Nafoora petroleums, the former showing an evident fieldwide variation. An implied source control was deduced to be operative here.

Crude oil maturity

As one control on gross compositional character, crude oil maturity was assessed via a range of molecular parameters, including the ethylcholestane

Table 1. *East Sirte Basin crude oil data for the 23-specimen study suite*

OIL	RESERVOIR	D13C(-)	D2H(-)	CV	PR/PY	29H/H	31H/H	S/H	%C28ST	%C29ST	C30X/H	19/23TRI	35/34H	24/23TRI	C27/17	%αααC29ST	C30/29ST	Ts/Tm	19/21TRI	22/21TRI	26/25TRI	26/27TRI	TET/C26TRI	C29ST(S/R)	MDBT
AUGILA-NAFOORA	AML	29.90	104	-2.65	2.14	0.83	0.29	0.29	34.0	30.0	0.22	0.21	0.56	0.66	0.44	32.73	0.07	1.4	0.13	0.25	1.30	0.37	0.45	0.76	2.30
GIALO	OLIGOCENE	27.90		-1.64	1.57	0.76	0.76	0.59	36.7	29.2	0.08	0.06	0.69	0.75	0.23	30.82	0.22	1.45	0.16	0.30	1.03	0.62	0.32	0.81	2.11
AMAL	AMAL	31.39	111	-0.54	1.84	0.76	0.38	0.28	38.3	28.1	0.15	0.09	0.61	0.64	0.37	31.56		1.39	0.16	0.27	1.04	0.45	0.44	0.69	2.46
TUAMA	SARIR	22.10	98	-2.48	2.34	0.78	0.45	0.18	20.0	55.0	2.17	0.56	2.56	0.51	0.61	36.36		7.67	0.48	0.27	1.54	1.04	0.19	1.50	24.00
JAKHIRA	TRIASSIC	34.67	111	2.25	1.91	0.88	0.28	0.07	37.0	30.0	0.49	0.24	0.44	0.67	0.51	48.84	0.00	2.17	0.16	0.25	1.60	0.36	0.53	2.39	6.25
AS SARAH	PUC	34.24	99	1.36	2.06	0.44	0.52	0.07	36.8	33.8	0.64		0.53	0.81	0.44	44.00		2.33	0.14	0.28	1.93	0.53	0.32	>2.00	>6.00
SARIR-L	SARIR	28.45	120	-0.17	1.81	0.79	0.31	0.93	39.0	27.0	0.13	0.13	0.60	0.81	0.30	30.54	0.18	2.11	0.13	0.32	0.95	0.88	0.20	1.01	2.70
MESSLA	SARIR	28.52	116	0.01	1.86	0.77	0.31	1.08	39.0	26.0	0.14	0.09	0.44	0.76	0.28	30.25	0.20	2.33	0.10	0.32	0.96	1.00	0.32	0.92	3.04
BU-ATTIFEL	SARIR	23.27	93	-4.57	2.33	1.45	0.26	0.12	16.0	41.0	1.07	0.60	0.39	0.73	0.64	50.00	0.00	6.29	0.25	0.22	1.83	0.68	0.17	2.02	9.60
NAKLA	PUC	21.65	107	-1.91	2.36	0.79	0.65	0.12	26.1	47.8	0.74		0.61	0.62	0.69	49.00		6.04	0.18	0.24	1.63	0.54	0.32	>2.00	>6.00
A1-129	ANTELAT	26.64	95	0.25	1.59	0.96	0.63	0.56	32.0	36.0	0.08	0.31	1.09	0.70	0.69	39.90		1.02	0.41	0.37	1.03	0.23	1.13	0.77	2.21
B1-152	U. CRETACEOUS	26.90	88	0.14	1.49	0.76	0.45	0.63	31.0	43.0	0.32	1.15	0.61	0.73	0.23	52.50	0.11	4.65	1.11	0.45	1.11	0.11	2.20	1.24	11.39
SARIR-C238	SARIR	27.68	110	-1.25	2.05	0.79	0.37	0.31	40.0	29.0	0.14	0.20	0.49	0.73	0.41	38.00	0.02	1.39	0.29	0.30	1.04	0.37	0.37	0.82	1.95
K2-51	U. CRETACEOUS	33.62	97	0.39	2.21	0.84	0.33	0.07	32.0	39.0	0.37	0.25	0.54	0.69	0.42	57.50	0.09	2.14	0.16	0.25	1.72	0.33	0.42	1.45	5.64
UU1-65	BUSAT-SARIR	26.71	93	-0.96	3.44	1.02	0.31	0.07	39.0	39.0	0.21	1.37	0.51	0.69	0.36	59.40	0.00	1.77	1.22	0.22	1.55	9.00	2.67	1.13	3.68
V1-51	TAGRIFET	34.87	96	0.16	2.44	0.83	0.33	0.07	30.0	38.0	0.29	0.22	0.56	0.66	0.44	42.50		1.53	0.14	0.22	1.72	0.26	0.41	1.30	1.22
AMAL-B2212	MESDAR	26.52	94	1.52	1.40	0.90	0.52	0.26	33.4	30.8	0.02	0.07	1.21	0.50	0.23	34.50	0.00	0.46	0.23	0.48	0.98	0.22	0.91	0.74	1.50
A1-NCI71	U.SABIL	27.70	98	2.69	1.57	0.91	0.57	0.89	27.0	27.0	0.20	0.18	0.83	0.79	0.23	30.40		1.94	0.27	0.33	1.08	0.36	0.75	0.75	2.48
B1-NCI71	U.SABIL	28.00	103	2.41	1.59	0.52	0.30	1.00	37.3	29.3	0.18	0.08	0.57	0.79	0.26	34.50		2.33	0.20	0.39	1.11	0.54	0.44	0.89	2.06
SHATIRAH	U.SABIL	27.94	107	1.58	1.77	0.49	0.84	0.69	38.2	28.8	0.12	0.08	0.83	0.81	0.13	29.00		1.90	0.19	0.39	1.11	0.56	0.33	0.91	
INTISAR-A103	U.SABIL	27.18	106	1.78	1.76	0.79	0.52	0.96	36.0	29.0	0.11	0.25	0.59	0.72	0.13	32.35		1.99	0.29	0.31	1.07	0.48	0.52		2.23
EIA-NCI71	GIALO	24.39	110	-4.83	2.13	0.80	0.41	0.39	31.6	35.5	0.05	0.35	1.26	0.72	0.04		0.21	1.25	0.94	0.31	0.92	0.92	1.27	0.77	5.39
K1-12	EOCENE	23.77	74	-3.29	0.60	1.75	0.63	0.12	28.0	37.0	0.00	0.19	2.06	0.31	0.54	40.70		0.25	0.63	1.24	0.94	0.02	12.05	0.55	1.64

Parameters are as follows: D13C ($\delta^{13}C$); D2H ($\delta^{2}H$); C_V (Canonical variable); PR/PY (pristane/phytane); 29H/H (norhopane/hopane); 31H/H (homohopane/hopane); S/H (sterane/hopane); %C28ST (%methylcholestane); %C29ST (%ethylcholestane); C30X/H (diahopane/hopane); 19/23TRI (19/23 tricyclic terpane); 35/34H (pentakis/tetrakishomohopane); 24/23 TRI (24/23 tricyclic terpane); C27/17 ($nC27/nC17$ wax factor); % $\alpha\alpha\alpha$ C29S (% $\alpha\alpha\alpha$ ethylcholestane of total $\alpha\alpha\alpha$ steranes); C30/29S (% $\alpha\alpha\alpha$ ethylcholestane of total $\alpha\alpha\alpha$ steranes); C30/29S (n-propyl/ethylcholestane); T_S/T_M; 19/21 TRI (19/21 tricyclic terpane); 22/21 TRI (22/21 tricyclic terpane); 26/25 TRI (26/25 tricyclic terpane); 26/27 TRI (26/27 tricyclic terpane); TET/26 (tetracyclic terpane/26 triterpane); C29S/R (ethylcholestane 20S/20R); and MDBT (4/1 methyldibenzothiopene) ratios.

($\alpha\alpha\alpha$ 20S/R), methylphenanthrene (MPI1) and methyldibenzothiophene (MDBT 4/1) ratio procedures (Radke *et al.* 1986; Bein & Sofer 1987; Radke & Willsch 1994). The majority of the petroleums showed main-phase generative status, with the implication that conventional thermal control and kerogen kinetics apparently controlled their generation and expulsion (Fig. 5). The predominance of regular gravity petroleums is thereby rationalized. A small population of low- and/or high-maturity crudes was corroborated by the methyldibenzothiophene procedure (Fig. 6), the existence of the more mature oils having implications relating to their deep-seated Maragh, Hameimat and Sarir trough provenances. Other than for these petroleums, it was assumed that maturity considerations would not necessarily complicate understanding of generic relationships between the oils studied.

Stable isotope segregation

Stable isotope values (δ^2H, $\delta^{13}C$) were assembled on whole oils, saturate and aromatic fractions. A $\delta^{13}C$ range of 13 ppt (parts per thousand) revealed a considerable diversity among the oils, with isotopically highly depleted (As Sarah) and enriched (Nakla) end-members embracing a broad mid-range, including the Augila–Nafoora, Messla, Sarir, Intisar, Gialo and Coastal Cyrenaica petroleums (Fig. 7). This diversity, previously recognized by El Alami *et al.* 1989, is accentuated in Figure 8; it is noteworthy that the disparate end-members include many of the waxier petroleums. The diagnostic value of this display is, however, of limited empirical use in that many of the waxy petroleums plot in the 'non-waxy' (i.e. marine-source depositional trend) domain as originally conceived by Sofer (1984). As subsequently demonstrated, a conclusion indicating a terrestrial-lacustrine provenance for these oils would have been more appropriate.

Further semi-empirical segregation of the oils was achieved by incorporation of deuterium isotope (^2H) information (Fig. 9). At least five domains can be delineated, the ^2H-depleted petroleums being more indicative of a lacustrine source depositional provenance (Burwood *et al.*

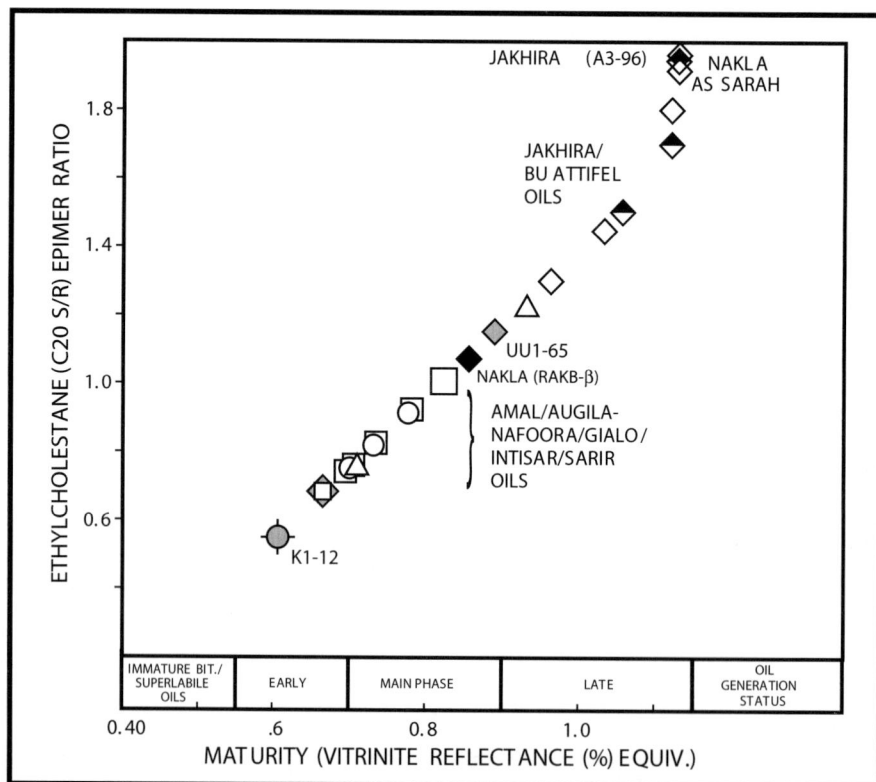

Fig. 5. East Sirte Basin crude oil inversion analysis: maturity assessment by ethylcholestane epimer (C20S/R) ratio procedure.

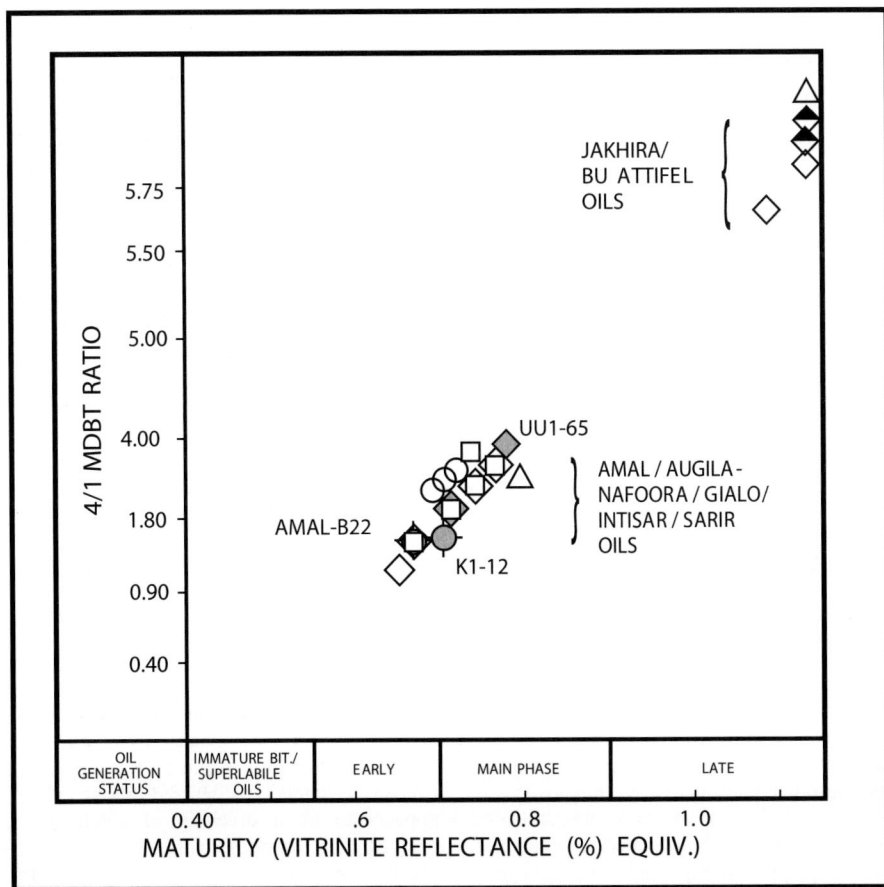

Fig. 6. East Sirte Basin crude oil inversion analysis: maturity assessment by methyldibenzothiophene (4/1 MDBT) ratio procedure.

1995), as could be anticipated for the As Sarah and Nakla end-members. Overall, isotopic composition suggests a large population of oils showing elements of a common, or similar, source provenance thought to be 'marine' in nature. Nevertheless, the existence of non-related disparate lacustrine-derived end-members is also evident, these being potentially available for incorporation into aggregate petroleums of a hybrid provenance.

Biomarker characteristics and inferences

From the biomarker dataset, sterane-based parameters have proved particularly useful in achieving a more intimate generic segregation of the petroleums. In an inversion sense, such oil-derived information can be highly diagnostic in revealing details of the progenic source-rock depositional palaeoenvironment and nature of the constituent organofacies.

As a generality, the sterane/hopane ratio v. $nC27/nC17$ cross-plot is informative, indicating a strong correlation between crude-oil waxiness and a non-marine (i.e. lacustrine) source-depositional provenance (Fig. 10).

Ethylcholestane content, as a measure of terrestrial/lacustrine organo-detrital input, is exploited as an environmental marker in Figure 11a and b. A cross-plot of the $\alpha\alpha\alpha$ C29 epimer content against dependent variable considerations of 'waxiness' and 'non-marineness', again represented by $nC27/nC17$ and sterane/hopane ratio parameters respectively, confirmed this strong correlation between lacustrine depositional conditions and wax context (e.g. As Sarah and Nakla end-member oil types).

A similar series of displays employing the abundance of n-propylcholestane as the strongly diagnostic marker of marine depositional inputs (Moldowan *et al.* 1990) again corroborates a con-

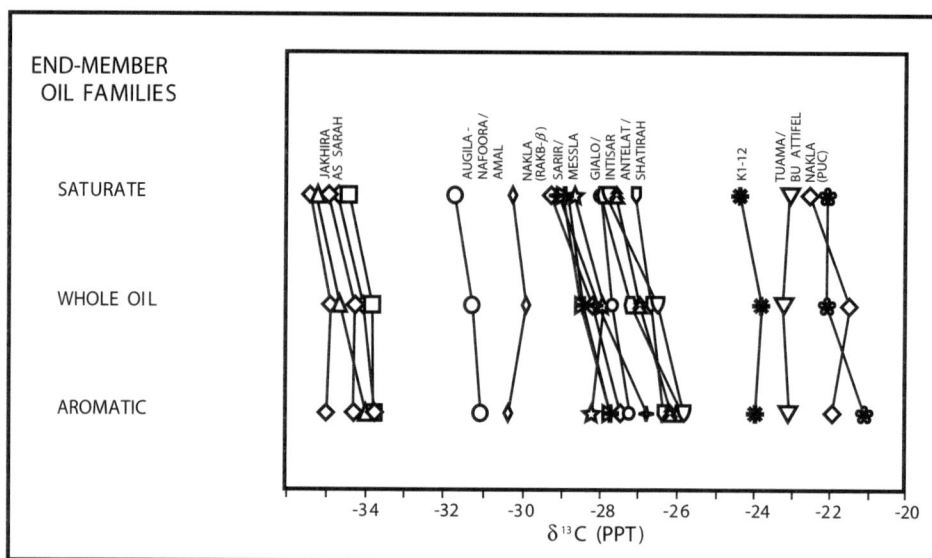

Fig. 7. East Sirte Basin crude oil inversion analysis: carbon-isotope Galimov-style profiles for a selection of 21 key oils, grouped as families (Table 1).

sistent segregation of the petroleums (Fig. 12a, b). On this occasion, oils such as Messla, Gialo, Intisar etc. are positively confirmed to be of marine-source provenance. Nevertheless, among output from the ethyl and *n*-propylcholestane-based diagnostic routines (Figs 11a, b & 12a, b), there appears to be ample evidence for oils of intermediate character, some of which presumably could be aggregates sharing in a hybrid source provenance (e.g. Sarir-C239, Augila–Nafoora, Amal etc.).

With the involvement and interplay between marine- and lacustrine-derived petroleum charge firmly established, other diagnostic markers can be used to advantage. Thus the observation of conspicuous β-carotane and gammacerane contents and high C26/C25 tricyclic terpane ratios provides abundant corroboration for the lacustrine-source provenance of the As Sarah, Jakhira, Nakla, Bu Attifel, UUI-65 petroleums and their derivatives (cf. Table 1).

The Coastal Cyrenaica (Concessions NC 129, 152 etc.) petroleums are exceptional in showing a discernible 18[α]oleanane content, this being interpretable in terms of a source age-dating of Late Cretaceous or younger. Additionally, prominent 2α-methylhopane, 30-norhopane and enhanced pentakishomohopane contents all testify to a highly anoxic, carbonate source-rock provenance. Assumed to have marine depositional affinities, this candidate system is clearly differentiable from that of the main marine source rock (i.e. the Rachmat, Tagrifet and Sirte Shale package), as illustrated in Figures 11a and 13. Further, and in this context,

these oils are exceptional, showing high wax contents. An unrelated Harash reservoired oil (K1–12) also shows strong carbonate-source affinities.

Thus, in addition to the lacustrine-derived petroleum system(s) recognized above, there also appears to be a diversity of candidate marine sources. This further expands the scope and complexity for emplacement of mixed hybrid petroleums.

For purposes of a graphic representation of a consensus of the contributory trends recognized from the data inversion processes employed, a cross-plot of the pristane/phytane ratio against carbon-isotopic composition achieves the most explicit results (Fig. 13). Five end-member petroleum systems can be delineated, including the As Sarah (!), UU1–65 (!) and Bu Attifel (!) lacustrine variants, plus the marine carbonate K1–12 (!) curiosity. These systems are well separated from a central domain embracing many of the assumed marine source derived oils (Messla, Intisar, Gialo, Shatirah etc.), which at this stage is provisionally referred to as the Upper Cretaceous (!) petroleum system (i.e. inclusive of the Rachmat, Tagrifet and Sirte Shale section). Subsequent application of source-oil correlation procedures and oil-data multivariate statistics allows further differentiation into the component systems, with recognition of the Sirte Shale (!) system as being the volumetrically predominant contributor.

Additionally, what is most evident from Figure 13 is the existence of hybrid oils and preliminary evidence for the mixing lines identifying their con-

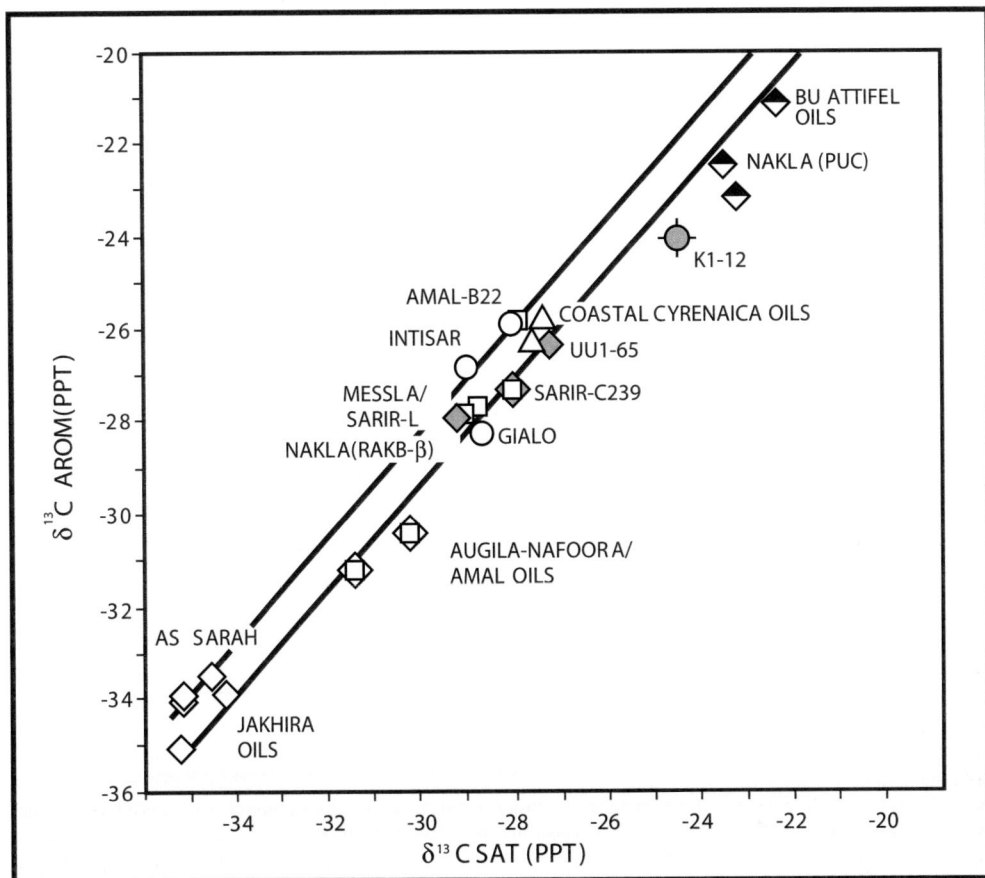

Fig. 8. East Sirte Basin crude oil inversion analysis: carbon-isotope Sofer-style cross-plot of saturate versus aromatic fraction δ¹³C values for 22 key petroleums (Table 1).

tributary end-member petroleum systems. In this context, the deeper-reservoired Augila–Nafoora and Amal oils derive from a hybrid Upper Cretaceous–As Sarah (!) system, with the waxy Sarir oils being ascribable to the Upper Cretaceous–U1–65 (!) analogue. Thus, as a result of the oil-data inversion enquiry, the complexity of the East Sirte Basin hydrocarbon habitat is revealed, with recognition of five unique, and at least two hybrid, petroleum systems. A minimum five-fold multiplicity in contributory source rocks is necessary to support this conclusion.

Candidate source-rock inventory

Screening of the 50 km of well section available to the study has permitted evaluation of previously accepted and new candidate sources of Silurian through Eocene age. The cumulative data are presented in terms of Palaeozoic, Triassic, Cretaceous and Paleogene sub-units. For each candi-

date, a type-well section is employed where feasible and features a comprehensive source rock classification covering intrinsic petroleum potential, kerogen transformation kinetics and carbon isotope signatures (Burwood *et al.* 1988).

Palaeozoic section

Early Palaeozoic. The Silurian is of widespread distribution throughout North Africa; however, the search for source quality sediments of this age in the East Sirte Basin was unsuccessful. An extensive shale section penetrated in A1–46, for instance, failed to show log characteristics suggestive of a basal Silurian (Gothlandian) high gamma-ray zone. Mature sediments with no suggestion of even residual hydrocarbon potential were observed. Although unpromising, these findings do not exclusively eliminate Silurian candidature, the relevant section being locally missing, having been eroded out as a result of the Hercynian orogeny.

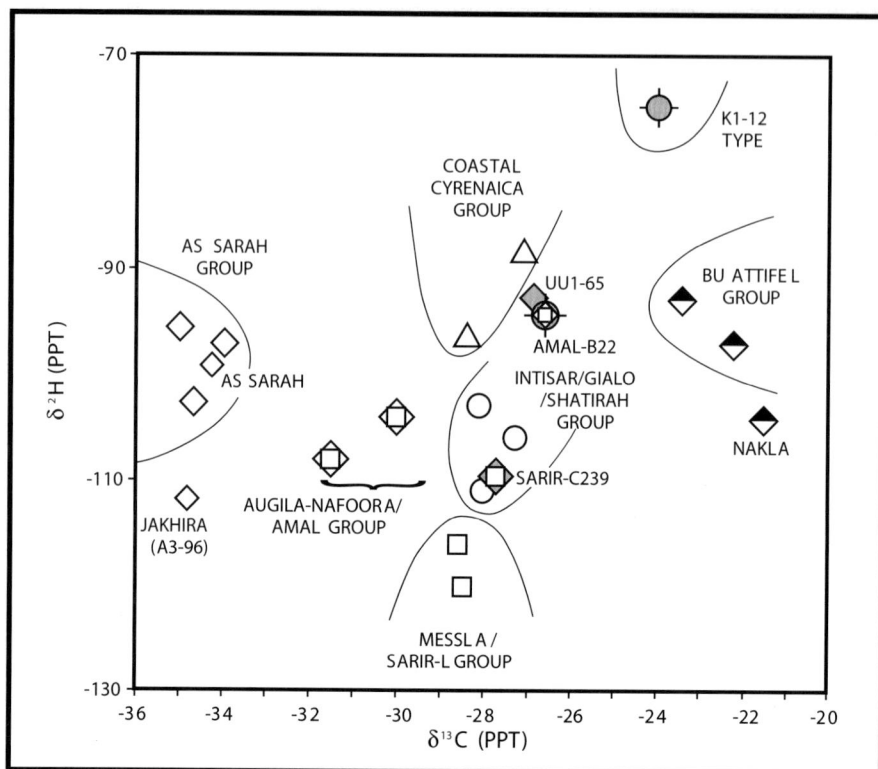

Fig. 9. East Sirte Basin crude oil inversion analysis: carbon and deuterium isotopic cross-plot for 21 key petroleums. Note the segregation into at least six discrete domains implying generic relationships.

Triassic section

Nubian (Triassic). Sediments with prolific source potential were penetrated in the Maragh Trough well L4–51. Dated to be of Anisian age (El Arnauti & Shelmani 1988), a ~30 m-thick effective source interval is characterized by a high gamma ray/low sonic velocity response zone.

Source evaluation data are summarized in Table 2a and identify presently immature sediments equating to high-activation energy Type I kerogen assemblages (Fig. 14). Although highly oil-prone, these kinetic attributes would dictate a forceful thermal regime in order to realize the hydrocarbon potential of these sediments. Progenic hydrocarbons with a highly depleted and diagnostic carbon isotope signature at $\delta^{13}C$ *c.*-34.0 ppt could be anticipated (Fig. 15).

Cretaceous section

Nubian (Sarir Formation). Screening of the Pre-Upper Cretaceous section in numerous Maragh, Hameimat and Sarir trough wells for anticipated

lacustrine-style source potential was of only limited success. Other than for some partially spent and intermittently developed potential in A1–96 (Maragh Trough) and A1-NC170 (Sarir Trough), dominantly sand-rich and/or organically lean, varicoloured sections were encountered. This was thought to imply that any lacustrine-source facies developments were of very localized and limited extent in what are collectively termed the 'variegated shales'.

Data for A1–96 are presented in Table 2b and identify the existence of residual primary source potential, these sediments being adjudged to be oil window mature and partially spent. A $\delta^{13}C$-depleted progenic oil signature of *c.*-30.5 ppt was observed (Fig. 15).

Analogous data for the Sarir Trough well (A1-NC170; Table 2c) again confirmed the development of thinly bedded potential, cumulatively equating to an effective source bed of *c.*30 m. The oil-prone assemblages comprised resistant, high activation energy Type I kerogens with an exceptionally $\delta^{13}C$-enriched progenic oil signature at −17.0 ppt (Figs 14 & 15).

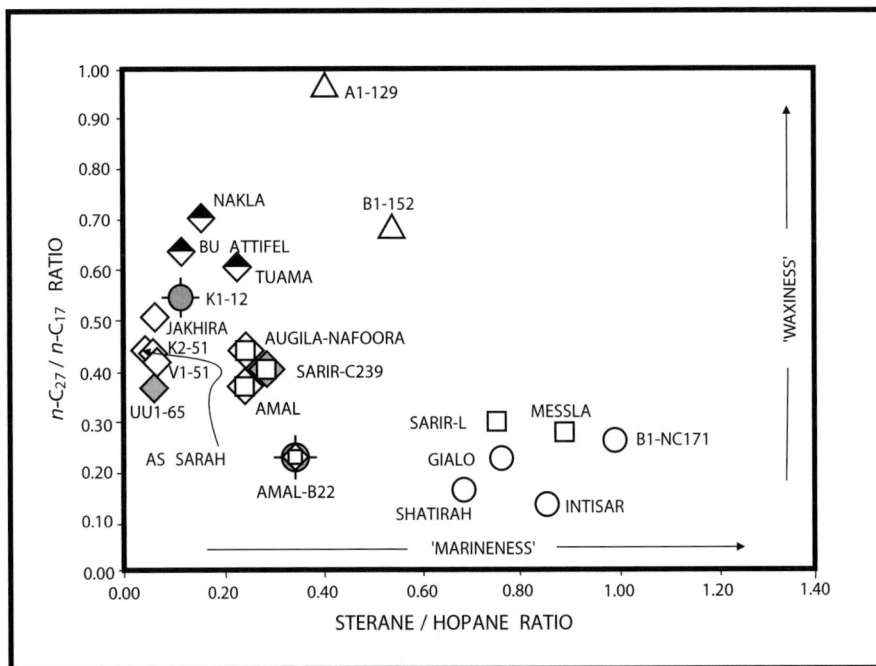

Fig. 10. East Sirte Basin crude oil inversion analysis: wax factor $nC27/nC17$ v. sterane/hopane ratio cross-plot showing a strong correlation between waxiness and non-marine (i.e. lacustrine) source depositional provenance.

Nubian (Calanscio Formation). Data for the Hameimat Trough well C5–97 revealed thinly bedded intervals of partially spent source-quality potential within the varicoloured shales (Table 2d). With instances of residual primary potential, these sediments were adjudged to have possessed attractive precursor source capacity. A highly $\delta^{13}C$-enriched formation mean progenic oil signature at -21.5 ppt (Fig. 15) characterized evident oil-prone potential with anticipated refractory kerogen kinetics.

The range in kerogen carbon isotope signature between the Nubian (Triassic) and Sarir plus Calanscio formation candidates (-34 ppt to -17 ppt) is noteworthy. This indicates an extensive organofacies variation and control as to the progenic hydrocarbon product that could be anticipated from these source rocks.

Etel Formation. Deposited as a result of the Cenorianian to Turonian marine incursions, source-quality potential has been recorded for intervals within the carbonate-evaporite sediments of the Etel Formation (El Alami 1996). In both the Agedabia (well 5P1–59) and Hameimat (well A1-LP4F) troughs, organic-rich sections in excess of 200 m with anticipated oil-prone potential were observed. A $\delta^{13}C$-enriched signature at -22.5 ppt for the source rock in well U1–82 has similarly been recorded.

Rachmat–Tagrifet formations. Comprising the lower part of the Upper Cretaceous source rock package, these marine sediments showed evidence of limited but widespread development. Control for the Rachmat Formation is provided by wells MM1–59 and D1-NC170 from the flanks of the Faregh-Messla High and identified a thick interval of sustained primary source potential (Table 2e). The oil-prone assemblages equated to Type II kerogens, with relatively labile kinetic transformation character (Fig. 14; E_a 52 kcal/mol), and gave a depleted progenic hydrocarbon signature at a $\delta^{13}C$ -28.0 ppt (Fig. 15).

The stratigraphically higher Tagrifet Formation in MM1–59 was less thickly developed but again showed evidence of primary source potential provided by oil-prone Type II kerogen assemblages (Table 2f). A progenic oil signature, comparable to the Rachmat Formation, at $\delta^{13}C$ -27.8 ppt was observed.

Sirte Shale Formation. Considered to be the archetypal source rock in the East Sirte Basin, the Sirte Shale is of variable thickness and richness. At its maximum development, in the Agedabia depocentre, up to 760 m of sediment with total organic carbon (TOC) contents of 1–5% have been recorded (El-Alami *et al.* 1989). In the present study, control was established for well Q1–59 with supporting

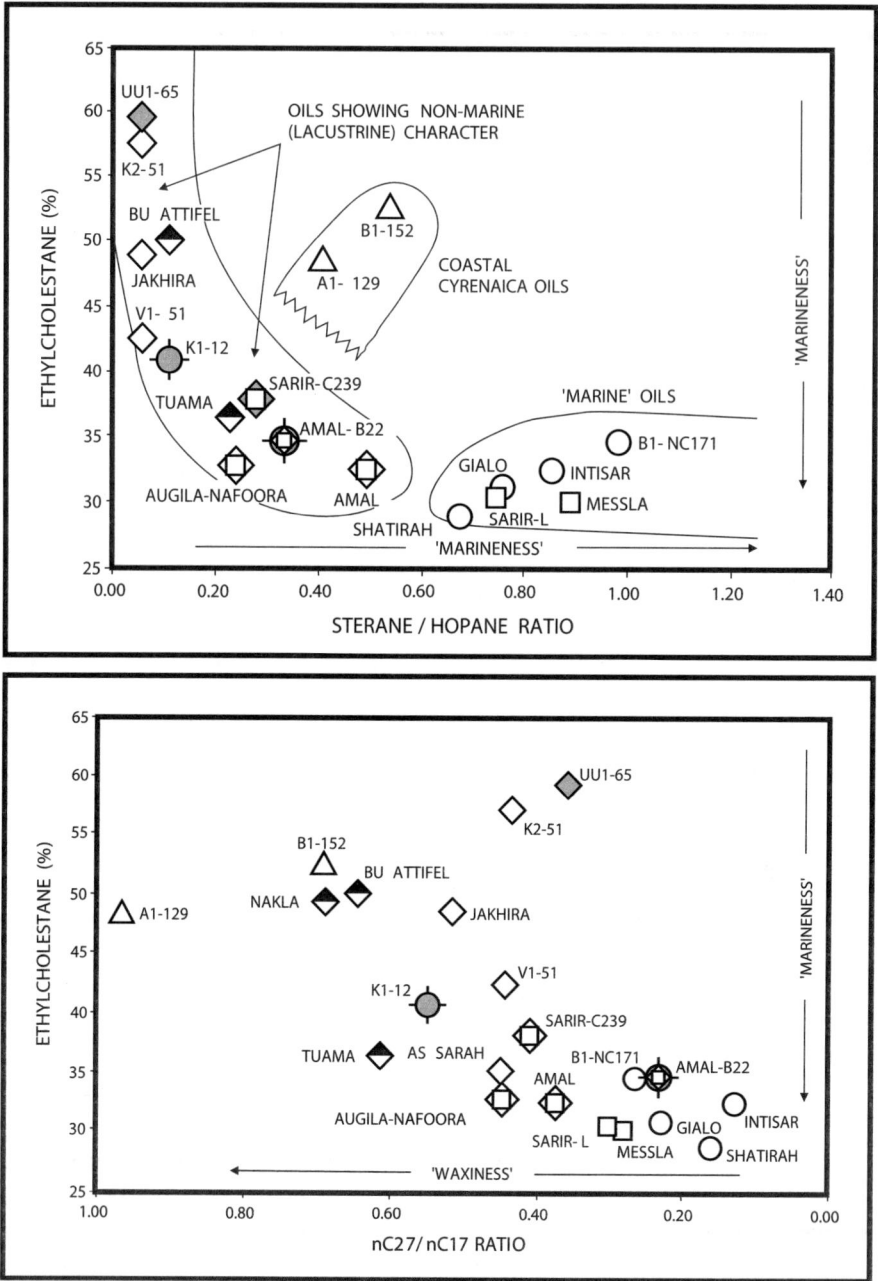

Fig. 11. East Sirte Basin crude oil inversion analysis: ααα ethylcholestane content parameter cross-plot with (**a**) sterane/hopane ratio and (**b**) *n*C27/*n*C17 wax factor confirming the strong direct correlation between the non-marine source detrital input parameters and providing *a priori* generic segregation of the oils into marine, non-marine (lacustrine) and those of intermediate character.

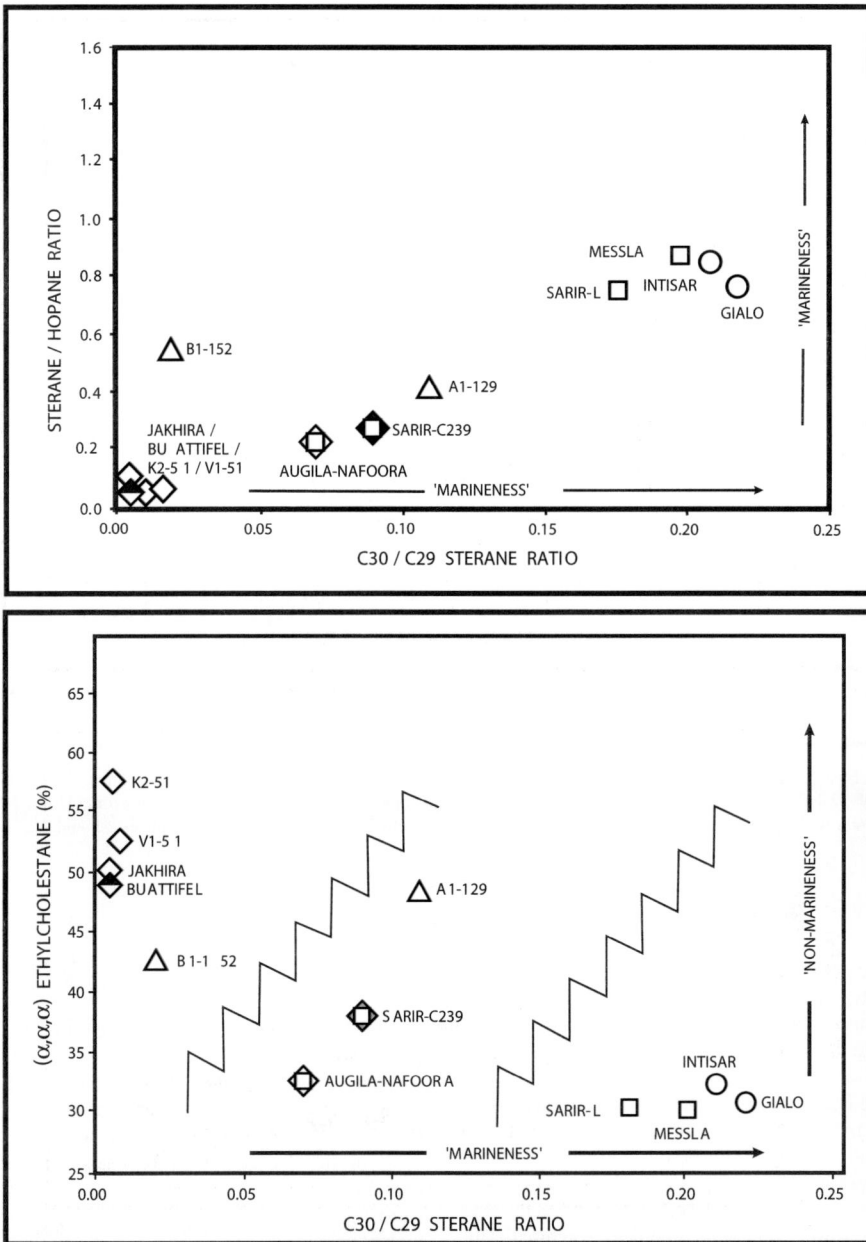

Fig. 12. East Sirte Basin crude oil inversion analysis: *n*-propylcholestane content 'marine' source parameter cross-plot with (**a**) sterane/hopane ratio and (**b**) ααα ethylcholestane content confirming the segregation into oils with non-marine (lacustrine), marine, and intermediate affinities.

data from D1-NC170, both these locations being on the flanks and south of the main Sirte Shale depocentre. A 380 m interval of sustained primary source potential, deriving from Type II kerogens, showed labile transformation kinetics (E_a 52 kcal/mol) and mixed oil-/associated gas-prone-

ness (Table 2g; Fig. 14). Kerogen carbon isotope values embraced a 3 ppt range, a potential-weighted average suggesting a mean progenic hydrocarbon signature at $\delta^{13}C$ −29.0 ppt (Fig. 15).

A cross-comparison of the three Upper Cretaceous candidate source units reveals comparable

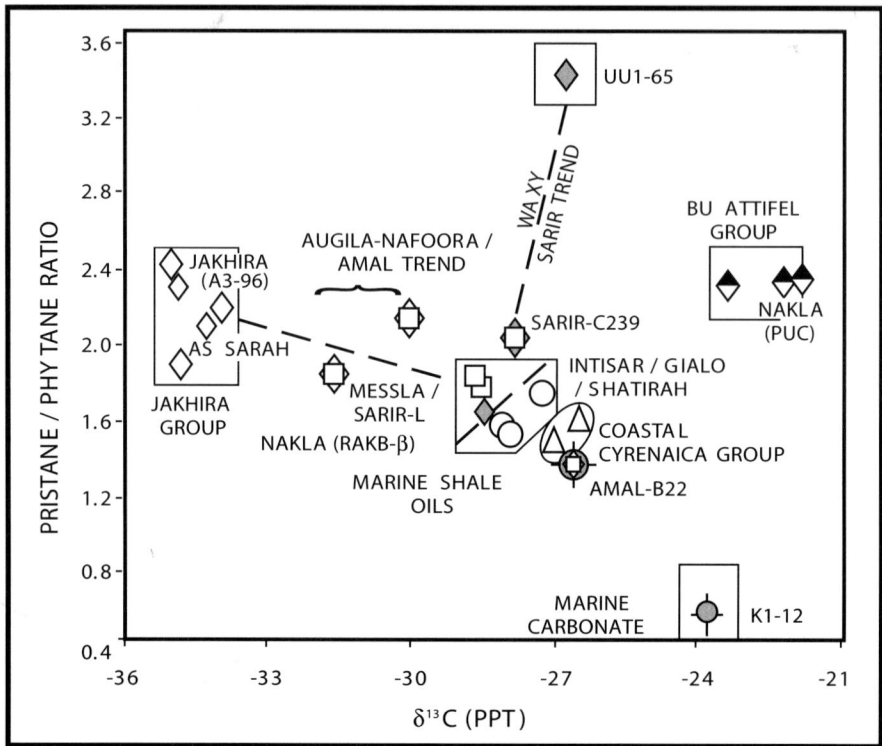

Fig. 13. East Sirte Basin crude oil inversion analysis: pristane/phytane ratio–carbon-isotope value cross-plot segregation of 22 key oils into generic groups. Note the mixing lines accounting for an aggregate composition for those oils of intermediate characteristics as previously identified by the sterane parameters in Figures 11 and 12.

intrinsic petroleum potential varying from 21 to 27 PP units, the main difference being in thickness and hence volumetric significance. In this context, the Sirte Shale is undoubtedly the premier candidate as and where fully developed. Nevertheless, both the Rachmat Formation and, to a lesser extent, the Tagrifet Formation offer substantial source potential that should not be overlooked. It is therefore important to note the subtle differences in carbon-isotopic signature between the candidates (Fig. 15). The lower and earlier-maturing Rachmat–Tagrifet formations show a more δ^{13}C-enriched signature (*c.* −28 ppt) than the stratigraphically higher Sirte Shale mean value (−29 ppt). Although only a narrow isotopic contrast, this observation proves to be invaluable in differentiating between the 'marine-sourced' oils.

In addition, isolated datasets attributable to the Kalash Formation have been obtained. With their source characteristics not dissimilar to that of the Sirte Shale Formation control, these data have not been individually flagged and quantified.

Paleogene section

Hagfa–Kheir formations. In the present study minor source-potential indications were observed for the Hagfa Shale. Again, these sediments showed source rock characteristics similar to that for the Upper Cretaceous candidates. Elsewhere, most of the Paleocene–earliest Eocene-aged section, including the Shetarat and Kheir formations, were adjudged to offer only insignificant to poor hydrocarbon potential.

Harash Formation. As the more basinal Kheir time-equivalent in the central Agedabia Trough, intermittent source indications have been observed for these dominantly carbonate sediments. Basal sediments from control well C1-NC171 revealed thinly developed horizons with secondary oil-prone potential (Table 2h). Kerogen assemblages with labile transformation kinetics (E_a 52 kcal/mol; Fig. 14) were characterized by a highly enriched, and

Table 2. *East Sirte Basin source-rock study: summary source-rock data for the more attractive candidate rock units recognized over the Triassic to Paleogene section*

(a)

Source Unit: Nubian (Triassic) Control well: L4–51 (Maragh Trough)					Progenic Oil ($\delta^{13}C_{KPY}$) Signature: -33.8 ppt			
Primary Source					Kinetic Data		Petroleum Potential	
Effective thickness (m)	TOC (%)	S2 (kg/t)	HI	GOPR	E_a (kcal/mol) [%]	A (s^{-1})	PP Units ($\times 10^6$m^3/km^3)	Bbl /acre-ft
+30	4.8	29.7	619	0.14	58 [99]	7.802 E+14	89	691

(b)

Source Unit: Nubian (Sarir$_M$) Control well: A1–96 (Maragh Trough)					Progenic Oil ($\delta^{13}C_{KPY}$) Signature: -30.5 ppt			
Primary Source					Kinetic Data		Petroleum Potential	
Effective thickness (m)	TOC (%)	S2 (kg/t)	HI	GOPR	E_a (kcal/mol) [%]	A (s^{-1})	PP Units ($\times 10^6$m^3/km^3)	Bbl /acre-ft
?	1.9	5.1	268	0.27	–	–	15	119

(c)

Source Unit: Nubian (Sarir$_S$) Control well: A1-NC170 (Sarir Trough)					Progenic Oil ($\delta^{13}C_{KPY}$) Signature: -17.0 ppt			
Primary Source					Kinetic Data		Petroleum Potential	
Effective thickness (m)	TOC (%)	S2 (kg/t)	HI	GOPR	E_a (kcal/mol) [%]	A (s^{-1})	PP Units ($\times 10^6$m^3/km^3)	Bbl /acre-ft
+33	2.5	11.2	455	0.16	58 [98]	7.135 E+14	34	261

(d)

Source Unit: Nubian (Calanscio Formation) Control well: C5–97 (Hameimat Trough)					Progenic Oil ($\delta^{13}C_{KPY}$) Signature: -21.4 ppt			
Primary Source					Kinetic Data		Petroleum Potential	
Effective thickness (m)	TOC (%)	S2 (kg/t)	HI	GOPR	E_a (kcal/mol) [%]	A (s^{-1})	PP Units ($\times 10^6$m^3/km^3)	Bbl /acre-ft
?	2.1	5.3			Partially Spent		>15	>115

Continued

Table 2. *Continued*

(e)

Source Unit: Rachmat Formation Control well: MM1–59/D1-NC170					Progenic Oil ($\delta^{13}C_{KPY}$) Signature:		-28.1 ppt	
Primary Source					Kinetic Data		Petroleum Potential	
Effective thickness (m)	TOC (%)	S2 (kg/t)	HI	GOPR	E_a (kcal/mol) [%]	A (s^{-1})	PP Units ($\times 10^6 m^3/km^3$)	Bbl /acre-ft
+260	2.1	8.6	420	0.21	52 [89]	2.181 E+13	26	200

(f)

Source Unit: Tagrifet Formation Control well: MM1–59/01–103					Progenic Oil ($\delta^{13}C_{KPY}$) Signature:		-27.8 ppt	
Primary Source					Kinetic Data		Petroleum Potential	
Effective thickness (m)	TOC (%)	S2 (kg/t)	HI	GOPR	E_a (kcal/mol) [%]	A (s^{-1})	PP Units ($\times 10^6 m^3/km^3$)	Bbl /acre-ft
+65	1.9	6.9	356	0.27	–	–	21	161

(g)

Source Unit: Sirte Shale Formation Control well: Q1–59/DI-NC170					Progenic Oil ($\delta^{13}C_{KPY}$) Signature:		-29.0 ppt	
Primary Source					Kinetic Data		Petroleum otential	
Effective thickness (m)	TOC (%)	S2 (kg/t)	HI	GOPR	E_a (kcal/mol) [%]	A (s^{-1})	PP Units ($\times 10^6 m^3/km^3$)	Bbl /acre-ft
+380	2.4	8.9	366	0.24	52 [63]	4.003 E+13	27	207

(h)

Source Unit: Lower Harash Formation Control well: C1-NC171					Progenic Oil ($\delta^{13}C_{KPY}$) Signature:		-23.4 ppt	
Primary Source					Kinetic Data		Petroleum Potential	
Effective thickness (m)	TOC (%)	S2 (kg/t)	HI	GOPR	E_a (kcal/mol) [%]	A (s^{-1})	PP Units ($\times 10^6 m^3/km^3$)	Bbl /acre-ft
1<5	0.9	4.2	467	0.19	52 [75]	3.433 E+13	13	97

Continued

Table 2. *Continued*

(i)

Source Unit: Gialo Formation (Gialo Member) Control well: D1-NC171					Progenic Oil ($\delta^{13}C_{KPY}$) Signature:	-25.5 ppt		
Primary Source					Kinetic Data		Petroleum Potential	
Effective thickness (m)	TOC (%)	S2 (kg/t)	HI	GOPR	E_a (kcal/mol) [%]	A (s$^{-1)}$)	PP Units ($\times10^6$m^3/km^3)	Bbl /acre-ft
25<40	0.9	5.5	577	0.13	49 [22]	0.950 E+13	16	125

(j)

Source Unit: Antelat Formation (Eocene) Control well: B1-NC129					Progenic Oil ($\delta^{13}C_{KPY}$) Signature:	-26.6 ppt		
Primary Source					Kinetic Data		Petroleum Potential	
Effective thickness (m)	TOC (%)	S2 (kg/t)	HI	GOPR	E_a (kcal/mol) [%]	A (s$^{-1)}$)	PP Units ($\times10^6$m^3/km^3)	Bbl /acre-ft
+20	5.4	30.3	558	0.18	52 [60]	2.557 E+13	91	705

$\delta^{13}C_{KPY}$ is the mean value for the formation weighted against S2.

Petroleum Potential (PP) units are in $\times10^6$ m$^3_{(oil\ equivalent)}$/km$^3_{(source\ rock)}$ reflecting a 35° API gravity crude oil and a rock density of 2.55 g/cm^3.

TOC (total organic carbon), S2 and HI are mean values for the effective source thickness.

GOPR, gas-oil production ratio, being a PGC parameter comprising the C1 to C5 component, normalized to total pyrolysate.

diagnostic, carbon-isotope signature at $\delta^{13}C$ -23.4 ppt (Fig. 15).

Intermittent minor potential is also developed higher in the Harash section. This shows a more gas-prone character (gas/associated light oil-condensate) with correspondingly less-labile kinetics and variable, but more ^{13}C-depleted, kerogen signatures.

Gialo Formation (Gialo Member). A persistent feature of the Middle Eocene in the central–southern Agedabia Trough is the development of a thin, highly oil-prone interval at the base of the Gialo Member. Of marginal primary source status in the control well D1-NC171, these sediments showed exceptionally labile kerogen kinetics (E_a 49 kcal/mol) and an enriched carbon isotope signature at $\delta^{13}C$ -25.5 ppt (Table 2i; Figs 14 &15). Although thoroughly immature in the type locality, such kinetics would facilitate transformation into hydrocarbons under greater burial depths, as perhaps prevalent in the present-day Agedabia depocentre.

Antelat Formation. Embraced within the alternative stratigraphic nomenclature applied to Coastal Cyrenaica, Eocene sediments penetrated in Concession NC129 wells showed excellent primary source potential (Table 2j). Whether these rocks are representative of a more basinal and effective development of those vestigial sediments recognized as the Gialo Member to the south is presently equivocal.

Source data for the type-section well (B1-NC129) suggested *c.*20 m of sustained, highly oil-prone potential deriving from Type II assemblages with labile transformation kinetics (E_a 52 kcal/mol; Fig. 14). A more ^{13}C-depleted, but diagnostically mid-range, progenic oil signature at -26.6 ppt applied (Fig. 15).

The systematic source-screening undertaken above has resulted in identification and quantification of eight authenticated candidate source rock systems. A synthesis of multivariate oil data analysis output, with carbon isotope-based, source to oil correlation assignments, now provides the basis for definition of the operative petroleum systems.

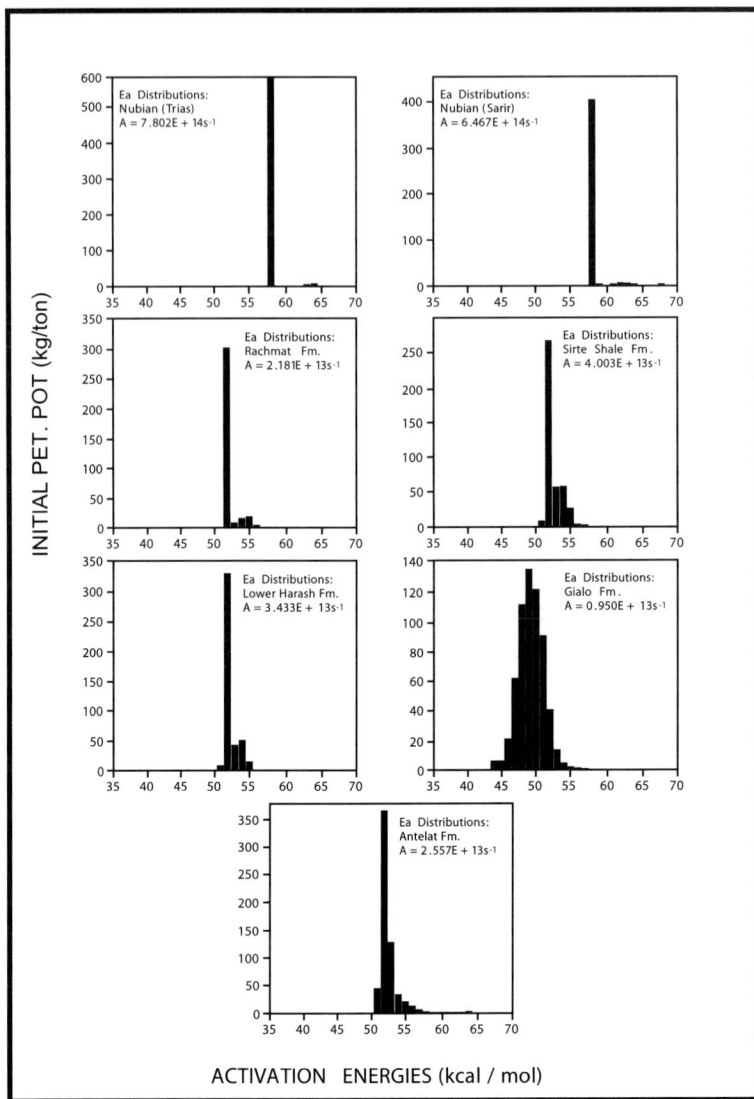

Fig. 14. Candidate source-rock study: kerogen transformation kinetic data for the repertoire of more attractive source rocks recognized. Data acquisition was by the Rock-Eval 5/Optkin procedure using pre-extracted (bitumen-free) rock powders employing a series of four heating rates (2, 5, 15 and 30 °C/min).

Multivariate oil analysis and source-oil correlation assignments

The various two-parameter inversion analysis treatment of the East Sirte oil data has provided evidence for the probable existence of eight end-member petroleum systems and of mixed-charge oils deriving from hybrid systems (Figs 9 & 13).

Corroboration of these observations, with extensions providing the basis for definitive source-oil correlation and petroleum system assignment, has been achieved using multivariate data processing. In this context a selection of parameters (Table 3), with strong palaeoenvironmental connotations and a low sensitivity to maturity effects, have been subject to PCA as illustrated in Figure 16a and b. The loadings-plot (Fig. 16a) reveals those critical source depositional traits that control the relative distribution of the oil families/sub-families in the attendant scores-plot (Fig. 16b). Note the discrimination between the end-member oil families and the evident mixing lines for the waxy Sarir (Sarir-

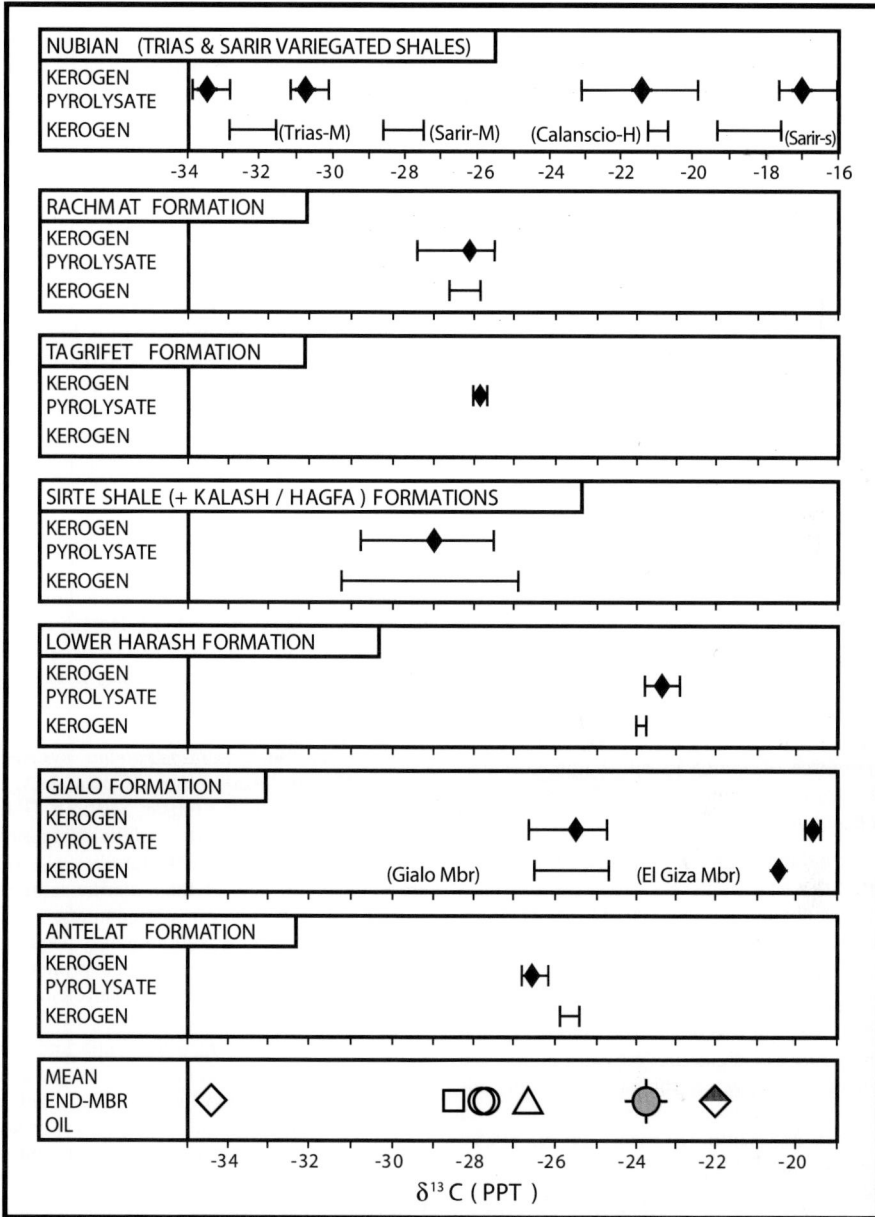

Fig. 15. Candidate source-rock study: kerogen and kerogen pyrolysate carbon isotope signatures for the more attractive source rock units recognized. Weighted mean (◆) pyrolysate values are calculated versus source potential as explained in the text. End-member oil family identities as given in Figure 16.

L to UU1–65) and Augila–Nafoora/Amal (Sarir-L to As Sarah) petroleums. The only equivocal group is that for the mixed Coastal Cyrenaica oils, the relative positioning of this domain suggesting tripartite charge inclusive of the Antelat, K1–12 and another (marine) contribution.

PCA can be extended, via an HCA routine, revealing additional fine structure as to the inter-relationships between the oils. Imposition of a correlation coefficient cut-off (≥80%) to the resulting dendrogram (Fig. 17) discriminated 12 groupings of sufficiently consistent composition to be

Table 3. *East Sirte Basin regional oil study: parameters employed for principal component and hierarchical cluster analysis in segregation of generic oil families*

1.	δ^2H whole oil
2.	$\delta^{13}C$ saturates
3.	$\delta^{13}C$ aromatics
4.	Canonical variable (C_V)
5.	nC_{27}/nC_{17} wax factor
6.	C35-hopane/C34-hopane ratio
7.	T_S/T_M ratio
8.	Sterane/Hopane ratio
9.	Pristane/Phytane ratio
10.	C24 tetracyclic terpane/C23 tricyclic terpane ratio
11.	C26/C25 tricyclic terpane ratio
12.	C19/C21 tricyclic terpane ratio
13.	C22/C21 tricyclic terpane ratio
14	Tetracyclic/C26 tricyclic terpane ratio
15.	Norhopane/Hopane ratio
16.	Homohopane (22S+R)/Hopane ratio
17.	Cholestane (% $\alpha\alpha\alpha$20R)
18.	Methycholestane (% $\alpha\alpha\alpha$20R)
19.	Ethylcholestane (% $\alpha\alpha\alpha$20R)

regarded as discrete generic families/sub-families. A specific provenance, as to charge, is thus required for each.

For purposes of source to oil assignments two general approaches are available based on petroleum-matching with either:

- The candidate source bitumen (biomarker route); or
- The candidate source kerogen assemblage (carbon isotope route).

The latter was used on this occasion in view of the extensive and discriminatory dataset that has been assembled. Carbon isotopic correlation via the kerogen pyrolysate signature provides a reliable tool in that it achieves a comparison of the oil-labile part of a precursor source assemblage with the progenic petroleum (Burwood *et al.* 1988). However, it should be noted that isotopic correlation alone does not necessarily or uniquely identify the source of a given oil. Rather, it establishes an *a priori* match, confirmation of which, by a second parameter, most notably diagnostic biomarker fingerprints, is desirable.

Carbon isotopic data for the candidate sources encountered in this study are summarized in Table 4 with their range and mean values illustrated in Figure 15. For the display of data, mean kerogen pyrolysate $\delta^{13}C$ values are weighted against source richness measured as S2 thus:

$$\overline{\delta^{13}C}\text{ (pyrolysate)} = [\Sigma\delta^{13}C\text{ (pyrolysate)} \times S2]/\Sigma S2.$$

Oil to source correlation was achieved by a whole oil/oil fraction v. kerogen pyrolysate profiling procedure as illustrated in Figures 15 and 18. Here, an acceptable correlation was deemed to exist when the whole oil-pyrolysate differential did not exceed ± 1 ppt, with a more stringent correspondence (± 0.5 ppt) conferring a greater degree of confidence.

Oil-to-source assignments, and hence the definition of the contributory petroleum system(s), can now be developed on the basis of the dendrogram and the supporting kerogen pyrolysate carbon isotope signatures, as summarized in Figure 17. An overview of the dendrogram reveals that the 12 generic groupings are distributed between a subordinate collection of oils with 'lacustrine'-source attributes and a major grouping of 'marine'-sourced petroleums. Within this 'marine' grouping, it is noteworthy that there is an evident fine structure between the Intisar, Shatirah and Gialo families and that these petroleums are quite distinct from the Messla genera.

Subtlety in source provenance can thus be anticipated. Elsewhere, the K1–12 Harash-reservoired oil appears to be of unique 'carbonate' provenance and three groupings, necessitating hybrid charge, are also evident.

End-member oil genera/petroleum system allocation. Family 1 oils embrace central Agedabia Trough fields including Intisar, Shatirah, Gialo and the shallow-reservoired Augila–Nafoora petroleums. Two differentiable families (1a and 1c) are evident. In view of their marine-source provenance and $\delta^{13}C$ mid-range signatures, Family 1a (Intisar) is assigned a predominantly Tagrifet Formation origin and constitutes the product of the Tagrifet Formation–Intisar (!) petroleum system (Fig. 19d).

Family 1c oils embrace the Gialo and shallow-reservoir Augila–Nafoora petroleums and are noteworthy in being the most sulphur-rich fluids of the eastern Sirte area. Marine, $\delta^{13}C$, mid-range kerogen signature sourcing again applies, with these petroleums being attributed to a dominantly Rachmat Formation provenance and constituting the Rachmat Formation–Gialo (!) petroleum system (Fig. 19c).

Hamid, Messla and some Sarir field production comprises Family 3, these oils again showing marine-source attributes, but with a somewhat more depleted $\delta^{13}C$ signature. A Sirte Shale Formation provenance, in which any Rachmat and/or Tagrifet Formation contribution is of very subordinate impact, can be attributed to these oils, this constituting the Sirte Shale Formation–Messla (!) petroleum system (Fig. 19a). Nakla, Bu Attifel and Tuama field petroleums comprise the Family 5

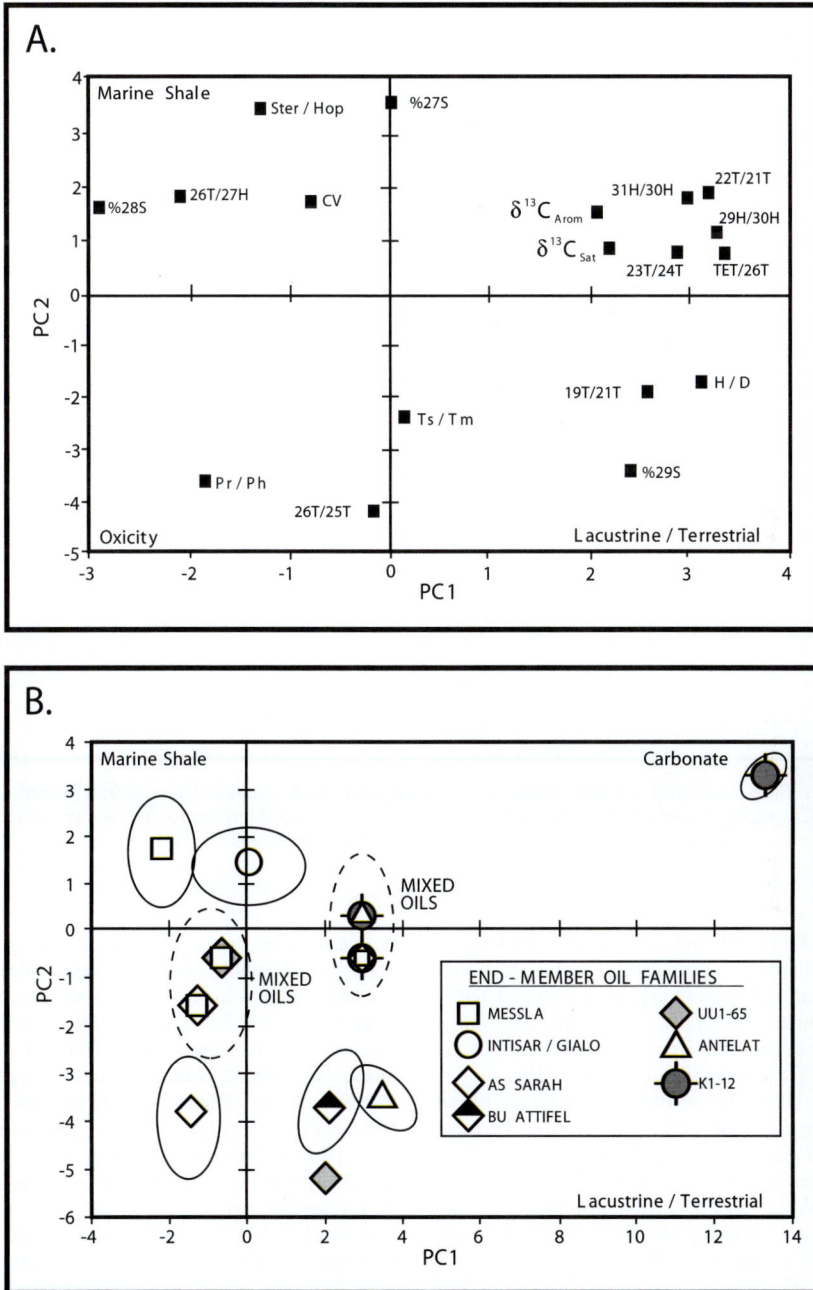

Fig. 16. East Sirte Basin crude oil multivariate principal component analysis (PCA): (**a**) PCA loadings plot using the palaeoenvironmental marker parameters listed in Table 3, and (**b**) PCA scores-plot illustrating the segregation into generically distinct families/sub-families. Note the mixing lines linking the end-member contributors to those petroleums of mixed, aggregate composition.

TYPE OIL (RESERVOIR)	GENERIC OIL FAMILY	SOURCE TYPE	OIL / SOURCE GENERATION	CORRELATION COEFFICIENT	PETROLEUM SYSTEM
INTISAR A-130 (U. SABIL)	1 / 1a	MARINE SHALE (M2/M3)	LOWEST UPPER CRETACEOUS SOURCE ROCK PACKAGE		TAGRIFET / INTISAR (!)
SHATIRAH (U. SABIL)	1b				TAGRIFET / RACHMAT-SHATIRAH (!!)
GIALO (OLIGOCENE)	1c				RACHMAT / GIALO (!)
A1-129 (ANTELAT)	2		HYBRID (M1 + M4 + N − M)		SIRTE SHALE / ANTELAT / HARASH - A1-129(.!.)
MESSLA (SARIR SST.)	3	MARINE SHALE (M1)	UPPERMOST U.CRET. SOURCE ROCK PACKAGE		SIRTE SHALE / MESSLA (!)
SARIR-C239 (SARIR SST.)	4 / 4a	MIXED MARINE / LACUSTRINE	(M1 + L3)		SIRTE SHALE / NUBIAN (PUCₛ)-SARIR-C(!.)
AUGILA - NAFOORA (AMAL)	4b		(M1 + L1)		SIRTE SHALE / NUBIAN (Trias.) - AMAL (!!)
AS SARAH (TRIASSIC SST.)	8	LAC.(L1)	NUBIAN (Trias.) LAC.		NUBIAN (Trias.) - AS SARAH (!)
BU ATTIFEL (SARIR SST.)	5	LAC.(L2)	NUBIAN (Pre-U.Cret.) LAC. (CALANSCIO FM.)		NU.(CALANS.) -BU ATTIFEL. (!)
UU (SARIR SST.)	6	LAC.(L3)			NU.(PU Cₛ)-UU1-65 (.)
B1-152 (ANTELAT)	7	RES. MAR.	EOC. MAR?		ANTELAT - B1-152(!)
K1-12 (HARASH)	9	MARINE (M4)	(M4)		HARASH - K1-1 2 (!)

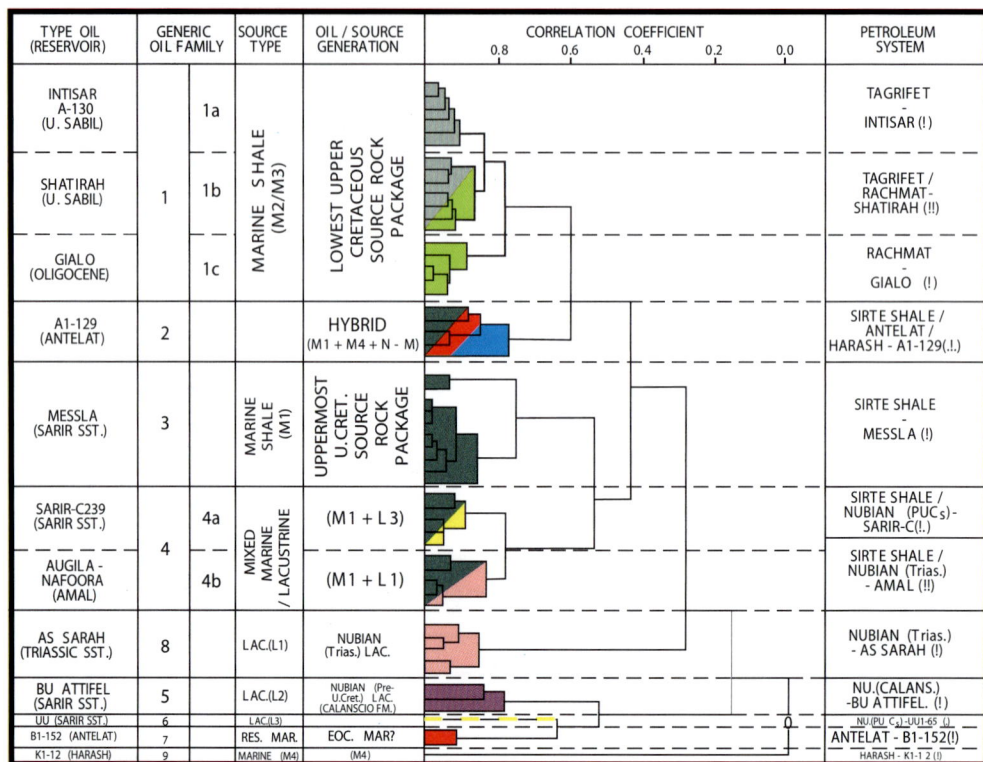

Fig. 17. East Sirte Basin crude oil and source-oil correlation assignment synthesis based upon a hierarchical cluster analysis (HCA) dendrogram. Note the additional generic fine structure revealed by the HCA procedure and the obvious internal diversity among those oils with a marine source provenance attribution.

members. These oils are of lacustrine provenance from the distinctive Pre-Upper Cretaceous Calanscio Formation organofacies, as developed within the Hameimat Trough, and constitute the Nubian Pre-Upper Cretaceous Hameimat Trough (PUC_H)–Bu Attifel (!) petroleum system (Fig. 19g).

The single member of Family 6 (UU1–65) again shows lacustrine attributes, but on this occasion appears to derive from a Pre-Upper Cretaceous organofacies developed within the Sarir Trough. This petroleum system is of considerable significance in contributing the high wax-modifying component to many of the Sarir-C oils and is defined as the Nubian Pre-Upper Cretaceous Sarir Trough (PUC_S)–UU1–65 (.) variant.

Family 7 petroleums comprise the end-member for the Coastal Cyrenaica oils (Concessions NC129, 152) and derive from the Eocene Antelat Formation (Antelat Formation–Antelat (!) petroleum system, Fig. 19e). The Antelat source derived oils show unusual palaeoenvironmental characteristics of both a highly carbonate/anoxic marine yet 'non-marine/terrestrial' waxy character.

A highly restricted, possibly speciality, marine depositional regime could be responsible.

Family 8 petroleums derive from the As Sarah, Jakhira, V1- and K2–51 fields. Of lacustrine provenance, these oils are sourced from an exceptionally $\delta^{13}C$-depleted Nubian (Triassic) organofacies specific to the Maragh Trough. The resulting Nubian (Triassic)–As Sarah (!) petroleum system (Fig. 19b) is of considerable importance in providing the co-charge to the Augila–Nafoora and Amal deep-reservoir fields.

A single oil (K1–12), of unique Harash Formation marine-carbonate source provenance, constitutes Family 9, being the representative of the Harash Formation–K1–12 (!) petroleum system (Fig. 19f). Again, this system is thought to be a subordinate contributor to some of the northern Agedabia Trough fields.

Hybrid petroleum system crude oils. Family 1b oils, embracing the Shatirah and NC171 discoveries, appear to be composite Rachmat–Tagrifet Formation derived petroleums and constitute the

Table 4. *East Sirte Basin source-rock study: kerogen and kerogen pyrolysate carbon isotope values for candidate source rock units*

Candidate Source Unit	Well	Kerogen $\delta^{13}C^*$ (-ppt)	Kerogen Pyrolysate $\delta^{13}C$ (-ppt)	Mean† $\delta^{13}C$ (-ppt)
Antelat Formation (Eocene)	B1-NC129	25.66	26.49	
		25.42	26.63	
		25.89	26.79	
		25.81	26.16	
		25.68	26.80	26.57
Gialo Formation (Gialo Member)	B1-NC171	24.84	25.59	
		24.68	25.42	
	C1-NC171	24.90	24.71	
	D1-NC171	25.26	24.84	
		25.49	25.27	
	A1-NC170	26.51	26.66	
		26.46	26.02	25.50
Gialo Formation (El Giza Member)	D1-NC171	20.46	19.60	19.60
Lower Harash Formation	C1-NC171	23.77	22.91	
		24.02	23.83	23.37
Hagfa Formation	A1-NC31B	27.36	27.91	
	KK1–65	27.07	27.56	27.61
Kalash Formation	A1-NC31B	28.34	28.48	
	KK1–65	27.37	28.04	28.27
Sirte Shale Formation	A1-NC31B	30.30	30.37	
		27.53	30.19	
	D1-NC170	27.18	28.19	
		27.46	28.34	
		26.89	28.66	
	HH1–80	28.12	27.83	
		30.12	30.82	
		28.60	29.53	
		31.25	28.25	
	JJ1–65	27.93	28.27	
	KK1–65	27.30	29.74	
	O1–1403	28.05	27.86	
		–	27.48	
	TT1–65	28.36	30.08	28.97
Tagrifet Formation	O1–103	–	27.85	27.85
Rachmat Formation	D1-NC170	27.83	29.34	
	01–103	28.06	29.41	28.10
Nubian (Sarir Formation)	A1–96	28.61	31.14	
Maragh Trough		27.47	30.10	30.75
Nubian (Sarir Formation)	A1-NC170	19.33	17.66	
Sarir Trough		18.87	17.36	
		17.58	15.89	16.97
Nubian (Calanscio Formation)	C5–97	20.68	20.22	
Hameimat Trough		20.86	23.16	
		20.92	19.80	
		21.00	22.58	21.44
Nubian (Triassic)	L4–51	32.57	33.80	
Maragh Trough				33.80

*$\delta^{13}C$ measurements referenced against NBS22 at -29.80 ppt
†See text for calculation of weighted mean kerogen pyrolysate values.

Rachmat/Tagrifet–Shatirah(!!) hybrid petroleum system (Fig. 20c).

Family 2 oils contain a variable collection of fluids from the northern Agedabia Trough, including the Coastal Cyrenaica fields, Amal B22–12 (Mesdar Formation reservoir) and an E1A-NC171 condensate, and proved the most equivocal in assigning a provenance. Plotting in the marine

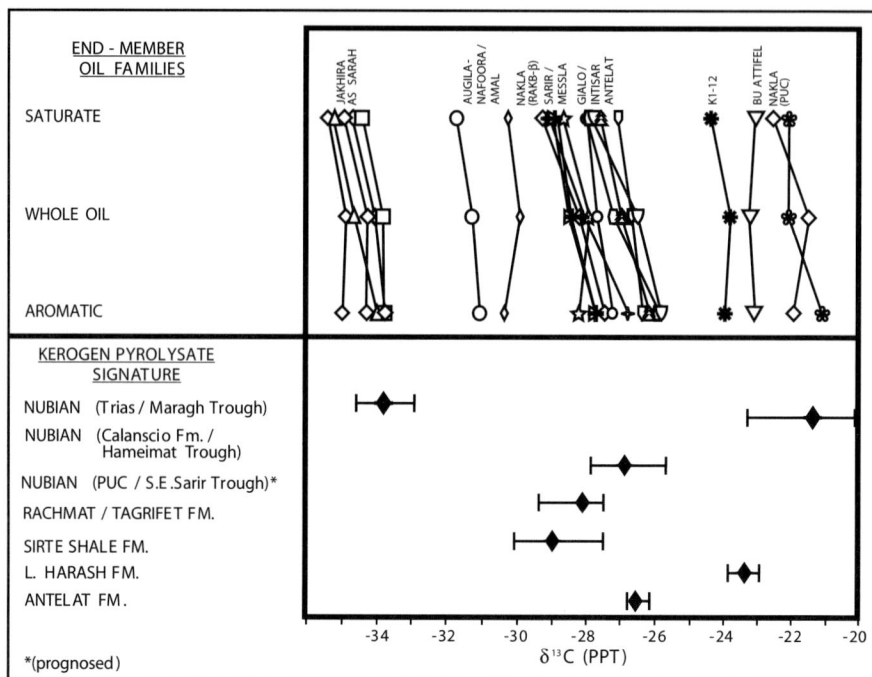

Fig. 18. Carbon isotope-based source-oil correlation assignment for the 21 key oils revealing an isotopic correspondence between the Nubian (Triassic)–As Sarah; Sirte Shale Formation–Messla; Rachmat–Tagrifet Formation/Intisar–Gialo; Antelat Formation–Antelat, Lower Harash Formation–K1–12 and Nubian (Calanscio Formation)–Bu Attifel end-member petroleum systems.

domain of the dendrogram, these petroleums can show both high wax and 'carbonate' source attributes. However, judging from their position within the scores-plot (Fig. 16a), they are not solely an admixture of the Antelat (!) and Kalash (!) systems, but some level of *bona fide* marine-source contribution is also involved.

Family 4 petroleums are essentially derived from the Sirte Shale Formation (!) system, but can be split into two sub-families according to the origin of their co-charge. Sub-family 4a oils comprise the waxy Sarir trend coming from various Sarir Sandstone reservoir locations over the extent of the Sarir-C field. The UU1–65 (!) system (see above) provides the co-charge on this occasion (Fig. 20b).

Sub-family 4b oils embrace the deep-reservoir Augila–Nafoora and Amal petroleums. Here, the highly distinctive As Sarah (!) system (see above) provides the modifying co-charge, these hydrocarbons currently being recognized as exclusive to a Maragh Trough provenance. This hybrid can thus be fully defined as the Sirte Shale Formation/Nubian (Triassic)–Amal (!!) petroleum system (Fig. 20a).

Petroleum systems: basin-modelling syntheses and exploration implications

The schematics and ranking of each defined petroleum system, with annotation as to oil type, has been quantified in terms of published original recoverable reserves information (GBOR) as illustrated in Figures 19a–g and 20a–c.

The petroleum systems identified within this study display distinct stratigraphic and areal distribution across the Sirte Basin. The recognition of a number of new source horizons, in addition to the traditional Upper Cretaceous marine shales, opens up further opportunities for exploration in the basin. In particular it suggests that a re-evaluation is required of areas previously downgraded because they are outside the main Upper Cretaceous kitchens or beyond reasonable migration distances. Early Cretaceous/Triassic-aged rifts are proven to contain effective and volumetrically significant source facies which may have considerable source potential in the southern and southwestern Sirte Basin. At present their genesis and distribution is not well defined.

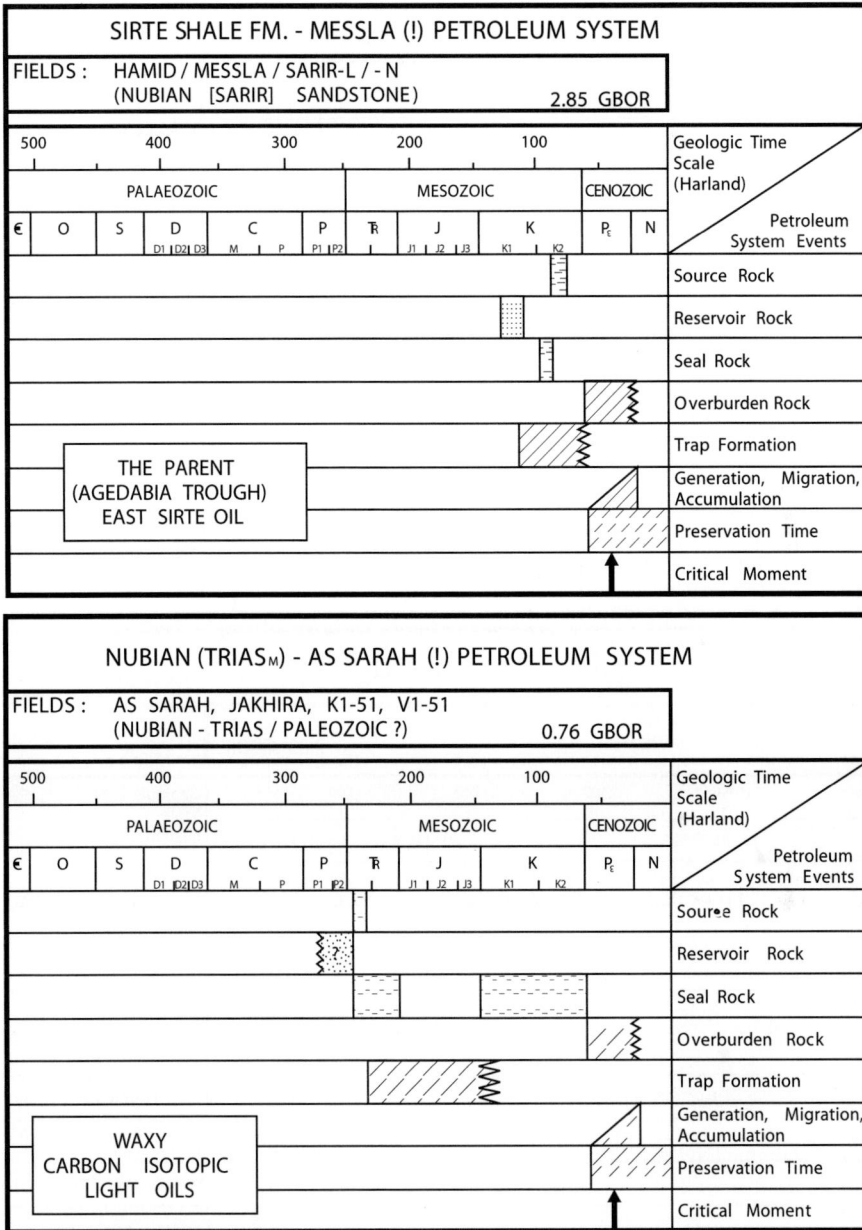

Fig. 19. East Sirte Basin petroleum system summary: proven unique systems involving a single source rock unit.(a) Sirte Shale Formation–Messla (!); (b) Nubian (Trias$_M$)–As Sarah (!); (c) Rachmat Formation–Gialo (!); (d) Tagrifet Formation–Intisar (!); (e) Antelat Formation–Antelat (!); (f) Harash Formation–K1–12 (!) and (g) Nubian (Calanscio Formation)–Bu Attifel (!) combinations.

Basin modelling

A number of well sections representative of more basinal locations within the main hydrocarbon kitchens, including the Hameimat, Agedabia and Sarir troughs, have been modelled (1-D TerraMod™ software) to assess the subsidence history and maturation profiles for the area (Figs 21, 22 & 23). The modelling utilized experimentally determined source rock kinetics (cf. Fig. 14), default Type IV kerogen kinetics (for vitrinite reflectance equivalent computation), present-day corrected geothermal

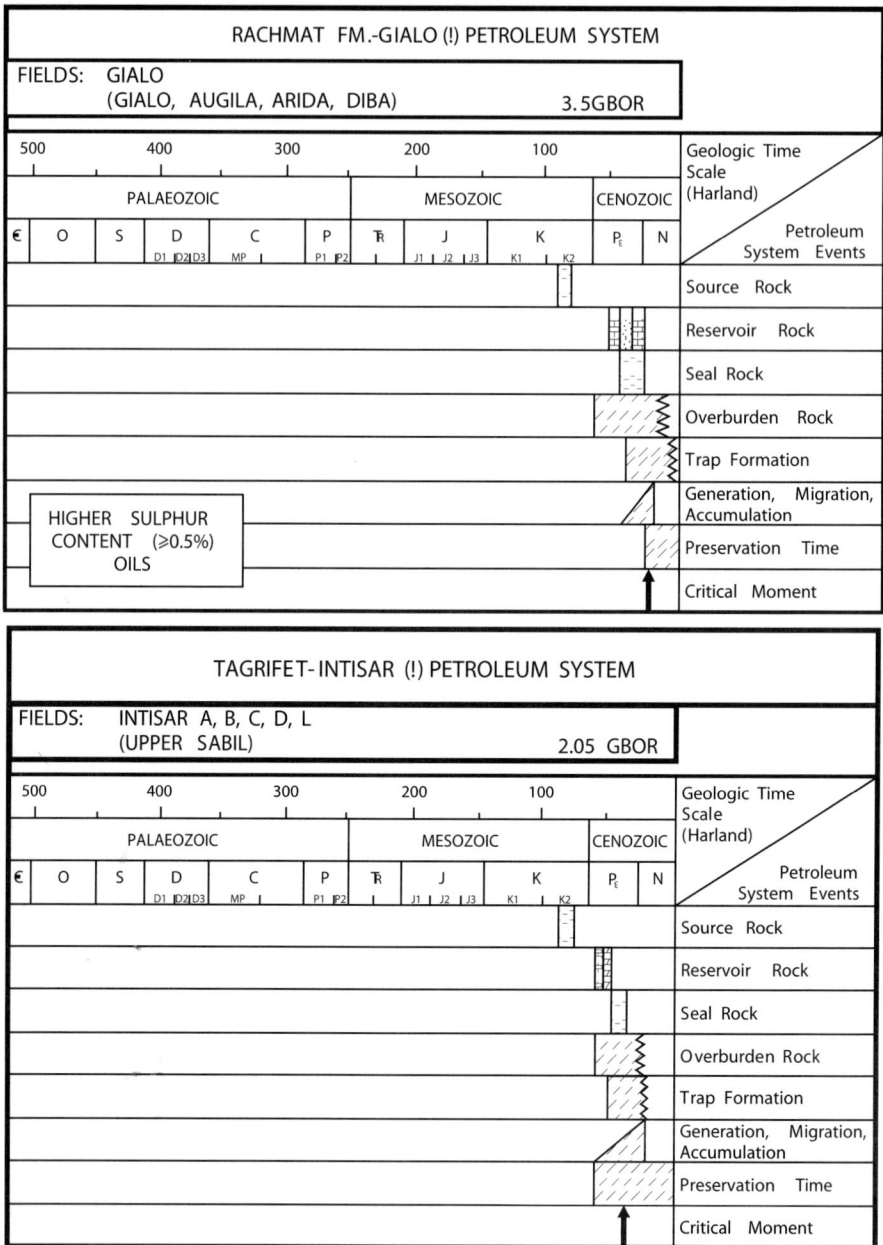

Fig. 19. Continued.

gradient data and available published information (Gumati & Schamel 1988; El Alami *et al.* 1989; Ghori & Mohammed 1996). Suleman and Roy (1987) suggested a range of heat flows in the Sirte Basin from 83 mW/m^2 during early rifting to 51 mW/m^2 at the present day. Bender *et al.* (2001) reported lower average heat flows of 55 mW/m^2 used in modelling of the Hameimat Trough.

This study calibrated the heat flow against available vitrinite reflectance data to obtain a reliable match, and results indicate average values ranging from 1.35 HFU (56 mW/m^2) during the early rift phase to 1.14–1.20 HFU (47–50 mW/m^2) over the later thermal sag phase. The results support the comments by Bender *et al.* (2001) that it is the significant subsidence during the Paleocene and

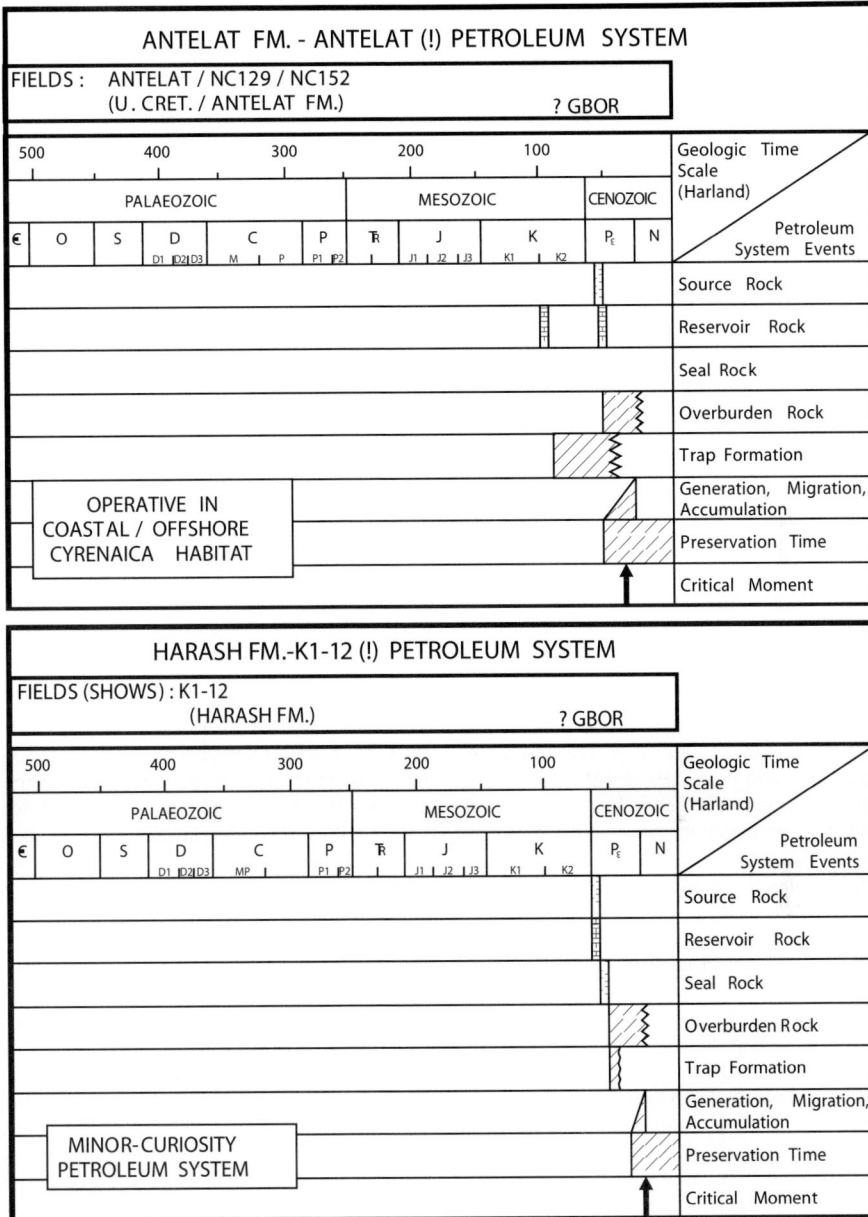

Fig. 19. Continued.

Eocene that drives the onset of oil generation in the basins, with the modelling less susceptible to moderate variations in heat flow through time, and in particular during the early rifting events.

Hameimat Trough

Basin modelling indicates that the Pre-Upper Cretaceous shales (Calanscio Formation) entered the

oil window in the Early Eocene, and are currently in the dry gas window (Fig. 21). The overlying Upper Cretaceous shales entered the oil window in the Late Eocene in the deeper parts of the trough. Present-day peak oil window (R_o 0.8%) is at 3600 m (11 800 ft) and the top gas window (R_o 1.3%) is at 4570 m (15 000 ft). Work by Bender *et al.* (2001) provides comparable results to this study, indicating the top mature zone in the Hamei-

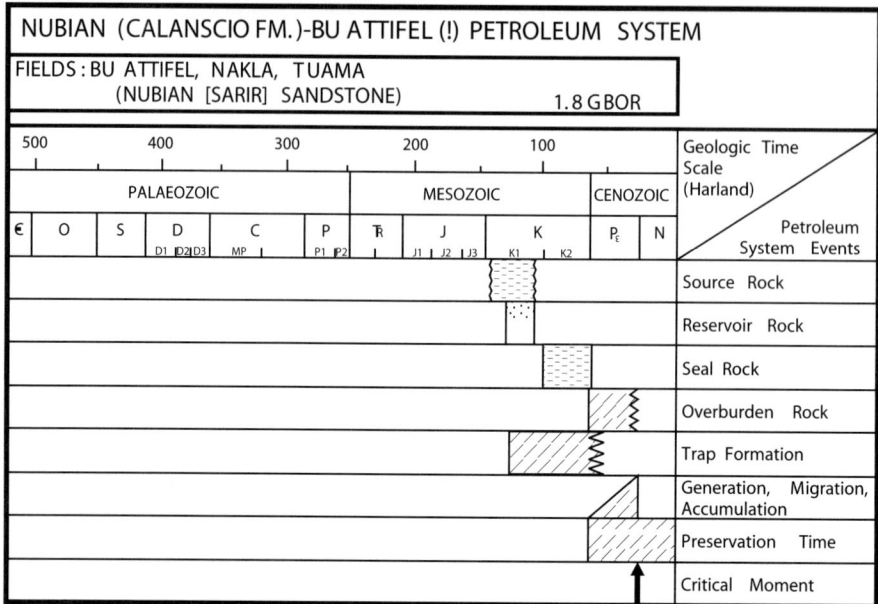

Fig. 19. Continued.

mat Trough in the range 3250–3500 m (10 660–11 480 ft), with top gas at 4570 m (15 000 ft). The Bu Attifel Field has a high GOR (gas–oil ratio; 230–600 m³/m³) and high API (Erba *et al.* 1984), which further supports the modelling.

A number of petroleum systems have been identified in the Hameimat Trough area, and a generalized play diagram is presented in Figure 24.

Nubian (PUC_H)–Bu Attifel (!) petroleum system. Within the Hameimat Trough the Nubian (Sarir) sandstones provide the main reservoir, with structural horsts and tilted fault blocks being the dominant trap type. Main fields include Bu Attifel, Nakla and Remel, with combined reserves estimated to be in excess of 1.8 GBOR.

This study confirms a non-marine source for the oils in the Nakla and Bu Attifel fields (Family 5), which is tentatively correlated with Pre-Upper Cretaceous Calanscio Formation shales. Reported restricted marine shales of source rock quality in the Upper Cretaceous Etel Formation (El Alami 1996) may also offer a potential candidate source for these fields, although this may be volumetrically subordinate to the thicker Pre-Upper Cretaceous shales.

Sirte Shale Formation–Messla (!) petroleum system. The thick Upper Cretaceous shales, while oil-mature and a prolific source, do not charge the main Nubian structures in the basin. This is interpreted to be due to a lack of an effective migration route, with the evaporitic Etel Formation forming a barrier (Fig. 24). Hydrocarbons generated from the mature Upper Cretaceous shales in the Hameimat Trough (Family 3) have migrated to the margins of the basin and are interpreted to have sourced the giant Messla and Sarir fields to the southwest, and possibly made a contribution to the Gialo Field to the west and the numerous smaller pools on the flanks of the Messla High.

Agedabia Trough

The Agedabia Trough is the main depocentre in the East Sirte Basin, with in excess of 6000 m (20 000 ft) of Mesozoic and Tertiary sediments. The potential Upper Cretaceous source interval is very thick, in excess of 1400 m (4500 ft) and thus there is a considerable variation in the maturity between the basal shales and the uppermost section, and consequently a range in timing for the onset of oil generation.

Basin modelling of the southern part of the basin, close to the Intisar oilfields, indicates present-day peak oil generation at 3750 m (12 250 ft) and gas generation at 4875 m (16 000 ft). The basal Upper Cretaceous shales (Rachmat Formation) entered the oil window in the Mid-Eocene (45 Ma BP) and the gas window by the end of the Oligocene. The younger and shallower Sirte Shales entered the oil window toward the end of the Eocene and are currently at peak oil maturity (Fig. 22).

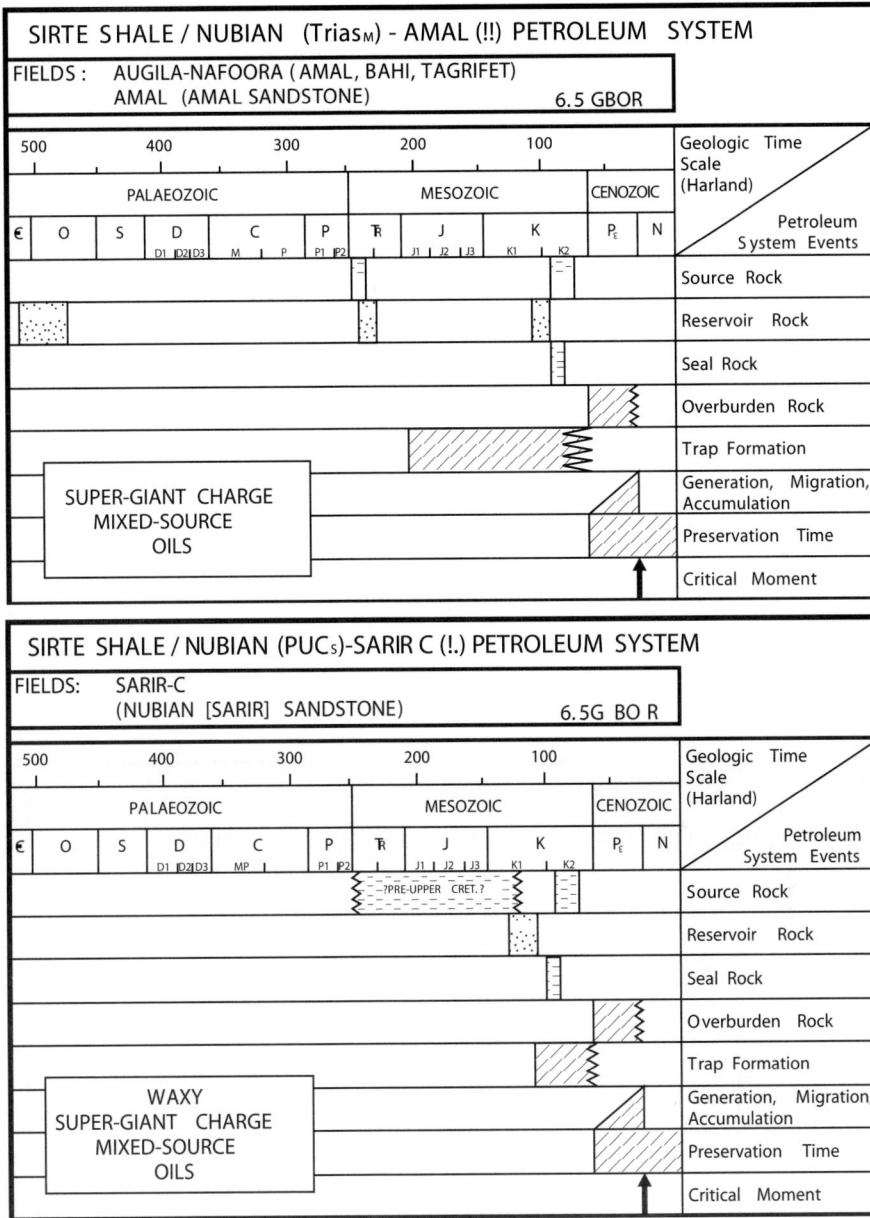

Fig. 20. East Sirte Basin petroleum system summary: hybrid systems involving multiple source rocks and aggregate charging. (**a**) the Sirte Shale Formation–Nubian (Trias$_M$)–Amal (!!); (**b**) the Sirte Shale Formation–Nubian (PLC$_S$)– Sarir C (!.) and (**c**) Rachmat/Tagrifet Formation–Shatirah (!!) variants.

The Agedabia Trough plunges steeply to the north, driving the entire section down into the gas window. This is demonstrated by the presence of gas discoveries at Sahl and Assoumoud on the northwest flank of the basin.

Tagrifet Formation–Intisar (!) petroleum system. The main play in the southern part of the Agedabia Trough is for Paleocene (Upper Sabil Formation) carbonates, within large reefal build-ups (Intisar pinnacle reefs) and smaller carbonate build-ups and

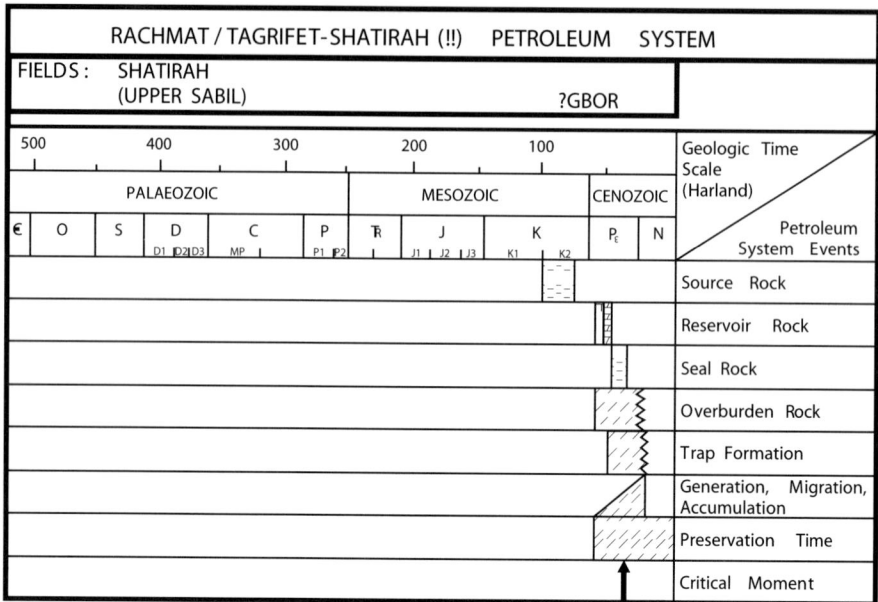

Fig. 20. Continued.

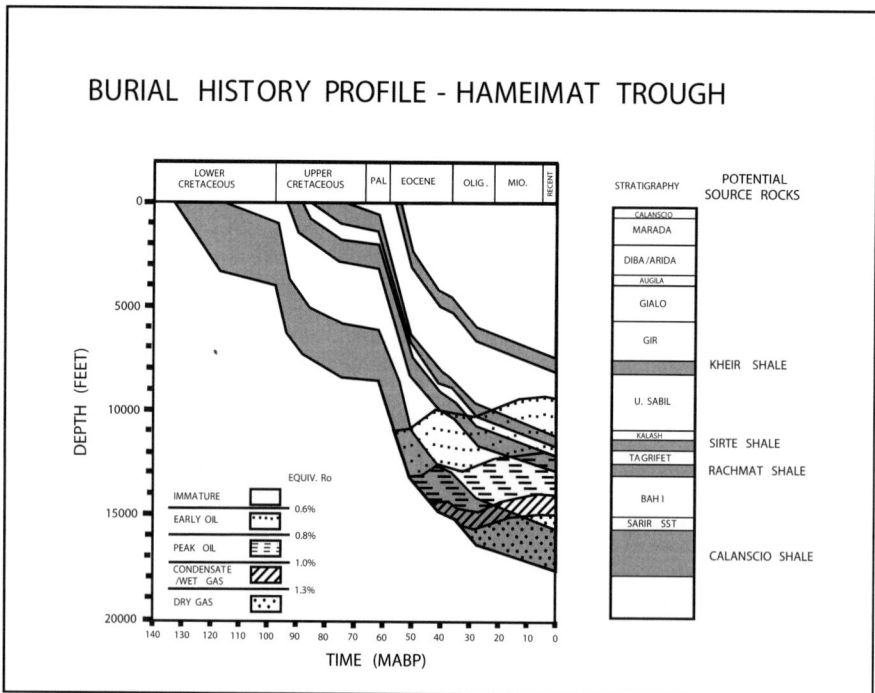

Fig. 21. Burial history profile and petroleum generation history for a near-depocentre location within the Hameimat Trough. 1-D TerraMod™ rifted model using heat flow of 1.35 HFU (Early Cretaceous rift phase) decaying to 1.20 HFU for present day. Vitrinite reflectance isopleths determined from Type IV kerogen kinetics.

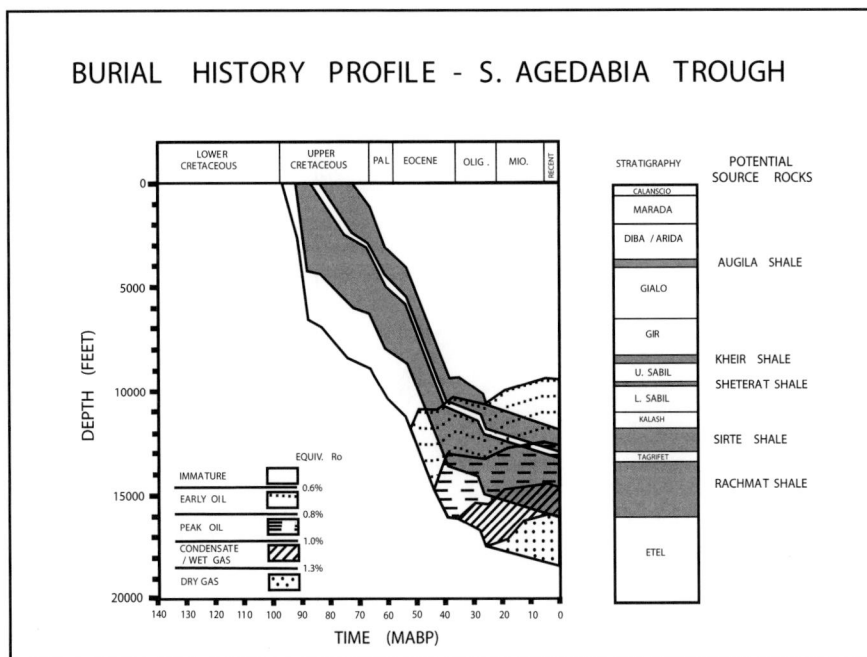

Fig. 22. Burial history profile and petroleum generation history for a near-depocentre location within the southern Agedabia Trough. Modelling as in Figure 21, with variable heat flow declining from 1.22 HFU, at Rachmat time, to 1.14 HFU present day.

structural traps. Reservoir development in the latter is primarily associated with higher energy carbonate facies, which rim the main depocentre. These reservoirs are charged from the underlying mature Upper Cretaceous shales via faults and fractures (Fig. 25).

Rachmat Formation–Gialo (!) petroleum system. The Gialo Field has a number of reservoirs (Paleocene Upper Sabil Formation, Eocene Gialo Formation, Eocene Augila Formation, Oligocene Chadra Member), but primary production is from the Eocene Gialo Formation carbonates. This very large regional structural high contains in excess of 3.5 GBOR. Hydrocarbon charge may have come from both the Hameimat Trough and Agedabia Trough and smaller subordinate basins on the margins of the structure. The main oil analysed was typed as Upper Cretaceous Rachmat-derived oil (Family 1c).

Harash Formation–K1–12 (!) petroleum system. The Paleocene Harash and Hagfa Shales (Family 9) have a relatively poor source quality and are not thought to be a significant source in the area. Deeper source facies have not been demonstrated by drilling, but may be present, although, due to

the excessive burial depths, these would be within the dry gas window.

Eocene Antelat Formation–Antelat (!) petroleum system. This petroleum system is developed on the northeast margin of the Agedabia Trough, and could possibly be more correctly defined geographically as being located on the Cyrenaica Platform. Oil analysis indicates a distinct petroleum system, with oils derived from the Eocene Antelat Formation (Family 7). It sources the Antelat Field, for which reserves and detailed reservoir information are currently unavailable. This appears to be a local source restricted to the Coastal Cyrenaica area and volumetrically limited. It may offer potential elsewhere in Cyrenaica or in similar depositional settings in other parts of Libya.

Additional petroleum systems. Additional petroleum systems may exist in the Agedabia Trough and adjacent highs. The Nubian (Sarir) sandstone reservoir has not to date been a major target in the area, and its distribution and quality is poorly known, primarily because of its excessive depth of burial. Recent discoveries on the margins of the Agedabia Trough, however, suggest that deeper Lower Cretaceous Nubian or Cambro-Ordovician

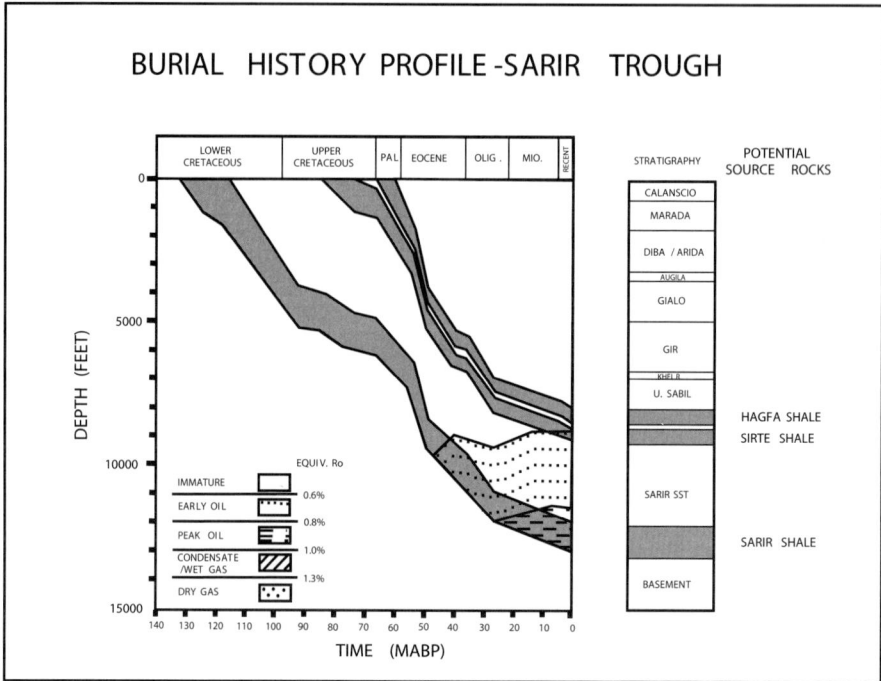

Fig. 23. Burial history profile and petroleum generation history for a near-depocentre location within the Sarir Trough. Modelling as in Figure 21 employing a Sarir rift-phase heat flow of 1.35 HFU declining to 1.20 HFU present day.

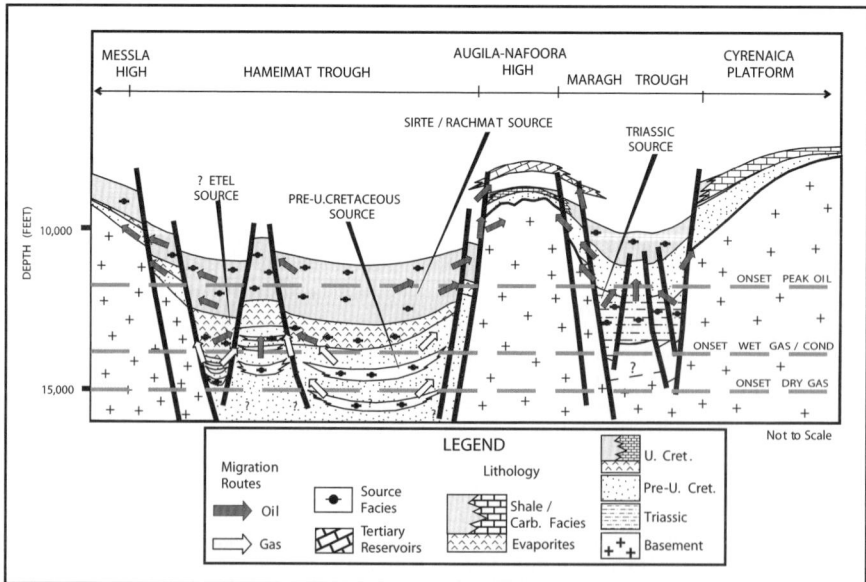

Fig. 24. Petroleum systems charge model for the Hameimat Trough.

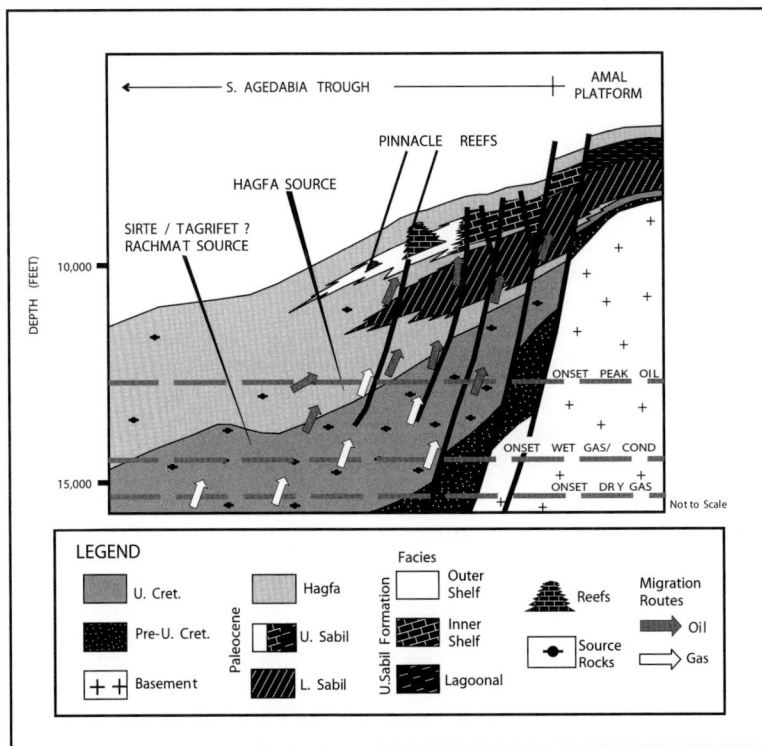

Fig. 25. Petroleum systems charge model for the southern Agedabia Trough.

Amal/Gargaf sandstone reservoir targets, within intra-basinal horsts or basin marginal fault blocks, may prove attractive. However, in such cases, there is clearly a risk of gas charge, especially to the north, and poor reservoir quality due to burial diagenesis.

Sarir Trough

Within the Sarir Trough, basin modelling suggests that the Upper Cretaceous marine shales are immature and have not generated significant volumes of oil (Fig. 23). Present-day peak oil generation is at 3450 m (11 250 ft), and the Early Cretaceous section entered the peak oil window (R_o 0.8%) in the Middle Oligocene (26 Ma BP)

Nubian (PUC$_S$)–UU1–65 (.) petroleum system. Support for an oil-mature Pre-Upper Cretaceous source in the Sarir Trough comes from the Agoco UU1–65 discovery. Located to the south of the Messla High, on the northern flank of the Sarir Trough, this field is a complex fault/stratigraphic pinchout play with a Nubian (Sarir) sandstone reservoir (Ambrose 2000). The structure is not on a migration route from the Hameimat or Agedabia

troughs and can only have been charged from the adjacent Sarir Trough. Analysis of the oil indicates a Pre-Upper Cretaceous source (Family 6), supporting the model for Pre-Upper Cretaceous lacustrine source-rock development within the Sarir Trough. This demonstrates the potential for Pre-Upper Cretaceous source rocks within early rift basins

Hybrid Sirte Shale–Nubian (PUC$_S$)–Sarir-C (!.) petroleum system. To the east, the giant Sarir Field oil has been demonstrated to have a complex hybrid geochemistry indicating migration from two source rocks (Family 4a). It has clearly received charge from both the Upper Cretaceous marine shales of the Hameimat Trough (by long-distance migration 'spill and fill' via Messla) and also a smaller contribution from the Sarir Trough of non-marine waxy crude.

Maragh Trough

Basin modelling described in Gruenwald (2001) suggests that the peak oil window is at 3350 m (11 000 ft), with generation from the Mid-Eocene, and that the Pre-Upper Cretaceous section

(Triassic) interval is currently within the peak oil window. This modelling assumed a constant heat flow in this actively rifted basin of 62 mW/m^2. The younger Upper Cretaceous shales are only marginally mature over most the trough and are not thought to provide a volumetrically significant charge.

Nubian (Triassic)–As Sarah (!) petroleum system. The Maragh Trough contains two significant discoveries, the As Sarah and Jakhira fields, which have confirmed oil sourced from Triassic-aged shales (Family 8). This petroleum system has generated in excess of 0.76 GBOR in the Maragh Trough and, in addition, contributed to the charge of the giant Augila-Nafoora and Amal fields. The presence of this prolific Triassic source rock demonstrates the potential for early rifts within the Sirte Basin.

Hybrid Sirte Shale Formation/Nubian (Triassic)– Amal (!!) petroleum system. The giant Amal and Augila-Nafoora fields have multi-play reservoirs ranging from Basement, Cambro-Ordovician quartzites, Triassic–Lower Cretaceous Nubian sandstones, Upper Cretaceous sandstones and Upper Cretaceous, Paleocene and Eocene carbonates. The bulk of the reserves are within the lower clastic reservoirs. Oil analysis suggests a complex hybrid charge, with mixed oils within the samples analysed (Family 4b).

Additional petroleum systems. Potential for additional petroleum systems in the Maragh Trough and immediate area has been recently proposed by Gruenwald (2001). He suggests that there is also potential for younger reservoirs of Upper Eocene Augila Formation carbonates and Lower Oligocene Arida Formation sandstones charged by both Triassic source rocks in the Maragh Trough and Upper Cretaceous source rocks from the adjacent Hameimat Trough.

Conclusions

Combination of oil data inversion and multivariate statistical analysis, with the results of a carbon isotope-based source-oil correlation exercise, has permitted segregation of the East Sirte petroleums into their component genera. From this process, the existence of both Upper Cretaceous and younger candidate marine source-rock units, in addition to Pre-Upper Cretaceous lacustrine analogues, were recognized. *In toto* 12 oil family/sub-family genera were discriminated, these deriving from one of eight unique petroleum systems or hybrid mixtures thereof.

The results of maturation modelling for the various basins within the eastern Sirte area are consistent with the generation and accumulation of unique and hybrid oils recognized from the multivariate statistical analysis.

Of greatest volumetric significance was the Sirte Shale Formation–Messla(!) petroleum system, this accounting for the lower wax content Hamid, Messla and Sarir accumulations. This system also provided the base charge for the waxy Sarir, Augila–Nafoora and Amal petroleums, these deriving from the hybrid Sirte Shale Formation/Nubian (PUC$_S$)–Sarir-C(!!) and Sirte Shale Formation/ Nubian (Triassic)–Amal(!!) combinations, respectively.

For the other marine-sourced oils, the Intisar and Gialo petroleums could be differentiated, being attributable to the Tagrifet Formation–Intisar(!) and Rachmat Formation–Gialo(!) systems, respectively. These sources combine to charge the Shatirah(!!) hybrid system.

Among the lesser accumulations, the Pre-Upper Cretaceous lacustrine Nubian (PUC$_H$)–Bu Attifel(!) and Nubian (Triassic)–As Sarah(!) petroleum systems, yielding high-wax oils, were deduced to be operative in the hinterlands to the Hameimat and Maragh troughs, respectively. The Nubian (PUC$_S$)– UU1–65(.) system was specific to the Sarir Trough and provides the high wax modifying charge to the waxy Sarir-C petroleums.

In the northern Agedabia Trough/coastal Cyrenaica area the Antelat Formation–Antelat(!) system was deduced to be locally productive, the contributory Eocene source facies exhibiting an unusual marine/non-marine, carbonatic character. Elsewhere, production from minor Lower Harash Formation carbonate source potential was evident in the case of the Harash Formation–K1–12(!) system.

Overall, these results, in revealing the complexity of the East Sirte Basin hydrocarbon habitat, should stimulate the search for hydrocarbons beyond the currently identified areas of mature Upper Cretaceous shales. An additional focus on other potential source horizons and on areas such as Pre-Late Cretaceous-aged restricted rift basins could promote the as yet unrealized potential of the area.

This paper draws on material originally presented at the First Magrebian Conference on Petroleum Exploration, Benghazi, Libya, November 1996 and subsequently at the American Association of Petroleum Geologists Annual Convention, Salt Lake City, May 1998. We thank the then Management of PetroFina SA and Fina Exploration (Libya) BV for their encouragement and permission to publish. Thanks are especially due to A. M. Bezan and the NOC, Agoco (A. I. Asbali and A. Mansouri), and M. Kuehn at Wintershall. Also thanked are the other companies who kindly contributed rock and oil materials including OMV, Veba, Waha and Zueitina.

Appendix

1. Source evaluation parameters

PP units: Petroleum Potential ($\times 10^6$ m$^3_{(oe)}$/km$^3_{(rock)}$), calculated from S2 value (oe, oil equivalent of 35° API fluid; rock, source rock with density of 2.55 g/cm^3). Conversion factor to barrels/acre-ft: ($\times 7.758 \times 10^{-6}$).

GOPR: Gas-oil production ratio (0.0–1.0). Pyrolysis-gas chromatographic measure of kerogen hydrocarbon product, zero and unity being exclusively oil-prone and gas-prone, respectively.

δ^{13}C$_{KPY}$ Kerogen pyrolysate (simulated oil) carbon isotope signature. Formation mean values are source-richness weighted against S2 and calculated thus:

$$\overline{\delta^{13}}C_{(pyrolysate)} = [\Sigma\delta^{13}C_{(pyrolysate)} \times S2/\Sigma S2]$$

Values reported v. NBS22 at -29.8 ppt (PDB).

E$_a$/A: Kerogen transformation kinetics determined on bitumen-free sediments using the Rock-Eval 5/Optkin procedure. GT$_{(MAX)}$: real (geological) time optimum kerogen transformation temperature at heating rate 1°C/Ma.

C$_V$: Canonical variable $[(-2.53\delta^{13}C_{sat}+2.22\delta^{13}C_{arom})-11.65]$

2. Petroleum system nomenclature*

Known/simple (proven source–oil correlation):	(!)
Known/complex (poly system):	(!*)
Known/complex (proven hybrid system):	(!!)
Hypothetical (prognosed with confidence):	(.)
Hybrid/part proven (major source known):	(!.)
Speculative:	(?)

*After Magoon & Dow (1994)

References

AMBROSE, G. 2000. The geology and hydrocarbon habitat of the Sarir Sandstone, SE Sirt Basin, Libya. *Journal of Petroleum Geology*, 23, 165–192.

BEIN, A. & SOFER, Z. 1987. Origins of oils in the Helez Region, Israel – implications for exploration in the Eastern Mediterranean. *Bulletin of the American Association of Petroleum Geologists*, 71, 65–75.

BENDER, A. A., COELHO, F. M. & BEDREGAL, R. P. 2001. A basin modelling study in the southeastern part of the Sirt Basin, Libya. *In*: SALEM, M. J., MOUZUGHI, A. J. & HAMMUDA, O. S. (eds) *The Geology of Sirt Basin*, vol. 1. Elsevier, Amsterdam, 139–155.

BISSADA, K. K., ELROD, L. W., DARNELL, L. M., SZYMEZYK, H. M. & TROSTLE, J. L. 1992. Geochemical inversion – a modern approach to inferring source-rock identity from characteristics of accumulated oil and gas. *Proceedings of the Indonesian Petroleum Association*, 21, 165–199.

BURWOOD, R., DROZD, R. J., HALPERN, H. I. & SEDIVY, R. A. 1988. Carbon isotopic variations of kerogen pyrolysates. *Organic Geochemistry*, 12, 195–205.

BURWOOD, R., DE WITTE, S.-M., MYCKE, B. & PAULET, J. 1995. Petroleum geochemical characterisation of the Lower Congo Coastal Basin Bucomazi Formation. In: Katz, B. (ed.) *Petroleum Source Rocks*. Springer-Verlag, Berlin, 235–263

BURWOOD, R. 1996. Geochemical evaluation of east Sirte Basin petroleum systems and oil provenance. *Proceed-ings of First Magrebian Conference on Petroleum Exploration, Benghazi, GSPLAJ, November 18–20*, 28.

EL ALAMI, M. 1996. Habitat of oil in Abu Attifel area, Sirt Basin, Libya. *In*: SALEM, M. J. *et al.* (eds) *The Geology of the Sirt Basin*, vol. 2. Elsevier, Amsterdam, 337–348.

EL ALAMI, M., RAHOUMA, S. & BUTT, A. A. 1989. Hydrocarbon habitat in the Sirte Basin, northern Libya. *Petroleum Research Centre Journal (Tripoli)*, 1, 17–28.

EL ARNAUTI, A. & SHELMANI, M. 1988. A contribution to the northeast Libyan subsurface stratigraphy with emphasis on pre-Mesozoic. *In*: EL ARNAUTI *et al.* (eds) *Subsurface Palynostratigraphy of Northeast Libya*. Garyounis University, Benghazi.

ERBA, M., ROVERE, A., TURRIANI, C. & SAFSAF, S. 1984. Bu Attifel Field – a synergenetic geological and engineering approach to reservoir management. *Proceedings of the 11th World Petroleum Congress, London, 1983*, 3, 89–99.

GRUENWALD, R. 2001. The hydrocarbon prospectivity of Lower Oligocene deposits in Maragh Trough, SE Sirte Basin. *Journal of Petroleum Geology*, 24, 213–231.

GHORI, A. & MOHAMMED, R. 1996. The application of petroleum generation modelling to the eastern Sirt Basin. *In*: SALEM, M. J. *et al.* (eds) *The Geology of the Sirt Basin*, vol. 2. Elsevier, Amsterdam, 529–539.

GUMATI, Y. D. & SCHAMEL, S. 1988. Thermal maturation history of the Sirte Basin, Libya. *Journal of Petroleum Geology*, 11, 205–218.

MACGREGOR, D. S. 1996. The hydrocarbon systems of North Africa. *Marine & Petroleum Geology*, 13, 329–340.

MAGOON, L. B. & DOW, W. G. 1994. The petroleum system. *In*: MAGOON, L. B. & DOW, W. G. (eds) *The Petroleum System – From Source to Trap*. American Association of Petroleum Geologists Memoirs, 60, 3–24.

MOLDOWAN, J. M., LEE, F. J., JACOBSON, Y., WATT, S. R., SLOUGUI, N. E., JEGANTHAN, A. & YOUNG, D. C. 1990. Sedimentary 24-*n*propylcholestanes, molecular fossils diagnostic of marine algae. *Science*, 247, 309–312.

PARSONS, M. G., ZAGAAR, A. M. & CURRY, J. J. 1980. Hydrocarbon occurrences in the Sirte Basin, Libya. *In*: MAILL, A. D. (ed.) *Facts and Principles of World Petroleum Occurrence*. Canadian Society of Petroleum Geologists Memoirs, **6**, 723–732.

PETROCONSULTANTS LTD 1981. Libya. *In*: GRUNAU, H. R. & MOSER, H. J. (compilers) *Overview of Source Rocks for Oil and Gas – A Worldwide Survey*. Petroconsulants Ltd, Dublin.

RADKE, M., WELTE, D. H. & WILLSCH, H. 1986. Maturity parameters based on aromatic hydrocarbons : influence of the organic matter type. *Organic Geochemistry,* **10**, 51–63.

RADKE, M. & WILLSCH, H. 1994. Extractable alkyldibenzothiophenes in Posidonia Shale (Toarcian) source rocks : Relationship of yields to petroleum formation and expulsion. *Geochimica et Cosmochimica Acta,* **58**, 5223–5244.

SOFER, Z. 1984. Stable carbon isotope compositions of crude oils : Application to source depositional environments and petroleum alteration. *Bulletin of the American Association of Petroleum Geologists,* **68**, 31–49.

SULEIMAN, I. S. & ROY, R. F. 1987. Heat-flow and basement heat production in the Sirt Basin, Libya. *3rd Symposium Geol. Libya (Tripoli), Abstracts,* 117.

Gravity signatures of sediment systems: predicting reservoir distribution in Angolan and Brazilian basins

W. G. DICKSON[1], A. DANFORTH[2] & M. ODEGARD[3]

[1]*Dickson International Geosciences (DIGs), 12503 Exchange Drive, Suite 510A, Stafford, Texas 77477, USA*

[2]*Consulting Exploration Geologist, Houston, Texas 77042, USA*

[3]*GETECH Inc. (Geophysical Exploration Technology), Stafford, Texas 77477, USA*

Abstract: The Lower Congo Basin and Congo Fan of Angola contain giant oil discoveries in Oligocene & Miocene deep-water sands; the Late Cretaceous and Tertiary fans in the Campos Basin of Brazil contain some 90% of Brazil's reserves. Petroleum exploration in these vast, deep-water regions is resource-hungry, increasing the value of techniques that facilitate quick and inexpensive extrapolation from discoveries into undrilled areas. Our work demonstrates a strong correlation in these basins between reservoir fairways and various attributes of the gravity signature. We provide comparisons between published interpretations based on well and seismic data and our interpretations of gravity attribute images. We demonstrate the use of geographic information systems (GIS) techniques to facilitate the comparisons, discuss the possible basis for these correlations and speculate on their application to other basins. We conclude with notes on current research to improve the resolution of the gravity data and future refinements to our processing and interpretation methods.

Since the mid-1990s, the Lower Congo Basin and Congo Fan of Angola have yielded numerous giant oil discoveries in Oligocene and Miocene deep-water sands. Although they have prominent seismic signatures (e.g. Alexander *et al.* 2000), the large areas of both fields and exploration licences require 3-D seismic surveys covering tens of thousands of square kilometres. The Late Cretaceous and Tertiary fans of the Campos Basin demand similar expenditures for both the seismic data and its interpretation. Despite the rapid advance of relevant technologies, in order to identify targets, the financial, staff and time budgets required remain large. Our techniques reduce those needs and permit quick and inexpensive extrapolation from discoveries into undrilled areas. A recent study, SAMBA, of the South Atlantic Margin Basins used geographic information systems (GIS) software to help implement an example of this new utility.

We demonstrated a strong correlation between reservoir fairways and the gravity signature on isostatic anomaly and other attribute maps. Combining knowledge of the underlying geology with enhanced gravity data from regional to sub-basin scale allowed recognition of inter-raft sediment pathways and depocentres in the Lower Congo and Kwanza basins of Angola. Similarly, we highlighted basement control of entry points for Oligocene fans and sediment bypass zones in the Greater Campos Basin. In summary, the displays allowed interpreters to relate reservoir controls to known points and to extrapolate across blocks and basins, leveraging inexpensive, existing knowledge and filling gaps in understanding the plays.

General procedure and results

Gravity data were compiled into regional displays and correlated with published geology using GIS software. First, the distal basin limits were redefined using detailed plate reconstructions of the South Atlantic as first-order controls. Next, more detailed work focused on the Greater Campos Basin of Brazil and the Aptian Salt Basin of West Africa. Striking correlations were observed between gravity anomalies and key reservoir distributions. These correlations were extended regionally, based on additional potential field displays and regional geologic knowledge. Final interpretations combined these datasets in a GIS-based report, showing extrapolated sediment distri-

From: ARTHUR, T. J., MACGREGOR, D. S. & CAMERON, N. R. (eds) *Petroleum Geology of Africa: New Themes and Developing Technologies.* Geological Society, London, Special Publications, **207**, 241–256. 0305-8719/03/$15

bution systems and the linked receiving basins or mini-basins.

Gravity compilation and mapping

To begin, gravity data were compiled from numerous sources at several resolutions, merged and processed (Fig. 1). The highest resolution data was on the continental margins (Fairhead *et al.* 1998, 2001; Green *et al.* 1998) and adjacent onshore areas (Fairhead & Watts 1989; Green & Fairhead 1992) where the main petroleum provinces were located. Original data points were on a 2 × 2 ft (4 × 4 km) grid offshore and 5 × 5 ft (11 × 11 km) grid onshore in the main study area. Elsewhere, the onshore grid (Green & Fairhead 1996; Lemoine *et al.* 1998) was about 50 × 50 km, while, in the oceans, the grid was still 4 × 4 km but from noisier, lower-resolution public domain data points (Sandwell & Smith 1997).

The resulting 4 × 4 km grids of the circum-Atlantic region were attached to and moved with plates based on the Cambridge Paleomap reconstructions (Green 1997). Six eras from 55 Ma to 165 Ma were restored, putting present-day grids and interpretation overlays into their palaeo-locations. Gravity maps included free air, Bouguer, isostatic residual (ISO), first vertical derivatives (1VD) and total horizontal derivatives (THD) and dip-azimuth (DAI) of the ISO plus topography/bathymetry.

Megaregional observations

The restorations showed the configuration of prerift Brazil and West Africa with sharply imaged transforms connecting corresponding points on opposite sides of the Atlantic (Fig. 2, window of THD gravity on the 95 Ma restoration). Work based on potential-fields methods has shown a steady progression in data coverages and resolution, from Cande & Rabinowitz (1978) to Fairhead (1988), through Mueller *et al.* (1997). On current imagery, the colour contrast between the cool, quiet oceanic and the hot, dissected continental terrains was displayed with much greater detail than typical bathymetric and magnetics-based maps, highlighting improvements in previously published interpretations of plate outlines and plate matches.

The Walvis Ridge and seamounts along the Brazilian margin north from the Vitoria–Trindade leaky transform have prominent expressions (Dickson & Fairhead 1998). The restorations emphasized the zipper-like separation of the continents, matching wide basins on one side with opposing narrow basins (e.g. Davison 1997). The

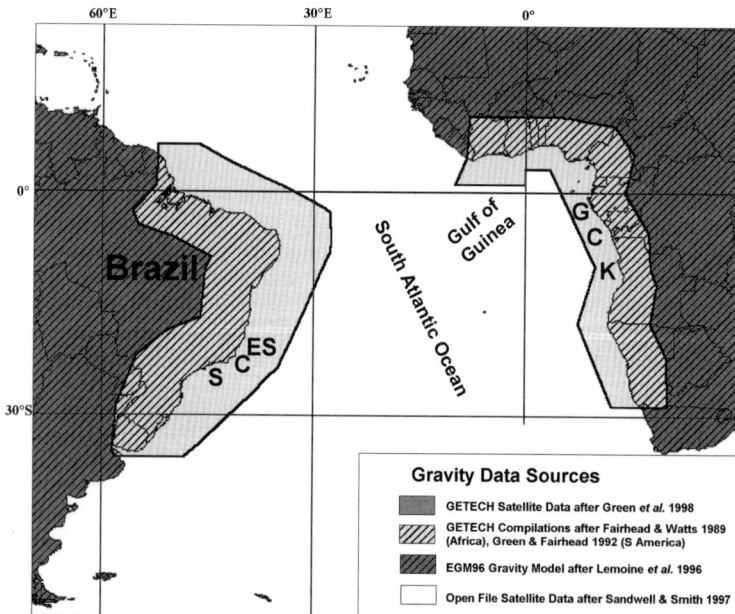

Fig. 1. South Atlantic Study Area with gravity data coverages after Fairhead and Watts (1989; Africa), Green and Fairhead (1992; South America), Sandwell and Smith (1997; oceanic regions), Lemoine *et al.* (1996; continental areas) and Green *et al.* (1998; continental margins). Note principal basin names: Brazil: ES, Espirito Santo; C, Campos; S, Santos. West Africa: G, Gabon; C, Congo; K, Kwanza.

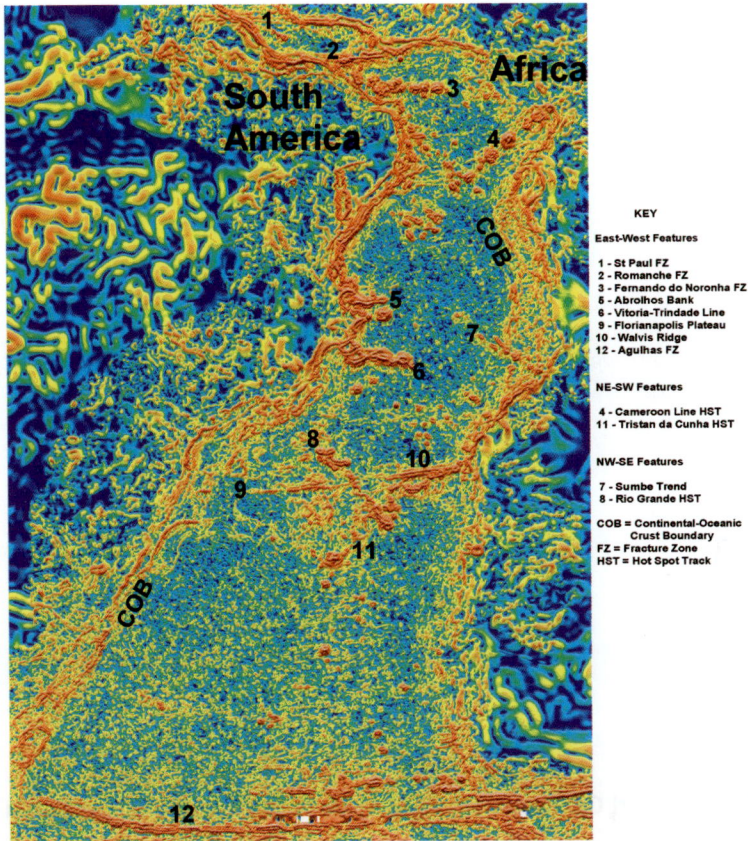

Fig. 2. Gravity total horizontal derivative (THD): South Atlantic region with plates restored to 95 Ma. Longitude is arbitrary; palaeolatitude based on Cambridge Paleomap work (after Green 1997).

continental margins are broadly partitioned at changes in structural strike into compartments with somewhat different geological histories. The boundaries offset generally coast-parallel trends both landward and seaward and these offsets often form entry points for major river systems, such as the Congo (Dickson *et al.* 1998 and in press) regarding the general implications for source and reservoir).

From regional to local

Our maps much improved the definition of the continental/oceanic crust boundary (COB) and seaward basin limits as well as redefining the basin segmentation and the nature of the crustal floor of some of the more seaward basins. With this better understanding of the margin-forming history and processes, we then turned our attention to specific examples of sediment controls in the prolific hydrocarbon provinces of Angola and Brazil.

Petroleum-related stratigraphy of the Congo Fan, West Africa

We refer the reader to the enormous body of literature (Belmonte *et al.* 1965; Brognon & Verrier 1966; Reyre 1966; Brink 1974; Vidal *et al.* 1977; Schlumberger 1983; Teisserenc & Villemin 1990; Duval *et al.* 1992; Lunde *et al.* 1992; Danforth *et al.* 1997; Marton *et al.* 2000) that describes the structure, stratigraphy and plays of the Aptian Salt Basin of West Africa. We only include a generalized stratigraphic column and illustrative cross-section as Figures 3 & 4.

Bypass zones have provided sediment transport across the shelf (Dickson & Macurda 2000) and coarser-grained sediments are expected far (at least 900 km) from shore, eventually forming base of slope fans (Evans 2001), and meaning that reservoirs can be found throughout the entire Congo Fan.

The Miocene–Mid-Oligocene Malembo (Angola), Landana (Cabinda) or Paloukou (Congo) sequence

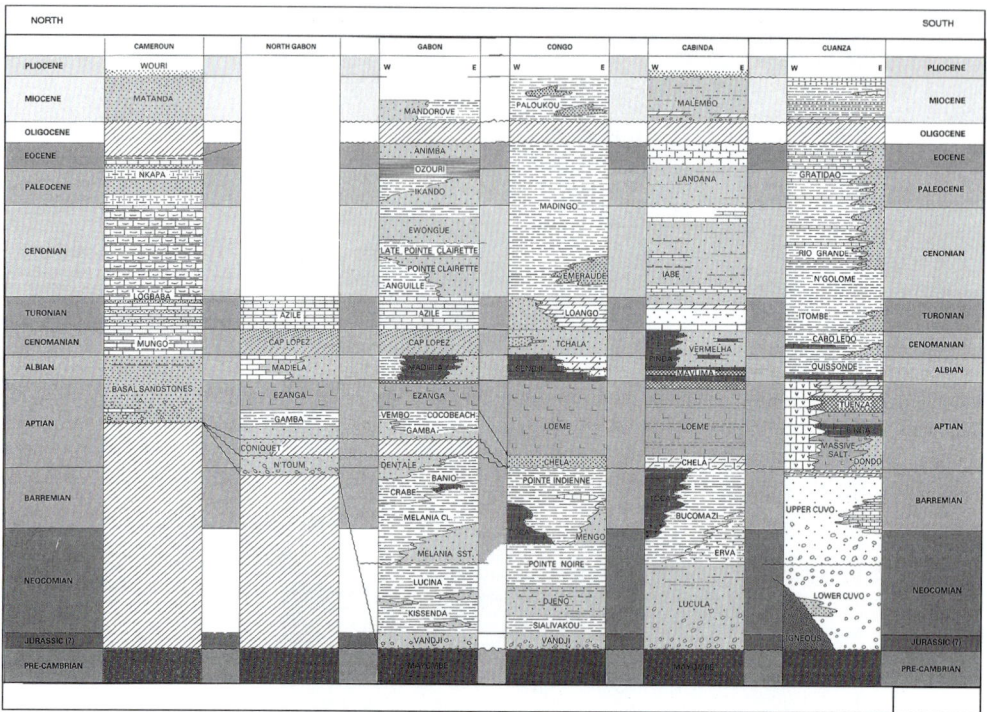

Fig. 3. Lithostratigraphic correlation, Aptian Salt Basin, West Africa (after Schlumberger 1983).

Generalized Geologic Profile: Aptian Salt Basin, West Africa

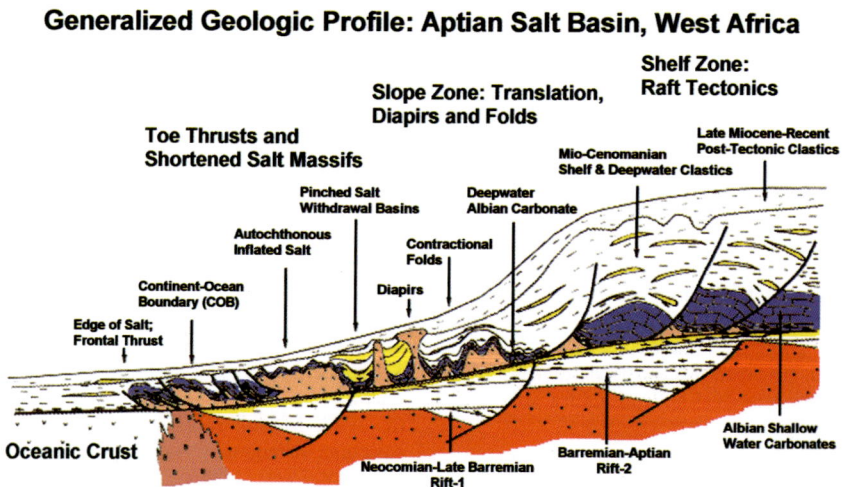

Fig. 4. Tectonostratigraphic elements of the Aptian Salt Basin of West Africa (after Norvick & Schaller 1998). Key Brazilian basins (Santos, Campos) in general mirror this profile.

contains the main oil reservoirs to date for the Congo Fan (see Fig. 5). Three deep-water reservoir facies are generally observed (e.g. Posamentier & Wisman 2000; Sikkema & Wojcik 2000) in deposits from gravity flows in a slope-and-basin environment:

- confined channel complexes, which appear as linear, frequently erosive features with internally discontinuous high-amplitude seismic character;
- overbank deposits, which exhibit more uniform concordant and internally continuous seismic

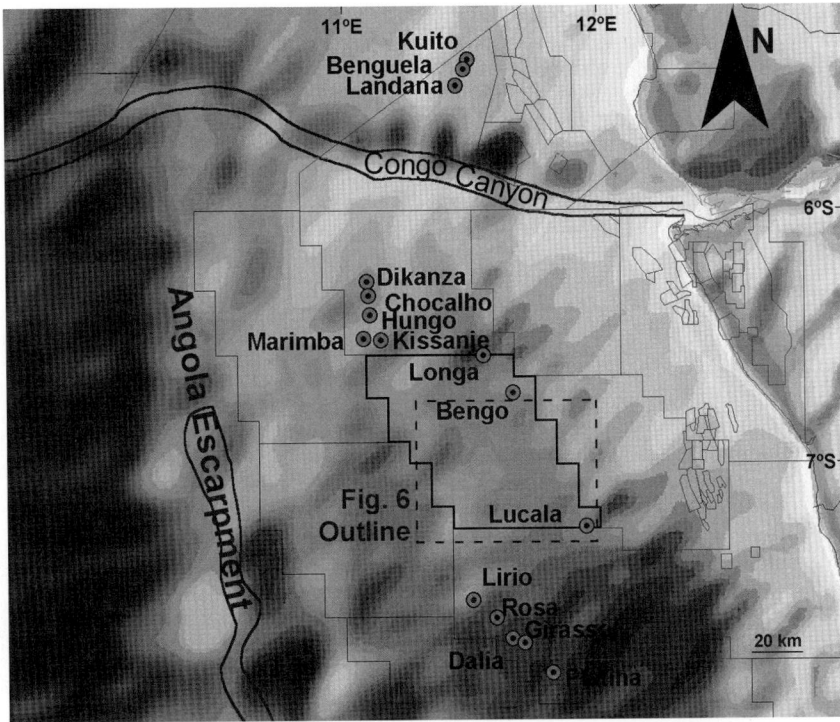

Fig. 5. Congo Fan bathymetry/topography, location of Oligocene–Miocene discoveries and Figure 6 detail area.

character. These strata were deposited as sediment gravity flows that escaped the banks of the confining channels and spread sand over the surrounding sea floor;

- levéed channels, which also exhibit conformable seismic reflections but with a constructive geometry.

The best reservoirs in these recent discoveries are in amalgamated or stacked channels and proximal levées. Frequent presentations at, for example, conferences of the Association of American Petroleum Geologists, the Society of Economic Paleontologists and Mineralogists, and the Geological Society of London since 1998 have illustrated these reservoir targets (e.g. Alexander et al. 2000). Identification from seismic data is feasible from bed geometries, amplitude variation with offset (AVO) signatures and direct hydrocarbon indicators (DHIs) such as flat-spots and phase changes. Especially on high-resolution or older academic seismic profiles, BSRs (bottom-simulating reflectors) from the shallow gas hydrates (Lindholm & Cunningham 1998; Cunningham & Lindholm 2000) may provide a quick-look indicator of shallow reservoirs.

The typical pre-salt lacustrine source rocks are prolific but require tortuous migration pathways to charge some of the post-salt reservoirs. The thickest Tertiary section drove most of the post-salt Late Cretaceous and Early Tertiary marine source rocks into the oil window. This simplified the hydrocarbon-charge pattern by eliminating the need for windows through the salt to access the mature pre-salt source. Extreme variability of the Tertiary thickness also had reservoir and trap implications. The advancing wedge of sediments squeezed upward both sediments and salt, creating bathymetric relief which trapped sands. As these mini-basins filled, they spilled sediments into the next downslope low as it formed, and so on. Beds uplifted by this movement were eroded and sands were redeposited further seaward, gradually improving their sorting. Mini-basin flanks and salt-related noses provided the structural relief to focus hydrocarbon migration into the reservoirs.

Structural controls on sedimentation

The early Angola–Congo–Gabon margin is typical of the South Atlantic. Major fracture systems, inherited from the Precambrian basement structural grain, were reactivated during the Early Cretaceous rifting that ultimately separated South America from Africa. These NE-trending fractures separated the margin into a fabric of generally N–S-trending horsts and grabens. The structural framework typically provided an eastern inner basin, a bounding horst block and a westward sag basin (e.g. Karner *et al.* 1997).

The inner basins were primarily filled with a syn-rift sequence and capped by a relatively thin drift sequence. The seaward sag basins had comparable syn-rift and drift but a great thickness of sag sequence deposited during late thermal subsidence. Between the Walvis Ridge/Florianopolis High in the south and the Romanche Fracture in the north (Fig. 2), the restricted syn-rift sequences typically were lacustrine and included important source beds.

Salt tectonics

The Angola–Congo–Gabon margin, although formed by rifting, was reshaped by multiple episodes of salt tectonics (Marton *et al.* 1998) related to thermal subsidence during sea-floor spreading and a later tilting. The syn-rift and earliest drift-age sequences were overlain by an Aptian salt layer, generally in turn overlain by a sequence of margin to shelf carbonates and basinal shales. The post-salt tectonic fabric directly reflected both the pre-salt morphology and sediment loading on the salt décollement layer. Major basement promontories bounded by transform faults segmented the margin into sub-basins. The resulting concave and convex shapes promoted convergent and divergent basinward salt flow and rafting, respectively (Spencer *et al.* 1998), showing widely varying structural styles among the basins.

There is a two-part history for salt tectonism. As the South Atlantic opened during the Aptian, salt, estimated locally up to 1500 m thick (Gyorgy *et al.* 1998), was deposited, probably reaching basinward to oceanic or rifted, volcanic continental crust. An Albian continental margin developed, supporting a carbonate platform overlying the Aptian salt that lubricated a regional décollement. Thermal subsidence caused westward tilting of the continental margin (Lunde *et al.* 1992), initiating regional, gravity-driven updip extension and associated downdip contraction. This produced rapid variations in salt thickness and the development of vertical salt structures (Spathopoulos 1996). Albian and Late Cretaceous sediments filled the lows created by extension and salt withdrawal.

A second stage of gravity-driven movement was triggered by the Neogene uplift of western Africa coupled with resultant rapid loading from clastics carried by the ancestral Congo, Kwanza and Ogooué river systems. Depositional loading by clastics shed from the raised areas enhanced downdip rafting. Lenses or slabs of post-salt section slid along the base-of-salt décollement, creating or amplifying ridges, diapirs and allochthonous sheets. In the southern Kwanza Basin, deformation was modified by buttressing against the Sumbe (Duarte *et al.* in press) volcanic chain. Elsewhere, the pattern of salt ridges and diapirs proceeded seaward to the COB (Dickson *et al.* in press), where salt nappes over-riding the abyssal plain and oceanic crust of the South Atlantic formed a prominent escarpment. Generally a series of mini-basins formed just behind (landward of) this escarpment.

There may have been strike-slip movement during the early extensional phases affecting the entire basin. However, displacement was largely during the two phases of gravity gliding across the salt that both deformed and provided a décollement layer for the overlying post-rift sequence. The salt was largely evacuated from the inner basins, only locally preserved in features such as pillows and isolated prisms. Salt reached great thicknesses in the seaward sag basins as diapirs, walls, ridges and allochthonous canopies. Consequently, the post-salt infill achieved great variations in thickness and structural complexity.

Extensional and contractional regimes of the Congo Fan

Within the Congo Fan, the updip extensional domain is separated from the downdip contractional domain by a complex structural zone. The extensional domain, located above continental crust, began at listric, mostly down-to-basin faults just west of the Atlantic Hinge. In the widest, central area of the fan, the Cretaceous series has translated oceanward up to 70 km, decreasing northward and southward to less than 10 km (e.g. Hudec *et al.* 2001; Jackson *et al.* 2001; Rowan *et al.* 2001). The middle complex zone was characterized by structural inversion, reactivation of older folds and thrusts in extension or by contraction of previous extensional diapir and listric faults.

Initially, Late Cretaceous contraction formed regularly spaced, variably verging, imbricate thrusts near the mid- to lower continental rise. Responding to Tertiary sedimentary loading, shortening has continued spasmodically until the present, forming distal toe-thrust complexes mostly within the Tertiary sequences nearer the COB (Dickson *et al.* in press). Contractional structures generally became younger seaward, climbing gradually to create a bumpy sea floor. The seaward younging indicated that an advancing sedimentary wedge caused most of the shortening. In contrast, only locally did contraction propagate in a partially landward direction as section piled up against a buttress created by a salt pinchout, a horst, or volcanoes.

Structural summary

The fabric of salt-induced rafting controlled sediment distribution in the Congo Fan. Raft blocks, cored with Albian carbonates, rested on a detachment surface at the Aptian salt (Duval *et al.* 1992; Demercian *et al.* 1993; Cobbold *et al.* 1995; Szatmari *et al.* 1996). The size, distribution and movement of these blocks controlled sediment distribution and salt deformation within the Congo Fan.

The key influences were:

- inter-raft lows near the shelf margin served as long-lived feeder channels that funnelled turbidites into seaward depocentres;
- depositional loading updip drove the rafting and thrusting downdip, in turn creating the bathymetric relief that caused ponding and fill-and-spill architectures.

Within the depocentres, bathymetric relief controlled the distribution of amalgamated levée channel complexes. Salt-cored highs deflected channel flows from the crests toward the flanks. Thrust-induced highs funnelled sands to an axial position along the thrust front. Subtle relief over previous channel levée complexes redirected younger systems. Several backstop basins were observed in the southern part of the area where multiple turbidite sequences were stacked in mini-basins between salt highs, caused by the interplay of high depositional rates and salt withdrawal. Subsequent migration of the salt-cored features has lifted regions containing reservoir into trapping configurations.

Rafts and inter-raft sediment transport fairways

The fabric of rafts that moved progressively seaward on the salt décollement controlled the distribution of Tertiary sands. Grabens which formed at the updip margin of each rafted area captured thick sections of clastics, as young as the Miocene, resting directly on pre-salt sediments. The grabens were bounded by listric faults that expanded as the underlying rafts moved seaward and sedimentation continued. Extensional lateral boundaries between sediment rafts provided avenues for basinward flow of sands, allowing bypass of parts of the shelf and upper slope.

In the downdip areas of the Kwanza/Congo/Ogooue fans, multiple phases of channel/fan sequences are reported in the Oligocene and Early Miocene, with the Tertiary thickness exceeding 8 km in the Congo Fan. In contrast, updip on the shelf, the channels appeared to represent a single event with much of the earlier section removed!

The orientation of these fill-and-spill reservoir systems varies from coast-parallel (NE–SW) to E–W to coast orthogonal as a function of the orientation of the underlying carbonate rafts and the steps from raft to raft (Fig. 4). These in turn were controlled by steps in the syn-rift architecture, i.e. steps in the basement caused steps in the syn-rift interval. Despite the bevelling at the break-up unconformity, slope changes or compaction effects tend to cause the carbonate rafts to break or stop moving at these older steps. Tertiary infill then deposited clastics or squeezed even lower-density salt into the gaps between the carbonate slabs.

Density contrasts and gravity expressions

Density contrasts arise from one of two causes: (1) compaction (and age) normally relating to loss of porosity, or (2) changes in lithology. The syn-rift section exhibits a horst-and-graben structural style with expanded wedges of sediments in the grabens. The grabenal sediments are infill, less compacted than on the adjacent horsts that have been partially eroded, leaving older, more-compacted (higher density) sediments at the same depth.

In the post-rift section, salt squeezed through gaps between the carbonate rafts. Where the rafts slid on the salt, clastic sequences filled the accommodation space. In each case, higher-density carbonates contrast laterally with lower-density clastics or salt.

All the way from surface to basement, therefore, density contrasts of the same sign tended to colocate. The cumulative stacking of all these density differences from basement to the near-surface resulted in strongly identifiable and correlatable anomalies in map-view. Our work to date has not attempted quantitative, profile-based modelling but simply checked that the sign of the density differences was correct for the interpreted depths and lithologies. The map-views provided similar checks to restrict the areal distribution of rafts, ridges and other features. Our interpretation was based on the strong map correlations between published features and our gravity attributes.

Observations from the Congo Fan region

Within the rafted province, individual rafts and the expanded younger sediment thicks that surround them are evident from regional gravity data. Through-going channels or sediment bypass fairways are inferred where gravity lows are interconnected. Transportation pathways of sediments into more distal salt mini-basins are interpreted.

Within the seaward allochthonous salt province, gravity highs correspond to salt canopies and diapirs. Gravity lows correspond to sediment thicks in mini-basins. Interconnected gravity lows observed on vertical residual maps of the regional gravity data correspond to sediment distribution fairways where long-lived bathymetric depressions have funnelled clastics in confined belts of expanded, stacked reservoirs that offer multiple exploration targets.

Dominey and Witte (1998) and Sikkema and Wojcik (2000) showed an interpretation of Block 16, offshore Angola. Figure 6 illustrates the correlation between their carbonate bank/rafts and inter-raft sediment systems v. our gravity attribute map. The blue lines indicate the eastward carbonate bank and westward rafts cored by dense Albian carbonates. Red lines indicate listric faults in the post-salt sequence. These faults created the roll-over traps that were tested by drilling at Longa, Bengo and Lucala (Fig. 5). The yellow/brown meander band represents the amalgamated channel complex of Miocene clastics as sedimentation moved seaward in the search for accommodation space. The underlying gravity map has blues in the ponds or catchment areas and reds in the highs or bypass areas. The channel complex of Figure 6 meanders through the connected lows of the gravity map. The gravity signatures associated with the channel complexes continue southwesterly into the next closed low.

Definition of the channel complexes helps target areas for specific seismic acquisition and interpret-

ation techniques. Seismic data is necessary since, although the defined channel complexes are 6–16 km in width, the individual channels at 10–200 m widths (Mayall & Stewart 2001) are well below resolution limits of these gravity maps. However, Figure 7 demonstrates a strong relationship between the recent discoveries and the gravity image. Armed with a thorough knowledge of the local geology, we were able to derive a gravity attribute specifically tuned to enhance the comparison to the sediment distribution systems for this area. This OG attribute (internal project name) is similar to a second vertical derivative but is based on depth-slicing the local gravity spectrum.

The areas of ponding highlighted from the closed gravity-map lows suggest where to look for high-reward fans. The yellow arrows demonstrate our interpreted channel systems winding downslope from basin to basin. The map lows (blues) represent mini-basins where the channel complexes have thickened; reds represent the bypass zones where underlying structures were raising the topography at the time of deposition. Intermediate colours are associated with the giant discoveries that required both a thick reservoir (better in the lows) and a trapping configuration (noses or turtles with reservoir drape across them). To our knowledge, no discoveries have yet been made in pure stratigraphic traps in the axial lows.

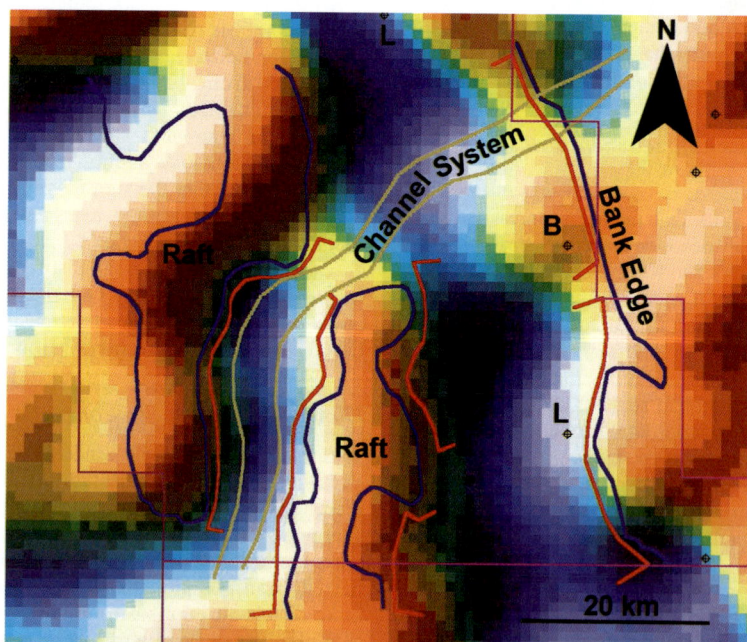

Fig. 6. Gravity derivative detail. Derivative maxima (red) and minima (blue) correspond to locations of carbonate bank (right), rafts (left) & inter-raft troughs that guided or captured the Oligocene–Miocene channel systems. Red, blue and tan lines trace the interpretation of Dominey & Witte (1998).

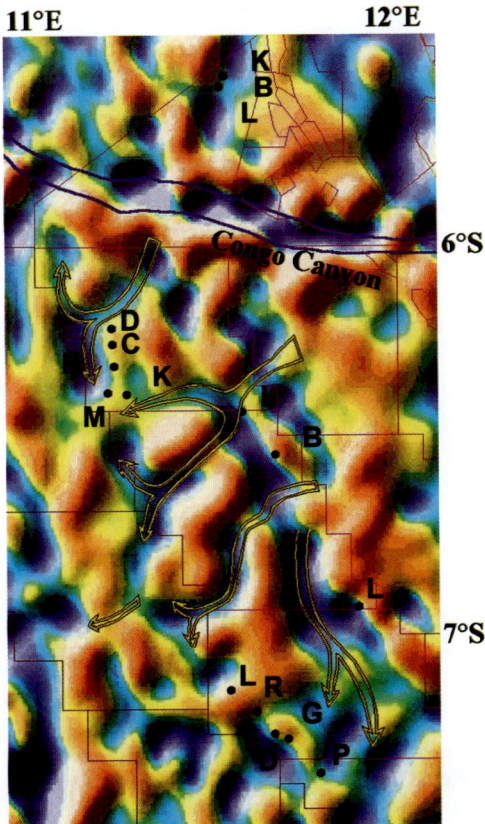

Fig. 7. Oligocene–Miocene sediment distribution pathways. Note discoveries (black dots) lie between maxima (red) and minima (blue). Selected pathways shown – see text for explanation. Block outlines in purple.

This may relate to the need for vertical migration paths from the source rocks – such pathways are often associated with faulting or microfracturing on structural flanks.

A further area of interest has been correlated with 3-D seismic results presented by Evans (2001) and at related conferences in 2000 and 2001. The westernmost arrow in Figure 7 indicates a transport path into the abyssal plain where we would predict true unconfined fans, in agreement with seismic signatures described by Evans.

Brazil

The Campos Basin alone holds more than 90% of Brazil's proved reserves (Bastos 1997), mostly in turbidite fan systems of the Late Cretaceous to Miocene (Fig. 8) (Bacoccoli & Toffoli 1988; Peres 1993). In our work here, we examined a simple concept of long-lived basement control on the

deposition of one of the main reservoir intervals in the Campos Basin fields. We observed on regional gravity imagery that, similarly to offshore West Africa, the syn-rift basins of Santos, Campos and Espírito-Santo (Figs 1 & 13) comprise a genetic whole that we refer to as the Greater Campos Basin, or SCES. We then extrapolated our inferences of similar basement influence and the identification of potential fan systems across the SCES and describe two examples.

Santos, Campos and Espírito Santo basin settings and sediment controls

Mirroring the position and development of the West African margin, the SCES of Brazil has a very similar history to that of the Aptian Salt Basin (Asmus & Ponte 1973; Campos 1974; Macedo 1989; Mohriak 1989; Rangel 1994; Davison 1997; Cainelli & Mohriak 1998). The overall stratigraphic successions and structural styles are very similar, although deposition generally youngs to the north, with the Santos Basin dominated by Cretaceous sediments (Fig. 9) while the Campos is infilled with a thick Tertiary succession. This region is structurally simpler than the Congo Fan, despite the similarity of geological history and stratigraphy because of the absence of the most intense salt-induced rafting and toe-thrusting.

Late Cretaceous to Holocene post-rift or drift stages reflect the different effects and dual influences of marginal to deep marine, mainly clastic, depositional systems and concurrent salt-tectonic structuring. Some of these halokinetic effects appear to be controlled by reactivation of the Early Cretaceous (rifting) normal faults. Most of the productive reservoirs of the SCES are found within the turbidites of the drift sequence.

The three SCES basins were limited on the western or cratonward side by outcrops and/or faulted hinge lines controlling sedimentation. These controls, such as the Campos Fault, Badejo High and 'pre-Aptian hinge' (e. g., Peres 1993; Mohriak *et al.* 1995) are particularly well imaged on the gravity maps (Figs 11–13; Cobbold 2001). Periodic uplift of the western shield areas exposed sediment sources for the basins; reservoirs were associated with both the upslope channels and the deeper, widespread turbidite facies.

Mirroring the West African margin, the margins stepped seaward into salt basins, often with a mid-basin ridge that coincides with major salt piercing. The Cretaceous Salt Province covered the main sedimentary areas of all three basins. The eastern side of the basins saw very active salt-dome structuring. The central portions contained salt-related or salt-mobilized features that were not actually piercements. Since the salt was mobilized almost

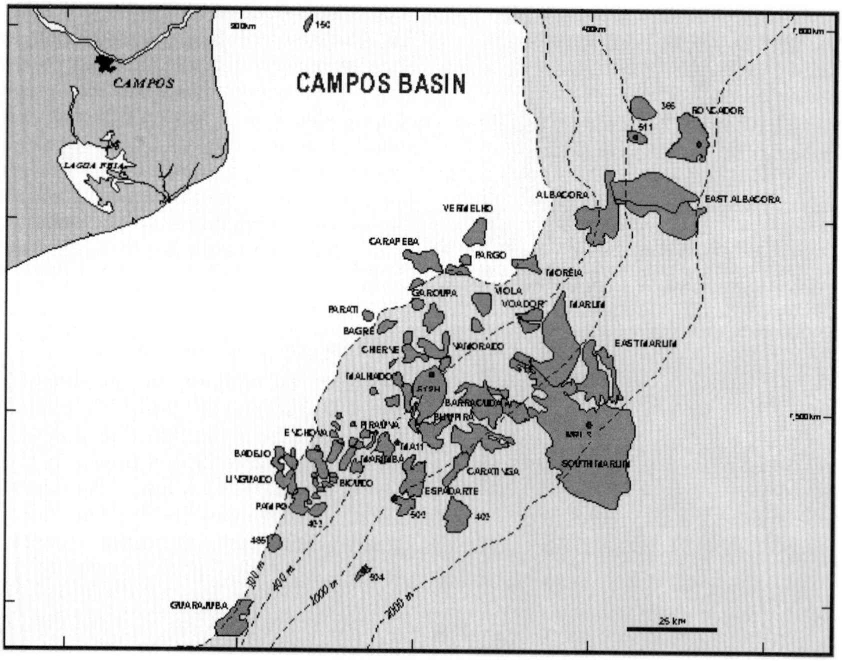

Fig. 8. Campos Basin bathymetry and fields (after Awad 1997).

directly after deposition, the interplay of fan deposition and salt movement formed both traps and reservoir trends.

While the structural style, environmental and gross sedimentary sequences are roughly similar in all three SCES basins, significant differences in basin subsidence rates have created important sedimentary and thickness contrasts. Also, the post-rift sequences generally young to the north so that there are much thicker Cretaceous sediments in the Santos Basin with thin or no Tertiary. This gives way to the reverse in the Campos and Espírito Santo basins.

The salt-involved structuring dies at the COB, similar to the limits seen on the opposing West African basins. Seaward across the oceanic crust, the section consists of a tapering wedge of relatively undisturbed sediments often overlying half-grabens floored with seaward-dipping reflectors (SDRs) (Abreu *et al.* 1997).

Inferred controls on sedimentation: build-ups, channels and fans

Our examples come specifically from the southern Campos and northern Santos basins. Our principal observations were the steps or offsets in the main landward basin-bounding faults (reservoir entry paths) and their locations v. the basin deeps (source rocks and migration paths) because steps in the

basin margins inferred sediment entry points. Their location with respect to the sediment provenances had implications for sand quality. Also, correspondence between published features and gravity signatures lead us to the recognition of other areas with similar patterns.

Channels (including those within the old Paraiba River system, Fig. 11) feeding the turbidites were trapped along the steps in the basin-bounding faults. The São João da Barra Low, just north of the Badejo High, shows a spectacular example of canyons v. the gravity signature. The northwestward convergence of shelf-edge canyons suggests a single (point) source representing the path of the Oligocene Paraiba River as it crossed the Badejo High at the interpreted basement offset. The Tertiary fans associated with the channels feeding through the São João da Barra Low are well documented (e.g. Peres 1993) and their correspondence with the ISO (Figs 11 & 12), 1VD and THD gravity maps is remarkable.

Similar channel/canyon features would be expected in two main offsets of the Badejo High just west of the Cabo Frio High in the northeastern corner of the Santos Basin (described by Cobbold *et al.* 2001). Fans are predicted in the northeastern Santos Basin fed through these steps in the bounding hinge line. Minor but economically significant channel/fan systems, such as at RJS-485 and Guarajuba Field in the Campos Basin, may be

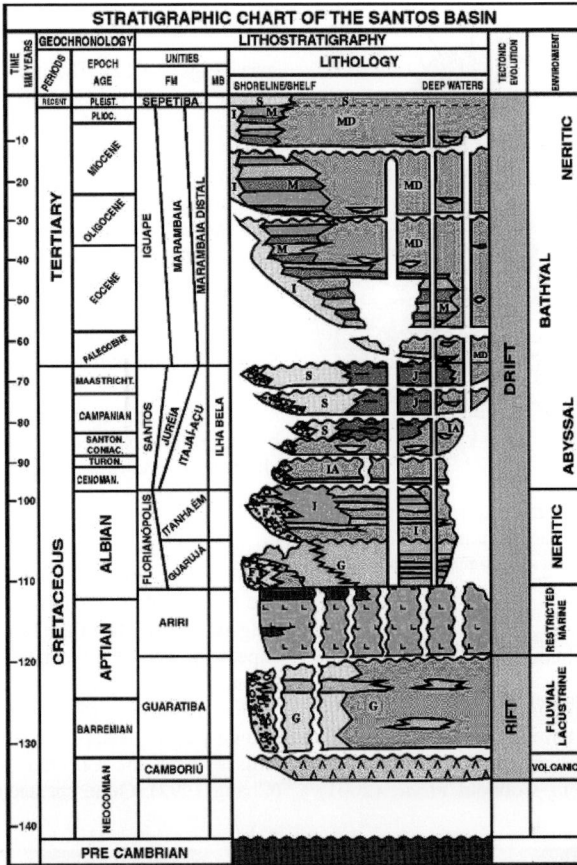

Fig. 9. Stratigraphic chart of the Santos Basin (after ANP 1999; modified from Pereira & Feijo 1994).

associated with other inflections in the coastal basin-bounding trend.

Campos Basin example

Many publications illustrated a series of Oligocene fans which reservoir giant reserves (Figs 8 & 10; Peres 1993) but, until Cobbold *et al.* (2001) and Meisling *et al.* (2001), most also indicated the landward hinge as a sinuous unbroken basin-bounding high (Badejo High–Campos Hinge). The modern Campos Basin features hundreds of shelf-edge canyons; simple extrapolation would suggest a similar system in the Tertiary. This would imply a line source of sediments with low prediction ability for fan locations. When we overlaid Oligocene fans on the isostatic gravity anomaly (Fig. 11) we saw something very different.

The gravity image reflects basement configuration, with reds being structurally high and blues, low. The NE-trending Badejo High is interrupted by a NW–SE fault (red line) that follows a major

onshore trend. The 40 km offset of the Badejo High was probably a long-lasting control on the Paraiba River and its outflow. The saddle linking the onshore sediment provenance to the fans encompasses the shale-filled canyons that comprise the bypass zone. The maximum fan isopach thick is cut off on its northeastern limit but the correlation with the gravity data shows maximum fan thicknesses in the ISO lows (blues). This reflects structural lows that tend to collect sands falling into the basin. Extrapolating fans across the basin lows becomes a simple exercise with this predictive method and the most promising areas can then be verified on selective seismic lines.

Santos Basin examples

We present two examples of trends in the Santos Basin: one relating to a set of basement offsets and the other simply to a trend of intrusions. Both have a gravitational signature but only the former

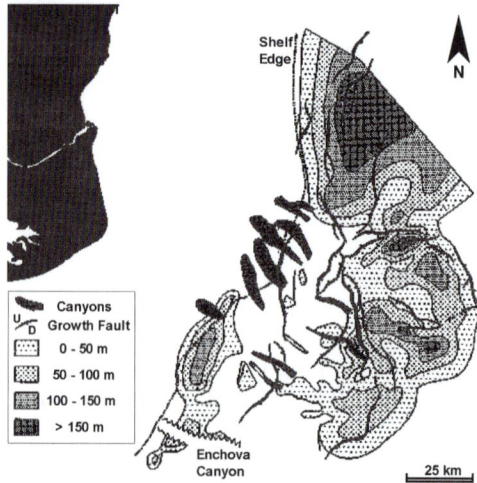

Fig. 10. Oligocene isopach, Campos Basin (after Peres 1993). Note termination of maximum isopach to the northeast and the northwestward convergence of shelf-edge canyons.

appears to have exerted control on reservoir deposition.

The structural configuration of a NE–SW-trending basin margin that is offset along NW–SE-trending lines is not described in the bulk of the literature but is clear in Cobbold *et al.* (2001).

Further, the offset trend follows the Rio Grande Rise and Hot Spot track, seen on the gravity imagery (Figs 2 &12). The much older Chaco–Paraná line, seen particularly well onshore (Chang 1992), takes a similar trend.

Two large offsets in the basin margin just west of Cabo Frio total about 100 km, as illustrated on the gravity image (Fig. 12) and in Cobbold (2001). The larger lateral displacement of 60 km greatly exceeds the step in the Badejo High of the Campos Basin. We expect that drainage capture of the Cretaceous system (Cainelli & Mohriak 1998) would funnel the main sediment volumes of the northern Santos through these offsets and into the basin. A semi-circular blob on the gravity image at the mouth of the predicted channel system is crossed by four or five NW–SE-trending lows and the overall effect is one of a channel feeding a fan system. The discovery of potentially commercial volumes of oil (Anon. 2001) in a suggested fan sequence in block BS-500 encourages the concept.

Certainly some sediments would derive from line sources along the basin margin and more from long-shore drift from the fan supply but we would expect the need for detailed seismic and well data to predict these lesser volumes of sand.

Our second example is the signature of the well-known pair of onshore dyke swarms at the Ponta Grossa Arch of Curitiba in the Paraná Basin (Chang 1992). Quite spectacular on the magnetics

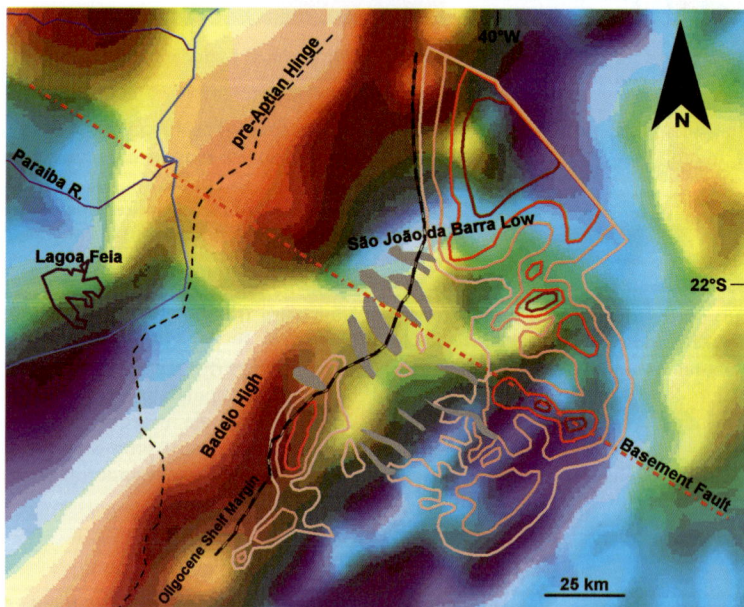

Fig. 11. Oligocene isopach and canyons of Figure 10 on gravity isostatic residual (ISO). Note correlation of isopach thicks to gravity lows (blue areas) and likely extension to northeast, and offset of Badejo High (red areas) and correlation of basement fault to canyon/fan system.

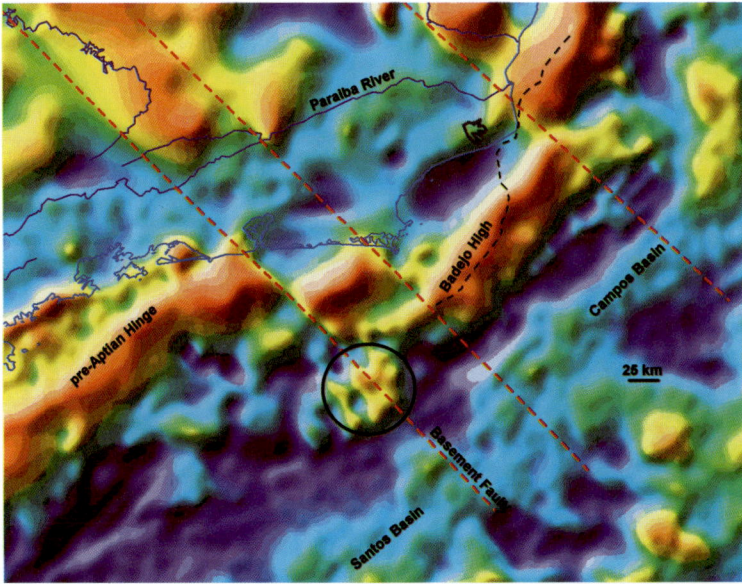

Fig. 12. Gravity isostatic residual (ISO) of the Campos and northern Santos basins showing three main NW–SE offsets in the basin margin and an inferred fan in the Santos Basin.

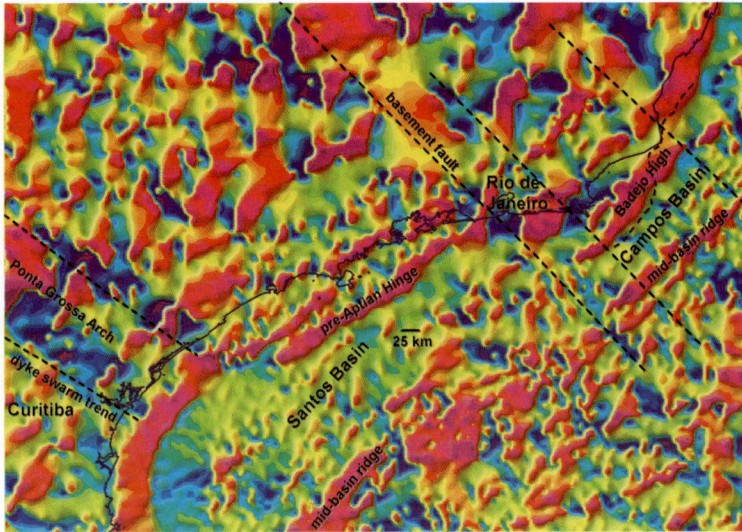

Fig. 13. Dip-azimuth display of gravity isostatic residual (ISO) for greater Campos Basin (Santos/Campos/Espírito Santo) showing principal basin margin offsets as Figure 12 and highlighting two WNW–ESE dyke swarm trends in the southern Santos Basin. Although the latter are clear on surface geology and magnetics maps (see Cainelli & Mohriak 1998), they do not offset the basin margin.

data (Anon. 2000), the same roughly NW–SE trend is seen on the gravity data (Fig. 13). This dip-azimuth display illustrates both dyke swarms but they do not offset the basin margin. Note the dyke trends are about 20° off the orientation of the basin-offsetting faults. While significant to the early opening of the South Atlantic, this trend appears not to have influenced the subsequent Cretaceous–Tertiary sedimentary infill of the Santos Basin.

Conclusions and implications for future exploration

We infer that the examination of multiple gravity attributes in map-view can thus provide a simple predictive technique for discriminating the likely influence on reservoir distribution of different structural features and trends. We stress the requirement to incorporate local geology in this analysis. Our examples confirmed predictions of fans in the Santos & Campos basins of Brazil and turbidites in the Congo Fan of Angola where seismic coverage was good, if relatively expensive. Using gravity signatures of known sedimentary features, such as reservoir fairways and depocentres, the locations and extents have also been projected into areas covered only by gravity data and (until 1999, per Evans 2001) some published research seismic lines. The result was a prediction of the locations of unconstrained sea-floor fans offshore Angola.

At its current level of development, this GIS-based potential-fields exploration technique has allowed teams to focus subsequent seismically oriented efforts quickly on areas of greatest potential. The method can be extended using higher-resolution gravity and bathymetry combined with additional gravity enhancements. These enhancements can be tuned to specific targets for each sub-basinal area as the OG filter (Fig. 7) in the Congo Fan example. The OG filter, named after Mark **O** de **G**ard, uses depth slicing techniques to enhance the gravity vertical derivative for the expected depth range of the sediments of interest.

Future work on gravity techniques

Despite the improvements from the first-level reprocessing employed for our project data, the resolution limit of current satellite-derived data is about 12–15 km wavelength anomalies, i.e. peak-to-peak. Half wavelengths or peak-to-trough limits would be 6–8 km on features with good spatial continuity. At the resolution limit, features are often seen with some spatial aliasing or mislocation, a phenomenon well known from work with satellite imagery. More precise 'pre-stack' reprocessing (Fairhead *et al.* 2001) improves on this limit to below 10 km full wavelength.

Conversely, poor bathymetry with resolution locally only 20 km degrades results with prominent expressions of seamount-associated artifacts, e.g. on THD maps (Fig. 2). Future work will include gravity inversion at the water–sediment interface to improve bathymetry, continued research to improve satellite gravity resolution, and means to separate surface and subsurface anomalies. Such improvements will reduce artefacts and increase

resolution to allow reliable interpretations down to perhaps half the feature size limits of the work in this paper.

The authors wish to thank GETECH for permission to publish the gravity images and the reviewers who greatly improved this paper.

References

ANON. 2000. Brazil's the place to be! *GETECH News*, Spring.

ANON. 2001. E & D action picks up in Santos Basin off Brazil. *Oil and Gas Journal*, **99**, 45–46.

ABREU, V., VAIL, P. R. & WILSON, E. 1997. Geologic evolution of conjugate volcanic passive margins: influence on the petroleum systems of the South Atlantic. *Bulletin of the Houston Geological Society*, **40**, 10–11.

ANP (AGÊNCIA NACIONAL DO PETRÓLEO). 1999. *Brasil Primera Rodada/Brazil Round 1.* [CD-ROM compilation *BrazilRnd1*]

ALEXANDER, C. S., MALEY, L. E., RAPOSO, A. & DOMINEY, J. 2000. The Plutonio discovery, Block 18, Angola – A 3D visualization and multi-attribute approach to exploration success. *American Association of Petroleum Geologists Bulletin*, **84**. A4. [Abstract]

ASMUS, H. E. & PONTE, F. C. 1973. The Brazilian marginal basins. *In*: NAIRN, A. E. M. & STEHLI, F. G. (eds) *The Ocean Basins and Margin. Vol. 1, The South Atlantic*. The Plenum Press, 102–113.

AWAD, S. P. 1997. Albacora Field FPS: Another deepwater development offshore Brazil. *Proceedings of the 1997 Offshore Technology Conference, OTC Paper 8468.*

BACOCCOLI, G. & TOFFOLI, L. C. 1988. The role of turbidites in Brazil's offshore exploration – a review. *Proceedings of the 1998 Offshore Technology Conference, OTC Paper 5659*, 379–388.

BASTOS, B. L. C. X. 1997. 20 years of drilling and completion experience in Campos Basin: a results review. *Proceedings of the 1997 Offshore Technology Conference*, SPE Paper 8488.

BELMONTE, Y., HIRTZ, P. & WENGER, R. 1965. The salt basins of the Gabon and the Congo (Brazzaville). *In*: *Salt Basins Around Africa*. Institute of Petroleum, London, 55–74.

BRINK, A. H. 1974. Petroleum geology of Gabon Basin. *American Association of Petroleum Geologists Bulletin*, **58**, 216–235.

BROGNON, G. P. & VERRIER, G. R. 1966. Oil and geology in Cuanza Basin of Angola. *American Association of Petroleum Geologists Bulletin*, **50**, 108–158.

CAINELLI, C. & MOHRIAK, W. U. 1998. *Geology of Eastern Atlantic Brazilian Basins: Brazilian Geology Part 2*. American Association of Petroleum Geologists Short Course.

CAMPOS, C. W. M., PONTE, F. C. & MIURR, K. 1974. Geology of the Brazilian Margin. *In*: BURKE, C. A. & DRAKE, C. L. (eds) *Geology of Continental Margins*, Springer-Verlag, New York, 447–463.

CANDE, S. & RABINOWITZ, P. D. 1978. Mesozoic sea-floor spreading bordering conjugate continental margins of Angola and Brazil. *Proceedings of the 1978*

Offshore Technical Conference, OTC Paper 3268, **78**, 1869–1876.

CHANG, H. K., KOWSMANN, R. O., FIGUEIREDO, A. M. F. & BENDER, A. 1992. Tectonics and stratigraphy of the East Brazil Rift System: an overview. *Tectonophysics*, **213**, 97–138.

COBBOLD, P. R., SZATMARI, P., DEMERCIAN, L. S., COELHO, D. & ROSSELLO, E. A. 1995. Seismic and experimental evidence for thin-skinned horizontal shortening by convergent radial gliding on evaporites, deep water Santos Basin, Brazil. *In*: JACKSON, M. P. A., ROBERTS, D. G. & SNELSON, S. (eds) *Salt Tectonics: A Global Perspective*. American Association of Petroleum Geologists Memoirs, **65**, 273–304.

COBBOLD, P. R., MEISLING, K. E. & MOUNT, V. S. 2001. Reactivation of an obliquely rifted margin, Campos and Santos Basins, southeastern Brazil. *American Association of Petroleum Geologists Bulletin*, **85**, 1925–1944.

CUNNINGHAM R. & LINDHOLM, R. M. 2000. Seismic Evidence for Widespread Gas Hydrate Formation, Offshore West Africa. *In*: MELLO, M. R. & KATZ, B. J. (eds) *Petroleum Systems of the South Atlantic Margins*. American Association of Petroleum Geologists Memoirs, **73**, 93–106.

DANFORTH, A., KONING, T. & DE DEUS, O. 1997. Petroleum systems of the coastal Kwanza and Benguela basins, Angola. *American Association of Petroleum Geologists International Conference/Exhibition, November 8–11, 1998, Rio de Janeiro, Brazil, Abstracts*.

DAVISON, I. 1997. Wide and narrow margins of the Brazilian South Atlantic. *Journal of the Geological Society, London*, **154**, 471–476.

DEMERCIAN, S. *et al*. 1993. Style and pattern of salt diapirs due to thin-skinned gravitational gliding, Campos and Santos Basins, offshore Brazil. *Tectonophysics*, **228**, 393–433.

DICKSON, W. G. & FAIRHEAD, J. D. 1998. Analysis of integrated satellite, land, marine and airborne gravity surveys using euler deconvolution and horizontal derivative processes. *American Association of Petroleum Geologists International Conference/Exhibition, November 8–11, 1998, Rio de Janeiro, Brazil, Abstracts*, 196.

DICKSON, W. G. & MACURDA, D. B. JR. 2000. Recognition and analysis of a major tributary of the Recent Congo Fan: Some Gully! (Abstract). *American Association of Petroleum Geologists Bulletin*, **84**, A39.

DICKSON, W. G., FRYKLUND, R. E. & GREEN, C. M. 1998. Constraints for plate reconstruction using gravity data: implications for source and reservoir distribution. *American Association of Petroleum Geologists International Conference/Exhibition, November 8–11, 1998, Rio de Janeiro, Brazil, Abstracts*, 248.

DICKSON, W. G., FRYKLUND, R. E., ODEGARD, M. E. & GREEN, C. M. (In press) Constraints for Plate Reconstruction using Gravity Data – Implications for Source and Reservoir Distribution or How To SAMBA: A GIS-Based Study Compares Reservoir Controls from Brazilian and West African Margin Basins.

DOMINEY, J. R. & WITTE, S. 1998. Salt tectonics and sedimentation: an integrated interpretation ultra deep water area, Lower Congo Basin, offshore Angola. *American Association of Petroleum Geologists International Conference/Exhibition, November 8–11, 1998, Rio de Janeiro, Brazil, Abstracts*, 590–591.

DUARTE MORAIS, M. L., MELLUSO L., MORAIS E., MORRA V. & SANTOGROSSO, I. 1998. *Cretaceous Magmatic Activity of the Sumbe Area (Angola)*.

DUVAL B., CRAMEZ C. & JACKSON, M. P .A. 1992. Raft tectonics in the Kwanza Basin, Angola. *Marine & Petroleum Geology*, **9**, 389–404.

EVANS, D. G. 2001. The depositional regime on the abyssal plain of the Congo Fan in Angola. *Proceedings of the Petroleum Geology of Deepwater Depositional Systems, Geological Society, London, March 2001*. [Abstract]

FAIRHEAD, J. D. 1988. Mesozoic plate tectonic reconstructions of the central South Atlantic Ocean: The role of the West and Central African rift system. *Tectonophysics*, **155**, 181–191.

FAIRHEAD, J. D. & WATTS, A. B. 1989. The African gravity project: academic, government and commercial data integrated for new map of continent and margins. *Lamont-Doherty Geological Observatory Newsletter*, **21**, 6–7.

FAIRHEAD, J. D., GREEN, C. M. & ODEGARD, M. E. 2001. Satellite-derived gravity having an impact on marine exploration. *Leading Edge*, **20**, 873–876.

FAIRHEAD, J. D., GREEN, C. M., MAUS, S. & WOOLLETT, R. 1998. Highest resolution satellite gravity data. *GETECH News*, Special Issue.

GREEN, C. M. 1997. *Global Gravity Compilation and its Application to Plate Tectonic Reconstruction*. Ph.D. thesis, University of Leeds.

GREEN, C. M. & FAIRHEAD, J. D. 1992. The South American gravity project. *In*: TÖRGE, W., GONZALES FLETCHER, A. & TANNER, J. G. (eds) *Recent Geodetic and Gravimetric Research in Latin America*. IAG Symposia, Springer-Verlag, Heidelberg, **111**, 82–95.

GREEN, C. M. & FAIRHEAD, J. D. 1996. New 5'x5' digital gravity and terrain models of the Earth. *In*: RAPP, R. H., CASENAVE, A. A. & NEREM, R. S. (eds) *Global Gravity Field and Its Temporal Variations*. IAG Symposia, Springer-Verlag, Heidelberg, **116**, 227–232.

GREEN, C. M., FAIRHEAD, J. D. & MAUS, S. 1998. Satellite-derived gravity: Where we are and what's next. *Leading Edge*, **17**, 77–79.

GYORGY, M., GABOR, T. & LEHMANN, C. 1998. Evolution of salt-related structures and their impact on the post-salt petroleum systems of the Lower Congo Basin, offshore Angola. *American Association of Petroleum Geologists International Conference/Exhibition, November 8–11, 1998, Rio de Janeiro, Brazil, Abstract. American Association of Petroleum Geologists Bulletin*, **82**, 1883–1984.

HUDEC, M. R., JACKSON, M. P. A., BINGA, ,L. F., DA SILVA, J. Q., FRAENK, R. & SIKKEMA, W. 2001. Regional restoration in the offshore Kwanza Basin, Angola: linked zones of extension, translation, and contraction. *American Association of Petroleum Geologists Bulletin*, **85**, A95. [Abstract].

JACKSON, M. P. A., HUDEC, M. R., FRAENK, R., SIKKEMA, W., BINGA, L. & DA SILVA, J. 2001. Minibasins translating down a basement ramp in the deepwater monocline province of the Kwanza Basin, Angola. *American Association of Petroleum Geologists Bulletin*, **85**, A89. [Abstract]

KARNER, G. D., DRISCOLL, N. W., McGINNIS, J. P., BRUMBAUGH, W. D. & CAMERON, N. 1997. Tectonic significance of syn-rift sedimentary packages across the Gabon–Cabinda continental margin. *Marine & Petroleum Geology*, **14**, 973–1000.

LEMOINE, F. G., KENYON, S. C. ET AL. 1998. *The Development of the Joint NASA GSFC and the National Imagery and Mapping Agency (NIMA) Geopotential Model EGM96, NASA/TP-1998-206861.*

LINDHOLM, ROSANNE M., & ROBERT CUNNINGHAM. 1998. Geologic controls on the gas hydrate distribution, offshore Congo. *American Association of Petroleum Geologists International Conference/Exhibition, November 8–11, 1998, Rio de Janeiro, Brazil, Abstracts*, 934–935.

LUNDE, G., AUBERT, K., LAURITZEN & LORANGE, E. 1992. Tertiary Uplift of the Kwanza Basin in Angola. *Geologie Africaine, Proceedings of the 1st Conference on the Stratigraphy and Palaeogeography of West African Sedimentary Basins, Libreville, 6–8 May 1991,* 99–117.

MACEDO, J. M. 1989. Tectonic evolution of the Santos Basin and adjacent continental areas. *Boletim de Geosciências da Petrobras*, **3**, 159–173.

MARTON, G., TARI, G. & LEHMANN, C. 1998. Evolution of salt-related structures and their impact on the postsalt petroleum systems of the Lower Congo Basin, offshore Angola. *American Association of Petroleum Geologists International Conference/Exhibition, November 8–11, 1998, Rio de Janeiro, Brazil, Abstracts.*

MARTON, G., TARI, G. & LEHMANN, C. 2000. Evolution of the Angolan Passive Margin, West Africa, with emphasis on Post-salt structural styles in Atlantic Basin evolution. *In*: MOHRIAK, W. & TALWANI, M. (eds) *Atlantic Rifts and Continental Margins*. American Geophysical Union, Geophysical Monograph Series, **115**, 129–149.

MAYALL, M. & STEWART, I. 2001. The architecture of turbidite slope channels. *Proceedings of the Petroleum Geology of Deepwater Depositional Systems, Geological Society, London, March 2001.* [Abstract]

MEISLING, K. E., COBBOLD, P. R. & MOUNT, V. S. 2001. Segmentation of an obliquely rifted margin, Campos and Santos basins, southeastern Brazil. *American Association of Petroleum Geologists Bulletin*, **85**, 1903–1924.

MOHRIAK, W. U., MELLO, M. R., KARNER, G. D., DEWEY, J. F. & MAXWELL, J. R. 1989. Structural and stratigraphic evolution of the Campos Basin, Offshore Brazil. *In*: TANKARD, A. J. & BALKWILL, H. R. (eds) *Extensional Tectonics and Stratigraphy of the North Atlantic Margins*. American Association of Petroleum Geologists Memoirs, **46**, 577–598.

MOHRIAK, W. U. et al. 1995. Salt tectonics and structural styles in the deep-water province of the Cabo Frio Region, Rio de Janeiro, Brazil. *In*: JACKSON, M. P. A., ROBERTS, D. G. & SNELSON, S. (eds) *Salt Tectonics: A Global Perspective*. American Association of Petroleum Geologists Memoirs, **65**, 273–304.

MUELLER, R. D., ROEST, W. R., ROYER, J.-Y., GAHAGAN, L. M. & SCLATER, J. G. 1997. Digital isochrons of the world's ocean floor, *Journal of Geophysical Research*, **102** (B2), 3211–3214.

NORVICK, M. S. & SCHALLER, H. 1998. The post-rift paleogeographic evolution of the South Atlantic Basins of Brazil and West Africa and the influence of hinterland uplift on drainage and sedimentary depocenters. *American Association of Petroleum Geologists International Conference/Exhibition, November 8–11, 1998, Rio de Janeiro, Brazil, Abstracts*, 36–7.

PEREIRA, M. J. & FEIJÓ. F. J. 1994. Bacia de Santos. *Boletim de Geosciências da Petrobras*, **8**, 219–234.

PERES, W. E. 1993. Shelf-fed turbidite system model and its application to the Oligocene deposits of the Campos Basin, Brazil. *American Association of Petroleum Geologists Bulletin*, **77**, 81–101.

POSAMENTIER, H. W. & WISMAN, P. S. 2000. Deep water depositional systems – ultra-deep Makassar Strait, Indonesia. *Proceedings of the Gulf Coast Section of the Society of Economic Paleontologists and Mineralogists Conference, Houston, December 3–6, 2000*, 806–816.

RANGEL, H. D. et al. 1994. Bacia de Campos. *Boletim de Geosciências da Petrobras*, **8**, 203–217.

REYRE, D. 1966. Evolution geologique du Bassin Gabonais. *In*: REYRE, D. *Sedimentary Basins of the African Coasts, Part 1, Atlantic Coast*. Association of African Geological Surveys, 171–189.

ROWAN, M. G., PEEL, F. J. & VENDEVILLE, B. C. 2001. Gravity-driven foldbelts on passive margins. *American Association of Petroleum Geologists Bulletin*, **85**, [Abstract]

SANDWELL, D. & SMITH, W. H. F. 1997. Marine gravity anomaly from GEOSAT and ERS-1 satellite altimetry. *Journal of Geophysical Research*, **102**, 10,039–10,054.

SIKKEMA, W. & WOJCIK, K. M. 2000. 3D visualization of turbidite reservoir architectures, Lower Congo Basin, offshore Angola. *Proceedings of the Gulf Coast Section of the Society of Economic Paleontologists and Mineralogists Conference, Houston, December 3–6, 2000*, 928–939.

SPATHOPOULOS, F. 1996. *An Insight on Salt Tectonics in the Angola Basin, South Atlantic, In*: ALSOP, G. I., BLUNDELL, D. J. & DAVISON, I. *Salt Tectonics*. Geological Society, London, Special Publications, **100**, 153–174.

SPENCER, J., TARI, G., JERONIMO, P. & HART, B. 1998. Comparison between offshore Angola and the Gulf of Mexico in terms of salt tectonics. *American Association of Petroleum Geologists International Conference/Exhibition, November 8–11, 1998, Rio de Janeiro, Brazil, Abstracts*, 968.

SZATMARI, P. et al. 1996. Genesis of large counterregional normal fault by flow of Cretaceous salt in the South Atlantic Santos Basin, Brazil. *In*: ALSOP, G. I., BLUNDELL, D. J. & DAVISON, I. *Salt Tectonics*. Geological Society, London, Special Publications, **100**, 259–264.

TEISSERENC, P. & VILLEMIN, J. 1990. Sedimentary Basin of Gabon – Geology and Oil Systems. *In*: EDWARDS, J. D. & SANTOGROSSI, P. A. (eds) *Divergent/Passive Margin Basins*. American Association of Petroleum Geologists Memoirs, **48**, 117–199.

VIDAL J. P., JOYES R. & VAN VEEN, J. 1977. L'Exploration petroliŁre au Gabon et Congo. *Proceedings. 9eme Congrés Mondial du Petrole, Tokyo, 1975. Vol. 3, Exploration and Transportation*. Applied Science Publishers, London, 149–165.

Optimizing 3-D seismic technologies to accelerate field development in the Berkine Basin, Algeria

J. M. DRUMMOND[1], R. KASMI[2], A. SAKANI[3], A. J. L. BUDD[4] & J. W. RYAN[4]

[1]Anadarko Petroleum Corporation, 17001 Northchase Drive, Houston, Texas, 77060, USA
[2]Sonatrach – Division PED, 8 Chemin du Reservoir, Hydra, Algiers, Algeria
[3]Lynx Geodata Sarl, Lotissement Alioua Fodil, Villa No. 94, Cheraga, Algeria (formerly Sonatrach Exploration)
[4]Target Finders Ltd, Unit 6B, Grays Farm Production Village, Grays Farm Road, Orpington, Kent, BR5 3BD, UK

Abstract: A radical approach to 3-D acquisition in country with large sand dunes was developed to improve seismic data quality. A methodology was devised which allowed accurate removal of near-surface statics and transmission effects giving enhanced depth prognosis. An adaptive noise-attenuation technique, based upon an image-processing procedure, was developed to improve the signal-to-noise ratio of 3-D data that were contaminated with 'acquisition footprint'. Finally, parallel survey geometries were devised that reduced the noise contamination of the next generation of 3-D surveys in the Berkine Basin. Most of the 3-D surveys acquired in the Berkine Basin are now designed with parallel templates rather than cross-spreads. This survey technique has been shown to provide better spatially sampled data, despite the interference caused by the huge sand dunes, which, in turn, respond more positively to standard pre-stack noise-attenuation processing. Attribute and coherency analysis, as well as stratigraphic interpretation, is now possible on these datasets with a much greater degree of confidence.

In 1995, after several successful exploration wells, Anadarko Algeria Corporation and partners Sonatrach, LASMO Oil Algeria and Maersk Olie Algeriet sought to accelerate their development programme in the Berkine Basin, Algeria (Fig. 1).

The total area encompassed by Anadarko's JV holdings was 2 023 500 ha (5 million acres). Prior to Anadarko's JV acquiring the exploration and development rights to the area, over 30,000 km of 2-D data had previously been acquired. These data were of many different vintages, with varying acquisition parameters, line lengths, azimuths and processing flows. The variability in data quality made an accurate interpretation of the regional geology very difficult and interpretation of potential subtle structures and reservoirs almost impossible. Anadarko therefore undertook to reprocess all of the vintage data.

Using the interpretation of the reprocessed 2-D data volume as a basis, Anadarko drilled an exploration well in the western portion of Block 404. This first well, BKW-1, shown in Figure 2, was targeted at the Frasnian and was plugged and aban-doned as a dry hole. The post-mortem analysis of the well results caused the exploration team to re-evaluate their approach from both a geological and geophysical standpoint.

It was determined that, to ensure success, the team had to formulate a geophysical strategy to acquire regional 2-D data. As a result a new exploration programme was instigated, with improved acquisition design based on analysis of the vintage data and field tests. As the exploration programme progressed, a new understanding of the regional geology was developed which highlighted prospective areas that would later be focused on for infill acquisition prior to exploration drilling. From 1990 to 1997 approximately 9000 km of 2-D seismic data was acquired with high-fold and increased spatial resolution. As a result of this new regional exploration programme the first successful discovery well (EME-1) was drilled in the south of Block 208.

Using successful drilling operations as a basis, the exploration team used their new knowledge and growing database as a springboard to plan and

From: Arthur, T. J., MacGregor, D. S. & Cameron, N. R. (eds) *Petroleum Geology of Africa: New Themes and Developing Technologies.* Geological Society, London, Special Publications, **207**, 257–273. 0305-8719/03/$15
© The Geological Society of London 2003.

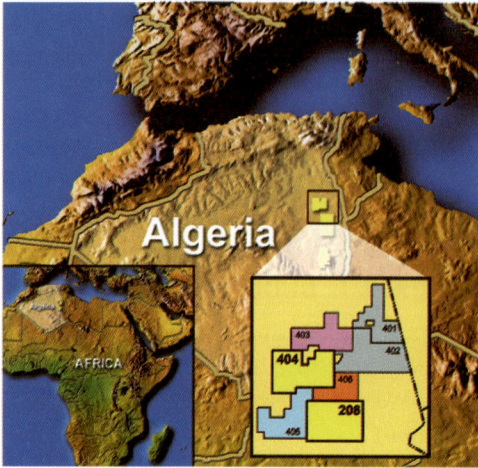

Fig. 1. Map of Algeria showing the Berkine Basin and Anadarko JV licensed Blocks.

because the factors controlling seismic data quality in the area were poorly understood, it was not clear how to optimize the acquisition and processing for interpretation. The approach to resolving these uncertainties was to launch a strategic 3-D project targeted at the Berkine Basin. The objectives of this project included: (1) understanding the near-surface mechanisms controlling seismic data quality in the group's acreage, (2) defining the seismic quality requirements for the development of 3-D surveys, (3) developing the 3-D acquisition and processing technologies to meet these requirements, and (4) implementing these technologies in order to minimize the 3-D cycle time.

The endgame of the 3-D strategic project was to produce high-quality geophysical data using a range of techniques, which would allow the definition of the reservoir architecture in the licence area. Here we describe how the project achieved its goals.

design the strategic goals and plans for field development through 3-D seismic acquisition and processing. The initial 3-D survey was planned over the HBNS field in the northern part of Block 404. To date, Anadarko and their partners have acquired *c.*3700 km² of 3-D seismic data (Fig. 2).

It was recognized that 3-D surveys should improve the positioning of development wells but,

The challenges

2-D seismic data from the Berkine Basin acquired in the early 1990s generally exhibit a poor signal-to-noise ratio. During the cycle of 2-D exploration, increasing levels of survey effort were employed to improve data quality, with 240-fold data becoming standard by 1995. Despite high field effort and extensive processing, direct interpretation of the

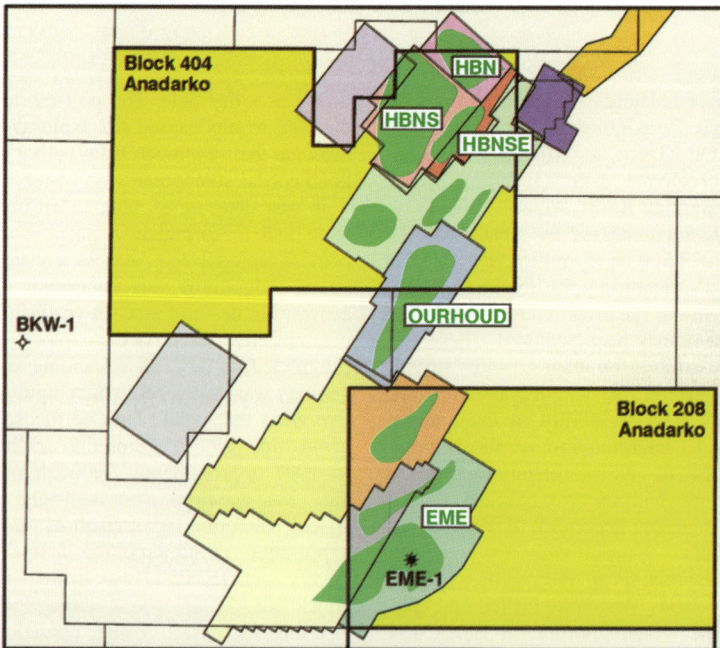

Fig. 2. Map showing detail of licensed blocks, discovered fields and 3-D surveys acquired to date.

reservoir formations was impossible. Lack of obvious seismic response of the reservoir, in conjunction with the low structural relief, represented the main technical challenges for 3-D survey design.

In addition, there were significant logistical issues to be considered. The Berkine Basin is home to some of the largest sand dunes in the world. Dunes such as those in Figure 3 can be around 450 m high and the subsequent problems of line access and cutting all influence the acquired geophysical data.

The Grand Erg Oriental extends over 50% of the Berkine Basin in Algeria and eastward into the western portion of Tunisia east of El Borma. The sand is very variable both in texture, colour and geometrical formation. In the western part of the basin the sand is formed into isolated star-shaped sand dunes with diameters up to 1200 m and heights up to 150 m. These dunes are fairly stable and are surrounded by plains characterized by sand-covered gypsum with sporadic sinuous sand ridges, or seifs, that impede travel through the valley areas.

As the sand progresses from the northwest of Block 404 to the south and east, the geomorphology of the sand changes dramatically. In Block 208 it appears that the star-dune density increases to the point where several dunes coalesce to form large dune ridges with a footprint in excess of 1.5 × 4–5 km. These larger footprints permit the sand to rise to heights in excess of 400 m and create impenetrable barriers. Figure 4 shows one of these large dunes with a WesternGeco seismic camp at its base.

Understanding the near-surface

There were two main motivations for studying the near-surface: (1) to understand the noise-generating mechanisms so that appropriate field parameters, such as source and receiver array lengths, could be optimized; and (2) to devise a near-surface model that provided the best statics solution and allowed the error in the solution to be estimated. This aspect of the project was considered necessary because of the general uncertainty on the quality of the statics solution in the area, and the possibility of long-period residual statics errors in the dataset which could adversely impact mapping of the field.

The near-surface layers range from 5 m of unconsolidated sand at the surface, through consolidated sand, to gypsum and high-velocity anhydrites in the first few hundred metres depth. An accurate picture of the near-surface layers was built from well logs and an extensive series of upholes drilled to a maximum of 300 m on a 1 km-square grid. Figure 5 shows a typical average velocity curve from an uphole. During the course of the 2-D acquisition in the area the depth of the upholes was increased from 100 m to 300 m. This increase in hole depth was in response to the prognosed depth of the refractor used in the statics derivation

Fig. 3. Star-shaped sand dunes coalesce into long ridges in Algeria's Eastern Sand Sea.

Fig. 4. A large sand dune in Block 208 in the Berkine Basin. A WesternGeco seismic camp is shown at the base of the sand dune.

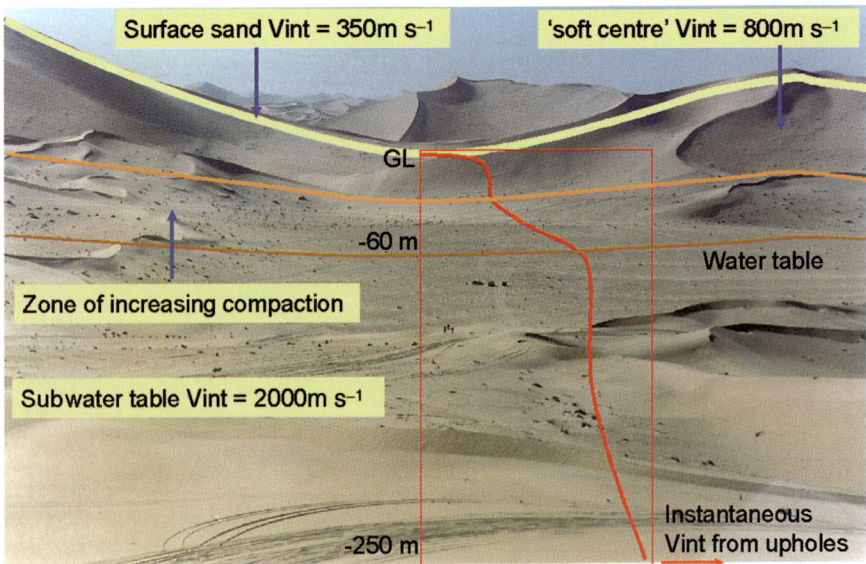

Fig. 5. Average velocity curve typical of near-surface sand compaction.

for the 2-D data. The holes were logged with a hammer and plate source. The wells were logged up to c.300 m below surface, enabling some integration of near- and subsurface measurements.

The uphole data were processed in a number of ways to gain insight into the near-surface. The time/depth curves were used to generate interval and average velocity functions. The data were also processed as vertical seismic profiles and used to generate corridor stacks.

In addition to the uphole data various types of 'dune' surveys were also recorded to improve the understanding of the internal architecture of the larger dunes. Access problems generally prohibited drilling upholes anywhere other than the valley floor. Reflection and refraction seismic surveys were acquired over large dunes and these were combined with global positioning system (GPS) mapping of the dunes. The results of these investigations led to the belief that the dunes had a 'soft' centre with some uplift of the water table also occurring (Fig. 5).

Near-surface and noise

The predominant noise mechanism involves multiple reflections of refraction events. The interfaces which generate the main refractions and multiples are generally within the Quaternary and Miocene–Pliocene section at depths up to c.400–500 m. The interface between the Senonian evaporite sequence and the softer overlying Tertiary sediments generally marks the last refractor of interest in the geophysical data. This refractor has velocities ranging from approximately 4000 m s^{-1} to 5000 m s^{-1}. Several slower refraction events are observed with velocities ranging down to approximately 900m/s. A velocity contrast from 900 m s^{-1} to approximately 2000 m s^{-1} is present and is identified as marking the contrast between dry and wet sand at the water table. This velocity contrast generates an important refraction event and most of the reverberant energy present in the seismic data is believed to be generated between this interface and the sand surface, in some respects resembling a leaky waveguide. In addition to the multiple refractions, which are generally non-dispersive, there is also at least one slower, dispersive event with a group velocity of approximately 400 m s^{-1}, which is considered to be a ground roll effect.

Near-surface velocities as determined by uphole surveys and refractor analyses change rapidly across the area. Poor seismic reflection data are obtained in locations where the near-surface velocity gradient increases, with a corresponding increase in shallow reflection coefficients and reduction in critical angle.

Seismic reflection quality is generally dependent on the velocity gradient in the near-surface. In some areas, P-wave velocities increase from 900 m s^{-1} to 2600 m s^{-1} at a depth of around 75 m below surface datum (an elevation of 125 m). This implies a P-wave reflection coefficient of 0.55 and a critical angle of 20 degrees. Subsurface penetration and angular resolution are reduced accordingly. Figure 6 shows one half of a seismic record up to a maximum offset of 4800 m. The pattern of refractors and reverberants generated in the near-surface is typical of the area and shows velocities ranging from approximately 400 m s^{-1} to 5000 m s^{-1}.

Near-surface and statics

The first step was to identify the boundaries that generate the main refractions in the area, which were relevant to the derivation of refraction statics. Three key events in the near-surface had been identified during the course of early work on 2-D data in the area. Figure 7 shows a schematic diagram of these events, the characteristics of which are described in Table 1.

The '2000 m s^{-1}' event was interpreted from uphole surveys as corresponding to the top of a consolidated sandstone column at the level of the water table. The uphole seismic recordings, which contained high levels of downgoing shear energy, enabled the water table to be identified with some confidence because of the characteristic change in Vp/Vs ratio between wet and dry sandstones. It is likely that the cementation process starts near the top of the water table. Above the level of the water table, the uphole surveys showed increasing velocity, characteristic of sand-grain compaction with a progressive reduction in pore space. Below the top of the water table a sandstone layer with a fairly constant velocity of approximately 2000 m s^{-1} overlies high-velocity evaporites. The water table varies in depth according to the surface elevation so the associated seismic event cannot

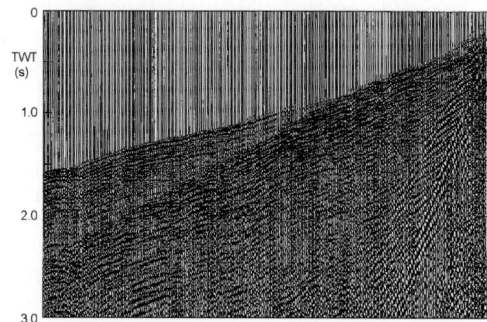

Fig. 6. Live shot record.

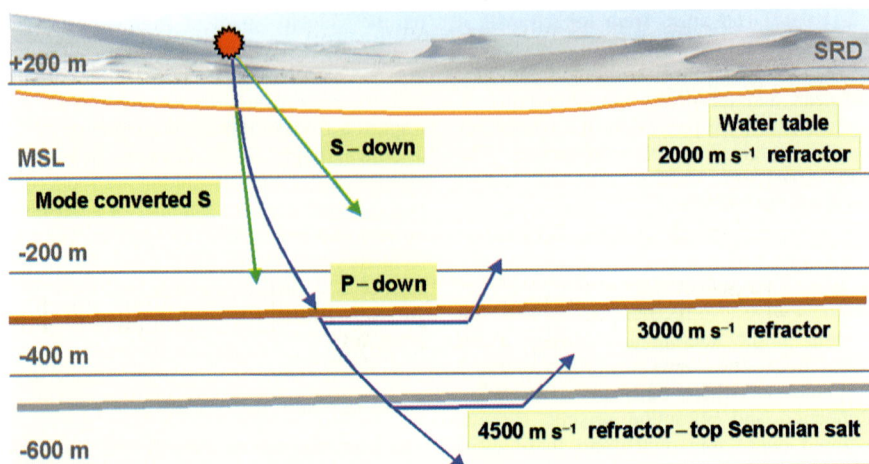

Fig. 7. Near-surface propagation – the three key events.

Table 1. *Summary of refractor parameters*

Approx. refractor velocity (m s⁻¹)	Uphole control	Pickable offset range (m)	'Pickability' of refractor
2000	Yes	100–400	Poor
3000	No	600–1200	Fair
4400	No	2600–4200	Good

conveniently be used as a basis for static corrections. The water table was logged during uphole recording.

The '3000 and 4400 m s⁻¹'refraction events proved more difficult to identify. A first estimate of the depth to these refractors was made using the delay times from the 2-D data statics analysis in conjunction with the time/depth curves from check-shot surveys in the exploration wells. The first check shots in the wells were generally around 75 m below the surface datum of 200 m.

The refractor depths estimated from this procedure were approximately 250–300 m below surface datum, just above total depth (TD) of the deeper upholes. This factor influenced the drilling of upholes to 300 m, but above the start of the sonic and density logs in the exploration wells. The computed velocities toward the bottom of the upholes were very erratic, probably due to the cable contacting the bottom of the hole and sagging. There was some slight evidence of high-velocity breaks near the bottom of some of the deeper upholes, but it was concluded at an early stage of the project that the upholes could not be used directly for calibration of the useful refractors. Fur-

thermore, owing to the absence of well logs in the initial zone of interest, the lithologies generating the refractions could not be positively identified.

Attempts were then made to generate a depth/velocity model over the zone of interest from the check-shot data. It became apparent that there was an anomaly in the refraction parameters, because the interval velocity profiles generated from check shots did not show sequences with the appropriate refraction velocities at the expected times. Either the refraction velocities or their delay times had to be in error. This anomaly was resolved by modelling a shot record based on the sonic log from the HBNS-3 well. Instantaneous velocities from the sonic log were combined with uphole and check-shot data to construct a continuous velocity profile from surface to TD. This velocity profile was used in finite-difference acoustic modelling to produce a shot record which showed the expected refraction arrivals.

A series of test models was then produced by progressively editing the velocity log from surface down to various velocity boundaries. For each test, the upper part of the log was replaced with a constant velocity designed to preserve the time/depth relationship at the boundary being investigated. Figure 8 shows a modelled record produced from this procedure. A constant velocity of 2000 m s⁻¹ has been inserted from surface to a depth of 500 m below surface. Progressively stripping the log in this fashion gave a good indication of the boundaries generating the refractions, which disappeared when the model had been stripped below that level.

The boundaries generating the 3000 m s⁻¹ and 4400 m s⁻¹ refractors were identified from this procedure. The 4400 m s⁻¹ refractor was identified as

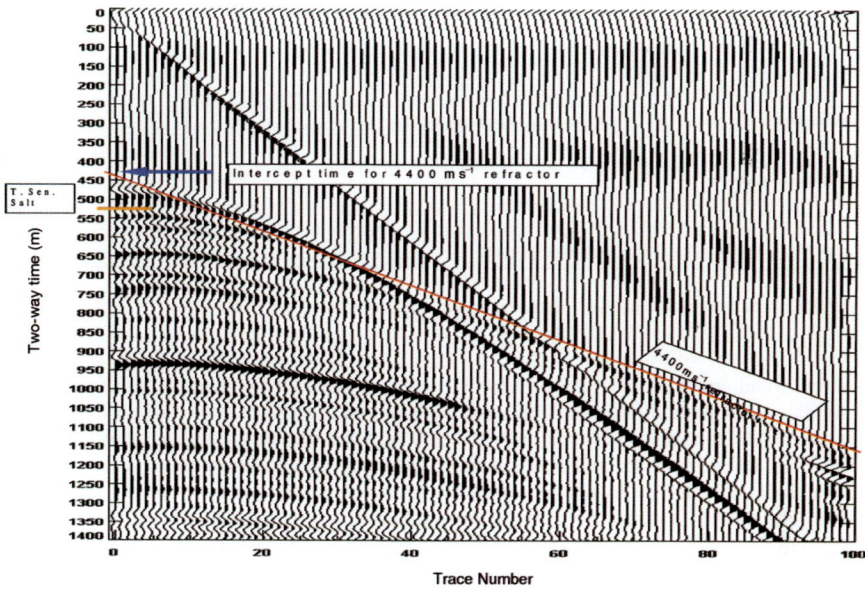

Fig. 8. Finite-difference acoustic model based on HBNS-3 logs.

the top of the Senonian Salt (SS). It has a very consistent expression on all the well logs in the area and can be picked very accurately by virtue of the characteristic kicks on both sonic and density logs. It thus transpired that the travel times, and therefore the computed depths to the refractors, had been considerably underestimated; the 4400 m s^{-1} refractor, for example, was approximately 650 m below surface datum rather than 300 m, as previously calculated from the 2-D refractor delay times. This is due to the zero-offset intercept time being a poor approximation to the true zero off-set time.

Once the refractors had been identified it remained to devise a technique for deriving statics corrections that would suit the near-surface model. Figure 9 shows the 'textbook' situation for refraction statics analysis and calibration using uphole surveys in relatively flat terrain. In this situation

the refractor is at the base of a relatively thin weathering layer that is a few tens of metres thick at most. The uphole surveys extend through the weathering layer and provide a good estimate of the time, depth and velocity from the surface to refractor. A single-layer refractor model provides an acceptable approximation to the true situation; in particular, a value of Vo, which is close to the true weathering velocity, can be used, and the zero-offset intercept time of the refractor is an acceptable approximation to the true zero-offset reflection time. Therefore the computed depth to refractor provides a good approximation to the true depth.

Figure 10 summarizes the general problems posed for refraction statics analysis in this area of the Berkine Basin in Algeria, compared to the ideal situation depicted in Figure 9. The most useful refractor (4400 m s^{-1}) is some 650 m below source receiver datum (SRD), underlying a complex geo-

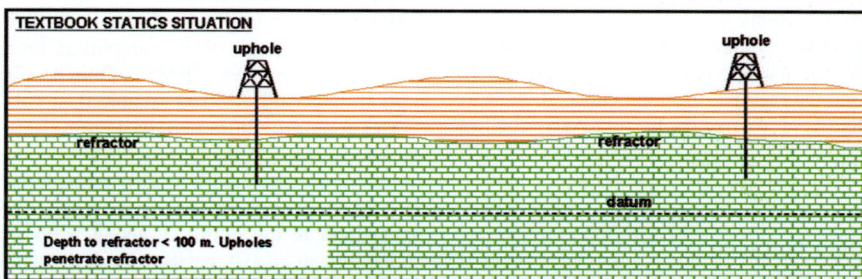

Fig. 9. Textbook statics situation (cf. Fig 10).

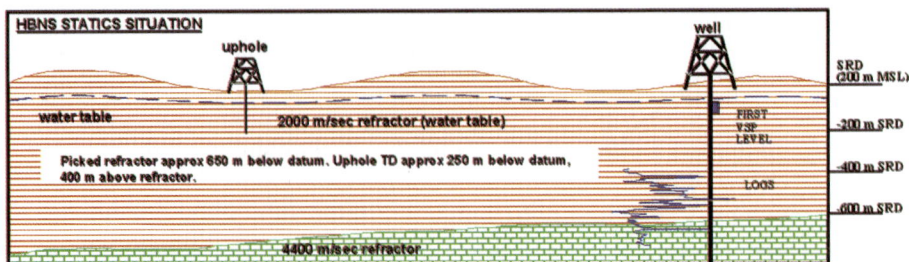

Fig. 10. Berkine Basin statics situation (cf. Fig. 9).

logical column which includes a 'weathering layer' of unconsolidated and partially consolidated sand. This layer varies in thickness between approximately 60 m and 160 m from ground level to the water table in Block 404 and up to 400 m in Block 208. The uphole surveys extend to some 300 m below SRD, but do not penetrate the most useful refractor. Wells are typically logged from approximately 300 m below SRD. However, uphole data below 250 m is unreliable and a proportion of the wells could only be logged below 400 m. So there is a 'grey zone' from approximately 250 m to 400 m below SRD for which vertical seismic profiling (VSP) surveys, sampled every 25 m in depth, provide the only source of regional time/depth information. In this case, a single-layer model was inadequate to describe the kinematics of the picked refractor. As previously mentioned, the zero-offset intercept time was a poor approximation to the true zero-offset time.

The technique used for the first 3-D survey in the block was to construct a depth model of the 4400 m s^{-1} refractor by fitting a spline surface to the logged depths of the top Senonian Salt. This model was then used to compute the long-period statics errors which might be present in the 3-D dataset. These modelled statics errors were largely verified by performing a mistie analysis between the 3-D seismic and borehole data on the Aptian. The modelling of the potential statics errors showed that the general effect over the HBNS field would be to underestimate the seismic time dips in a NE–SW direction.

The method of statics calibration used on the 3-D survey was feasible due to the density of well control and the fact that logs and check shots were run in most of the wells to relatively shallow depths.

The first 3-D survey: discovery of the acquisition footprint

After analysis of 2-D surface seismic and borehole data, an 80-fold cross-spread template design was

chosen for the inaugural 800 km^2 3-D survey acquired in 1996 over the HBNS field. At the time, it was felt that this template gave the best balance between offset distribution, fold of stack and required field logistics. Planning of this 3-D was focused on traditional thinking and logistical management of the 3-D survey. The severity of the terrain and previous experience of the bulldozer effort required for line clearance on 2-D surveys in the area was a major influence on the choice of field technique. Some simulations of the expected 3-D data quality were performed with 2-D data using a simulation technique that was later developed to more accurately model the 3-D system response.

Early fast-track 3-D data from the survey showed that, although the data quality was high (better than the 2-D data acquired to date), there was significant amplitude striping at all levels.

Many 3-D datasets contain variations in amplitude and phase related to the source and receiver geometry. These effects are often referred to under the catch-all term of 'acquisition footprint', although, in fact, several types of footprint may be distinguished. One type could be loosely classified as arising from 'signal-processing problems', such as residual normal moveout (NMO) caused through incorrect velocities. A typical example of the NMO footprint is shown in Figure 11. Systematic errors in computed offsets or amplitude variations caused by inadequate 3-D dip moveout (DMO) formulation (Walker *et al.* 1995; Budd *et al.* 1995) also give rise to periodic disturbances. Figure 12 shows DMO-generated amplitude fluctuations viewed on a root mean square (RMS) amplitude time slice. Due to their sensitivity, RMS amplitude slices are a good diagnostic for 3-D seismic data volumes.

The footprint illustrated in Figure 13, a time slice from the 80-fold cross-spread 3-D survey, is caused by coherent noise leaking through the stack. The footprint appears as an obvious amplitude modulation in the in-line direction, with a period approximately equal to the nominal source line spacing, as seen in the in-line display of Figure 14. A more subtle modulation in the cross-line direc-

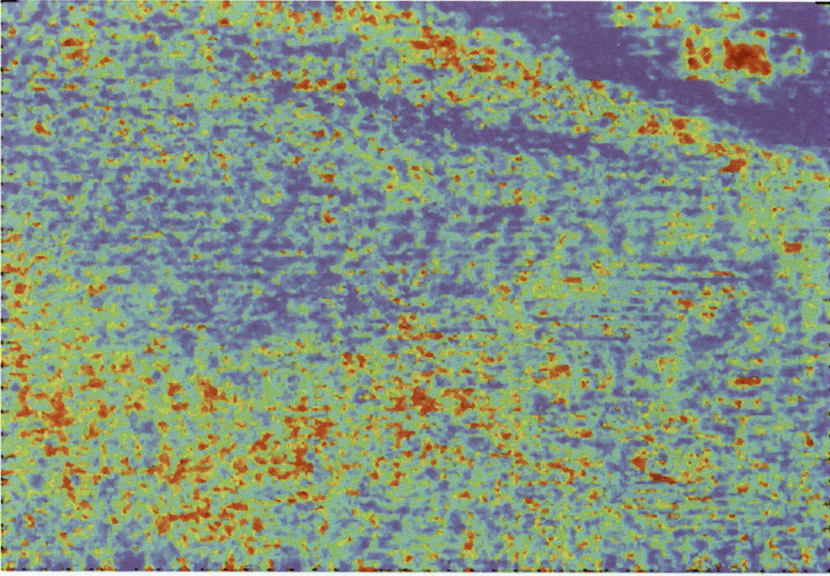

Fig. 11. Time slice from a 3-D cross-spread survey showing NMO footprint effects created through inappropriate velocities.

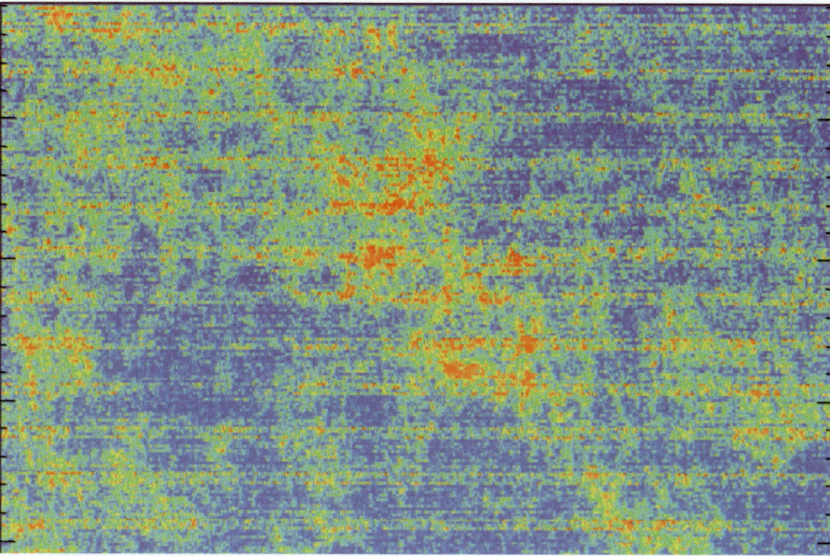

Fig. 12. RMS amplitude slice showing 3-D DMO footprint as a result of inadequate formulation of 3-D DMO operator.

tion, with a period equivalent to the nominal receiver line spacing, is also evident. This noise contaminated automatic reflection time and amplitude picking, and so affected structural mapping as well as reservoir attribute studies (Marfurt *et al.* 1995). A way to understand and then attenuate this noise footprint at several depths in the 3-D dataset was therefore required.

Devising a modelling technique to understand the noise footprint

To understand the effects of the noise footprint and produce both field and processing solutions, a novel technique was developed for prognosis of final processed 3-D data quality. A volume of 3-D data was simulated, starting from elastically mod-

|↔| Source line spacing 600 m

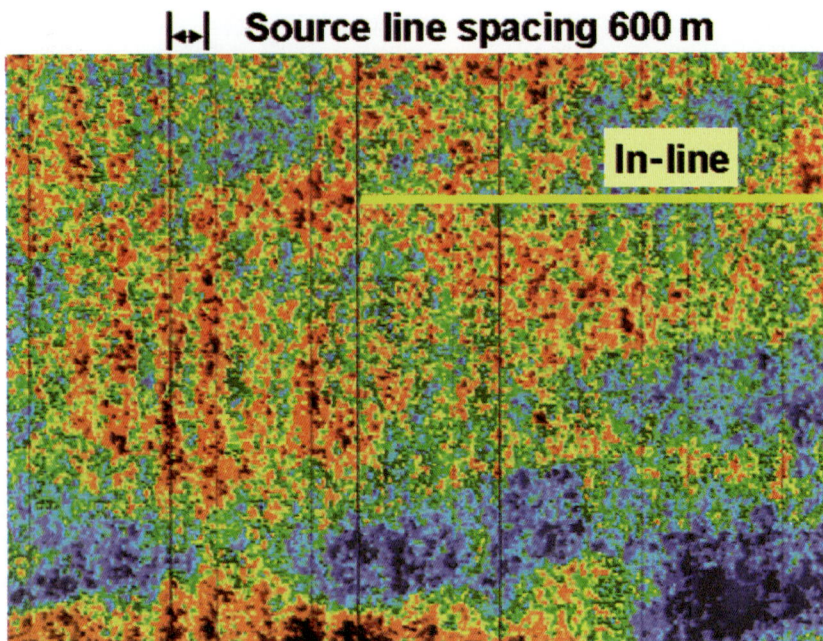

Fig. 13. Time slice from an initial stack volume of 80-fold cross-spread 3-D survey showing the effect.

Source lines

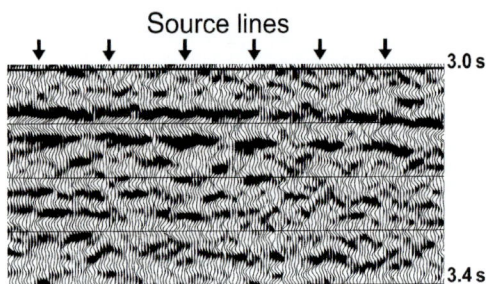

Fig. 14. In-line DMO stack from a cross-spread 3-D survey.

elled shot records, although previous 2-D pre-stack data could also have been used. This volume accurately predicted the relative performance of different 3-D acquisition schemes and demonstrated the severity of the acquisition footprint.

The elastically modelled shot records were generated using the modified wavenumber summation method of Kerner (1990). The input to the elastic modelling process was a 1-D log of P- and S-wave velocities and density. To predict the 'footprint' a realistic estimate of the shot-generated noise field was required. To achieve this, well log parameters were augmented with surface information derived from uphole data. The modelling was performed at sufficiently small trace spacing to permit digital

source and receiver array simulation in processing. The volume was also used to test various noise attenuation algorithms in processing prior to production processing of the first 3-D survey.

To simulate the required stack response, the modelled record was input to a trace-selection process driven by the theoretical survey geometry. The source-receiver offsets derived from the survey geometry were used to select traces from the modelled record in order to build common mid-point (CMP) gathers. An example of an in-line taken from one of the modelled test volumes is shown in Figure 15. This example also includes effects due to surface topography and geophone/source layout, factors that may also be included in the data generation. Travel time and amplitude perturbations seen

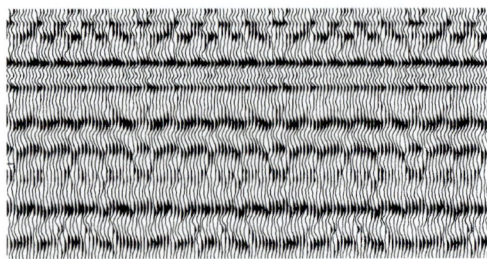

Fig. 15. In-line 1 of 24 from an elastically modelled 3-D cross-spread geometry.

very clearly in this figure proved to be a more accurate prognosis of one aspect of the 3-D volume than was realized at the time. It was felt that the severity of the perturbations was considerably greater in the modelled data than would be seen in an acquired 3-D volume. In reality this did not prove to be the case, demonstrating that simulations can be an accurate representation of the noise expected in a final processed 3-D data volume.

Defining the noise

The data obtained by stacking CMP gathers contained residual coherent noise whose arrival time was determined by the noise velocities and the pattern of offsets in the gather. Every bin within the basic repeating CMP pattern of the survey had a different combination of offsets, and therefore a different pattern of residual coherent noise in the stack trace. The net result was a complex residual noise field that repeats with the same spatial periodicity as the CMP offset distribution. Multiples would also show periodic variations in amplitude and stack time, as would primary events that have not been properly corrected for NMO. Sampling theory indicates that the footprint contains a significant aliased noise component.

The wavenumber response of the 3-D geometry can be decomposed into in-line and cross-line components and treated using 2-D sampling theory in Kx and Ky (Hampson 1994, 1998). This is an approximation, but it does suggest that the spatial bandwidth for cross-spread 3-D data depends on the source and receiver line spacings, which are analogous to shot-point intervals in x and y for the 2-D decomposition. The main characteristics of the noise field as relevant to post-stack noise attenuation are that the noise field contains components on K = 0, due to the wrapping of aliased noise across K = 0 and also due to direct current (DC) bias contained within the noise field of individual time slices. The noise field contains complex spectral components, e.g. 'harmonics' resulting from complex amplitude modulation and spatial frequency variations caused by deviations in surface geometry. The data contain signals dispersed over a wide area of wavenumber space due to faulting and, more importantly, the spectral components of the noise vary from one time slice to another.

Testing traditional approaches to removing the footprint

In-line f-k notch filtering is routinely applied to marine 3-D data after stack to remove the sampling fringes resulting from non-stack array geometries (Hampson 1994). This is an acceptable substitute for the much more expensive pre-stack process of wavefield reconstruction (Jacubowicz 1994). In the case of marine 3-D acquisition, the data are typically recorded, and f-k filtered, at 12.5 m group interval, adjacent trace summed and then stacked on a 4 CMP per shot-point geometry (25 m group interval, 50 m shot-point interval). In this case, the sampling fringes occur at the Ky Nyquist and half Nyquist wavenumbers, and are well segregated from the signal.

An analogous approach to post-stack wavenumber filtering has been advocated for land 3-D processing (Gulunay et al. 1994). There is a better chance of attenuating flat-lying reflection signal by notch filtering in this case because the spatial periods involved are significantly larger than in marine 3-D, and the spectral peaks of the noise therefore lie much closer to signal in the region of K = 0.

In principle, f-k filtering in either two or three dimensions would permit frequency-dependent noise attenuation, for example, to selectively attenuate lower temporal frequency noise. However, the output from low-frequency band-limited f-k_y reject tests generally contained residual higher and lower (swath width) frequency footprint components. It would therefore be necessary to apply f-k filtering over all frequencies, in which case the process is very similar to t-k_x-k_y filtering. Filtering may be performed in other domains. Marfurt et al. (1995) used a discrete radon transform in τ-p-q space, primarily, it appears, for economy of implementation on workstations.

Notch-filtering tests were performed in t-k_x-k_y by filtering individual time slices using a production 'k_x-k_y' filter program, by f-k_x-k_y one-pass filtering, as described by Gulunay et al. (1994), and by f-k_y notch filtering in the in-line direction, which contained the stronger footprint due to the source line spacing.

Some attenuation of the footprint was obtained in all three domains, usually in conjunction with a slight loss of low-frequency signal content. The t-k_x-k_y and f-k_y trials showed more improvement in data quality in the domain in which the filter was applied. Therefore t-k_x-k_y, or some such time-slice domain, was preferred for noise attenuation because of the extensive use made of time slices and areal horizon attributes in 3-D interpretation.

Modelling studies predicted that the statistics of the footprint noise should change significantly over time and this was confirmed by inspection of the noise field from the 3-D data volume. One problem with applying t-k_x-k_y filtering was that, ideally, one would want to change the filter definition over time to track these changes. It would not be practical to pick a large number of filters manually, so a technique needed to be devised to adapt the filter coefficients automatically. This line of thought led to

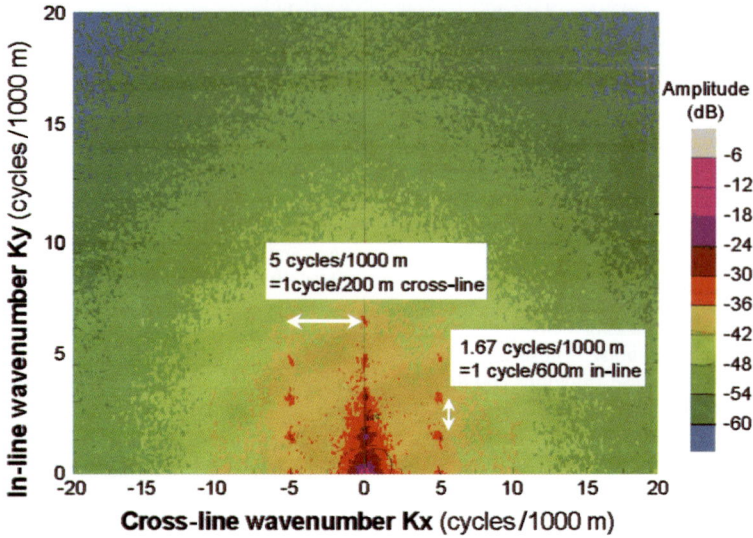

Fig. 16. The kx-ky transform of a 3-D time slice showing periodic components of the 3-D noise field from a cross-spread survey.

the development of an adaptive filtering process as there was no available application in the industry.

Footprint filtering by adaptive noise estimation

The problem of the acquisition footprint is well known in the image-processing world because the images obtained from optical scanners usually contain pseudoperiodic noise. Wavenumber notch filtering is used for image processing in cases where the noise is highly periodic and well segregated from the signal. When this is not the case, the inverse 2-D transform of the easily identifiable periodic components is used as an initial noise model in an adaptive filtering technique. In image processing terms, the forward problem is described as:

$$g(x,y) = f(x,y) + n(x,y) \qquad (1)$$

where $g(x,y)$ is the given image, consisting of the desired image $f(x,y)$ contaminated by additive noise $n(x,y)$ (Gonzales & Woods 1993).

The inverse process is formulated as:

$$F(x,y) = g(x,y) - w(x,y).p(x,y) \qquad (2)$$

where $F(x,y)$ is an estimate of the uncontaminated image obtained by subtracting the periodic component $p(x,y)$ multiplied by a weighting function $w(x,y)$ from the input image $g(x,y)$. The periodic component $p(x,y)$ is obtained by taking the inverse 2-D transform of the easily identifiable noise components in the t-k_x-k_y plane, as shown in Figure 16 for example. The simplest way to calculate the

weighting function is to minimize the variance of $F(x,y)$ over a specified region around x,y (i.e. to assume that the image is locally 'flat'). An additional requirement for processing a 3-D data volume, which is not necessary for processing single images, is that the weighting function $w(x,y)$ should vary smoothly over the temporal axis. The flow diagram of a simple adaptive footprint filtering of this type is shown in Figure 17.

A prototype program based on the Gonzales and Woods (1993) algorithm was tested to see whether it provided any advantages over t-k_x-k_y notch filtering. Figure 18 is a time slice from the first 3-D survey, which illustrates the noise footprint.

A k_x-k_y transform of a stack time slice, which clearly illustrates the wavenumber components in the footprint, is shown in Figure 16. This transform contains peaks at the spatial frequencies corresponding to the source and receiver line spacings:

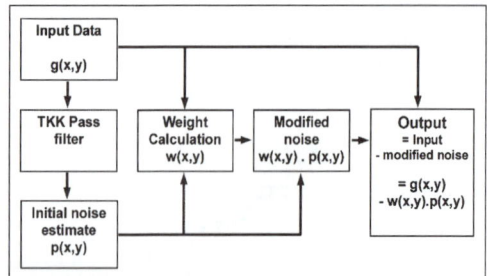

Fig. 17. Flowchart for a 3-D adaptive footprint filtering process developed to attenuate acquisition noise.

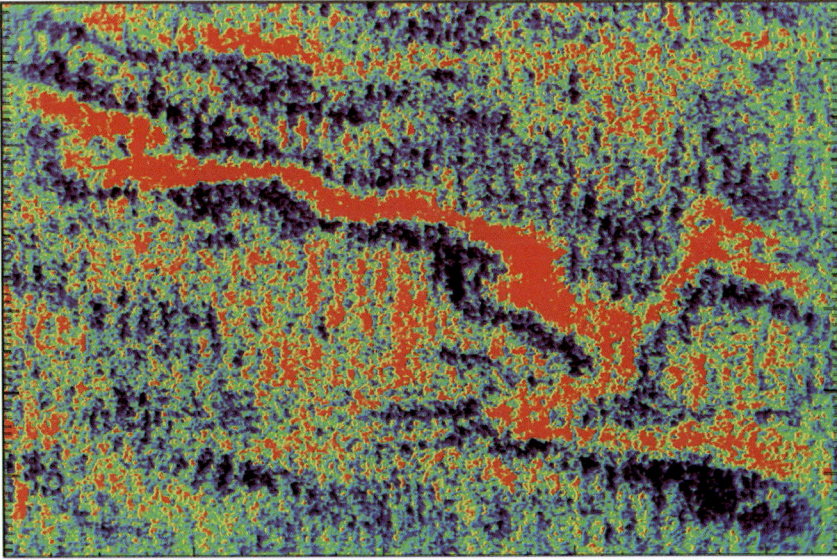

Fig. 18. Raw DMO stack time slice at 3.2 s from a 3-D cross-spread survey.

in the cross-line direction of one cycle per 200 m receiver line spacing (5 cycles/1000 m) and, in the in-line direction, one cycle per 600 m source line spacing (1.67 cycles/1000 m).

The data after t-k_x-k_y notch filtering and adaptive filtering can be seen in Figures 19 and 20 respectively. Both techniques have attenuated, but not completely removed, the source line footprint running from left to right. The difference in data quality between the two filter techniques is clearly seen, even on these low-noise time slices. In this case, the t-k_x-k_y filter was designed at the two-way time of the slice and was therefore optimum. At other times within the volume, the adaptive filter showed

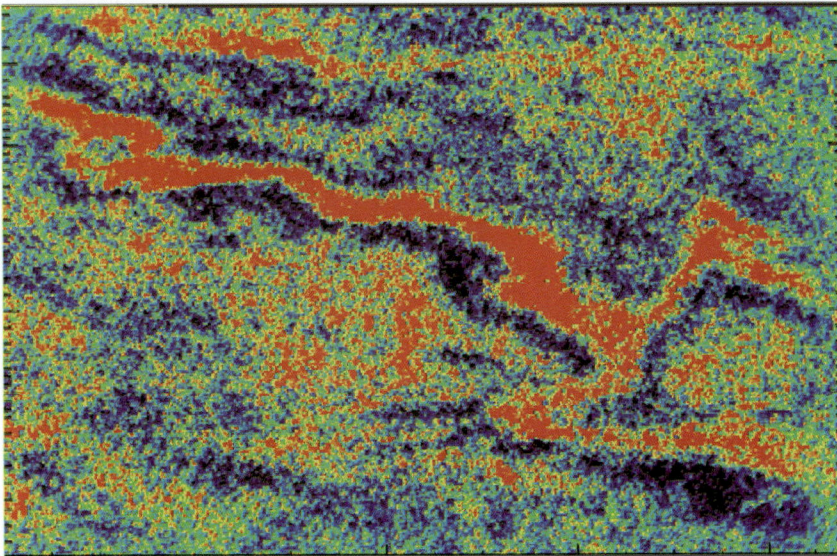

Fig. 19. Time slice at 3.2 s after t-k_x-k_y notch filtering optimized for this slice.

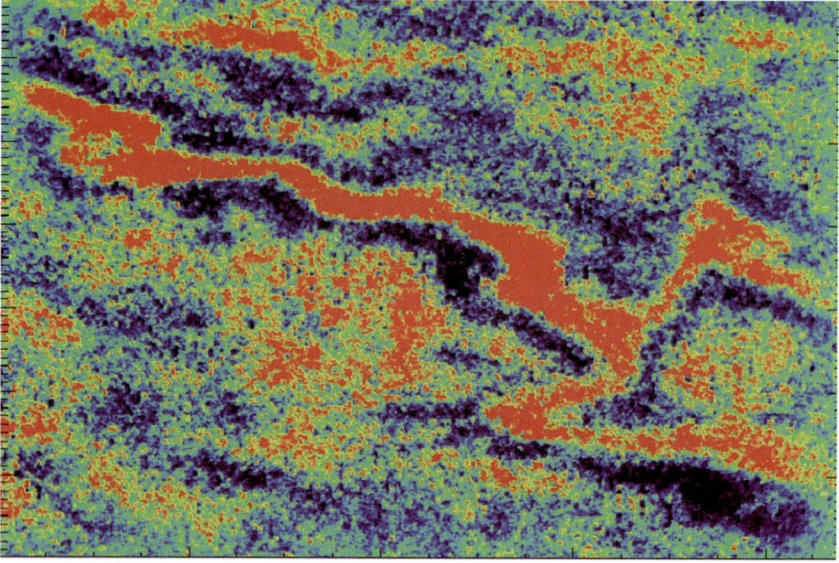

Fig. 20. Time slice at 3.2 s after adaptive filtering.

even more marked superiority as the statistics of the noise changed.

The noise fields rejected by the two processes were radically different, as shown in Figures 21 and 22. The noise rejected by t-k_x-k_y filtering, as shown in Figure 21, was predominantly low-angled, highly coherent noise, corresponding to the limited number of low wavenumber components that were rejected and were predominantly associated with the source line spacing. The noise field rejected by the adaptive process and shown in Figure 22 was less organized and contained higher spatial frequency components. Both noise fields show the complexity generated in the wavenumber

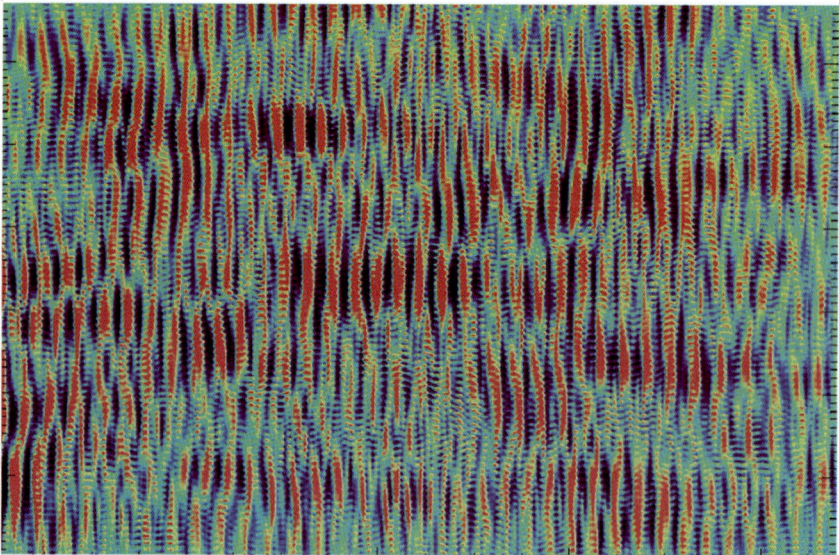

Fig. 21. Noise estimate at 3.2 s of t-k_x-k_y notch filter process.

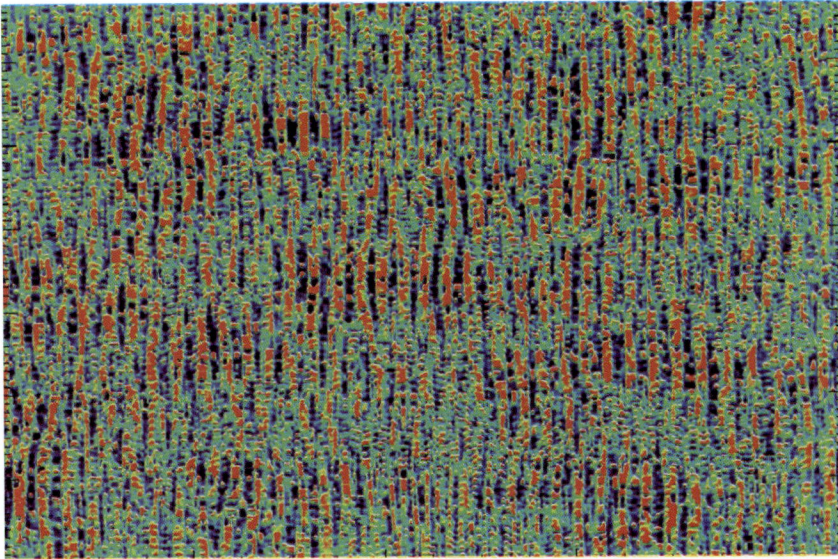

Fig. 22. Noise estimate at 3.2 s from adaptive-filtering application showing estimate of source and higher-frequency receiver components (cf. Fig. 21).

domain due to deviations in the source lines caused by the geomorphology of the sand conditions.

Development of field techniques to improve data quality

In order to enhance the acquired data for future surveys, two aspects of 3-D survey design were addressed following the acquisition of the Vibroseis 80-fold cross-spread 3-D dataset. The first was to investigate ways of regularizing the offset sampling in the CMP domain, and the second to see if there were any advantages to be gained using explosive sources.

The first cross-spread survey, acquired over HBNS, had more acquisition footprint contamination (Figs 13 & 18) than was desirable, despite the use of a high-fold cross-spread design with reasonable offset sampling. Subsequent modelling showed that the stack response would be improved and the noise-related footprint better attenuated by providing a better-sampled offset distribution in the CMP domain. The logical conclusion of this approach was to use a parallel acquisition geometry designed on the stack-array principles described by Anstey (1986) and Morse and Hildebrandt (1989). Although compromises due to logistics and cost may not allow the stack-array effect to be fully utilized in general terms, parallel geometries that do not fulfill the stack-array requirements still allow more effective application of pre- and post-stack noise attenuation techniques than cross-spread

data. Using the modelling techniques described above, parallel survey geometries were devised that reduced the noise contamination of the next generation of 3-D surveys in the Berkine Basin. The proposal to acquire parallel surveys met with initial resistance from traditional thinking, which was more predisposed to cross-spread recording. The first parallel template data were recorded in 1997 and fulfilled the predictions of reduced acquisition footprint and improved data quality as shown in Figures 23 and 24. Parallel template acquisition is now the accepted technique in the Berkine Basin for 3-D acquisition.

Field testing carried out as part of the strategic project showed that dynamite charges located at or near the base of the unconsolidated sand layer, approximately 5–6 m, would excite different noise and reverberant systems than a surface Vibroseis source.

Of the source types evaluated by 2-D testing, lines acquired with dynamite buried about 3 m below the surface had the best signal-to-noise ratio and resolution compared to both the surface dynamite and Vibroseis data. Events could be identified in the shot gathers throughout the offset range, in contrast to the Vibroseis data, where noise contamination made identification of target reflections in shot records almost impossible.

At 60-fold, the buried dynamite data showed significantly higher resolution than the Vibroseis stack at 240-fold, especially at reservoir levels. The potential of acquiring higher-quality 3-D data with

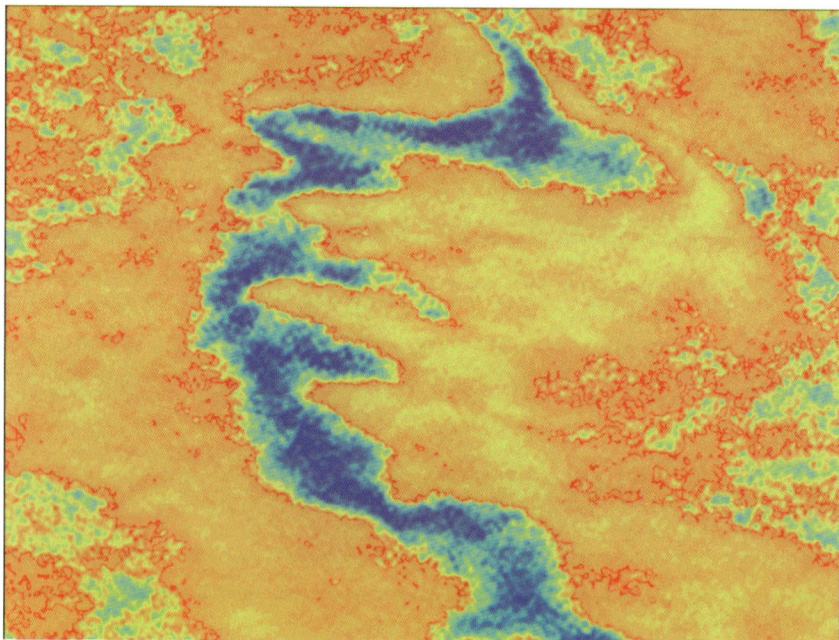

Fig. 23. A time slice from around 3 s from a 3-D survey acquired with a parallel template (cf. Fig. 18).

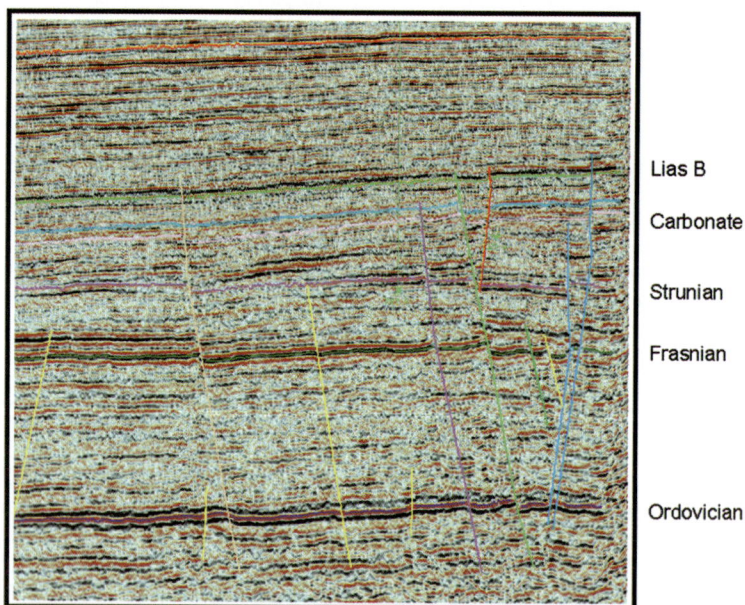

Fig. 24. 3-D parallel template migration.

much lower fold, albeit using explosives, was very appealing.

Parallel template 3-D volumes were modelled using the 2-D buried dynamite shots as the seed

data and suggested that the fold could be reduced to as low as 40 with acceptable data quality. The Berkine Groupement agreed to use dynamite for the next EME 3-D survey in an area with the highest dunes (*c*.400 m) encountered so far in the exploration area. The acquisition of this programme has recently been completed and the data is currently being processed and is showing encouraging results.

Conclusions

It is possible, through a combination of research and development, and innovation, to develop new technologies that address problems specific to an area. Software to examine and solve area-specific problems can be written in time to make a significant difference. Most of the 3-D surveys acquired in the Berkine Basin are now designed with parallel templates rather than cross-spreads. They have been shown to provide better spatially sampled data, despite the interference of the huge sand dunes, which, in turn, respond more positively to standard pre-stack noise attenuation processing. The 3-D volumes still contain some acquisition footprint and often include the DMO footprint, but both have been attenuated using the adaptive filter approach. Of a total of 3700 km² of 3-D, 610 km² were acquired with a double off-end cross-spread template and the remaining 3090 km² with multi-source parallel templates. Vibroseis accounted for 3200 km² whereas a dynamite energy source was used for the last survey of 500 km². This is the first dynamite source 3-D acquired in Algeria.

The adaptive footprint filter has to date been applied successfully to almost all the 3-D surveys in the Berkine Basin, and beyond. It can now be implemented easily during the standard processing so does not interfere with the project deadlines. Significantly, it has allowed other post-stack processes to be successfully applied, such as spectral shaping, as the signal-to-noise ratio has been improved. Attribute and coherency analysis, as well as stratigraphic interpretation, is now possible on these datasets with a much greater degree of confidence.

The authors wish to thank Anadarko Algeria Corporation and partners Sonatrach, LASMO Oil Algeria and Maersk Olie Algeriet for permission to publish this paper.

References

ANSTEY, N. A. 1986. Whatever happened to ground roll? *Leading Edge*, **5**, 40.

BUDD, A. J. L., HAWKINS, K., MACKEWN, A. R. & RYAN, J. W. 1995. Marine geometry design for optimum 3-D seismic imaging. *Proceedings of the 57th European Association of Geophysicists and Engineers Conference, Glasgow.*

GONZALES, R. C. & WOODS, R. E. 1993. *Digital Image Processing.* Addison-Wesley Inc., Massachusetts.

GULUNAY, N., MARTIN, E. & MARTINEZ, R. 1994. 3-D Data acquisition artifacts removal by spot editing in the spatial-temporal frequency domain. *Proceedings of the 56th European Association of Geophysicists and Engineers Conference, Vienna.*

HAMPSON, G. 1994. Relationships between wavefield sampling and coherent noise attenuation. *Proceedings of the 56th European Association of Geophysicists and Engineers Conference, Vienna.*

HAMPSON, G. 1998. 3-D wavefield sampling in the CMP method. *Proceedings of the 60th European Association of Geophysicists and Engineers Conference, Leipzig.*

JACUBOWICZ, H. 1994. Wavefield reconstruction. *Proceedings of the 56th European Association of Geophysicists and Engineers Conference, Vienna.*

KERNER, C. 1990. Modelling of soft sediments and liquid-solid interfaces: Modified wavenumber summation method and application. *Geophysical Prospecting*, **38**, 111.

MARFURT, K. J., SCHEET, R. M., SHARP, J. A., CAIN, G.J. & HARPER, M. G. 1995. Suppression of the acquisition footprint for seismic sequence attribute mapping. *Proceedings of the 65th Annual International Meeting of the SEG*, 949–952.

MORSE, P. F. & HILDEBRANDT, G. F. 1989. Ground-roll suppression by the stack-array. *Geophysics*, **54**, 290.

WALKER, C. D. T., MACKEWN, A. R., BUDD, A. J. L. & RYAN, J. W. 1995. Marine 3-D geometry design for optimum acquisition system response. *Proceedings of the 57th European Association of Geophysicists and Engineers Conference, Glasgow.*

Application of fluid inclusion studies to understanding oil charge, Pre-Salt succession, offshore Angola

J. PARNELL & H. CHEN

Department of Geology and Petroleum Geology, University of Aberdeen, King's College, Aberdeen, AB24 3UE, UK

Abstract: Fluid inclusion studies in the Pre-Salt succession of the Kwanza Basin show that valuable information can be obtained regarding fluid migration, including oil charge. A pilot study demonstrates that oil inclusions are widespread in both cements and healed microfractures, and that measurements can be made of their entrapment temperature, fluorescence characteristics, API gravity determination and bulk organic geochemistry. Successive populations of inclusions show changes of temperature with time, which give an insight into thermal history and emphasize the importance of hot fluids in the region. Multiple populations of oil inclusions offer the potential of reconstructing the evolution of oil chemistry and, in particular, the origin of mixed oils.

Detailed studies of fluid inclusions in mineral cements and healed microfractures allow integrated reconstruction of the fluid migration and thermal histories of sedimentary basins, including oil migration.

Fluid-inclusion analyses are widely undertaken in studies of oilfields in order to help relate the thermal history of the sequence to the cementation history and to place oil migration within the diagenetic sequence. Most such studies are undertaken in 'mature' regions where there is a very detailed existing context in which to interpret the inclusion data, such as in the North Sea (e.g. Burley *et al.* 1989; Walderhaug 1994; Munz *et al.* 1999) and the Arabian Gulf (e.g. Horsfield & McLimans 1984; Neilson *et al.* 1998). These studies tend to address very specific issues, such as reservoir compartmentalization and palaeopressure determination. However, there is a role for fluid-inclusion studies in less developed regions, even 'frontier' regions, particularly through recording of hydrocarbon inclusions. The simple observation of oil inclusions in a core or cuttings sample could be the first real evidence for a hydrocarbon system. Microthermometric data can provide fundamental information on the burial history of a region and support interpretations from other approaches, such as vitrinite-reflectance measurement and fission-track analysis.

As part of a long-term project on fluid migration and thermal histories of all Atlantic margins, pilot studies have been undertaken in West Africa. The data reported here is from the Kwanza Basin, offshore Angola, but the approaches are equally applicable elsewhere on the West Africa margin. This study is not intended to be comprehensive, but to illustrate the potential for fluid-inclusion analyses in this region, using a limited number of core and cuttings samples from old wells (Fig. 1).

We show examples of five approaches to obtaining and using data from fluid inclusions in the basin, all of which are relevant to an understanding of the oil charge.

Fluid-inclusion analysis

Fluid inclusions are micron-scale volumes of ambient fluid entrapped during mineral growth and rehealing of microfractures. Hence, analysis of the inclusions can provide direct evidence of the fluid composition and physicochemical conditions pertaining to cementation and fluid migration, including oil charge. They are identified and studied in doubly polished wafers (thick sections, typically 100–130 µm thick) of well cuttings or core. Inclusions of aqueous fluid are subjected to heating and cooling in order to determine temperatures of homogenization into a single fluid (minimum temperature of entrapment) and temperature of ice-melting (which reflects the fluid salinity). The theory of fluid-inclusion analysis is described in detail by Roedder (1984) and Shepherd *et al.* (1985). Inclusions of oil are recognized by fluorescence in ultraviolet light (McLimans 1987; Guil-

From: ARTHUR, T. J., MACGREGOR, D. S. & CAMERON, N. R. (eds) *Petroleum Geology of Africa: New Themes and Developing Technologies.* Geological Society, London, Special Publications, **207**, 275–283. 0305-8719/03/$15

haumou *et al.* 1990), and sometimes by yellow to brown colour in normal light. Their thermal behaviour is also recorded, although generally they do not 'freeze'.

Many samples contain multiple populations of inclusions, which may yield different homogenization temperatures and salinities or involve different fluid types (aqueous, oil, gas). Study of their paragenesis allows reconstruction of an evolving fluid-migration history, linked to an evolving thermal history.

Kwanza Basin succession

The Kwanza Basin succession is divisible into Pre-Salt and Post-Salt sequences, formerly separated by an Aptian salt layer. The Pre-Salt sequence includes Neocomian fluviatile and lacustrine sandstones passing upward into lacustrine shales and carbonates in fault-bound sub-basins: a Barremian fluviatile–lacustrine clastic wedge and thin early Aptian sands. Marine transgression resulted in Aptian salt deposition. The Post-Salt sequence consists of Albian–Cenomanian predeformation sediments, Turonian–Maastrichtian synextension sediments, then a Tertiary succession deposited during gravity gliding on a décollement of salt. Raft movement above the salt started no later than the Eocene, but was enhanced by cratonic uplift that commenced in the Late Oligocene. Tertiary sedimentation was focused in grabens between the separating rafts. The stratigraphy and its relationships with deformation are described by Brice *et al.* (1982), McHargue (1990), Duval *et al.* (1992) and Lundin (1992).

Samples

The samples used are cuttings or core from the Barremian–Early Aptian (Pre-Salt) succession in several wells in the Kwanza Basin (Fig. 1). Most are from the Falcão Formation (Bate *et al.* 2001) and equivalents (Second Lake Cycle of Bate 1999), and also from the overlying Aguia Formation. Doubly polished wafers were made from cemented sandstones and from limited skeletal carbonates that occur in the succession (Uncini *et al.* 1998).

Microthermometric measurements on wafers were made using a Linkam THM600 heating-freezing stage attached to a Nikon Optiphot2-POL microscope. Detailed methodology is described by Shepherd *et al.* (1985). Fluorescence under ultraviolet light was studied using a Nikon Eclipse 600 microscope with a UV-2A filter block (excitation 330–380 nm, barrier filter 420 nm, dichroic mirror 400 nm). Organic chemicals of known melting point were used as standards. Quantitative measurements of API gravity in oil inclusions were

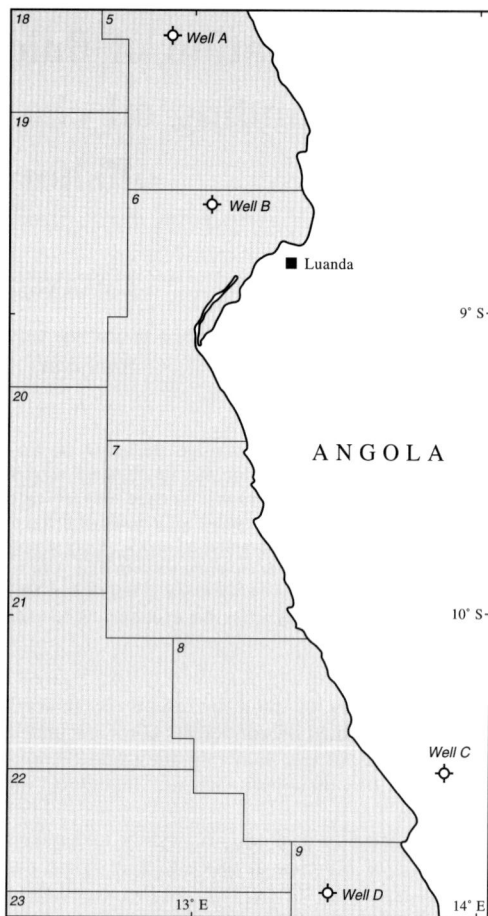

Fig. 1. Approximate location of wells sampled in fluid inclusion study, Kwanza Basin. Licence blocks also shown.

made by Fluid Inclusion Technology Inc., using a patented in-house technique based upon inclusion behaviour during heating and cooling. Gas chromatograms were prepared by Torkelson Geochemistry, using a proprietary technique in which inclusion oils are liberated by mechanical crushing after careful removal of hydrocarbons from mineral surfaces. The oils are analysed by a combined solvent and thermal extraction procedure.

Application of fluid-inclusion studies

Inclusion oils in the Pre-Salt succession

Our pilot studies have shown that offshore well samples of Pre-Salt sandstones and carbonates are suitable for study of oil migration. Oil inclusions

(a)
(b)

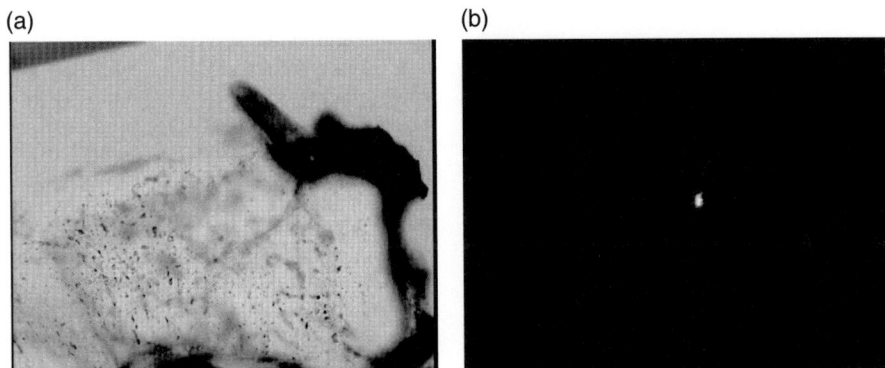

Fig. 2. (a) Inclusion trails in quartz grain, well B, depth about 3300 m. (b) Close-up of A, showing micron-scale oil inclusion fluorescing in ultraviolet light.

occur both in cements and rehealed microfractures (Figs 2 & 3). Oil inclusions have been identified previously in cements in Post-Salt reservoirs in the offshore Angola region (Walgenwitz *et al.* 1990; Eichenseer *et al.* 1999), but these are the first published records in the Pre-Salt succession. They are amenable to the study of entrapment temperature, fluorescence characteristics, API gravity determination and, in some cases, to bulk organic geochemistry.

Thermal histories from inclusion populations

In many samples, more than one population of fluid inclusions can be identified. From petrographic relationships, most can be placed in a relative time sequence. The entrapment temperature data from each population can thus contribute to a trend of changing temperature through time. Typical patterns are: (1) progressively increasing temperature during burial; (2) increasing temperature interrupted by cooling events, which reflect uplift; and (3)

'normal' burial temperatures with superimposed events of hot fluid migration. Fluid-inclusion temperature data, in combination with apatite fission-track data, can provide a powerful insight into regional thermal history (e.g. O'Brien *et al.* 1996; Parnell *et al.* 1999), and the recent availability of fission-track data for offshore Angola (Harris *et al.* 2002) makes this integration an exciting future possibility. Where temperatures recorded are higher than those predicted from the burial history, the passage of anomalously hot fluids is implied (e.g. Middleton *et al.* 2001; Parnell *et al.* 2001b). Examples from Angola are as follows.

Well C, sample about 1900 m Progressively higher temperatures are shown by successive quartz, feldspar and calcite cements, reflecting increasing temperature during burial, but cross-cutting trails through the calcite show a later cooling event (Fig. 4). Cooling probably reflects uplift, which is known to have been experienced by the Kwanza and other basins (Norvick & Schaller 1998).

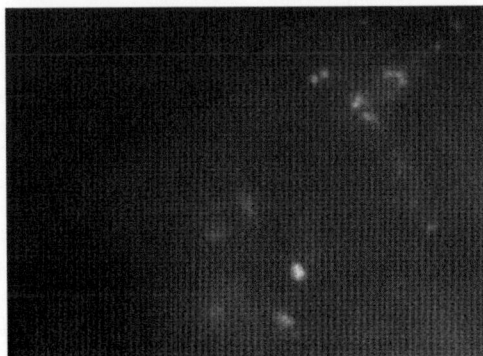

Fig. 3. Oil inclusions (micron-scale) fluorescing bluish white in ultraviolet light, calcite cement, well A, depth about 2250 m. Bluish white fluorescence reflects a relatively mature oil composition.

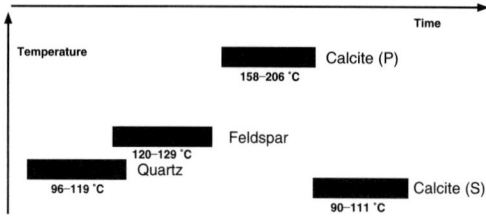

Fig. 4. Paragenesis of fluid inclusion populations (P, primary; S, secondary) from cements in sample from well C, showing increasing temperature with burial, then later cooler temperature (probable uplift) in secondary trails.

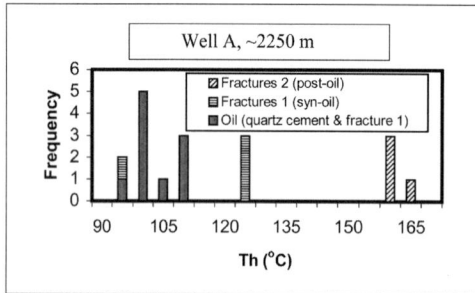

Fig. 5. Histogram of homogenization temperature data in quartz for distinct inclusion populations, well A, depth about 2250 m. Oil was present during first stage of fracturing, then experienced later high-temperature event.

Well A, sample about 2250 m Quartz grains exhibit two sets of rehealed microfractures (Fig. 5). The first set contains both oil and aqueous inclusions. The oil inclusions also occur along cleavage surfaces in calcite cement, so this oil was probably present during cementation. A second set of hotter aqueous inclusions occurs in the fractured quartz, i.e. oil emplacement was followed by a hot fracturing event, which could have allowed remigration.

Well D, sample about 3300 m A similar trend is exhibited in well D, where an oil-bearing fracture-fill stage is followed by another, hotter, fracturing event (Fig. 6). Other inclusion populations suggest

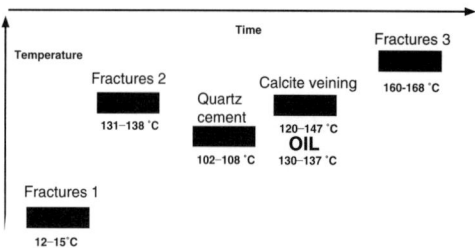

Fig. 6. Paragenesis of fluid inclusion populations, well D, depth about 3300 m, showing oil-bearing fracture-fill stage followed by hotter event.

that hot fluid pulses were superimposed on a cooler (burial) thermal history.

Evidence for hot fluids has also been obtained from mineralogical studies in the north of the Kwanza Basin by Crossley *et al.* (1993), who reported a remarkable alteration assemblage including epidote and garnet. As indicated by the inclusion data, the occurrence of bitumen post-dating this high-temperature assemblage shows that the first hydrothermal activity did not destroy the oil potential of the succession. Liquid hydrocarbons can survive temperatures up to 200°C and higher, as high pressures at depth confer stability to oils which would be broken down under surface conditions. Thus these high temperatures may be recorded from oil inclusions (e.g. Parnell *et al.* 2001*a*).

Fracture-bound oil

In several cases oil inclusions occur particularly in rehealed fractures through detrital grains, i.e. fracturing/healing occurred while oil was present, and may have been the predominant means by which oil passed through the rock. For example, in well A, oil inclusions occur in healed fractures through quartz grains.

Fracturing events can reflect a build-up of anomalous pore-fluid pressure, which is released by fracturing and fluid expulsion. In cases where oil inclusions occur in the fractures, overpressuring may have been caused/enhanced by liquid hydrocarbon generation/accumulation. If coeval aqueous and oil inclusions are present and oil compositional data can be obtained from microanalysis, it may be possible to calculate the palaeopressure (e.g. Swarbrick *et al.* 2000). This has been successfully accomplished at other Atlantic margin sites, where pressures calculated from inclusions exceed the hydrostatic pressure (i.e. they are overpressured) and appear to reflect the fracture pressure in the rock (Middleton, D., pers. comm.). The particular observation of fracture-bound oil in well A is significant, because this sample comes from a sandstone interval in a thick (nearly 1 km) shale section: Although oil may migrate through the pore network in shales (e.g. Düppenbecker *et al.* 1991), in such a thick low-permeability section fractures may be the predominant means of charging isolated sand intervals.

Characterization of oil

Oils are characterized by their temperature of entrapment and their API gravity, and in some circumstances by their fluorescence behaviour. Figure 7 shows a cross-plot of homogenization temperature and API gravity for the two examples in Fig-

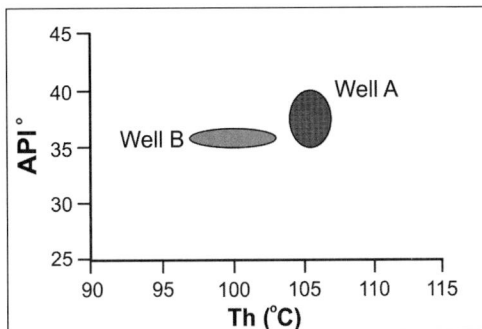

Fig. 7. Cross-plot of API gravity against homogenization temperature for two sets of oil inclusions shown in Figures 2 and 3.

ures 2 and 3. In samples with more than one population of oil inclusions, the change in oil character could be plotted in this manner. Characterization of oil inclusions at different levels in a well and between wells may help to determine trends in maturation and migration through space and time. Interpretation of fluorescence colour under ultraviolet light must be undertaken with caution but, in general, inclusions with reddish fluorescence are of low maturity or are degraded, while those with bluish white fluorescence tend to be of higher maturity (Bodnar 1990). Other factors that may influence fluorescence include the chemistry of the source rock, thermal alteration and fractionation of oil during trapping (see George *et al.* 2001), but the general relationship between colour and maturity serves as a useful comparison between different populations of oil inclusions within the same succession.

Mixing of oils and sequence of oil charge

Where sandstones are rich in oil inclusions, there is potential to obtain organic geochemical data to characterize the inclusion oil and compare it with oil in the current pore space (i.e. production oil). There may be a slight difference between the entrapped inclusion oil and the parent oil, due to differential trapping, and accompanying aqueous inclusions may contain water-soluble organic compounds which contribute to the analysis (Ruble *et al.* 1998), but major differences between the chemistry of the inclusion oil and the production oil show how the oil composition has evolved between inclusion-fluid entrapment and the present day. This can be very valuable for several reasons:

- information may be obtained about the relative

degree of oil maturity, which can be used in the progressive modelling of oil expulsion;
- where production oil shows evidence of biodegradation, study of the inclusion oil provides evidence to constrain whether the degradation was an event in the geologic past (i.e. it is recorded in inclusions) or reflects recent (e.g. near-surface) alteration (i.e. it is not recorded in inclusions);
- oils in some fields are found to be mixtures; either mixtures from two distinct sources, mixtures of two charges of different maturity from the same source, or mixtures of degraded and fresh oil. Study of the inclusion oil may constrain when mixing occurred and, where the inclusion oil is a single component, it can show which component in a mixture arrived in the reservoir first;
- in some circumstances, the production oil is completely different from the inclusion oil, which can reflect displacement of an earlier charge or refilling of an evacuated reservoir.

An example from well D (Fig. 8) shows a major difference in composition between the inclusion oil, which is relatively fresh, and the pore oil, which is degraded. The chromatogram of the pore oil shows a large 'hump' (unresolved complex mixture), typical of degradation, but also a set of superimposed well-defined high-end *n*-alkane peaks which may reflect the addition of a fresher oil component. The chromatogram of the inclusion oil lacks the 'hump', but also shows a different suite of *n*-alkanes. Further assessment of the difference should involve a larger dataset.

Mixing of oils, including oils from Pre-Salt and Post-Salt sources, occurred in the offshore West Africa basins (Burwood 1998, 1999). Thus geochemical studies of inclusion oils can help to understand when mixing occurred and which oil first passed through a migration pathway. Where Pre-Salt oils occur in Post-Salt successions (e.g. in the Lower Congo Basin, Angola block 4, Anderson *et al.* 1998; in the Kwanza Basin, Angola block 5, Marton *et al.* 2002), it may be possible to relate petrographical/geochemical evidence for the charge history to conclusions about the timing of fluid mixing from the structural history.

Potential for other analyses

In addition to the basic approaches described here, there is potential to undertake more sophisticated analyses on oil inclusions at various scales (Fig. 9). Organic geochemistry can be made more specific by coupling gas chromatography to mass spectrometry, which allows determination of maturation parameters and source typing, and

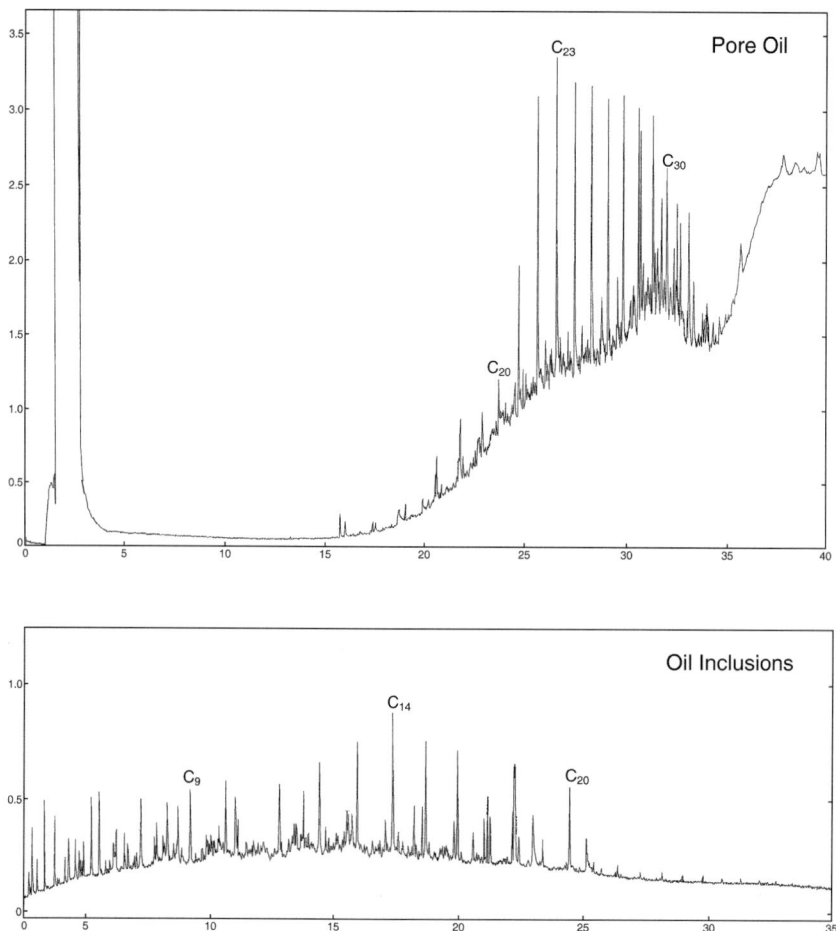

Fig. 8. Gas chromatograms of pore oil (degraded) and inclusion oil in same sample, showing different composition as evolved from entrapment in cement (selected *n*-alkanes highlighted), well D, depth about 3300 m.

correlation with source rocks and other oils (e.g. George *et al.* 1997; Isaksen *et al.* 1998). Oil composition can also be constrained by non-destructive and destructive techniques directed at individual inclusions, including Raman spectroscopy (Wopenka *et al.* 1990), infrared spectroscopy (Barres *et al.* 1987), confocal scanning microscopy (Aplin *et al.* 1999), micropyrolysis (Greenwood *et al.* 1998), and time-of-flight secondary ion mass spectrometry (Mazzini *et al.* 2002). Screening of cuttings and core from entire well profiles by bulk volatile analysis using mass spectrometry allows focusing on intervals where oil inclusions are prevalent (Barclay *et al.* 2000; Parnell *et al.* 2001*b*).

Conclusions

This demonstration study shows that fluid inclusion analysis can contribute to an understanding of several aspects of the thermal and oil charge histories in the Kwanza Basin:

- oil inclusions can be identified readily, allowing detection of palaeomigration;
- oil inclusions are sufficiently abundant to allow measurement of oil chemistry, which can be used to determine differences between reservoirs, and progressive differences through time;
- evidence for multiple populations of oil inclusions shows that a detailed history of oil

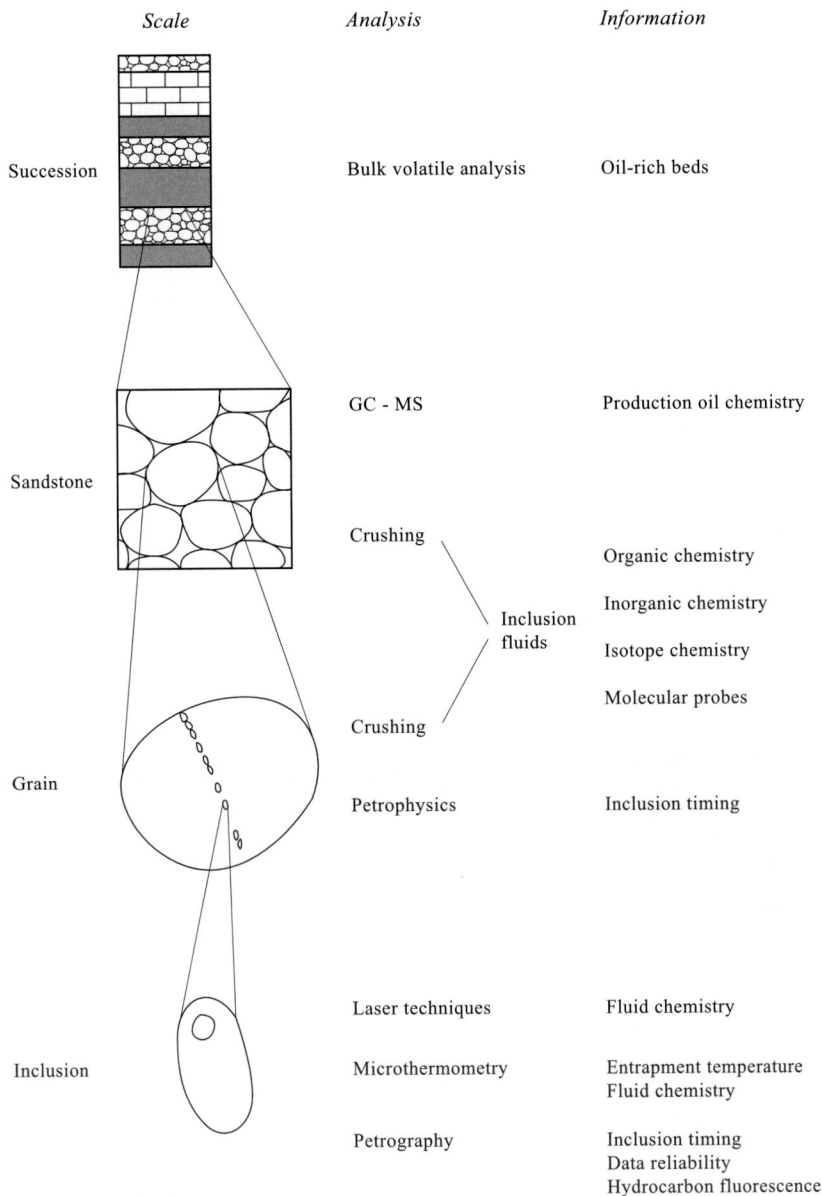

Fig. 9. Range of analyses that can be applied at various scales to obtain data from fluid inclusions. Screening of bulk volatile data (mass spectrometry after crushing to release inclusion fluids) allows selection of oil inclusion-rich sections.

migration and evolution could be developed, including the understanding of how oil mixtures are formed;

- the context of oil inclusions in the diagenetic sequence helps to constrain porosity-permeability conditions during oil charge;
- the relative timing of oil-inclusion entrapment and fracturing episodes constrains the role of such features in oil migration;

- the trends in temperature between successive populations of inclusions allows their timing to be integrated with basin burial histories.

We are grateful to R. Bate for providing essential background details to the study, B. Fulton for skilled technical assistance, Fluid Inclusion Technologies Inc. for API determination, and A. Carr and an anonymous referee for critical review.

References

ANDERSON, J., CRAIK, D., DRYSDALL, S. & VIVIAN, N. 1998. The Pre-Salt geology of Block 4, Angola and implications for the Post-Salt play. *American Association of Petroleum Geologists International Conference, November 8–11, 1998, Rio de Janeiro, Brazil, Extended Abstracts*, 352–353.

APLIN, A. C., MACLEOD, G., LARTER, S. R., PEDERSEN, K. S., SØRENSEN, H. & BOOTH, T. 1999. Combined use of confocal laser scanning microscopy and PVT simulation for estimating the composition and physical properties of petroleum in fluid inclusions. *Marine & Petroleum Geology*, **16**, 97–110.

BARCLAY, S. A., WORDEN, R. H., PARNELL, J., HALL, D. L. & STERNER, S. M. 2000. Assessment of fluid contacts and compartmentalization in sandstone reservoirs using fluid inclusions: an example from the Magnus Oil Field, North Sea. *American Association of Petroleum Geologists Bulletin*, **84**, 489–504.

BARRES, O., BURNEAU, A., DUBESSY, J. & PAGEL, M. 1987. Application of micro-FT-IR spectroscopy to individual hydrocarbon inclusion analysis. *Applied Spectroscopy*, **41**, 1000–1008.

BATE, R. H. 1999. Non-marine ostracod assemblages of the Pre-Salt rift basins of West Africa and their role in sequence stratigraphy. *In*: CAMERON, N. R., BATE, R. H. & CLURE, V. S. (eds) *The Oil and Gas Habitats of the South Atlantic*. Geological Society, London, Special Publications, **153**, 283–292.

BATE, R. H., CAMERON, N. R. & BRANDÃO, M. 2001. The Lower Cretaceous (Pre-Salt) lithostratigraphy of the Kwanza Basin, Angola. *Newsletters on Stratigraphy*, **38**, 117–127.

BODNAR, R. J. 1990. Petroleum migration in the Miocene Monterey Formation, California, USA: constraints from fluid-inclusion studies. *Mineralogical Magazine*, **54**, 295–304.

BRICE, S. E., COCHRAN, M. D., PARDO, G. & EDWARDS, A. D. 1982. Tectonics and sedimentation of the South Atlantic rift sequence: Cabinda, Angola. *In*: WATKINS, J. S. & DRAKE, C. L. (eds) *Studies in Continental Margin Geology*. American Association of Petroleum Geologists Memoirs, **34**, 5–18.

BURLEY, S. D., MULLIS, J. & MATTER, A. 1989. Timing diagenesis in the Tartan Reservoir (UK North Sea): constraints from combined cathodoluminescence microscopy and fluid inclusion studies. *Marine & Petroleum Geology*, **6**, 98–120.

BURWOOD, R. 1998. Angolan Atlantic-Margin petroleum systems: Pre- and Post-Salt source-rock control. *American Association of Petroleum Geologists International Conference, November 8–11, 1998, Rio de Janeiro, Brazil, Extended Abstracts*, 814–815.

BURWOOD, R. 1999. Angola: source rock control for Lower Congo coastal and Kwanza Basin petroleum systems. *In*: CAMERON, N. R., BATE, R. H. & CLURE, V. S. (eds) *The Oil and Gas Habitats of the South Atlantic*. Geological Society, London, Special Publications, **153**, 181–194.

CROSSLEY, R., CAMERON, N. & RAFALSKA, J. K. 1993. Hydrothermal fluids, organic geochemical maturation and hydrocarbon migration in Pre-Salt sequences of Angola. *In*: PARNELL, J., RUFFELL, A. H. & MOLES, N. R. (eds) *Geofluids '93: Contributions to an International Conference on Fluid Evolution, Migration and Interaction in Rocks*. Geological Society Publishing House, Bath, 111–114.

DÜPPENBECKER, S. J., DOHMEN, L. & WELTE, D. H. 1991. Numerical modelling of petroleum expulsion in two areas of the Lower Saxony Basin, Northern Germany. *In*: ENGLAND, W. A. & FLEET, A. J. (eds) *Petroleum Migration*. Geological Society, London, Special Publications, **59**, 57–64.

DUVAL, B., CRAMEZ, C. & JACKSON, M. P. A. 1992. Raft tectonics in the Kwanza Basin, Angola. *Marine & Petroleum Geology*, **9**, 389–404.

EICHENSEER, H. TH., WALGENWITZ, F. R. & BIONDI, P. J. 1999. Stratigraphic control on facies and diagenesis of dolomitized oolitic siliciclastic ramp sequences (Pinda Group, Albian, offshore Angola). *American Association of Petroleum Geologists Bulletin*, **83**, 1729–1758.

GEORGE, S. C., KRIEGER, F. W. *et al.* 1997. Geochemical comparison of oil-bearing fluid inclusions and produced oil from the Toro Sandstone, Papua New Guinea. *Organic Geochemistry*, **26**, 155–173.

GEORGE, S. C., RUBLE, T. E., DUTKIEWICZ, A. & EADINGTON, P. J. 2001. Assessing the maturity of oil trapped in fluid inclusions using molecular geochemistry data and visually-determined fluorescence colours. *Applied Geochemistry*, **16**, 451–473.

GREENWOOD, P. F., GEORGE, S. C. & HALL, K. 1998. Applications of laser micropyrolysis-gas chromatography-mass spectrometry. *Organic Geochemistry*, **29**, 1075–1089.

GUILHAUMOU, N., SZYDLOWSKII, N., & PRADIER, B. 1990. Characterization of hydrocarbon fluid inclusions by infra-red and fluorescence microspectrometry. *Mineralogical Magazine*, **54**, 311–324.

HARRIS, N. B., HEGARTY, K. A., GREEN, P. F. & DUDDY, I. R. 2002. Distribution, timing and intensity of major tectonic events on the west African margin from Gabon to Namibia: Results of a regional apatite fission track study. *American Society of Petroleum Geologists Annual Meeting, Houston, 2002, Abstracts*. [CD-ROM]

HORSFIELD, B. & McLIMANS, R. K. 1984. Geothermometry and geochemistry of aqueous and oil-bearing fluid inclusions from Fateh Field, Dubai. *Organic Geochemistry*, **6**, 733–740.

ISAKSEN, G. H., POTTORF, R. J. & JENSSEN, A. I. 1998. Correlation of fluid inclusions and reservoired oils to infer trap fill history in the South Viking Graben, North Sea. *Petroleum Geoscience*, **4**, 41–55.

LUNDIN, E. R. 1992. Thin-skinned extensional tectonics on a salt detachment, northern Kwanza Basin, Angola. *Marine & Petroleum Geology*, **9**, 405–411.

McHARGUE, T. R. 1990. Stratigraphic development of proto-South Atlantic rifting in Cabinda, Angola: a petroliferous lake basin. *In*: KATZ, B. J. (ed) *Lacustrine Basin Exploration: Case Studies and Modern Analogs*. American Association of Petroleum Geologists Memoirs, **50**, 307–326.

McLIMANS, R. K. 1987. The application of fluid inclusions to migration of oil and diagenesis in petroleum reservoirs. *Applied Geochemistry*, **2**, 585–603.

MARTON, G., HAUN, D. *et al.* 2002. Petroleum geology and exploration potential of Block 5, offshore Angola,

West Africa. *American Association of Petroleum Geologists Annual Meeting, Houston, 2002, Abstracts.* [CD-ROM]

MAZZINI, A., LI, R. & PARNELL, J. 2002. Spectroscopic methods for analyzing organic compounds in fluid inclusions during planetary exploration. *Lunar and Planetary Science XXXIII, Houston. Extended Abstracts.* Lunar and Planetary Institute Contribution, **1109**, 1645. [CD-ROM]

MIDDLETON, D. W. J., PARNELL, J., GREEN, P. F., XU, G. & MCSHERRY, M. 2001. Hot fluid flow events in Atlantic margin basins: an example from the Rathlin Basin. *In*: SHANNON, P. M., HAUGHTON, P. D. W. & CORCORAN, D. V. (eds) *The Petroleum Exploration of Ireland's Offshore Basins.* Geological Society, London, Special Publications, **188**, 91–105.

MUNZ, I. A., JOHANSEN, H., HOLM, K. & LACHARPAGNE, J.-C. 1999. The petroleum characteristics and filling history of the Frøy field and the Rind discovery, Norwegian North Sea. *Marine & Petroleum Geology,* **16**, 633–651.

NEILSON, J. E., OXTOBY, N. H., SIMMONS, M. D., SIMPSON, I. R. & FORTUNATOVA, N. K. 1998. The relationship between petroleum emplacement and carbonate reservoir quality: examples from Abu Dhabi and the Amu Darya Basin. *Marine & Petroleum Geology,* **15**, 57–72.

NORVICK, M. S. & SCHALLER, H. 1998. The Post-Rift paleogeographic evolution of the South Atlantic basins of Brazil and West Africa and the influence of hinterland uplift on drainage and sedimentary depocenters. *American Association of Petroleum Geologists International Conference, November 8–11, 1998, Rio de Janeiro, Brazil, Extended Abstracts*, 36–37.

O'BRIEN, G. W., LISK, M., DUDDY, I., EADINGTON, P. J., CADMAN, S. & FELLOWS, M. 1996. Late Tertiary fluid migration in the Timor Sea: A key control on thermal and diagenetic histories? *Australian Petroleum Production and Exploration Association Journal,* **36**, 399–424.

PARNELL, J., CAREY, P. F., GREEN, P. & DUNCAN, W. 1999. Hydrocarbon migration history, West of Shetland: integrated fluid inclusion and fission track studies. *In*: FLEET, A. J. & BOLDY, S. A. R. (eds) *Petroleum Geology of Northwest Europe: Proceedings of the 5th Conference.* Geological Society, London, 613–625.

PARNELL, J., CHEN, H. & KLUBOV, B. 2001*a*. Hot oil in the Russian Arctic: Precipitation of vanadiferous bitumens, Novaya Zemlya. *In*: PIESTRZYNSKI, A. (ed.) *Mineral Deposits at the Beginning of the 21st Century.* Balkema, Lisse, 71–74.

PARNELL, J., MIDDLETON, D., CHEN, H. & HALL, D. 2001*b*. The use of integrated fluid inclusion studies in constraining oil charge history and reservoir compartmentation: examples from the Jeanne d'Arc Basin, offshore Newfoundland. *Marine & Petroleum Geology,* **18**, 535–549.

ROEDDER, E. 1984. *Fluid Inclusions.* Mineralogical Society of America, Reviews in Mineralogy, **12**.

RUBLE, T. E., GEORGE, S. C., LISK, M. & QUEZADA, R. A. 1998. Organic compounds trapped in aqueous fluid inclusions. *Organic Geochemistry,* **29**, 195–205.

SHEPHERD, T. J., RANKIN, A. H. & ALDERTON, D. H. 1985. *A Practical Guide to Fluid Inclusion Studies.* Blackie, London.

SWARBRICK, R. E., OSBORNE, M. J. *et al.* 2000. Integrated study of the Judy Field (block 30/7a) – an overpressured Central North Sea oil/gas field. *Marine & Petroleum Geology,* **17**, 993–1010.

UNCINI, G., BRANDÃO, M & GIOVANNELLI, A. 1998. Neocomian–Upper Aptian Pre-Salt sequence of southern Kwanza Basin: a regional view. *American Association of Petroleum Geologists International Conference, November 8–11, 1998, Rio de Janeiro, Brazil, Extended Abstracts*, 346–347.

WALDERHAUG, O. 1994. Temperatures of quartz cementation in Jurassic sandstones from the Norwegian continental shelf – evidence from fluid inclusions. *Journal of Sedimentary Research,* **A64**, 311–323.

WALGENWITZ, F, PAGEL, M., MEYER, A., MALUSKI, H. & MONIE, P. 1990. Thermo-chronological approach to reservoir diagenesis in the offshore Angola basin: a fluid inclusion, 40Ar-39Ar and K-Ar investigation. *American Association of Petroleum Geologists Bulletin,* **74**, 547–563.

WOPENKA, B., PASTERIS, J. D. & FREEMAN, J. J. 1990. Analysis of individual fluid inclusions by Fourier transform infrared and Raman microspectroscopy. *Geochimica et Cosmochimica Acta,* **54**, 519–533.

Index

Note: Page references in *italics* refer to Figures; those in **bold** refer to Tables